Information Theory and Statistical Learning

Frank Emmert-Streib • Matthias Dehmer

Information Theory and Statistical Learning

Springer

Frank Emmert-Streib
University of Washington
Department of Biostatistics
and Department of Genome Sciences
1705 NE Pacific St.,
Box 357730
Seattle WA 98195, USA
and
Queen's University Belfast
Computational Biology
and Machine Learning
Center for Cancer Research
and Cell Biology
School of Biomedical Sciences
97 Lisburn Road, Belfast BT9 7BL, UK
v@bio-complexity.com

Matthias Dehmer
Vienna University of Technology
Institute of Discrete Mathematics
and Geometry
Wiedner Hauptstr. 8–10
1040 Vienna, Austria
and
University of Coimbra
Center for Mathematics
Probability and Statistics
Apartado 3008, 3001–454
Coimbra, Portugal
matthias@dehmer.org

ISBN: 978-1-4419-4650-8
e-ISBN: 978-0-387-84816-7
DOI: 10.1007/978-0-387-84816-7

Printed on acid-free paper

springer.com

Preface

This book presents theoretical and practical results of information theoretic methods used in the context of statistical learning. Its major goal is to advocate and promote the importance and usefulness of information theoretic concepts for understanding and developing the sophisticated machine learning methods necessary not only to cope with the challenges of modern data analysis but also to gain further insights into their theoretical foundations. Here *Statistical Learning* is loosely defined as a synonym, for, e.g., Applied Statistics, Artificial Intelligence or Machine Learning. Over the last decades, many approaches and algorithms have been suggested in the fields mentioned above, for which information theoretic concepts constitute core ingredients. For this reason we present a selected collection of some of the finest concepts and applications thereof from the perspective of information theory as the underlying *guiding principles*. We consider such a perspective as very insightful and expect an even greater appreciation for this perspective over the next years.

The book is intended for interdisciplinary use, ranging from Applied Statistics, Artificial Intelligence, Applied Discrete Mathematics, Computer Science, Information Theory, Machine Learning to Physics. In addition, people working in the hybrid fields of Bioinformatics, Biostatistics, Computational Biology, Computational Linguistics, Medical Bioinformatics, Neuroinformatics or Web Mining might profit tremendously from the presented results because these data-driven areas are in permanent need of new approaches to cope with the increasing flood of high-dimensional, noisy data that possess seemingly never ending challenges for their analysis.

Many colleagues, whether consciously or unconsciously, have provided us with input, help and support before and during the writing of this book. In particular we would like to thank Shun-ichi Amari, Hamid Arabnia, Gökhan Bakır, Alexandru T. Balaban, Teodor Silviu Balaban, Frank J. Balbach, João Barros, Igor Bass, Matthias Beck, Danail Bonchev, Stefan Borgert, Mieczyslaw Borowiecki, Rudi L. Cilibrasi, Mike Coleman, Malcolm Cook, Pham Dinh-Tuan, Michael Drmota, Shinto Eguchi, B. Roy Frieden, Bernhard Gittenberger, Galina Glazko, Martin Grabner, Earl Glynn, Peter Grassberger, Peter Hamilton, Kateřina Hlaváčková-Schindler, Lucas R. Hope, Jinjie Huang, Robert Jenssen, Attila Kertész-Farkas, András Kocsor,

Elena Konstantinova, Kevin B. Korb, Alexander Kraskov, Tyll Krüger, Ming Li, J.F. McCann, Alexander Mehler, Marco Möller, Abbe Mowshowitz, Max Mühlhäuser, Markus Müller, Noboru Murata, Arcady Mushegian, Erik P. Nyberg, Paulo Eduardo Oliveira, Hyeyoung Park, Judea Pearl, Daniel Polani, Sándor Pongor, William Reeves, Jorma Rissanen, Panxiang Rong, Reuven Rubinstein, Rainer Siegmund Schulze, Heinz Georg Schuster, Helmut Schwegler, Chris Seidel, Fred Sobik, Ray J. Solomonoff, Doru Stefanescu, Thomas Stoll, John Storey, Milan Studeny, Ulrich Tamm, Naftali Tishby, Paul M.B. Vitányi, José Miguel Urbano, Kazuho Watanabe, Dongxiao Zhu, Vadim Zverovich and apologize to all those who have been missed inadvertently. We would like also to thank our editor Amy Brais from Springer who has always been available and helpful. Last but not least we would like to thank our families for support and encouragement during all the time of preparing the book for publication.

We hope this book will help to spread the enthusiasm we have for this field and inspire people to tackle their own practical or theoretical research problems.

Belfast and Coimbra
June 2008

Frank Emmert-Streib
Matthias Dehmer

Contents

Contributors

Frank J. Balbach, University of Waterloo, Waterloo, ON, Canada, fbalbach@uwaterloo.ca

João Barros, Instituto de Telecomunicações, Universidade do Porto, Porto, Portugal, barros@dcc.fc.up.pt

Rudi L. Cilibrasi, CWI, Kruislaan 413, 1098 SJ Amsterdam, The Netherlands, cilibrar@cilibrar.com

Pham Dinh-Tuan, Laboratory Jean Kuntzmann, CNRS-INPG-UJF BP 53, 38041 Grenoble Cedex, France, Dinh-Tuan.Pham@imag.fr

Shinto Eguchi, Institute of Statistical Mathematics, 4-6-7 Minami-Azabu, Minato-ku, Tokyo 106-8569, Japan, eguchi@ism.ac.jp

B. Roy Frieden, College of Optical Sciences, University of Arizona, Tucson, AZ 85721, USA, roy.frieden@optics.Arizona.edu

Peter Grassberger, Department of Physics and Astronomy and Institute for Biocomplexity and Informatics, University of Calgary, 2500 University Drive NW, Calgary AB, Canada T2N 1N4, pgrassbe@ucalgary.ca

Lucas R. Hope, Bayesian Intelligence Pty. Ltd., lhope@bayesian-intelligence.com

Kateřina Hlaváčková-Schindler, Commission for Scientific Visualization, Austrian Academy of Sciences and Donau-City Str. 1, 1220 Vienna, Austria and Institute of Information Theory and Automation of the Academy of Sciences of the Czech Republic, Pod Vodárenskou věží 4, 18208 Praha 8, Czech Republic, katerina.schindler@assoc.oeaw.ac.at

Jinjie Huang, Department of Automation, Harbin University of Science and Technology, Xuefu Road 52 Harbin 150080, China, jinjiehyh@yahoo.com.cn

Robert Jenssen, Department of Physics and Technology, University of Tromsø, 9037 Tromso, Norway, robert.jenssen@phys.uit.no

Attila Kertész-Farkas, Research Group on Artificial Intelligence, Aradi vértanúk tere 1, 6720 Szeged, Hungary, kfa@inf.u-szeged.hu

András Kocsor, Research Group on Artificial Intelligence, Aradi vértanúk tere 1, 6720 Szeged, Hungary, kocsor@inf.u-szeged.hu

Kevin B. Korb, Clayton School of IT, Monash University, Clayton 3600, Australia, kevin.korb@infotech.monash.edu.au

Alexander Kraskov, UCL Institute of Neurology, Queen Square, London WC1N 3BG, UK, akraskov@ion.ucl.ac.uk

Ming Li, University of Waterloo, Waterloo, ON, Canada, mli@uwaterloo.ca

Marco Möller, Adaptive Systems Research Group, School of Computer Science, University of Hertfordshire, Hatfield, UK XXX@herts.ac.uk

Noboru Murata, Waseda University, Tokyo 169-8555, Japan, noboru.murata@eb.waseda.ac.jp

Erik P. Nyberg, School of Philosophy, University of Melbourne, Parkville 3052, Australia, e.nyberg@pgrad.unimelb.edu.au

Hyeyoung Park, Kyungpook National University, Daegu 702-701, Korea, hypark@knu.ac.kr

Daniel Polani, Adaptive Systems Research Group, School of Computer Science, University of Hertfordshire, Hatfield, UK, d.polani@herts.ac.uk

Sándor Pongor, Protein Structure and Bioinformatics Group, International Centre for Genetic Engineering and Biotechnology, Padriciano 99, 34012 Trieste, Italy and Bioinformatics Group, Biological Research Centre, Hungarian Academy of Sciences, Temesvári krt. 62, 6701 Szeged, Hungary, pongor@icgeb.org

Jorma Rissanen, Helsinki Institute for Information Technology, Technical Universities of Tampere and Helsinki, and CLRC, Royal Holloway, University of London, London, UK, jorma.rissanen@hiit.fi

Panxiang Rong, Department of Automation, Harbin University of Science and Technology, Xuefu Road 52 Harbin 150080, China, pxrong@hrbust.edu.cn

Reuven Rubinstein, Faculty of Industrial Engineering and Management, Technion, Israel Institute of Technology Haifa 32000, Israel, ierrr01@ie.technion.ac.il

Ray J. Solomonoff, Visiting Professor, Computer Learning Research Centre, Royal Holloway, University of London, London, UK, rjsolo@ieee.org

Paul M. B. Vitányi, CWI, Kruislaan 413, 1098 SJ Amsterdam, The Netherlands, paulv@cwi.nl

Chapter 1
Algorithmic Probability: Theory and Applications

Ray J. Solomonoff

Abstract We first define Algorithmic Probability, an extremely powerful method of inductive inference. We discuss its completeness, incomputability, diversity and subjectivity and show that its incomputability in no way inhibits its use for practical prediction. Applications to Bernoulli sequence prediction and grammar discovery are described. We conclude with a note on its employment in a very strong AI system for very general problem solving.

1.1 Introduction

Ever since probability was invented, there has been much controversy as to just what it meant, how it should be defined and above all, what is the best way to predict the future from the known past. Algorithmic Probability is a relatively recent definition of probability that attempts to solve these problems.

We begin with a simple discussion of prediction and its relationship to probability. This soon leads to a definition of Algorithmic Probability (ALP) and its properties. The best-known properties of ALP are its *incomputibility* and its *completeness* (in that order). *Completeness* means that if there is any regularity (i.e. property useful for prediction) in a batch of data, ALP will eventually find it, using a surprisingly small amount of data. The *incomputability* means that in the search for regularities, at no point can we make a useful estimate of how close we are to finding the most important ones. We will show, however, that this incomputability is of a very benign kind, so that in no way does it inhibit the use of ALP for good prediction. One of the important properties of ALP is *subjectivity*, the amount of personal experiential information that the statistician must put into the system. We will show that this

R.J. Solomonoff
Visiting Professor, Computer Learning Research Centre, Royal Holloway, University of London, London, UK
http://world.std.com/ rjs, e-mail: rjsolo@ieee.org

F. Emmert-Streib, M. Dehmer (eds.), *Information Theory and Statistical Learning*,
DOI: 10.1007/978-0-387-84816-7_1,

is a desirable feature of ALP, rather than a "Bug". Another property of ALP is its diversity – it affords *many* explanations of data giving very good understanding of that data.

There have been a few derivatives of Algorithmic Probability – Minimum Message Length (MML), Minimum Description Length (MDL) and Stochastic Complexity – which merit comparison with ALP.

We will discuss the application of ALP to two kinds of problems: Prediction of the Bernoulli Sequence and Discovery of the Grammars of Context Free Languages. We also show how a variation of Levin's search procedure can be used to search over a function space very efficiently to find good predictive models.

The final section is on the future of ALP – some open problems in its application to AI and what we can expect from these applications.

1.2 Prediction, Probability and Induction

What is Prediction?

"An estimate of what is to occur in the future" – But also necessary is a measure of confidence in the prediction: As a negative example consider an early AI program called "Prospector". It was given the characteristics of a plot of land and was expected to suggest places to drill for oil. While it did indeed do that, it soon became clear that without having any estimate of confidence, it is impossible to know whether it is economically feasible to spend $100,000 for an exploratory drill rig. Probability is one way to express this confidence.

Say the program estimated probabilities of 0.1 for 1,000-gallon yield, 0.1 for 10,000-gallon yield and 0.1 for 100,000-gallon yield. The expected yield would be $0.1 \times 1{,}000 + 0.1 \times 10{,}000 + 0.1 \times 100{,}000 = 11{,}100$ gallons. At $100 per gallon this would give $1,110,000. Subtracting out the $100,000 for the drill rig gives an expected profit of $1,010,000, so it *would* be worth drilling at that point. The moral is that predictions by themselves are usually of little value – it is necessary to have confidence levels associated with the predictions.

A strong motivation for revising classical concepts of probability has come from the analysis of human problem solving. When working on a difficult problem, a person is in a maze in which he must make choices of possible courses of action. If the problem is a familiar one, the choices will all be easy. If it is not familiar, there can be much uncertainty in each choice, but choices must somehow be made. One basis for choices might be the probability of each choice leading to a quick solution – this probability being based on experience in this problem and in problems like it. A good reason for using probability is that it enables us to use Levin's Search Technique (Sect. 1.11) to find the solution in near minimal time.

The usual method of calculating probability is by taking the ratio of the number of favorable choices to the total number of choices in the past. If the decision to use integration by parts in an integration problem has been successful in the past 43%

of the time, then its present probability of success is about 0.43. This method has very poor accuracy if we only have one or two cases in the past, and is undefined if the case has never occurred before. Unfortunately it is just these situations that occur most often in problem solving.

On a very practical level: If we cross a particular street 10 times and we get hit by a car twice, we might estimate that the probability of getting hit in crossing that street is about $0.2 = 2/10$. However, if instead, we only crossed that street twice and we didn't get hit either time, it would be unreasonable to conclude that our probability of getting hit was zero! By seriously revising our definition of probability, we are able to resolve this difficulty and clear up many others that have plagued classical concepts of probability.

What is Induction?

Prediction is usually done by finding inductive models. These are deterministic or probabilistic rules for prediction. We are given a batch of data – typically a series of zeros and ones, and we are asked to predict any one of the data points as a function of the data points that precede it.

In the simplest case, let us suppose that the data has a very simple structure:

$$0101010101010\ldots$$

In this case, a good inductive rule is "zero is always followed by one; one is always followed by zero". This is an example of deterministic induction, and deterministic prediction. In this case it is 100% correct every time!

There is, however, a common kind of induction problem in which our predictions will not be that reliable. Suppose we are given a sequence of zeros and ones with very little apparent structure. The only apparent regularity is that zero occurs 70% of the time and one appears 30% of the time. Inductive algorithms give a probability for each symbol in a sequence that is a function of any or none of the previous symbols. In the present case, the algorithm is very simple and the probability of the next symbol is independent of the past – the probability of zero seems to be 0.7; the probability of one seems to be 0.3. This kind of simple probabilistic sequence is called a "Bernoulli sequence". The sequence can contain many different kinds of symbols, but the probability of each is independent of the past. In Sect. 1.9 we will discuss the Bernoulli sequence in some detail.

In general we will not always be predicting Bernoulli sequences and there are many possible algorithms (which we will call "models") that tell how to assign a probability to each symbol, based on the past. Which of these should we use? Which will give good predictions in the future?

One desirable feature of an inductive model is that if it is applied to the *known* sequence, it produces good predictions. Suppose R_i is an inductive algorithm. R_i predicts the probability of an symbol a_j in a sequence $a_1, a_2 \cdots a_n$ by looking at the previous symbols: More exactly,

$$p_j = R_i(a_j|a_1.a_2 \cdots a_{j-1})$$

a_j is the symbol for which we want the probability. $a_1, a_2 \cdots a_{j-1}$ are the previous symbols in the sequence. Then R_i is able to give the probability of a particular value of a_j as a function of the past. Here, the values of a_j can range over the entire "alphabet" of symbols that occur in the sequence. If the sequence is a binary, a_j will range over the set 0 and 1 only. If the sequence is English text, a_j will range over all alphabetic and punctuation symbols. If R_i is a good predictor, for most of the a_j, the probability it assigns to them will be large – near one.

Consider S, the product of the probabilities that R_i assigns to the individual symbols of the sequence, $a_1, a_2 \cdots a_n$. S will give the probability that R_i assigns to the sequence as a whole

$$S = \prod_{j=1}^{n} R_i(a_j|a_1, a_2 \cdots a_n) = \prod_{j=1}^{n} p_j \,. \tag{1.1}$$

For good prediction we want S as large as possible. The maximum value it can have is one, which implies perfect prediction. The smallest value it can have is zero – which can occur if one or more of the p_j are zero – meaning that the algorithm predicted an event to be impossible, yet that event occurred!

The "Maximum Likelihood" method of model selection uses S only to decide upon a model. First, a set of models is chosen by the statistician, based on his experience with the kind of prediction being done. The model within that set having maximum S value is selected.

Maximum Likelihood is very good when there is a lot of data – which is the area in which classical statistics operates. When there is only a small amount of data, it is necessary to consider not only S, but the effect of the likelihood of *the model itself* on model selection. The next section will show how this may be done.

1.3 Compression and ALP

An important application of symbol prediction is text compression. If an induction algorithm assigns a probability S to a text, there is a coding method – *Arithmetic Coding* – that can re-create the entire text without error using just $-\log_2 S$ bits.

More exactly: Suppose x is a string of English text, in which each character is represented by an 8-bit ASCII code, and there are n characters in x. x would be directly represented by a code of just $8n$ bits. If we had a prediction model, R, that assigned a probability of S to the text, then it is possible to write a sequence of just $-\log_2 S$ bits, so that the original text, x, can be recovered from that bit sequence without error.

If R is a string of symbols (usually a computer program) that describes the prediction model, we will use $|R|$ to represent the length of the shortest binary sequence that describes R. If $S > 0$, then the probability assigned to the text will be in two parts: the first part is the code for R, which is $|R|$ bits long, and the second part is the code for the probability of the data, as given by R – it will be just $-\log_2 S$ bits in length. The sum of these will be $|R| - \log_2 S$ bits. The compression ratio achieved by R would be

$$\frac{8N}{|R| - \log_2 S}$$

PPM, a commonly used prediction algorithm, achieves a compression of about three for English text. For very large strings of data, compressions of as much as six have been achieved by highly refined prediction algorithms. We can use $|R| - \log_2 S$ bits, the length of the compressed code, as a "figure of merit" of a particular induction algorithm with respect to a particular text.

We want an algorithm that will give good prediction, i.e. large S, and small $|R|$, so $|R| - \log_2 S$, the figure of merit, will be as small as possible and the probability it assigns to the text will be as large as possible. Models with $|R|$ larger than optimum are considered to be *overfitted.* Models in which $|R|$ are smaller than optimum are considered to be *underfitted*. By choosing a model that minimizes $|R| - \log_2 S$, we avoid both underfitting and overfitting, and obtain very good predictions. We will return to this topic later, when we tell how to compute $|R|$ and S for particular models and data sets.

Usually there are many inductive models available. In 1960, I described Algorithmic Probability – ALP [5–7], which uses all possible models in parallel for prediction, with weights dependent upon the figure of merit of each model.

$$P_M(a_{n+1}|a_1, a_2 \cdots a_n) = \sum 2^{-|R_i|} \, S_i \, R_i(a_{n+1}|a_1, a_2 \cdots a_n) \tag{1.2}$$

$P_M(a_{n+1}|a_1, a_2 \cdots a_n)$ is the probability assigned by ALP to the $(n+1)$th symbol of the sequence, in view of the previous part of the sequence.

$R_i(a_{n+1}|a_1, a_2 \cdots a_n)$ is the probability assigned by the ith model to the $(n+1)$th symbol of the sequence, in view of the previous part of the sequence.

S_i is the probability assigned by R_i, (the ith model) to the known sequence, $a_1, a_2 \cdots a_n$ via (1.1).

$2^{-|R_i|} S_i$ is 1/2 with an exponent equal to the figure of merit that R_i has with respect to the data string $a_1, a_2 \ldots a_n$. It is the weight assigned to $R_i(\,)$. This weight is large when the figure of merit is good – i.e. small.

Suppose that $|R_i|$ is the shortest program describing the ith model using a particular "reference computer" or programming language – which we will call M. Clearly the value of $|R_i|$ will depend on the nature of M. We will be using machines (or languages) that are "Universal" – machines that can readily program any conceivable function – almost all computers and programming languages are of this kind. The subscript M in P_M expresses the dependence of ALP on choice of the reference computer or language.

The universality of M assures us that the value of ALP will not depend *very* much on just which M we use – but the dependence upon M is nonetheless important. It will be discussed at greater length in Sect. 1.5 on "Subjectivity".

Normally in prediction problems we will have some time limit, T, in which we have to make our prediction. In ALP what we want is a set of models of maximum total weight. A set of this sort will give us an approximation that is as close as possible to ALP and gives best predictions. To obtain such a set, we devise a search technique that tries to find, in the available time, T, a set of Models, R_i, such that the total weight,

$$\sum_i 2^{-|R_i|} S_i \tag{1.3}$$

is as large as possible.

On the whole, ALP would seem to be a complex, time consuming way to compute probabilities – though in fact, if suitable approximations are used, these objections are not at all serious.

Does ALP have any advantages over other probability evaluation methods? For one, it's the only method known to be *complete*. The completeness property of ALP means that if there is any regularity in a body of data, our system is guaranteed to discover it using a relatively small sample of that data. More exactly, say we had some data that were generated by an *unknown* probabilistic source, P. Not knowing P, we use instead, P_M to obtain the Algorithmic Probabilities of the symbols in the data. How much do the symbol probabilities computed by P_M differ from their true probabilities, P? The Expected value with respect to P of the total square error between P and P_M is bounded by $-1/2\, lnP_0$.

$$E_P\left[\sum_{m=1}^{n}(P_M(a_{m+1}=1|a_1,a_2\cdots a_m)-P(a_{m+1}=1|a_1,a_2\cdots a_m))^2\right]\le -\frac{1}{2}\, lnP_0$$

$$lnP_0 \approx kln2 \tag{1.4}$$

P_0 is the a priori probability of P. It is the probability we would assign to P if we knew P.

k is the *Kolmogorov complexity* of the data generator, P. It's the shortest binary program that can describe P, the generator of the data.

This is an extremely small error rate. The error in probability approaches zero more rapidly than $1/n$. Rapid convergence to correct probabilities is a most important feature of ALP. The convergence holds for any P that is describable by a computer program and includes many functions that are formally *incomputable*. Various kinds of functions are described in the next section. The convergence proof is in Solomonoff [8].

1.4 Incomputability

It should be noted that in general, it is impossible to find the truly best models with any certainty – there is an infinity of models to be tested and some take an unacceptably long time to evaluate. At any particular time in the search, we will know the best ones so far, but we can't ever be sure that spending a little more time will not give *much* better models! While it is clear that we can always make approximations to ALP by using a limited number of models, we can never know how close these approximations are to the "True ALP". ALP is indeed, *formally incomputable.*

In this section, we will investigate how our models are generated and how the incomputability comes about – why it is a *necessary, desirable feature* of any high performance prediction technique, and how this incomputability *in no way* inhibits its use for practical prediction.

How Incomputability Arises and How We Deal with It

Recall that for ALP. we added up the predictions of all models, using suitable weights:

$$P_M = \sum_{i=1}^{\infty} 2^{-|R_i|} S_i R_i \ . \tag{1.5}$$

Here R_i gives the probability distribution for the next symbol as computed by the ith model. Just what do we mean by these models, R_i?

There are just four kinds of functions that R_i can be:

1. Finite compositions of a finite set of functions
2. Primitive recursive functions
3. Partial recursive functions
4. Total recursive functions

Compositions are combinations of a small set of functions. The finite power series

$$3.2 + 5.98 * X - 12.54 * X^2 + 7.44 * X^3$$

is a composition using the functions *plus* and *times* on the real numbers. Finite series of this sort can approximate any continuous functions to arbitrary precision.

Primitive Recursive Functions are defined by one or more *DO* loops. For example to define $Factorial(X)$ we can write

$Factorial(0) \leftarrow 1$
$DO \quad I = 1, X$
$Factorial(I) \leftarrow I * Factorial(I-1)$
$EndDO$

Partial Recursive Functions are definable using one or more *WHILE* loops. For example, to define the factorial in this way:

$Factorial(0) \leftarrow 1$
$I \leftarrow 0$
$WHILE\ I \neq X$
$I \leftarrow I+1\ Factorial(I) \leftarrow I * Factorial(I-1)$
$EndWHILE$

The loop will terminate if X is a non negative integer. For all other values of X, the loop will run forever. In the present case it is easy to tell for which values of X the loop will terminate.

A simple *WHILE* loop in which it is *not* so easy to tell:
$WHILE\ X > 4$
$IF\ X/2$ is an integer $THEN\ X \leftarrow X/2$
$ELSE\ X \leftarrow 3*X+1$
EndWHILE

This program has been tested with X starting at all positive integers up to more than sixty million. The loop has always terminated, but no one yet is certain as to whether it terminates for *all* positive integers!

For any *Total Recursive Function* we know *all* values of arguments for which the function has values. Compositions and primitive recursive functions are all total recursive. Many partial recursive functions are total recursive, but some are not. As a consequence of the insolvability of Turing's "Halting Problem", it will sometimes be impossible to tell if a certain *WHILE* loop will terminate or not.

Suppose we use (1.2) to approximate ALP by sequentially testing functions in a list of all possible functions – these will be the partial recursive functions because this is the only *recursively enumerable* function class that includes *all* possible predictive functions. As we test to find functions with good figures of merit (small $(|R_i| - \log_2 S_i)$) we find that certain of them don't converge after say, a time T, of 10 s. We know that if we increase T enough, eventually, all converging trials will converge and all divergent trials will still diverge – so eventually we will get close to true ALP – but we cannot recognize when this occurs. Furthermore for any finite T, we cannot *ever* know a useful upper bound on how large the error in the ALP approximation is. That is why this particular method of approximating ALP is called "incomputable". Could there be another *computable* approximation technique that *would* converge? It is easy to show that any computable technique cannot be "complete" – i.e. having very small errors in probability estimates.

Consider an arbitrary *computable* probability method, R_0. We will show how to generate a sequence for which R_0's errors in probability would *always* be 0.5 or more. We start our sequence with a single bit, say zero. We then ask R_0 for the most probable next bit. If it says "one is more probable", we make the continuation zero, if it says "zero is more probable", we make the next bit one. If it says "both are equally likely" we make the next bit zero. We generate the third bit in the sequence in the same way, and we can use this method to generate an arbitrarily long continuation of the initial zero.

For this sequence, R_0 will always have an error in probability of at least one half. Since *completeness* implies that prediction errors approach zero for all finitely describable sequences, it is clear that R_0 or any other *computable* probability method *cannot* be *complete*. Conversely, any *complete* probability method, such as ALP, *cannot* be *computable*.

If we cannot compute ALP, what good is it? It would seem to be of little value for prediction! To answer this objection, we note that from a practical viewpoint, we *never* have to calculate ALP exactly – we can *always* use approximations. While it is impossible to know how close our approximations are to the true ALP, *that information is rarely needed for practical induction.*

What we actually need for practical prediction:

1. Estimates of how good a particular approximation will be in future problems (called "Out of Sample Error")
2. Methods to search for good models
3. Quick and simple methods to compare models

For 1., we can use *Cross Validation* or *Leave One Out* – well-known methods that work with most kinds of problems. In addition, because ALP does not overfit or underfit there is usually a better method to make such estimates.

For 2. In Sect. 1.11 we will describe a variant of Levin's Search Procedure, for an efficient search of a very large function space.

For 3., we will always find it easy to compare models via their associated "Figures of Merit", $|R_i| - \log_2(S_i)$.

In summary, it is clear that all computable prediction methods have a serious flaw – they cannot ever approach completeness. On the other hand, while approximations to ALP *can* approach completeness, we can never know how close we are to the final, incomputable result. We can, however, get good estimates of the future error in our approximations, and *this is all that we really need* in a practical prediction system.

That our approximations approach ALP assures us that if we spend enough time searching we will eventually get as little error in prediction as is possible. *No computable probability evaluation method can ever give us this assurance.* It is in this sense that the incomputability of ALP is a *desirable* feature.

1.5 Subjectivity

The subjectivity of probability resides in a priori information – the information available to the statistician before he sees the data to be extrapolated. This is independent of what kind of statistical techniques we use. In ALP this a priori information is embodied in M, our "Reference Computer". Recall our assignment of a $|R|$ value to an induction model – it was the length of the program necessary to describe R. In general, this will depend on the machine we use – its instruction set. Since the machines we use are Universal – they can imitate one another – the length of description of programs will not vary widely between most reference machines we might consider. But nevertheless, using small samples of data (as we often do in AI), these differences between machines can modify results considerably.

For quite some time I felt that the dependence of ALP on the reference machine was a serious flaw in the concept, and I tried to find some "objective" universal device, free from the arbitrariness of choosing a particular universal machine. When I thought I finally found a device of this sort, I realized that I really didn't want it – that I had no use for it at all! Let me explain:

In doing inductive inference, one begins with two kinds of information: First, the data itself, and second, the a priori data – the information one had before seeing the data. It is possible to do prediction without data, but one cannot do prediction without a priori information. In choosing a reference machine we are given the opportunity to insert into the a priori probability distribution any information about the data that we know before we see it.

If the reference machine were somehow "objectively" chosen for all induction problems, we would have no way to make use of our prior information. This lack of an objective prior distribution makes ALP very subjective – as are all Bayesian systems.

This certainly makes the results "subjective". If we value objectivity, we can routinely reduce the choice of a machine and representation to certain universal "default" values – but there is a tradeoff between objectivity and accuracy. To obtain the best extrapolation, we must use whatever information is available, and much of this information may be subjective.

Consider two physicians, *A* and *B*: *A* is a conventional physician: He diagnoses ailments on the basis of what he has learned in school, what he has read about and his own experience in treating patients. *B* is not a conventional physician. He is "objective". His diagnosis is entirely "by the book" – things he has learned in school that are universally accepted. He tries as hard as he can to make his judgements free of any bias that might be brought about by his own experience in treating patients. As a lawyer, I might prefer defending *B*'s decisions in court, but as a patient, I would prefer *A*'s intelligently biased diagnosis and treatment.

To the extent that a statistician uses objective techniques, his recommendations may be easily defended, but for accuracy in prediction, the additional information afforded by subjective information can be a critical advantage.

Consider the evolution of a priori in a scientist during the course of his life. He starts at birth with minimal a priori information – but enough to be able to learn to walk, to learn to communicate and his immune system is able to adapt to certain hostilities in the environment. Soon after birth, he begins to solve problems and incorporate the problem solving routines into his a priori tools for future problem solving. This continues throughout the life of the scientist – as he matures, his a priori information matures with him.

In making predictions, there are several commonly used techniques for inserting a priori information. First, by restricting or expanding the set of induction models to be considered. This is certainly the commonest way. Second, by selecting prediction functions with adjustable parameters and assuming a density distribution over those parameters based on past experience with such parameters. Third, we note that much of the information in our sciences is expressed as definitions – additions to our language. ALP, or approximations of it, avails itself of this information by using these definitions to help assign code lengths, and hence a priori probabilities to models. Computer languages are usually used to describe models, and it is relatively easy to make arbitrary definitions part of the language.

More generally, modifications of computer languages are known to be able to express any conceivable a priori probability distributions. This gives us the ability to incorporate whatever a priori information we like into our computer language. It is certainly more general than any of the other methods of inserting a priori information.

1.6 Diversity and Understanding

Apart from accuracy of probability estimate, ALP has for AI another important value: Its multiplicity of models gives us many different ways to understand our data.

A very conventional scientist understands his science using a *single* "current paradigm" – the way of understanding that is most in vogue at the present time. A more creative scientist understands his science in *very many* ways, and can more easily create new theories, new ways of understanding, when the "current paradigm" no longer fits the current data.

In the area of AI in which I'm most interested – Incremental Learning – this diversity of explanations is of major importance. At each point in the life of the System, it is able to solve with acceptable merit, all of the problems it's been given thus far. We give it a new problem – usually its present Algorithm is adequate. Occasionally, it will have to be modified a bit. But every once in a while it gets a problem of real difficulty and the present Algorithm has to be *seriously revised*. At such times, we try using or modifying *once sub-optimal algorithms*. If that doesn't work we can use parts of the sub-optimal algorithms and put them together in new ways to make new trial algorithms. It is in giving us a broader basis to learn from the past, that this value of ALP lies.

1.6.1 ALP and "The Wisdom of Crowds"

It is a characteristic of ALP that it averages over all possible models of the data: There is evidence that this kind of averaging may be a good idea in a more general setting. "The Wisdom of Crowds" is a recent book by James Serowiecki that investigates this question. The idea is that if you take a bunch of very different kinds of people and ask them (independently) for a solution to a difficult problem, then a suitable average of their solutions will very often be better than the best in the set. He gives examples of people guessing the number of beans in a large glass bottle, or guessing the weight of a large ox, or several more complex, very difficult problems.

He is concerned with the question of what kinds of problems can be solved this way as well as the question of when crowds are wise and when they are stupid. They become very stupid in mobs or in committees in which a single person is able to strongly influence the opinions in the crowd. In a wise crowd, the opinions are individualized, the needed information is shared by the problem solvers, and the individuals have great diversity in their problem solving techniques. The methods of combining the solutions must enable each of the opinions to be voiced. These conditions are very much the sort of thing we do in ALP. Also, when we *approximate* ALP we try to preserve this diversity in the *subset* of models we use.

1.7 Derivatives of ALP

After my first description of ALP in 1960 [5], there were several related induction models described, minimum message length (MML) Wallace and Boulton [13], Minimum Description Length (MDL) Rissanen [3], and Stochastic Complexity, Rissanen [4]. These models were conceived independently of ALP – (though Rissanen had read Kolmogorov's 1965 paper on minimum coding [1], which is closely related to ALP). MML and MDL select induction models by minimizing the figure of merit, $|R_i| - \log_2(S_i)$ just as ALP does. However, instead of using a weighted sum of models, they use only the single best model.

MDL chooses a space of computable models then selects the best model from that space. This avoids any incomputability, but greatly limits the kinds of models that it can use. MML recognizes the incomputability of finding the best model so it is in principle much stronger than MDL. Stochastic complexity, like MDL, first selects a space of computable models – then, like ALP it uses a weighted sum of all models in the that space. Like MDL, it differs from ALP in the limited types of models that are accessible to it. MML is about the same as ALP when the best model is much better than any other found. When several models are of comparable figure of merit, MML and ALP will differ. One advantage of ALP over MML and MDL is in its diversity of models. This is useful if the induction is part of an ongoing process of learning – but if the induction is used on one problem only, diversity is of much less value. Stochastic Complexity, of course, does obtain diversity in its limited set of models.

1.8 Extensions of ALP

The probability distribution for ALP that I've shown is called "The Universal Distribution for *sequential* prediction". There are two other universal distributions I'd like to describe:

1.8.1 A Universal Distribution for an Unordered Set of Strings

Suppose we have a corpus of n unordered discrete objects, each described by a finite string a_j: Given a new string, a_{n+1}, what is the probability that it is in the previous set? In MML and MDL, we consider various algorithms, R_i, that assign probabilities to strings. (We might regard them as *Probabilistic Grammars*). We use for prediction, the grammar, R_i, for which

$$|R_i| - \log_2 S_i \tag{1.6}$$

is minimum. Here $|R_i|$ is the number of bits in the description of the grammar, R_i. $S_i = \prod_j R_i(a_j)$ is the probability assigned to the entire corpus by R_i. If R_k is the best stochastic grammar that we've found, then we use $R_k(a_{n+1})$ as the probability of a_{n+1}. To obtain the ALP version, we simply sum over all models as before, using weights $2^{-|R_i|}S_i$.

This kind of ALP has an associated convergence theorem giving very small errors in probability. This approach can be used in linguistics. The a_j can be examples of sentences that are grammatically correct. We can use $|R_i| - \log_2 S_i$ as a likelihood that the data was created by grammar, R_i. Section 1.10 continues the discussion of Grammatical Induction.

1.8.2 A Universal Distribution for an Unordered Set of Ordered Pairs of Strings

This type of induction includes almost all kinds of prediction problems as "special cases". Suppose you have a set of question answer pairs, $Q_1, A_1; Q_2, A_2; \ldots Q_n, A_n$: Given a new question, Q_{n+1}, what is the probability distribution over possible answers, A_{n+1}? Equivalently, we have an unknown analog and/or digital transducer, and we are given a set of input/output pairs $Q_1, A_1; \ldots$ – For a new input Q_i, what is probability distribution on outputs? Or, say the Q_i are descriptions of mushrooms and the A_i are whether they are poisonous or not.

As before, we hypothesize operators $R_j(A|Q)$ that are able to assign a probability to any A given any Q: The ALP solution is

$$\begin{aligned} P(A_{n+1}|Q_{n+1}) &= \sum_j 2^{-|R_j|} \prod_{i=1}^{n+1} R_j(A_i|Q_i) \\ &= \sum_j a_n^j R_j(A_{n+1}|Q_{n+1}), \\ a_n^j &= 2^{-|R_j|} \prod_{i=1}^{n} R_j(A_i|Q_i). \end{aligned} \tag{1.7}$$

$2^{-|R_j|}$ are the a priori probabilities associated with the R_j.

a_n^j is the weight given to R_j's predictions in view of it's success in predicting the data set $Q_1, A_1 \ldots Q_n, A_n$.

This ALP system has a corresponding theorem for small errors in probability. As before, we try to find a set of models of maximum weight in the available time. Proofs of convergence theorems for these extensions of ALP are in Solomonoff [10]. There are corresponding MDL, MML versions in which we pick the single model of maximum weight.

1.9 Coding the Bernoulli Sequence

First, consider a binary Bernoulli sequence of length n. It's only visible regularity is that zeroes have occurred n_0 times and ones have occurred n_1 times. One kind of model for this data is that the probability of 0 is p and the probability of 1 is $1-p$. Call this model R_p. S_p is the probability assigned to the data by R_p

$$S_p = p^{n_0}(1-p)^{n_1} \,. \tag{1.8}$$

Recall that ALP tells us to sum the predictions of each model, with weight given by the product of the a priori probability of the model ($2^{-|R_i|}$) and S_i, the probability assigned to the data by the model ..., i.e.:

$$\sum_i 2^{-|R_i|} \, S_i \, R_i(\,) \,. \tag{1.9}$$

In summing we consider all models with $0 \le p \le 1$.

We assume for each model, R_p, precision Δ in describing p. So p is specified with accuracy, Δ. We have $\frac{1}{\Delta}$ models to sum so total weight is 1

$$2^{-|R_i|} = \Delta,$$

$$S_i = S_p = p^{n_0}(1-p)^{n_1},$$

$$R_i(\,) = R_p = p.$$

Summing the models for small Δ gives the integral

$$\int_0^1 p^{n_0}(1-p)^{n_1} p \, dp = \int_0^1 p^{n_0+1}(1-p)^{n_1} dp. \tag{1.10}$$

This integral can be evaluated using the Beta function, $B(\,,\,)$

$$\int_0^1 p^x(1-p)^y dy = B(x+1, y+1) = \frac{x!\,y!}{(x+y+1)!} \,. \tag{1.11}$$

So our integral of (1.10) equals $\frac{(n_0+1)!\,n_1!}{(n_0+n_1+2)!}$.

We can get about the same result another way: The function $p^{n_0}(1-p)^{n_1}$ is (if n_0 and n_1 are large), narrowly peaked at $p_0 = \frac{n_0}{n_0+n_1}$. If we used MDL we would use the model with $p = p_0$. The a priori probability of the model itself will depend on how accurately we have to specify p_0. If the "width" of the peaked distribution is Δ, then the a priori probability of model M_{p_0} will be just $\Delta \cdot p_0^{n_0}(1-p_0)^{n_1}$.

It is known that the width of the distribution is just $2\sqrt{\frac{p_0(1-p_0)}{n_0+n_1+1}}$.[1] As a result the probability assigned to this model is $\sqrt{\frac{p_0(1-p_0)}{n_0+n_1+1}} \cdot p_0^{n_0}(1-p_0)^{n_1} \cdot 2$. If we use Sterling's approximation for $n!$ ($n! \approx e^{-n}n^n\sqrt{2\pi n}$), it is not difficult to show that

$$\frac{n_0!n_1!}{(n_0+n_1+1)!} \approx p_0^{n_0}(1-p_0)^{n_1}\sqrt{\frac{p_0(1-p_0)}{n_0+n_1+1}} \cdot \sqrt{2\pi}, \tag{1.12}$$

$\sqrt{2\pi} = 2.5066$ which is roughly equal to 2.

To obtain the probability of a zero following a sequence of n_0 zeros and n_1 ones: We divide the probability of the sequence having the extra zero, by the probability of the sequence without the extra zero, i.e.:

$$\frac{(n_0+1)!n_1!}{(n_0+n_1+2)!} \Big/ \frac{n_0!n_1!}{(n_0+n_1+1)!} = \frac{n_0+1}{(n_0+n_1+2)}. \tag{1.13}$$

This method of extrapolating binary Bernoulli sequences is called "Laplace's Rule". The formula for the probability of a binary sequence, $\frac{n_0!n_1!}{(n_0+n_1+1)!}$ can be generalized for an alphabet of k symbols.

A sequence of k different kinds of symbols has a probability of

$$\frac{(k-1)!\prod_{i=1}^{k} n_i!}{(k-1+\sum_{i=1}^{k} n_i)!}. \tag{1.14}$$

n_i is the number of times the ith symbol occurs.

This formula can be obtained by integration in a $k-1$ dimensional space of the function $p_1^{n_1}p_2^{n_2}\cdots p_{k-1}^{n_{k-1}}(1-p_1-p_2\cdots-p_{k-1})^{n_k}$.

Through an argument similar to that used for the binary sequence, the probability of the next symbol being of the jth type is

$$\frac{n_j+1}{k+\sum_{i=1}^{k} n_i}. \tag{1.15}$$

A way to visualize this result: the body of data (the "corpus") consists of the $\sum n_i$ symbols. Think of a "pre-corpus" containing one of each of the k symbols. If we think of a "macro corpus" as "corpus plus pre-corpus" we can obtain the probability of the next symbol being the jth one by dividing the number of occurrences of that symbol in the macro corpus, by the total number of symbols of all types in the macro corpus.

[1] This can be obtained by getting the first and second moments of the distribution, using the fact that $\int_0^1 p^x(1-p)^y dp = \frac{x!y!}{(x+y+1)!}$.

It is also possible to have different numbers of each symbol type in the pre-corpus, enabling us to get a great variety of “a priori probability distributions” for our predictions.

1.10 Context Free Grammar Discovery

This is a method of extrapolating an unordered set of finite strings: Given the set of strings, $a_1, a_2, \cdots a_n$, what is the probability that a new string, a_{n+1}, is a member of the set? We assume that the original set was generated by some sort of probabilistic device. We want to find a device of this sort that has a high a priori likelihood (i.e. short description length) and assigns high probability to the data set. A good model R_i, is one with maximum value of

$$P(R_i) \prod_{j=1}^{n} R_i(a_j). \qquad (1.16)$$

Here $P(R_i)$ is the a priori probability of the model R_i. $R_i(a_j)$ is the probability assigned by R_i to data string, a_j.

To understand *probabilistic* models, we first define *non-probabilistic* grammars. In the case of context free grammars, this consists of a set of *terminal* symbols and a set of symbols called *nonterminals*, one of which is the initial starting symbol, S.

A grammar could then be:

$$S \to Aac$$
$$S \to BaAd$$
$$A \to BAaS$$
$$A \to AB$$
$$A \to a$$
$$B \to aBa$$
$$B \to b$$

The capital letters (including S) are all *nonterminal* symbols. The lower case letters are all *terminal* symbols. To generate a legal string, we start with the symbol, S, and we perform either of the two possible substitutions. If we choose $BaAd$, we would then have to choose substitutions for the nonterminals B and A. For B, if we chose aBa we would again have to make a choice for B. If we chose a terminal symbol, like b for B, then no more substitutions can be made.

An example of a string generation sequence:

$S, BaAd, aBaaAd, abaaAd, abaaABd, abaaaBd, abaaabd$.

The string $abaaabd$ is then a legally derived string from this grammar. The set of all strings legally derivable from a grammar is called the *language* of the grammar. The language of a grammar can contain a finite or infinite number of strings. If we replace the deterministic substitution rules with probabilistic rules, we have a *probabilistic* grammar. A grammar of this sort assigns a probability to every string it can generate. In the deterministic grammar above, S had two rewrite choices, A had three, and B had two. If we assign a probability to each choice, we have a probabilistic grammar.

Suppose S had substitution probability 0.1 for Aac and 0.9 for $BaAd$. Similarly, assigning probabilities 0.3, 0.2 and 0.5 for A's substitutions and 0.4, 0.6 for B's substitutions.

$$
\begin{aligned}
S\ &0.1\ Aac\\
&0.9\ BaAd\\
A\ &0.3\ BAaS\\
&0.2\ AB\\
&0.5\ a\\
B\ &0.4\ aBa\\
&0.6\ b
\end{aligned}
$$

In the derivation of *abaaab* of the previous example, the substitutions would have probabilities 0.9 to get *BaAd*, 0.4 to get *aBaaAd*, 0.6 to get *abaaAd*, 0.2 to get *abaaABd*, 0.5 to get *abaaaBd*, and 0.6 to get *abaaabd*. The probability of the string *abaabd* being derived this way is $0.9\times0.4\times0.6\times0.2\times0.5\times0.6 = 0.01296$. Often there are other ways to derive the same string with a grammar, so we have to add up the probabilities of all of its possible derivations to get the total probability of a string.

Suppose we are given a set of strings, *ab*, *aabb*, *aaabbb* that were generated by an unknown grammar. How do we find the grammar?

I wouldn't answer that question directly, but instead I will tell how to find a sequence of grammars that fits the data progressively better. The best one we find may not be the true generator, but will give probabilities to strings close to those given by the generator.

The example here is that of A. Stolcke's, PhD thesis [12]. We start with an ad hoc grammar that *can* generate the data, but it $overfits$ ... it is too complex:

$$
\begin{aligned}
S &\to ab\\
&\to aabb\\
&\to aaabbb
\end{aligned}
$$

We then try a series of modifications of the grammar (*Chunking* and *Merging*) that increase the total probability of description and thereby decrease total description length. *Merging* consists of replacing two nonterminals by a single nonterminal. *Chunking* is the process of defining new nonterminals. We try it when a string or substring has occurred two or more times in the data. *ab* has occurred three times so we define $X = ab$ and rewrite the grammar as

$$\begin{aligned} S &\to X \\ &\to aXb \\ &\to aaXbb \\ X &\to ab \end{aligned}$$

aXb occurs twice so we define $Y = aXb$ giving

$$\begin{aligned} S &\to X \\ &\to Y \\ &\to aYb \\ X &\to ab \\ Y &\to aXb \end{aligned}$$

At this point there are no repeated strings or substrings, so we try the operation *Merge* which coalesces two nonterminals. In the present case merging S and Y would decrease complexity of the grammar, so we try:

$$\begin{aligned} S &\to X \\ &\to aSb \\ &\to aXb \\ X &\to ab \end{aligned}$$

Next, merging S and X gives

$$\begin{aligned} S &\to aSb \\ &\to ab \end{aligned}$$

which is an adequate grammar. At each step there are usually several possible *chunk* or *merge* candidates. We chose the candidates that give minimum description length to the resultant grammar.

How do we calculate the length of description of a grammar and its description of the data set?

Consider the grammar

$$\begin{aligned} S &\to X \\ &\to Y \\ &\to aYb \\ X &\to ab \\ Y &\to aXb \end{aligned}$$

There are two kinds of terminal symbols and three kinds of nonterminals. If we know the number of terminals and nonterminals, we need describe only the right hand side of the substitutions to define the grammar. The names of the nonterminals (other than the first one, S) are not relevant. We can describe the right hand side by the string $Xs_1Ys_1aYbs_1s_2abs_1s_2aXbs_1s_2$. s_1 and s_2 are punctuation symbols. s_1 marks the end of a string. s_2 marks the end of a sequence of strings that belong to the same nonterminal. The string to be encoded has seven kinds of symbols. The number of times each occurs: X, 2; Y, 2; S, 0; a, 3; b, 3; s_1, 5; s_2, 3. We can then use the formula

$$\frac{(k-1)!\prod_{i=1}^{k} n_i!}{(k-1+\sum_{i=1}^{k} n_i)!} \tag{1.17}$$

to compute the probability of the grammar: $k = 7$, since there are seven symbols and $n_1 = 2$, $n_2 = 2$, $n_3 = 0$, $n_4 = 3$, etc. We also have to include the probability of 2, the number of kinds of terminals, and of 3, the number of kinds of nonterminals.

There is some disagreement in the machine learning community about how best to assign probability to integers, n. A common form is

$$P(n) = A2^{-\log_2^* n}\,, \tag{1.18}$$

where $\log_2^* n = \log_2 n + \log_2 \log_2 n + \log_2 \log_2 \log_2 n \cdots$ taking as many positive terms as there are, and A is a normalization constant. There seems to be no good reason to choose 2 as the base for logs, and using different bases gives much different results. If we use natural logs, the sum diverges.

This particular form of $P(n)$ was devised by Rissanen. It is an attempt to approximate the shortest description of the integer n, e.g. the Kolmogorov complexity of n. Its first moment is infinite, which means it is very biased toward large numbers. If we have reason to believe, from previous experience, that n will not be very large, but will be about λ, then a reasonable form of $P(n)$ might be $P(n) = Ae^{-n/\lambda}$, A being a normalization constant.

The forgoing enables us to evaluate $P(R_i)$ of (1.16). The $\prod_{j=1}^{n} R_i(a_j)$ part is evaluated by considering the choices made when the grammar produces the data corpus. For each nonterminal, we will have a sequence of decisions whose probabilities can be evaluated by an expression like (1.14), or perhaps the simpler technique of (1.15) that uses the "pre-corpus". Since there are three nonterminals, we need the product of three such expressions.

The process used by Stolcke in his thesis was to make various trials of chunking or merging in attempts to successively get a shorter description length – or to increase (1.16). Essentially a very greedy method. He has been actively working on Context Free Grammar discovery since then, and has probably discovered many improvements. There are many more recent papers at his website.

Most, if not all CFG discovery has been oriented toward finding a *single best grammar*. For applications in AI and genetic programming it is useful to have large sets of *not necessarily best* grammars – giving much needed diversity. One way to implement this: At each stage of modification of a grammar, there are usually several different operations that can reduce description length. We could pursue such paths in parallel ... perhaps retaining the best 10 or best 100 grammars thus far.

Branches taken early in the search could lead to very divergent paths and much needed diversity. This approach helps avoid *local optima* in grammars and *premature convergence* when applied to Genetic Programming.

1.11 Levin's Search Technique

In the section on incomputability we mentioned the importance of good search techniques for finding effective induction models. The procedure we will describe was inspired by Levin's search technique [2], but is applied to a different kind of problem.

Here, we have a corpus of data to extrapolate, and we want to search over a function space, to find functions ("models") $R_i(\,)$ such that $2^{-|R_i|}S_i$ is as large as possible. In this search, for some R_i, the time needed to evaluate S_i, (the probability assigned to the corpus by R_i), may be unacceptably large – possibly infinite.

Suppose we have a (deterministic) context free grammar, G, that can generate strings that are programs in some computer language. (Most computer languages have associated grammars of this kind.) In generating programs, the grammar will have various choices of substitutions to make. If we give each substitution in a k-way choice, a probability of $1/k$, then we have a *probabilistic* grammar that assigns *a priori probabilities* to the programs that it generates. If we use a functional language (such as LISP), this will give a probability distribution over all functions it can generate. The probability assigned to the function R_i will be denoted by $P_M(R_i)$. Here M is the name of the functional computer language. $P_M(R_i)$ corresponds to what we called $2^{-|R_i|}$ in our earlier discussions. $|R_i|$ corresponds to $-\log_2 P_M(R_i)$. As before, S_i is the probability assigned to the corpus by R_i. We want to find functions $R_i(\,)$ such that $P_M(R_i)S_i$ is as large as possible.

Next we choose a small initial time T – which might be the time needed to execute 10 instructions in our Functional Language. The initial T is not critical. We then compute $P_M(R_i)S_i$ for all R_i for which $t_i/P_M(R_i) < T$. Here t_i is the time needed to construct R_i and evaluate its S_i.

There are only a finite number of R_i that satisfy this criterion and if T is very small, there will be very few, if any. We remember which R_i's have large $P_M(R_i)S_i$.

$t_i < T \cdot P_M(R_i)$, so $\sum_i t_i$, the total time for this part of the search takes time $< T \cdot \sum_i P_M(R_i)$. Since the $P_M(R_i)$ are a priori probabilities, $\sum_i P_M(R_i)$ must be less than or equal to 1, and so the total time for this part of the search must be less than T.

If we are not satisfied with the R_i we've obtained, we double T and do the search again. We continue doubling T and searching until we find satisfactory R_i's or until we run out of time. If T' is the value of T when we finish the search, then the total time for the search will be $T' + T'/2 + T'/4 \cdots \approx 2T'$.

If it took time t_j to generate and test one of our "good" models, R_j, then when R_j was discovered, T' would be no more than $2t_j/P_M(R_j)$ – so we would take no more time than twice this, or $4t_j/P_M(R_j)$ to find R_j. Note that this time limit depends on R_j only, and is independent of the fact that we may have aborted the S_i evaluations of many R_i for which t_i was infinite or unacceptably large. This feature of Levin Search is a mandatory requirement for search over a space of partial recursive functions. Any weaker search technique would seriously limit the power of the inductive models available to us.

When ALP is being used in AI, we are solving a sequence of problems of increasing difficulty. The machine (or language) M is periodically "updated" by inserting subroutines and definitions, etc., into M so that the solutions, R_i to problems in the past result in larger $P_M(R_j)$. As a result the $t_j/P_M(R_j)$ are smaller – giving quicker solutions to problems of the past – and usually for problems of the future as well.

1.12 The Future of ALP: Some Open Problems

We have described ALP and some of its properties:

First, its *completeness*: Its remarkable ability to find any irregularities in an apparently small amount of data.

Second: That any *complete* induction system like ALP must be formally *incomputable*.

Third: That this *incomputability* imposes no limit on its use for practical induction. This fact is based on our ability to estimate the future accuracy of any particular induction model. While this seems to be easy to do in ALP without using Cross Validation, more work needs to be done in this area.

ALP was designed to work on difficult problems in AI. The particular kind of AI considered was a version of "Incremental Learning": We give the machine a simple problem. Using Levin Search, it finds one or more solutions to the problem. The system then updates itself by modifying the reference machine so that the solutions found will have higher a priori probabilities. We then give it new problems somewhat similar to the previous problem. Again we use Levin Search to find solutions – We continue with a sequence of problems of increasing difficulty, updating after each solution is found. As the training sequence continues we expect that we will need less and less care in selecting new problems and that the system will eventually be able to solve a large space of very difficult problems. For a more detailed description of the system, see Solomonoff [11].

The principal things that need to be done to implement such a system:

* We have to find a good reference language: Some good candidates are APL, LISP, FORTH, or a subset of Assembly Language. These languages must be augmented with definitions and subroutines that we expect to be useful in problem solving.
* The design of good training sequences for the system is critical for getting much problem-solving ability into it. I have written some general principles on how to do this [9], but more work needs to be done in this area. For early training, it might be useful to learn definitions of instructions from Maple or Mathematica. For more advanced training we might use the book that Ramanujan used to teach himself mathematics – "A Synopsis of Elementary Results in Pure and Applied Mathematics" by George S. Carr.

* We need a good update algorithm. It is possible to use PPM, a relatively fast, effective method of prediction, for preliminary updating, but to find more complex regularities, a more general algorithm is needed. The universality of the reference language assures us that any conceivable update algorithm can be considered. APL's diversity of solutions to problems maximizes the information that we are able to insert into the a priori probability distribution. After a suitable training sequence the system should know enough to usefully work on the problem of updating itself.

Because of ALP's *completeness* (among other desirable properties), we expect that the *complete* AI system described above should become an extremely powerful general problem solving device – going well beyond the limited functional capabilities of current *incomplete* AI systems.

References

1. Kolmogorov, A.N.: Three approaches to the quantitative definition of information. Problems of Information Transmission **1**(1), 1–7 (1965)
2. Levin, L.A.: Universal search problems. Problemy Peredaci Informacii **9**, 115–116 (1973); Translated in Problems of Information Transmission **9**, 265–266
3. Rissanen, J.: Modeling by the shortest data description. Automatica **14**, 465–471 (1978)
4. Rissanen, J.: Stochastical Complexity and Statistical Inquiry. World Scientific, Singapore (1989)
5. Solomonoff, R.J.: A preliminary report on a general theory of inductive inference. (Revision of Report V–131, Feb. 1960), Contract AF 49(639)–376, Report ZTB–138. Zator, Cambridge (Nov, 1960) (http://www.world.std.com/~rjs/pubs.html)
6. Solomonoff, R.J.: A formal theory of inductive inference, Part I. Information and Control **7**(1), 1–22 (1964)
7. Solomonoff, R.J.: A formal theory of inductive inference, Part II. Information and Control **7**(2), 224–254 (1964)
8. Solomonoff, R.J.: Complexity-based induction systems: comparisons and convergence theorems. IEEE Transactions on Information Theory **IT–24**(4), 422–432 (1978)
9. Solomonoff, R.J.: A system for incremental learning based on algorithmic probability. In: Proceedings of the Sixth Israeli Conference on Artificial Intelligence, Computer Vision and Pattern Recognition 515–527 (Dec. 1989)
10. Solomonoff, R.J.: Three kinds of probabilistic induction: universal distributions and convergence theorems. Appears in Festschrift for Chris Wallace (2003)
11. Solomonoff, R.J.: Progress in incremental machine learning. TR IDSIA-16-03, revision 2.0. (2003)
12. Stolcke, A.: On learning context free grammars. PhD Thesis (1994)
13. Wallace, C.S and Boulton, D.M.: An information measure for classification. Computer Journal **11**, 185–195 (1968)

Further Reading

Cover, T. and Thomas, J.: Elements of Information Theory. Wiley, New York (1991) – Good treatments of statistics, predictions, etc.

Li, M. and Vitányi, P.: An Introduction to Kolmogorov Complexity and Its Applications. Springer, New York (1993) (1997) – Starts with elementary treatment and development. Many sections very clear, very well written. Other sections difficult to understand. Occasional serious ambiguity of notation (e.g. definition of "enumerable"). Treatment of probability is better in 1997 than in 1993 edition.

Shan, Y., McKay, R.I., Baxter, R., et al.: Grammar Model-Based Program Evolution. (Dec. 2003) A recent review of work in this area, and what looks like a very good learning system. Discusses mechanics of fitting Grammar to Data, and how to use Grammars to guide Search Problems.

Solomonoff, R.J.: The following papers are all available at the website: world.std.com/ rjs/ pubs.html.

Stolcke, A., Omohundro, S.: Inducing Probabilistic Grammars by Bayesian Model Merging. ICSI, Berkeley (1994) This is largely a summary of Stolcke's: On Learning Context Free Grammars [12].

A Preliminary Report on a General Theory of Inductive Inference. (1960).

A Formal Theory of Inductive Inference. Information and Control, Part I. (1964).

A Formal Theory of Inductive Inference, Part II. (June 1964) – Discusses fitting of context free grammars to data. Most of the discussion is correct, but Sects. 4.2.4 and 4.3.4 are questionable and equations (49) and (50) are incorrect.

A Preliminary Report ... and A Formal Theory ... give some intuitive justification for the way ALP does induction.

The Discovery of Algorithmic Probability. (1997) – Gives heuristic background for discovery of ALP. Page 27 gives a time line of important publications related to development of ALP.

Progress in Incremental Machine Learning; Revision 2.0. (Oct. 30, 2003) – A more detailed description of the system I'm currently working on. There have been important developments since, however.

The Universal Distribution and Machine Learning. (2003) – Discussion of irrelevance of incomputability to applications for prediction. Also discussion of subjectivity.

Chapter 2
Model Selection and Testing by the MDL Principle

Jorma Rissanen

Abstract This chapter is an outline of the latest developments in the *MDL* theory as applied to the selection and testing of statistical models. Finding the number of parameters is done by a criterion defined by an *MDL* based *universal model*, while the corresponding optimally quantized real valued parameters are determined by the so-called structure function following Kolmogorov's idea in the algorithmic theory of complexity. Such models are *optimally distinguishable*, and they can be tested also in an optimal manner, which differs drastically from the Neyman–Pearson testing theory.

2.1 Modeling Problem

A data generating physical machinery imposes restrictions or properties on data. In statistics we are interested in statistical properties, describable by distributions as models that can be fitted to a set of data $x^n = x_1,\ldots,x_n$ or $(y^n,x^n) = (y_1,x_1),\ldots,(y_n,x_n)$, in the latter case conditional models to data y^n given other data x^n. This case adds little to the discussion, and to simplify the notations we consider only the first type of data with one exception on regression.

Parametric models

$$\mathcal{M}_k = \{f(x^n;\theta,k) : \theta = \theta_1,\ldots,\theta_k \in \Omega^k\} \tag{2.1}$$

$$\mathcal{M} = \{\mathcal{M}_k : k \geq 0\} \tag{2.2}$$

capture almost every type of statistical property that can be described in a constructive manner. These include also the usual so-called nonparametric models when they

J. Rissanen
Helsinki Institute for Information Technology, Technical Universities of Tampere and Helsinki, and CLRC, Royal Holloway, University of London, London, UK
e-mail: jorma.rissanen@hiit.fi

F. Emmert-Streib, M. Dehmer (eds.), *Information Theory and Statistical Learning*,
DOI: 10.1007/978-0-387-84816-7_2,

can be fitted to data. Typically, we have a set of n parameters $\theta_1, \theta_2, \ldots, \theta_n$, but we wish to fit sub collections of these – not necessarily the k first. Each sub collection would define a structure. To simplify the notations we consider the structure defined by the first k parameters in some sorting of all the parameters. We also write $\theta = \theta^k$ when the number of the parameters needs to be emphasized, in which case $f(x^n; \theta, k)$ is written as $f(x^n; \theta^k)$. Finally, the class $\mathcal{M}$ can be made nested if we identify two models $f(x^n; \theta^k)$ and $f(x^n; \theta^{k+1})$ if $\theta^{k+1} = \theta^k, 0$. This can be sometimes useful.

There is no unique way to express the statistical properties of a physical machinery as a distribution, hence, no unique "true" model. As an example, a coin has a lot of properties such as its shape and size. By throwing it we generate data that reflect statistical properties of the coin together with the throwing mechanism. Even in this simple case it seems clear that these are not unique for they depend on many other things that we have no control over. All we can hope for is to fit a model, such as a Bernoulli model, which gives us some information about the coin's statistical behavior. At any rate, to paraphrase Laplace' statement that he needed no axiom of God in his celestial mechanics, we want a theory where the "true" model assumption is not needed.

In fitting models to data a yardstick is needed to measure the fitting error. In traditional statistics, where a "true" model is hypothesized, the fitting error can be defined as the mean difference of the observed result from the true one, which however must be estimated from the data, because the "true" model is not known. A justification of the traditional view, however vague, stems from the confusion of statistics with probability theory. If we construct problems in probability theory that resemble statistical problems, the relevance of the results clearly depends on how well our hypothesized "true" model captures the properties in the data. In simple cases like the coin tossing the resemblance can be good, and useful results can be obtained. However, in realistic more complex statistical problems we can be seriously misled by the results found that way, for they are based on a methodology which is close to circular reasoning. Nowhere is the failure of the logic more blatant than in the important problem of hypothesis testing. Its basic assumption is that one of the hypotheses is true even when none exists, which leads to a dangerous distorted theory. We discuss below model testing without such an assumption.

In absence of the "true" model a yardstick can be taken as the value $f(x^n; \theta^k)$; a large probability or density value represents a good fit and a small one a bad fit. Equivalently, we can take the number $\log 1/f(x^n; \theta^k)$, which by the Kraft inequality can be taken as an ideal code length, ideal because a real code length must be integer valued. But there is a well-known difficulty, for frequently $\log 1/f(x^n; \theta^k) \to 0$ as $k \to n$. We do get a good fit, but the properties learned include too much, "noise" if we fit too complex models. To overcome this without resorting to ad hoc means to penalize the undefined complexity we must face the problem of defining and measuring "complexity".

"Complexity" and its close relative "information" are widely used words with ambiguous meaning. We know of only two formally defined and related notions of complexity which have had significant implications, namely, *algorithmic complexity* and *stochastic complexity*. Both are related to Shannon's entropy and hence fundamentally to Hartley's information: the logarithm of the number of elements in a finite set; that is, the *shortest code length* as the number of binary digits in the coded string, with which any element in the set can be encoded. Hence, the amount of "complexity" is measured by the unit, *bit*, which will also be used to measure "information", to be defined later. In reality, the set that includes the object of interest is not always finite, and the optimal code length is taken either literally, whenever possible, or the shortest in a probability sense, or the shortest in the worst case. In practice, however, the sense of optimality is often as good as literally the shortest code length.

The formal definition of complexity means that it is always relative to a framework within which the description or coding of the object is done. A conceptually supreme way, due to Solomonoff [23], is to define it as the length of the shortest program of a universal computer that delivers it as the output. Hence, the framework is the programming language of the computer. Such a programming language can describe the set of partial recursive functions for integers and tuples of them, and if we identify a finite or constructive description with a computer program we can consider program defined sets and their elements as the objects of interest. With a modified version of Solomonoff's definition due to Kolmogorov and Chaitin, Kolmogorov defined a data string's model as a finite set that includes the string, and the best such model gets defined in a manner which amounts to the *MDL* principle. The best model we take to give the algorithmic information in the string, given by the complexity of the best model. The trouble with all this is the fact that such an "algorithmic complexity" itself is noncomputable and hence cannot be implemented. However, Kolmogorov's plan is so beautiful and easy to describe without any involved mathematics just because it is noncomputable, and we discuss it below.

We wish to imitate Kolmogorov's plan in statistical context in such a way that the needed parts are computable, and hence that they can at least be approximated with an estimate of the approximation error. The role of the set of programs will be played by a set of parametric models, (2.1), (2.2). Such a set is simply selected, perhaps in a tentative way, but we can compare the behavior of several such suggestions. Nothing can be said of how to find the very best set, which again is a noncomputable problem, and consequently we are trying to do only what can be done, rather than something that cannot be done – ever.

In this chapter we discuss the main steps in the program to implement the afore outlined ideas for a theory of modeling. We begin with a section on the *MDL* principle and the Bayesian philosophy, and discuss their fundamental differences. These two are often confused. Next, we describe three universal models, the first of which, the Normalized Maximum Likelihood model, makes the definition of stochastic complexity possible and gives a criterion for fitting the number of parameters. This is followed by an outline of Kolmogorov's structure function in the algorithmic theory of complexity, and its implementation within probability models. The resulting optimally quantized parameters lead into the idea of optimally distinguishable models, which are also defined in an independent manner thereby providing confidence in the constructs. These, in turn, lead naturally to model testing, in which the inherent lopsidedness in testing a null-hypothesis against a composite hypothesis in the Neyman–Pearson theory is removed.

2.2 The MDL Principle and Bayesian Philosophy

The outstanding feature of Bayesian philosophy is the use of distributions not only for data but also for the parameters, even when their values are non repeating. This widens enormously the use of probability models in applications over the orthodox statistics. To be specific, consider a class of parametric models $\mathcal{M} = \{f(x^n;\mu)\}$, where μ represents any type of parameters, together with a "prior" distribution $Q(\mu)$ for them. By Bayes' formula, then, the prior distribution is converted in light of observed data into the posterior

$$P(\mu|x^n) = \frac{f(x^n;\mu)Q(\mu)}{\int f(x^n;\mu)Q(\mu)d\mu},$$

where the integral is a sum in case of discrete parameter values. The posterior may then be maximized over the parameters for estimation. More generally, the posterior is taken to play the role of a "true" distribution in further developments of Bayesian analysis including decision and risk theory, hypothesis testing and other statistical problems.

For the Bayesians the distribution Q for the parameters represents prior knowledge, and its meaning is the probability of the event that the value μ is the "true" value. This causes some difficulty if no value is "true", which is often the case. An elaborate scheme has been invented to avoid this by calling $Q(\mu)$ the "degree of belief" in the value μ. The trouble comes from the fact that any rational demand of the behavior of "degrees of belief" makes them to satisfy the axioms for probability, which apparently leaves the original difficulty intact.

A much more serious difficulty is the selection of the prior, which obviously plays a fundamental role in the posterior and all the further developments. One attempt is to try to fit it to the data, but that clearly not only contradicts the very foundation of Bayesian philosophy but without restrictions on the priors disastrous outcomes can be prevented only by ad hoc means. A different and much more worthwhile line of endeavor is to construct "noninformative" priors, even though there is the difficulty in defining the "information" to be avoided. Clearly, a uniform distribution appeals to intuition whenever it can be defined, but as we see below there is a lot of useful knowledge to be gained even with such "noninformative" distributions!

The MDL principle, too, permits the use of distributions for the parameters. However, the probabilities used are defined in terms of code lengths, which not only gives them a physical interpretation but also permits their optimization thereby removing the anomalies and other difficulties in the Bayesian philosophy. Because of the difference in the aims and the means to achieve them the development of the *MDL* principle and the questions it raises differ drastically from Bayesian analysis. It calls for information and coding theory, which are of no particular interest nor even utility for the Bayesians.

The *MDL* principle was originally aimed at obtaining a criterion for estimating the number of parameters in a class of ARMA models [13], while a related method was described earlier in [29] for an even narrower problem. The criterion was arrived at by optimization of the quantification of the real-valued parameters. Unfortunately the criterion for the ARMA models turned out to be asymptotically equivalent with *BIC* [21], which has given the widely accepted wrong impression that the *MDL* principle *is BIC*. The acronym "BIC" stands for Bayesian Information Criterion, although the "information" in it has nothing to do with Shannon information nor information theory.

The usual form of the *MDL* principle calls for minimization of the ideal code length

$$\log 1/f(x^n;\mu) + L(\mu),$$

where $L(\mu)$ is a prefix code length for the parameters in order to be able to separate their code from the rest. Because a prefix code length defines a distribution by Kraft inequality we can write $L(\mu) = \log 1/Q(\mu)$. We call them "priors" to respect the Bayesian tradition even though they have nothing to do with anybody's prior knowledge. The fact that the meaning of the distribution is the (ideal) code length $\log 1/Q(\mu)$ with which the parameter value can be encoded imposes restrictions on these distributions, and unlike in the Bayesian philosophy we can try to optimize them.

The intent with the *MDL* principle is to obtain the shortest code length of the data in a self containing manner; i.e., by including in the code length all the parts needed. But technically this amounts to a prefix code length for the data to be calculated, ideally, using only the means provided by the model class, although in some cases this may have to be slightly augmented. Otherwise, regardless of the model class given we could get a shorter coding by Kolmogorov complexity, which we want to exclude. Hence, not only are the minimizing parameters and the prior itself to be included when the code length is calculated but also the probability model for the priors, and so on. This process stops when the last model for a model for a model ... is found which either assigns equal code length to its arguments or is common knowledge and need not be encoded. Since each model teaches a property of the data, we may stop when no more properties of interest can be learned. Usually, two or three steps in this process suffice. For more on the *MDL* principle we refer to [7].

Before applying the *MDL* principle to the model classes (2.1) and (2.2) we illustrate the process with an example. Take an integer n as the object of interest without any family of distributions. As the first "model" we take the set $\{1,2,\ldots,2^m\}$, where m is the smallest integer such that n belongs to the set or that $\log_2 n \le m$. We need to encode the model, or the number m. We repeat the argument and get a model for the first model, or the smallest integer k such that $\log_2 m \le k$. This process ends when the last model has only one element $1 = 2^0$. Such an actual coding system was described in [6], and the total code length is about $L(n) = \log_2^*(n) = \log_2 n + \log_2 \log_2 n + \cdots$, the sum ending with the last positive iterated logarithm value. It was shown in [14] that $P^*(n) = C^{-1} 2^{-L(n)}$ for $C = 2.865\ldots$ defines a universal prior for the integers. We can now define $\log_2 1/P^*(n)$ as the complexity of n and $\log_2 1/P^*(n) - \log_2 n$, or, closely enough, the sum $\log_2^*(\log_2) = \log_2 \log_2 n + \cdots$, as the "information" in the number n that we learn with the models given. In other words, the amount of information is the code length for encoding the models needed to encode the object n.

2.3 Complexity and Universal Models

A universal model is a fundamental construct borrowed from coding theory and little known in ordinary statistics. Its roots, too, are in the algorithmic theory of complexity. In fact, it was the very reason for Solomonoff's query for the shortest program length, because he wanted to have a universal prior for the integers as binary strings; see the section on Kolmogorov complexity below.

Given a class of models (2.1) we define a model $\widehat{f}(y^n;k)$ to be *universal* for the class if

$$\frac{1}{n}\log\frac{f(y^n;\theta,k)}{\widehat{f}(y^n;k)} \to 0 \tag{2.3}$$

for all parameters $\theta \in \Omega^k$, and *optimal universal* if the convergence is the fastest possible for almost all θ; the convergence is in a probabilistic sense, either in the mean, taken with respect to $f(y^n;\theta,k)$, in probability, or almost surely. The mean sense is of particular interest, because we then have the convergence in Kullback–Leibler distance between the two density functions and hence consistency. The qualification "almost all θ" may be slightly modified; see [15, 16]. If we extend these definitions to the model class (2.2), we can talk about consistency in the number of parameters, and ask again for fastest convergence. In the literature studies have been made of the weakest criteria under which consistency takes place [8]. Such criteria cannot measure consistency in terms of the Kullback–Leibler distance and do not seem to permit query for optimality.

2.3.1 The NML Universal Model

Let $y^n \mapsto \widehat{\theta}(y^n)$ be the Maximum Likelihood (ML) estimate which minimizes the ideal code length $\log 1/f(y^n;\theta,k)$ for a fixed k. Consider the maximized joint density or probability of the data in the minimized negative logarithm form

$$L(y^n,\widehat{\theta}) = -\log f(y^n;\widehat{\theta},k)/h(\widehat{\theta},\widehat{\theta})) - \log w(\widehat{\theta})\,, \tag{2.4}$$

$$h(\widehat{\theta};\theta) = \int_{y^n:\widehat{\theta}(y^n)=\widehat{\theta}} f(y^n;\theta,k)dy^n\,,$$

$$w(\widehat{\theta}) = h(\widehat{\theta};\widehat{\theta})/C_{n,k}\,, \tag{2.5}$$

$$C_{n,k} = \int_{\widehat{\theta}\in\Omega^k} h(\widehat{\theta};\widehat{\theta})d\widehat{\theta}\,,$$

where Ω^k is such that the integral is finite. We have rederived the Normalized ML or *NML* model

$$\widehat{f}(y^n;k) = \frac{f(y^n;\widehat{\theta}(y^n),k)}{C_{n,k}} \tag{2.6}$$

by an application of the *MDL* principle. It was originally obtained as the solution to Shtarkov's minmax problem

$$\min_q \max_{y^n} \log \frac{f(y^n;\widehat{\theta}(y^n),k)}{q(y^n)} \tag{2.7}$$

[2, 17, 22]. Indeed, the *NML* model gives the unique prefix code which is closest to the shortest possible code length $\log 1/f(y^n;\widehat{\theta}(y^n);k)$, itself not a prefix code length. It is also the unique solution to the related maxmin problem [19]

$$\max_g \min_q E_g \log \frac{f(Y^n;\widehat{\theta}(Y^n),k)}{q(Y^n)} = \max_g \min_q D(g\|q) - D(g\|\widehat{f}(Y^n;k)) + \log C_{n,k},$$

where g ranges over any set that includes $\widehat{f}(y^n;k)$, as well as the associated minmax problem – although the maximizing distribution is then nonunique.

For the special prior $w(\widehat{\theta})$ the joint density function equals the marginal $p(y^n) = \int_{\widehat{\theta}\in\Omega^k} f(y^n;\widehat{\theta},k)w(\widehat{\theta})d\widehat{\theta}$, and the maximized posterior is given by the $\delta(\widehat{\theta},\theta)$-functional, whose integral over any subset of Ω^k of volume Δ is unity. This means that the posterior probability of the ML parameters, quantized to any precision, is unity. We might call it the *canonical* prior.

We have defined [17]

$$\log 1/\widehat{f}(y^n;k) = \log 1/f(y^n;\widehat{\theta}(y^n),k) + \log C_{n,k} \tag{2.8}$$

as the *stochastic complexity* of the data y^n, given the class of models (2.1). If the model class satisfies the central limit theorem and other smoothness conditions the stochastic complexity is given by the decomposition [17]

$$\log \frac{f(y^n;\widehat{\theta}(y^n),k)}{\widehat{f}(y^n;k)} = \frac{k}{2}\log\frac{n}{2\pi} + \log \int_{\Omega^k} |J(\theta)|^{1/2} d\theta + o(1), \tag{2.9}$$

where $J(\theta)$ is the Fisher information matrix

$$J(\theta) = \lim n^{-1} E\{\frac{\partial^2 \log 1/f(y^n;\theta,k)}{\partial\theta_i \partial\theta_j}\}.$$

There is a further asymptotic justification for *stochastic complexity* by the theorems in [15] and [16], which imply, among other things, that $\widehat{f}(y^n;k)$ is optimal universal in the mean sense. This means that there is no density function which converges to the data generating density function $f(y^n;\theta,k)$ faster than $\widehat{f}(y^n;k)$. In particular, no other estimator $y^n \mapsto \bar{\theta}(y^n)$ can give a smaller mean length for the normalized density function $f(y^n;\bar{\theta}(y^n),k)$ than the ML estimator.

We mention in conclusion that for discrete data the integral in the normalizing coefficient becomes a sum, for which efficient algorithms for reasonable size of n have been developed in a number of special cases [9, 10, 24–27]. For such cases the structure of the models can be learned much better than by minimization of the asymptotic criterion (2.9).

2.3.1.1 NML Model for Class $\mathcal{M}$

The minimization of the stochastic complexity (2.8) with respect to k is meaningful both asymptotically and non-asymptotically. For instance, for discrete data $\log 1/\widehat{f}(y^n;n)$ can equal $\log C_{n,n}$, which is larger than the minimized stochastic complexity $\log 1/\widehat{f}(y^n;\widehat{k})$. This is a bit baffling, because $\log 1/\widehat{f}(y^n;\widehat{k})$ is not a prefix code length; i.e., $\widehat{f}(y^n;\widehat{k})$ is not a model. To get a logical explanation as well as to be able at least to define an optimal universal model for the class (2.2) let $y^n \mapsto \widehat{k}(y^n) = \widehat{k}$ denote the estimator that maximizes $\widehat{f}(y^n;k)$, and hence minimizes the joint code length

$$L(y^n,\widehat{k}) = -\log \widehat{f}(y^n;\widehat{k})/g(\widehat{k};\widehat{k})) - \log v(\widehat{k}), \tag{2.10}$$

$$g(\widehat{k};k) = \int_{y^n:\widehat{k}(y^n)=\widehat{k}} \widehat{f}(y^n;k)dy^n ,$$

$$v(\widehat{k}) = g(\widehat{k};\widehat{k})/C_n , \tag{2.11}$$

$$C_n = \sum_1^n g(\widehat{k};\widehat{k}).$$

The model

$$\widehat{f}(y^n) = \frac{\widehat{f}(y^n;\widehat{k})}{C_n} \tag{2.12}$$

is optimal universal for the class $\mathcal{M}$, and it solves Shtarkov's type of minmax problem

$$\min_q \max_{y^n} \log \frac{\widehat{f}(y^n;\widehat{k}(y^n))}{q(y^n)}$$

and the associated maxmin problem

$$\max_g \min_q E_g \log \frac{\widehat{f}(Y^n;\widehat{k}(Y^n))}{q(Y^n)}.$$

The difficulty is to calculate the probabilities $g(\widehat{k};\widehat{k})$, but for our purpose here we do not need that, for we see that $\widehat{k}$, which minimizes the non-prefix code length $\log 1/\widehat{f}(y^n;\widehat{k}(y^n))$, also minimizes the prefix code length $\log 1/\widehat{f}(y^n)$.

Although $\widehat{f}(y^n;k)$ provides an excellent criterion for the structure and the number of parameters in the usual cases where $\widehat{k}$ is not too large in comparison with n, it is not good enough for denoising, where $\widehat{k}$ is of the order of n. Then we cannot calculate $h(\widehat{\theta},\widehat{\theta})$, $C_{n,k}$, nor the joint density, accurately enough. One way in the regression problems, which include denoising problems, is to approximate the maximum joint density by two-part coding thus

$$\log 1/f(y^n|X_n;\widehat{\theta}^k,k) + L(k), \tag{2.13}$$

where X_n is an $n \times n$ regressor matrix and $L(k)$ is a prefix code length of k. For instance, when X_n is defined by wavelets we need to specify which k rows correspond to the same number of the largest wavelet coefficients [19], which can be done with $L(k) = \log \binom{n}{k}$ bits.

2.3.2 Mixture Model

The second universal model for the class (2.1) is the mixture model together with a prior $\mu(\theta)$

$$f_\mu(y^n;k) = \int_\Omega f(y^n;\theta,k)\mu(\theta)d\theta. \tag{2.14}$$

Such a mixture model $\widehat{q} = f_\mu(y^n;k)$ satisfies the problem

$$\min_q \int \mu(\theta) D(f(Y^n;\theta,k)\|q(Y^n))d\theta, \tag{2.15}$$

where $D(f(Y^n;\theta,k)\|q(Y^n))$ is the Kullback–Leibler distance between the two density functions shown. There is a special prior $\bar{\mu}$ maximizing the minimized result

$$\max_\mu \int \mu(\theta) D(f(Y^n;\theta,k)\|f_\mu(Y^n;k))d\theta = K_{n,k},$$

which is often called the capacity of the channel $\Theta \to Y^n$, even though the analogy with Shannon's capacity of a noisy channel for sending a finite number of messages is somewhat obscure. With this prior the mixture lies at the center of a hyperball with the model class $\mathcal{M}_k$ as the surface and the radius given by the constant distance $D(f(Y^n;\theta,k)\|f_{\bar{\mu}}(Y^n;k))$. Such a prior is not easy to calculate.

However, asymptotically one can show, [4], that with Jeffreys' prior

$$w(\theta) = \frac{|J(\theta)|^{1/2}}{\int_\Omega |J(\eta)|^{1/2} d\eta},$$

we get

$$E_\theta \log \frac{f(Y^n;\widehat{\theta}(Y^n),k)}{f_w(Y^n,k)} = \frac{k}{2}\log\frac{n}{2\pi} + \log \int_\Omega |J(\eta)|^{1/2} d\eta + o(1). \tag{2.16}$$

Since the right-hand side does not depend on θ the channel capacity achieving prior coincides in the limit with Jeffreys' prior. We see that with Jeffreys' prior the mixture universal model, too, is optimal.

The negative logarithm of the mixture universal model provides also a criterion for the selection of the number of parameters, at least asymptotically. For small and moderate amounts of data the prior must be taken so that the mixture can be computed. Hence, the result will depend on the prior.

2.3.3 Sequentially Normalized Universal Model

The *NML* model, while providing often more information about the data than any other model, has two drawbacks: the difficulty in calculating the normalizing coefficient, and it does not define a random process, which prevents its use for prediction. When the normalization integral does not exist, as in the important linear-quadratic regression problem, we need to restrict the range by hyper parameters, which however means that the resulting universal model depends on their selection. A way to solve this for cases where the ML estimate $\widehat{k}$ is small is discussed in [18]. We can create a random process defining universal model by sequential normalization, which, moreover, simplifies the normalization. In the important linear-quadratic regression problem the normalization difficulty disappears completely without hyper parameters.

Let $\widehat{\theta}_t = \widehat{\theta}(y^t)$ denote the maximum likelihood estimate with maximum number of components, not exceeding a selected k, which can be uniquely solved from the data y^t. Consider the sequentially maximized likelihood

$$f(y^n;\widehat{\theta}^n) = \prod_{t=0}^{n-1} f(y_{t+1} \mid y^t;\widehat{\theta}_{t+1}), \tag{2.17}$$

where $\widehat{\theta}^n = (\widehat{\theta}_1, \widehat{\theta}_2, \ldots, \widehat{\theta}_{m+1}, \ldots, \widehat{\theta}_n)$, with repetitions allowed. We take $f(y_1|y_0, \widehat{\theta}_1)$ $= f(y_1)$ as a suitable density function with $\widehat{\theta}_1$ empty. Also, for $t \le m$, the index after which k parameters can be solved, we may have $\widehat{\theta}_t = \widehat{\theta}_{t-1}$. One can show that in general $f(y^n;\widehat{\theta}^n) > f(y^n;\widehat{\theta}_n)$ so that it is bigger than the ordinary maximized likelihood!

We wish to normalize the conditionals (2.17) to generate recursively a density function $\widehat{f}(y^t) = \widehat{f}(y^{t-1})\widehat{f}(y_t \mid y^{t-1})$:

$$\widehat{f}(y_t \mid y^{t-1}) = \frac{f(y_t \mid y^{t-1};\widehat{\theta}(y^t))}{K(y^{t-1})}, \tag{2.18}$$

$$K(y^{t-1}) = \int f(y_t \mid y^{t-1};\widehat{\theta}(y^t))dy_t\,, \tag{2.19}$$

where the range of the integral may have to be restricted to be finite. This gives the universal sequentially normalized maximum likelihood, *SNML*, model as a random process

$$\widehat{f}_S(y^n;k) = \prod_{t=1}^{n} \widehat{f}(y_t \mid y^{t-1}). \tag{2.20}$$

In the linear quadratic regression case, [20], consider the representation

$$y_t = \sum_{i=1}^{k} \widehat{a}_{t,i} x_{t,i} + e_t,$$

where the coefficients $\widehat{a}_{t,i}$ are the least squares estimates from the data $y^t|X_t$, and $X_t = [\bar{x}_1,\ldots,\bar{x}_t]$ is the regressor matrix defined by the columns $\bar{x}_i = \{x_{i,j}\}$. Notice that these depend on the current data point y_t, which means that the sum is not its prediction even in the AR case, where $x_{t,i} = y_{t-i}$. These estimates minimize the sum of the updated squared errors $\widehat{s}_n = \sum_{t=1}^{n} e_t^2$. If we normalize the ratio

$$\frac{\widehat{s}_t^{\,-t/2}}{\widehat{s}_{t-1}^{\,-(t-1)/2}} = \widehat{s}_{t-1}^{\,-1/2}\left(1+\frac{(y_t-\widehat{y}_t)^2}{\widehat{s}_{t-1}}\right)^{-t/2}$$

we obtain conditional density functions with their product defining a universal sequentially normalized least squares, *SNLS*, model in terms of its negative logarithm

$$\widehat{f}(y^n \mid X_n) = \prod_{t=1}^{n} \widehat{f}(y_t \mid y^{t-1}, X_t)$$

$$\ln 1/\widehat{f}(y^n \mid X_n) = \frac{n}{2}\ln(\widehat{s}_n) - \ln\Gamma(n/2) + \ln\prod_t \frac{\sqrt{\pi}}{1-d_t},$$

where $d_t = \bar{x}_t'[X_t X_t']^{-1}\bar{x}_t$, the prime denoting the transposition. Put $\widehat{f}(y_1|y^0, x_1) = 1$.

We mention to this end that if the estimate $\widehat{a}_{t,i}$ is replaced by $\widehat{a}_{t-1,i}$ we get Dawid's prequential model, [5]. Such a model is also universal. It would be worse than the *SNML* model except for data where y_t adds little new information and is easy to predict.

2.4 Kolmogorov's Structure Function

Kolmogorov never published his work on the structure function, but it is discussed extensively in [28] along with elaborations. We begin with the current definition of the Kolmogorov complexity, as it is now called. It was given originally by Solomonoff [23], in a defective form, which was later fixed by Kolmogorov, see [11], and Chaitin [3]. Consider the set of programs as binary strings for any universal computer U such that when a program p is executed the output is a binary string, which we write as $U(p) = x$. Any finite binary string x has actually a countable number of programs that can deliver it. The programs then are codewords for binary strings, and with a preamble we can make them to occupy a leaf in a countable prefix free tree. Write dp as the so adjusted self-delimiting program.

Kolmogorov complexity relative to the programming language of U is defined as the shortest prefix free program that generates the string x,

$$K(x) = \min\{|dp| : U(dp) = x\},$$

where $|y|$ denotes the length of the binary string y. The reason for the requirement that the programs are leaves is that they satisfy the Kraft inequality,

$$\sum_x 2^{-K(x)} < 1,$$

and by normalization we get a superb universal model for the set of finite binary strings

$$P(x) = 2^{-K(x)} / \sum_y 2^{-K(y)}.$$

This can, in fact, imitate any computable distribution $Q(x)$ by assigning almost as large a probability to strings as this distribution; i.e., for a constant C_Q, which does not depend on the strings x,

$$P(x) \geq C_Q Q(x).$$

We follow here [19]. Consider next any finite set S that includes x defining a *model* of data x; i.e., it describes some properties of x, perhaps not all of them. This corresponds to intuition as follows:

- All strings in S share the defining property of the set S.
- Size $|S|$ is an inverse measure of the amount of properties extracted from string with S.
- Large $|S| \Leftrightarrow$ few properties (restrictions) of x extracted.
- $S = \{x\}$, ($|S| = 1$), captures all conceivable properties of x.

It is clearly pointless to claim that one model (set $S \ni x$) is "true" and others "false"! However, we can define an *optimal* model by the *structure function*:

$$h_x(\alpha) = \min_{S \in \mathscr{S}} \{\log |S| : K(S) \leq \alpha\}, \tag{2.21}$$

where α is a parameter and $\mathscr{S}$ a set restricting the properties we are after. Kolmogorov himself did not have such a restriction, but without it one can get into trouble, which we do not go into. When we implement this procedure in statistics we automatically restrict the properties desired. Clearly any finite set has its computable membership function and hence its complexity. The idea is to have the shortest code length of whatever remains when properties of maximum amount α are extracted from x. The structure function then may be interpreted to measure the amount of "noise".

It is clear that $\alpha > \alpha' \Rightarrow h_x(\alpha) < h_x(\alpha')$. The *structure line* $K(x) - \alpha$ represents the least possible amount of unexplained "noise" on level α. We also have the two-part code length for the string x and the best model on the level α

$$L(x, \alpha) = L(x|S_\alpha) + L(S_\alpha) = h_x(\alpha) + \alpha.$$

The optimum level $\bar{\alpha}$ and model $S_{\bar{\alpha}}$ was defined by Kolmogorov as follows

$$\min\{\alpha : h_x(\alpha) + \alpha \doteq K(x)\},$$

where the notation $\doteq$ means equality to within a constant not dependent on the length of the string x. But since $K(x)$ is a lower bound for the two-part code length we get the same by an application of the *MDL* principle with the twist that the conditional code length $L(x|S_\alpha)$ is not the shortest length but the shortest for the worst case $x \in S_\alpha$. The reason for this is to regard the code length for all "noise" sequences as the same, namely, the worst case code length, so that we could not take advantage of a special string in S_α that happens to be easy to encode. For the same reason we do not take $K(x|S_\alpha)$ as the conditional code length, because it might take advantage of other properties than those in the set S_α.

The optimal set $S_{\bar{\alpha}}$ represents all learnable properties of x that can be captured by finite sets of the desired kind. However, the structure function cannot be applied as such for the Kolmogorov complexity is noncomputable.

2.5 Structure Function and Information

The stochastic complexity provides an excellent criterion for finding the structure of the data by the models fitted to the data, and as the first try we may take the code length of $\widehat{k}$ to be the amount of structure information in the number of parameters gained from the data. Indeed, intuitively we regard the properties of the optimal model to represent the "information" we are after, but when the data include something that we do not capture with our models and what we call "noise" the issue gets more complex. We would like to define "information" as the properties we learn in the presence of "noise"; i.e., when the effect of noise is removed to the extent it can be removed. For instance, in case of the structure parameter $\widehat{k}$, consisting of the first $\widehat{k}$ components, there may be so much noise that specifying $\widehat{k}$ only to the precision $\widehat{k}+1$ or $\widehat{k}-1$ would better represent the learnable information. This is particularly striking when we ask for the "information" in the real-valued parameters $\widehat{\theta}$, which without quantization would be infinite. And clearly, we cannot expect to gain more information from the data than the finite code length needed to encode all them.

We next follow the plan in Kolmogorov's structure function to find the optimal quantization of the real-valued parameters; for details we refer to [19]. As a result we can describe a finite subset of the compact parameter space which consists of models that can be optimally distinguished from a finite amount of data. The other models in a neighborhood of each of them add no useful learnable information.

For the model class $\mathcal{M}_k = \{f(y^n;\theta)\}$, where we drop the index k in the models, consider a finite partition $\Lambda_n = \{B_i : i = 1,2,\ldots,N_n\}$ of a compact parameter space Ω^k and the representatives $\theta^i \in B_i$ such that the Kullback–Leibler distance $D(f_i \| f_{i+1})$ between the adjacent models $f(y^n;\theta_i) = f_i$ is constant. This can be achieved to an increasing accuracy as n grows by the following construct, [19]: Take $B_i = B_{d/n}(\theta^i)$ as the maximal rectangle within the hyper ellipsoid $(\theta-\theta^i)'J(\theta^i)(\theta-\theta^i) = d/n$, centered at θ^i, where $J(\theta)$ is the Fisher information matrix satisfying

$$0 < J(\theta) = \lim n^{-1} E_\theta \left\{ \frac{\partial^2 \log 1/f(Y^n;\theta)}{\partial\theta_i \partial\theta_j} \right\} < \infty;$$

the expectation is with respect to the distribution $f(y^n;\theta)$. The volume of $B_{d/n}(\theta^i)$ is given by

$$V = \left(\frac{4d}{kn}\right)^{k/2} |J(\theta^i)|^{-1/2}.$$

We do not need to actually construct such a partition for we only need an approximation of a rectangle that includes the ML estimate $\widehat{\theta}$. In fact, we'll take the center θ_i of it as a quantized version of $\widehat{\theta}$, defined by the size parameter d. Let $N_{d,n}$ be the number of the rectangles.

We need the following analogies of the algorithmic notions:

- A set of programs is to be replaced by the model class $\mathscr{M}_k$.
- A set S is to be replaced by a quantized model $f(x^n;\theta^i)$.
- Kolmogorov complexity is to be replaced by stochastic complexity.
- $K(S)$ is to be replaced by the shortest code length $L(\theta^i)$.
- $\log|S|$, which is the maximum code length of $y \in S$, is to be replaced by the maximum or the mean code length of typical strings of $f(x^n;\theta^i)$, which we take to be the strings for which $\widehat{\theta} \in B_{d/n}(\theta^i)$.

The reason to measure the code length of the typical strings by the maximum or the mean is the same as in Kolmogorov's structure function. We consider the typical strings as noise, and we should take their code lengths as equal, either as the maximum or the mean. It is true that if we order the strings in some fashion and encode them by their ordinal, then a small fraction of the strings could be encoded with a short code length. But this does not correspond to intuition about noise strings, of which none should be treated differently from the others.

The density function for the ML estimates in (2.4) induces the probability distribution for the equivalence classes

$$Q_{d/n}(\theta^i) = \int_{\widehat{\theta}\in B_{d/n}(\theta^i)} h(\widehat{\theta};\widehat{\theta})d\widehat{\theta} \to \left(\frac{2d}{\pi k}\right)^{k/2}. \tag{2.22}$$

This gives asymptotically the uniform distribution for the centers $W(\theta^i)$ given by $Q_{d/n}(\theta^i)/C_{n,k}$ and the code length $L(\theta^i) = \log 1/W(\theta^i)$. Under the conditions for which the formula (2.9) holds, we get

$$L_d(\theta^i) = \ln 1/W(\theta^i) \cong \frac{k}{2}\ln\frac{\pi k}{2d} + \ln C_{n,k}.$$

We can now define two structure functions. The first is obtained by measuring the amount of noise by the maximum code length of strings $\ln 1/f(x^n;\theta^i)$ in the rectangle $B_{d/n}(\theta^i)$, which is given by

$$\ln 1/f(x^n;\widehat{\theta}(x^n)) + d/2.$$

Indeed, the lengths $\ln 1/f(x^n;\widehat{\theta}(x^n))$ are nearly constant for all strings in $B_{d/n}(\theta^i)$, which is small for large n, and hence if we fix the center θ^i, then by Taylor's series the maximum code length $\ln 1/f(x^n;\theta^i)$ of strings in $B_{d/n}(\theta^i)$ is when the ML estimate falls in a corner of the rectangle. Hence, $\ln[f(x^n;\widehat{\theta}(x^n))/f(x^n;\theta^i)] = d/2$. We then get the first structure function

$$h^1_{x^n}(\alpha) = \min_d\{\ln 1/f(x^n;\widehat{\theta}(x^n)) + d/2 : L_d(\theta^i) \leq \alpha\}, \tag{2.23}$$

where $\widehat{\theta}(x^n) \in B_{d/n}(\theta^i)$. The minimizing d is the one that minimizes the two-part code length $\ln 1/f(x^n;\widehat{\theta}(x^n)) + d/2 + L_d(\theta^i)$, or asymptotically $\widehat{d} = k$.

The other structure function, which we need later, results from taking the average code length of the typical strings. We get

$$\begin{aligned} h^2_{x^n}(\alpha) &= \ln 1/f(x^n;\widehat{\theta}(x^n)) \\ &+ \min_d\{\frac{1}{|B_{d/n}(\theta^i)|}\int_{\widehat{\theta}(y^n)\in B_{d/n}(\theta^i)} \ln\frac{f(x^n;\widehat{\theta}(x^n))}{f(x^n;\theta^i)}dy^n : L_d(\theta^i)\leq\alpha\}. \end{aligned} \tag{2.24}$$

The average of the integral gives asymptotically $d/6$, and the optimum size parameter is $\widehat{d} = 3k$. With the second structure function we get the *universal sufficient statistics decomposition*

$$h^2_{x^n}(\bar{\alpha}) + \bar{\alpha} = \ln 1/f(x^n;\widehat{\theta}(x^n)) + k/2 + \ln C_{n,k} + (k/2)\ln(\pi/6), \tag{2.25}$$

where

- The amount of learnable information in data is given by

$$\bar{\alpha} = \ln C_{n,k} + (k/2)\ln(\pi/6).$$

- Leaving the amount of unexplained noise as

$$h^2_{x^n}(\bar{\alpha}) = -\ln f(x^n;\widehat{\theta}(x^n)) + k/2.$$

Figure 2.1 illustrates these.

2.6 Optimal Distinguishability

There is another property in the set of the models quantized by the size parameter $\widehat{d} = 3k$, which changes the way one can do hypothesis testing. First, we need an index measuring the separation between the set of the quantized models. Let $P_{i|j}$ denote the probability of $B_{d/n}(\theta^i)$ under the model $f(x^n;\theta^j)$. Then take

$$\frac{1}{N_{d,n}}\sum_{i,j} P_{i|j} = 1 - \frac{1}{N_{d,n}}\sum_i P_{i|i}, \tag{2.26}$$

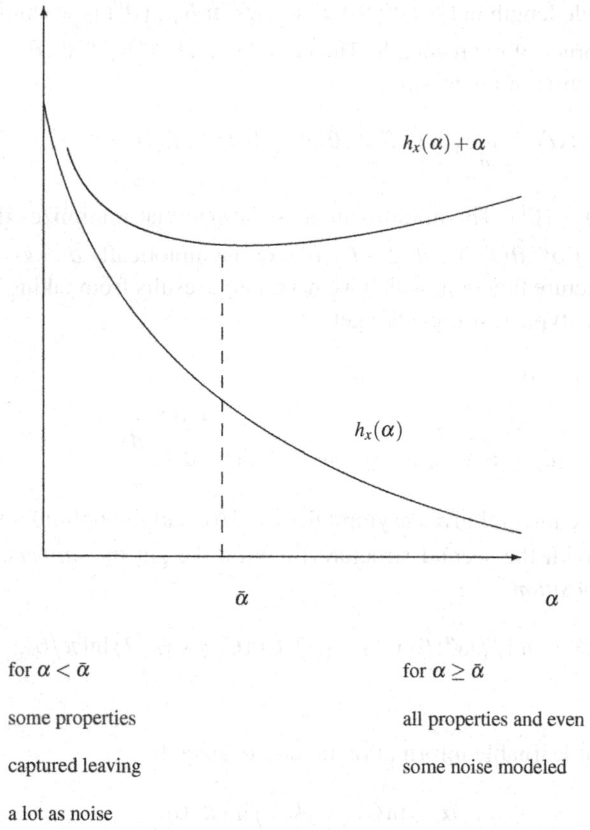

Fig. 2.1 The universal sufficient statistics decomposition

as the measure of separation between the models in the family. It is seen to generalize the index of separation between two models. We see that the smaller this index is the better the models in the family are separated. One can show that a similarly defined index of separation for any partition of size $N_{d,n}$ cannot be smaller than that of the partition $\Lambda_{d,n} = \{B_{d/n}(\theta^i)\}$.

The persistent problem with the Neyman–Pearson hypothesis theory is that one cannot optimize the level of the test of a null-hypothesis against a composite hypothesis, simply because the power of an opposite hypothesis grows with an increasing separation between it and the null-hypothesis. Hence, all one can do is set arbitrarily the level to, say 0.05, and reject the null hypothesis only for data where the test statistic, such as the ML estimate, falls in the so-selected tail. The trouble with such a lopsided test is that it rejects almost all reasonable opposing hypotheses, because if we bother to test at all we certainly expect the data to be such that the test statistic of the opposing hypotheses does not fall very far in the tail of the null-hypothesis.

It is clear that we cannot minimize the index of separation, which shrinks to zero as the distance between two hypotheses grows; i.e., d in our family grows. How then can we optimize it? The answer requires us to examine hypothesis testing more closely. First, the hypotheses should be tested as models rather than testing when one of them is "true", which never is the case. There are two desired properties of a well separated family of models: The first is that the density functions $f(y^n;\theta)$ for $\theta \in B_{d/n}(\theta^i)$ in each equivalence class should be close to its representative $f(y^n;\theta^i)$ so that they could be collapsed to it, and the properties they capture about the data differ only due to noise. The second is that each representative should assign a large probability mass to its equivalence class to make the adjacent models defined by the representatives different.

If the CLT holds, these are conflicting properties for the family constructed above, but ideally satisfied by the family

$$\widehat{f}(y^n|\theta^i) = \begin{cases} f(y^n;\widehat{\theta}(y^n))/Q_{d/n}(\theta^i), & \text{if } \widehat{\theta}(y^n) \in B_{d/n}(\theta^i), \\ 0, & \text{otherwise.} \end{cases}$$

This suggests that we get an optimal separation by asking for the size index d for which the models are as close as they can be to the corresponding perfectly separated models in the Kullback–Leibler distance

$$\min_d D(\widehat{f}(Y^n|\theta^i)\|f(Y^n;\theta^i)). \tag{2.27}$$

We see that for very small values of d the values of the density function $f(y^n;\theta^i)$ within its equivalence class vary little so that the ideal flatness is nearly reached, but the peak value of $\widehat{f}(y^n|\theta^i)$ is much greater than that of $f(y^n;\theta^i)$, and the KL distance is considerable. In the other extreme for large values of d the peaks differ less but $f(y^n;\theta^i)$ varies a lot and again the KL distance is large. Hence, for the minimizing d the two desired requirements are balanced and an optimal separation reached.

Perhaps not too surprisingly the optimal separation asymptotically is reached for $\widehat{d} = 3k$, which agrees with the previous value that removes the average noise in (2.24). The same value is close to that found by Balasubramanian [1], by very different means without partitions, who was the first to study distinguishability.

The test of the null hypothesis, say $f_0(y^n)$, one of the optimally distinguishable models, against the others, is simply to accept the null hypothesis if and only if $\widehat{\theta}(x^n) \in B_{3k/n}(\theta^0)$. The confidence in the result as measured by the error probability in case of acceptance, is $1 - P_{0|0}$, and $P_{0|i}$ in case of rejection when $\widehat{\theta}(x^n)$ falls in $B_{3k/n}(\theta^i)$ for $i \neq 0$. We mention that this differs from the clumsy way the confidence was calculated in [19].

Normally distributed data in an frequently studied signal processing application were studied in [12], where $\widehat{d} = 3k$ even non asymptotically.

We mention to this end that there is a different way to define the optimally distinguishable models, which has the powerful property that if and only if the growth rate of the number of the models is below $n^{k/2}$ each can be estimated without error in the limit. This is close to Shannon's channel capacity.

2.7 Conclusions

This chapter describes an approach to statistics, above all model selection and testing, without the customary assumption that the data have been generated by a "true" distribution, parametric or nonparametric. Instead of measuring a model's performance by the nearness to the imagined "true" distribution the yardstick is the number of bits with which the data and the model can be encoded when advantage is taken of the constraints the model prescribes to the data. The objective then is to find a minimizing code for a class of distributions as models, which is the *minimum description length* principle, or, equivalently, a global maximum likelihood principle, global because it permits finding an optimal model, including its structure and the number of parameters.

References

1. Balasubramanian, V. (1996), Statistical inference, Occam's razor and statistical mechanics on the space of probability distributions, *Neural Comput.*, **9**(2), 349–268
2. Barron, A.R., Rissanen, J., and Yu, B. (1998), The MDL principle in modeling and coding, *IEEE Trans. Inform. Theory* **IT-44**(6), 2743–2760 (special issue to commemorate 50 years of information theory)
3. Chaitin, G.J. (1969), On the length of programs for computing finite binary sequences: statistical considerations *JACM*, **16**, 145–159
4. Clarke, B.C. and Barron, A.R. (1990), Information-theoretic asymptotics of Bayes methods, *IEEE Trans. Inform. Theory*, **IT-36**, 453–471
5. Dawid, A.P. (1984), Present position and potential developments: some personal views, statistical theory, the prequential approach, *J. R. Stat. Soc. A*, **147**(Part 2), 278–292
6. Elias, P. (1975), Universal codeword sets and representations of the integers, *IEEE Trans. Inform. Theory*, **IT-21**(2), 194–203
7. Grünwald, Peter, D. (2007), *The Minimum Description Length Principle*, MIT, Cambridge, 701 pages
8. Hannan, E.J. and Quinn, B.G. (1979), The determination of the order of an autoregression, *J. R. Stat. Soc. B*, **41**, 190–195
9. Korodi, G. and Tabus, I. (2005), An efficient normalized maximum likelihood algorithm for DNA sequence compression, *ACM Trans. Inform. Syst.*, **23**(1), 3–34
10. Korodi, G. and Tabus, I. (2007), Normalized maximum likelihood model of order-1 for the compression of DNA sequences, in *Proc. IEEE Data Compression Conference, DCC'07*, pp. 33–42, Snowbird, 27–29 March 2007.
11. Li, M. and Vitanyi, P.M.B. (1997), *An Introduction to Kolmogorov Complexity and its Applications*, Second Edition, Springer, New York, 637 pages
12. Razavi, S.A. and Giurcaneanu, C.D. (2008), Composite hypothesis testing by optimally distinguishable distributions, in *Proceedings of 2008 IEEE International Conference on Acoustics, Speech, and Signal Processing, ICASSP*, Las Vegas, Nevada, USA, 30 Mar – 4 Apr (accepted for publication)
13. Rissanen, J. (1978), Modeling by shortest data description, *Automatica*, **14**, 465–471
14. Rissanen, J. (1983), A universal prior for integers and estimation by minimum description length, *Ann. Stat.*, **11**(2), 416–431
15. Rissanen, J. (1984), Universal coding, information, prediction, and estimation, *IEEE Trans. Inform. Theory*, **IT-30**(4), 629–636

16. Rissanen, J. (1986), Stochastic complexity and modeling, *Ann. Stat.*, **14**, 1080–1100
17. Rissanen, J. (1996), Fisher information and stochastic complexity, *IEEE Trans. Inform. Theory*, **IT-42**(1), 40–47
18. Rissanen, J. (2000), MDL Denoising, *IEEE Trans. Inform. Theory*, **IT-46**(7), 2537–2543
19. Rissanen, J. (2007), *Information and Complexity in Statistical Modeling*, Springer, Berlin, 142 pages
20. Rissanen, J., Roos, T., and Myllymaki, P. (2008), Sequentially Normalized Least Squares Models, (to appear), http://www.mdl-research.org/jorma.rissanen
21. Schwarz, G. (1978), Estimating the dimension of the model. *Ann. Stat.* **6**, 461–464
22. Shtarkov, Y. (1987), Universal sequential coding of single messages, *Prob. Inform. Transmission*, **23**(3), 175–186
23. Solomonoff, R.J. (1960), A preliminary report on a general theory of inductive inference, Report ZTB-135, Zator, Cambridge, November 1960
24. Tabus, I. and Korodi, G. (2008), Genome compression using normalized maximum likelihood models for constrained Markov sources, in *IEEE Information Theory Workshop*, Porto, Portugal, May 5–9, 2008
25. Tabus, I., Korodi, G., and Rissanen J. (2003), DNA sequence compression using the normalized maximum likelihood model for discrete regression, in *Data Compression Conference 2003, DCC '03*, pp. 253–262, 2003
26. Tabus, I. and Rissanen, J., (2006), Maximum likelihood model for logit regression, Festschrift for T. Pukkila, University of Tampere, June 2006
27. Tabus, I., Rissanen, J., and Astola, J. (2006), Classification and feature gene selection using the normalized maximum likelihood model for discrete regressions, *Signal Process.*, Special issue on genomic signal processing; also available at http://sigwww.cs.tut.fi/ tabus/GSP.pdf
28. Vereshchagin, N.K. and Vitanyi P.M.B. (2004), Kolmogorov's Structure functions and model selection, *IEEE Trans. Inform. Theory*, **IT-50**(12), 3265–3290
29. Wallace, C.S. and Boulton, D.M. (1968), An information measure for classification, *Comput. J.*, **11**(2), 185–195

Chapter 3
Normalized Information Distance

Paul M. B. Vitányi, Frank J. Balbach, Rudi L. Cilibrasi, and Ming Li

Abstract The normalized information distance is a universal distance measure for objects of all kinds. It is based on Kolmogorov complexity and thus uncomputable, but there are ways to utilize it. First, compression algorithms can be used to approximate the Kolmogorov complexity if the objects have a string representation. Second, for names and abstract concepts, page count statistics from the World Wide Web can be used. These practical realizations of the normalized information distance can then be applied to machine learning tasks, especially clustering, to perform feature-free and parameter-free data mining. This chapter discusses the theoretical foundations of the normalized information distance and both practical realizations. It presents numerous examples of successful real-world applications based on these distance measures, ranging from bioinformatics to music clustering to machine translation.

3.1 Introduction

The typical data mining algorithm uses explicitly given features of the data to assess their similarity and discover patterns among them. It also comes with many parameters for the user to tune to specific needs according to the domain at hand.

P.M.B. Vitányi (✉)
CWI, Kruislaan 413, 1098 SJ Amsterdam, The Netherlands
e-mail: paulv@cwi.nl

F.J. Balbach
University of Waterloo, Waterloo, ON, Canada
e-mail: fbalbach@uwaterloo.ca
(supported by a postdoctoral fellowship of the German Academic Exchange Service (DAAD))
R.L. Cilibrasi
CWI, Kruislaan 413, 1098 SJ Amsterdam, The Netherlands
e-mail: cilibrar@cilibrar.com

M. Li
University of Waterloo, Waterloo, ON, Canada
e-mail: mli@uwaterloo.ca

F. Emmert-Streib, M. Dehmer (eds.), *Information Theory and Statistical Learning*,
DOI: 10.1007/978-0-387-84816-7_3,

In this chapter, by contrast, we are discussing algorithms that neither use features of the data nor provide any parameters to be tuned, but that nevertheless often outperform algorithms of the aforementioned kind. In addition, the methods presented here are not just heuristics that happen to work, but they are founded in the mathematical theory of Kolmogorov complexity. The problems discussed in this chapter will mostly, yet not exclusively, be clustering tasks, in which naturally the notion of distance between objects plays a dominant role.

There are good reasons to avoid parameter laden methods. Setting the parameters requires an intimate understanding of the underlying algorithm. Setting them incorrectly can result in missing the right patterns or, perhaps worse, in detecting false ones. Moreover, comparing two parametrized algorithms is difficult because different parameter settings can give a wrong impression that one algorithm is better than another, when in fact one is simply adjusted poorly. Comparisons using the optimal parameter settings for each algorithm are of little help because these settings are hardly ever known in real situations. Lastly, tweaking parameters might tempt users to impose their assumptions and expectations on the algorithm.

There are also good reasons to avoid feature based methods. Determining the relevant features requires domain knowledge, and determining how relevant they are often requires guessing. Implementing the feature extraction in an algorithm can be difficult, error-prone, and is often time consuming. It also limits the applicability of an algorithm to a specific field.

How can an algorithm perform well if it does not extract the important features of the data and does not allow us to tweak its parameters to help it do the right thing? Of course, parameter and feature free algorithms cannot mind read, so if we a priori know the features, how to extract them, and how to combine them into exactly the distance measure we want, we should do just that. For example, if we have a list of cars with their color, motor rating, etc. and want to cluster them by color, we can easily do that in a straightforward way.

Parameter and feature free algorithms are made with a different scenario in mind. In this *exploratory data mining* scenario we are confronted with data whose important features and how to extract them are unknown to us (perhaps there are not even features). We are then striving not for a certain similarity measure, but for *the* similarity measure between the objects. Does such an absolute measure of similarity exist at all? Yes, it does, in theory. It is called the information distance, and the idea behind it is that two objects are similar if there is a simple description of how to transform each one of them into the other one. If, however, all such descriptions are complex, the objects are deemed dissimilar. For example, an image and its negative are very similar because the transformation can be described as "invert every pixel." By contrast, a description of how to transform a blank canvas into da Vinci's *Mona Lisa* would involve the complete, and comparably large, description of that painting.

The latter example already points to some issues one has to take care of, like asymmetry and normalization. Asymmetry refers to the fact that, after all, the inverse transformation of the *Mona Lisa* into a blank canvas can be described rather simply. Normalization refers to the fact that the transformation description size must be seen in relation to the size of the participating objects. Section 3.2 details how these and other issues are dealt with and explains in which sense the resulting *information distance* measure is universal. The formulation of this distance measure will involve the mathematical theory of Kolmogorov complexity, which is generally concerned with shortest effective descriptions.

While the definition of the information distance is rather theoretical and cannot be realized in practice, one can still use its theoretical idea and approximate it with practical methods. Two such approaches are discussed in subsequent sections. They differ in which property of the Kolmogorov complexity they use and to what kind of objects they apply. The first approach, presented in Sect. 3.3, exploits the relation between Kolmogorov complexity and data compression and consequently employs common compression algorithms to measure distances between objects. This method is applicable whenever the data to be clustered are given in a compressible form, for instance, as a text or other literal description.

The second approach, presented in Sect. 3.4, exploits the relation between Kolmogorov complexity and probability. It uses statistics generated by common Web search engines to measure distances between objects. This method is applicable to non-literal objects, names and concepts, whose properties and interrelations are given by common sense and human knowledge.

3.2 Normalized Information Distance

Kolmogorov complexity measures the absolute information *content* of individual objects. For the purpose of data mining, especially clustering, we would also like to be able to measure the absolute information *distance* between individual objects. Such a notion should be universal in the sense that it contains all other alternative or intuitive notions of computable distances as special cases. Such a notion should also serve as an absolute measure of the informational, or cognitive, distance between discrete objects x and y. Such a notion of universal informational distance between two strings is the minimal quantity of information sufficient to translate between x and y, generating either string effectively from the other. As a result of the universality requirement, this information distance is uncomputable. However, the study of the abstract properties of such an absolute information distance leads to formulas and approaches applicable in practice, as we will demonstrate in subsequent sections.

In this section, we first give a brief introduction to the theory of Kolmogorov complexity, providing definitions and fundamental results. We then derive the unnormalized information distance and show its universality with respect to unnormalized distance measures. Finally we discuss the normalized information distance.

3.2.1 Kolmogorov Complexity

To provide some formal framework, we have to give a brief introduction to the theory of Kolmogorov complexity (a comprehensive treatment of this theory is [19]). In order to give a mathematically rigorous definition of Kolmogorov complexity and related terms, we need a few notations and definitions beforehand. By $\mathscr{N}$, $\mathscr{Q}$, $\mathscr{R}$, and $\mathscr{R}^+$ we denote the set of all natural, rational, real, and non-negative real numbers, respectively. For the cardinality of a set S we write $|S|$. In the following we will only consider binary strings $x \in \{0,1\}^*$. All other objects that we might consider can be encoded in a natural way as such strings. We write ε for the empty string, and $\ell(x)$ for the length of string x. All binary strings can be totally ordered according to their length and within the same length lexicographically. We implicitly identify every string with its number in this ordering. Using this identity, we have $\ell(x) = \lfloor \log(x+1) \rfloor$.

For any string x we denote by $\bar{x}$ the string $\bar{x} = 1^{\ell(x)}0x$, called the self-delimiting encoding of x. The set $\{\bar{x} : x \in \{0,1\}^*\}$ then is a prefix set, that is, no element of it is prefix of another element. Prefix sets have an important property, namely they satisfy the *Kraft inequality*:

Lemma 3.1. *Let $S \subset \{0,1\}^*$ be a prefix set. Then*

$$\sum_{x \in S} 2^{-\ell(x)} \leq 1 .$$

Partial functions whose domain is a prefix set are called prefix functions. They play a major role in the theory of Kolmogorov complexity.

Using $\bar{x}$ one can define a pairing function $\langle x,y \rangle = \bar{x}y$, which can be extended to k strings: $\langle x_1,\ldots,x_k \rangle = \langle x_1, \langle x_2, \langle \ldots \langle x_{k-1}, x_k \rangle \ldots \rangle\rangle\rangle = \bar{x}_1 \ldots \bar{x}_{k-1} x_k$. Functions with more than one argument can be realized in the usual way via the pairing function as $\varphi(x_1,\ldots,x_k) = \varphi(\langle x_1,\ldots,x_k \rangle)$. We use an effective enumeration $\varphi_1, \varphi_2, \ldots$ of all partial recursive functions with one argument, and also an effective enumeration $\psi_1, \psi_2, \ldots$ of all partial recursive prefix functions.

Now everything is in place to formulate the fundamental definition of Kolmogorov theory. Consider a function φ and two strings p and x such that $\varphi(p) = x$. The string p can be interpreted as a description of x by means of the *description language* φ. Of course, the string x can have many such descriptions, among which the shortest ones are special. The length of a shortest description p is called the complexity of the string x with respect to φ. A slightly more general version of complexity takes into account an additional input y. In this generalization the description of x is conditional to y:

Definition 3.1. Let φ be a partial recursive function. The conditional complexity (with respect to φ) of x given y is defined by

$$C_\varphi(x|y) = \min\{\ell(p) : \varphi(y,p) = x\} ,$$

the unconditional complexity of x by $C_\varphi(x) = C_\varphi(x|\varepsilon)$.

Clearly the complexity of a string depends on the choice of φ. There is, however, a distinguished function φ_0, which essentially assigns the lowest possible values, over all partial recursive functions φ.

Theorem 3.1. *There is a partial recursive function φ_0 such that for all partial recursive functions φ there is a constant c with*

$$C_{\varphi_0}(x|y) \le C_{\varphi}(x|y) + c$$

for all x and y.

It is sufficient for φ_0 to satisfy $\varphi_0(y,n,p) = \varphi_n(y,p)$. Intuitively the behavior of φ_0 is to take one of its arguments, n, and simulate the n-th partial recursive function on the input comprised of φ_0's other two arguments. In other words, φ_0 is a universal function for our enumeration $(\varphi_i)_{i\ge 1}$. The function φ_0 is not unique, but every such function defines essentially the same complexity C_{φ_0}, that is, up to an additive constant (this follows from Theorem 3.1). Instead of C_{φ_0} one typically simply writes C.

A helpful intuition for understanding C is to regard $C(x|y)$ as the length of a shortest computer program, in any popular language, that outputs x on input y.

While simple and elegant, the notion of $C(\cdot)$ has some oddities. For instance, $C(xy)$ is in general not upper-bounded by the sum of $C(x)$ and $C(y)$. This is one of the reasons that in many cases it is beneficial not to consider *all* partial recursive functions, but only the prefix functions. A result very similar to Theorem 3.1 holds for these functions.

Theorem 3.2. *There is a partial recursive prefix function ψ_0 such that for all partial recursive prefix functions ψ there is a constant c with*

$$C_{\psi_0}(x|y) \le C_{\psi}(x|y) + c$$

for all x and y.

Instead of $C_{\psi_0}(x|y)$ and $C_{\psi_0}(x|\varepsilon)$ it is customary to write $K(x|y)$ and $K(x)$, respectively. Also the expression $K(\langle x,y\rangle)$ is usually written simply as $K(x,y)$. We refer to K as the Kolmogorov complexity. Our intuition about values of $K(x|y)$ is essentially the same as for $C(x|y)$, the length of the shortest program that outputs x on input y. But in contrast to C, all shortest programs that "occur" in K constitute a prefix set. This implies, among other important things, that two such programs can be concatenated and still recognized as two distinct programs. This in turn allows the construction of a program that simulates two other programs and combines their output, at the same time being only a constant number of bits larger than the concatenation of the original two programs. A consequence of this is that for K the subadditivity holds, that is, $K(xy) \le K(x) + K(y) + O(1)$. The next theorem summarizes some more properties of K.

Theorem 3.3. *1. K is not partial recursive.*
2. $K(x) \leq \ell(x) + 2\log\ell(x) + O(1)$ for all x.
3. $K(x,y) \leq K(x) + K(y|x) + O(1)$ for all x, y.
4. Up to an additional term of $O(1)$:

$$K(x,y) = K(x) + K(y|\langle x, K(x)\rangle) = K(y) + K(x|\langle y, K(y)\rangle)$$

for all x, y.
5. Up to an additional term of $O(\log K(xy))$:

$$K(x,y) = K(x) + K(y|x) = K(y) + K(x|y)$$

for all x, y.

Item 1 does not come as a surprise since, intuitively, in order to find $K(x)$ one has to verify that no program under a certain length outputs x, a task that conflicts with the undecidability of the halting problem.

Item 2 gives an upper bound of $K(x)$ in terms of the length of x. In order to describe x prefix free, one can use an advanced self-delimiting encoding of x namely $\bar{\bar{x}} = \overline{\ell(x)}x$, which has the length $\ell(x) + 2\ell(\ell(x)) + 1$.

Items 3, 4, and 5 elaborate on the subadditivity property. While Item 3 only provides a better *upper bound*, which can be easily understood via the intuition of program lengths, the other two items state *equalities* and require sophisticated proofs. These results go by the name *Symmetry of Information* and will prove useful later in Sect. 3.3.1.

All programs that can be identified as shortest ones by K form a prefix set. It follows by the Kraft inequality that $\sum_x 2^{-K(x)} \leq 1$. This means that the values $2^{-K(x)}$ can be regarded as quantities very similar to probabilities, because they sum up to at most 1. In this assignment of values, less complex objects receive a higher probability than more complex objects. This can be seen as a "smooth" compromise between the contrary views of *Occam's Razor*, which advises to consider the simplest explanation only, and Epicurus's *Principle of Multiple Explanations*, which advises to consider all explanations. Indeed, this *algorithmic probability* $R(x) = 2^{-K(x)}$ is universal in a sense made clear in the remainder of this section.

Since in the realm of probabilities we are dealing with real valued functions, we first need to introduce some notions of computability for them.

Definition 3.2. A real valued function $f\colon \mathscr{N} \to \mathscr{R}$ is called *lower semicomputable* if there is a recursive function $g\colon \mathscr{N} \times \mathscr{N} \to \mathscr{Q}$ such that for all x the series $(g(x,k))_{k\in\mathscr{N}}$ is nondecreasing and $f(x) = \lim_{k\to\infty} g(x,k)$. The function f is called *upper semicomputable* if $-f$ is lower semicomputable.

A semimeasure assigns a non-negative real number to every string (or, equivalently, natural number). It differs from a probability measure in that the sum of all these values can be less than 1. In the same way as there are conditional probabilities, we can also consider conditional semimeasures.

Definition 3.3. A *discrete conditional semimeasure* is a function $P: \mathcal{N} \times \mathcal{N} \rightarrow \mathcal{R}^+$ such that for all y:

$$\sum_x P(x|y) \leq 1 .$$

The large class of *lower semicomputable* semimeasures has a universal element **m** that dominates every other lower semicomputable semimeasure up to a multiplicative constant.

Theorem 3.4. *There is a lower semicomputable discrete semimeasure* **m** *such that for all lower semicomputable discrete semimeasures P:*

$$P(x|y) = O(\mathbf{m}(x|y)) .$$

This universal semimeasure is intimately related to the Kolmogorov complexity via the *Conditional Coding Theorem* [19]:

Theorem 3.5. $-\log \mathbf{m}(x|y) = K(x|y) + O(1)$.

3.2.2 Information Distance

Intuitively, the minimal information distance between x and y is the length of the shortest program for a universal computer to transform x into y and y into x. This program then functions in a "catalytic" manner, being retained in the computer before, during, and after the computation. This measure will be shown to be, up to a logarithmic additive term, equal to the *maximum* of the conditional Kolmogorov complexities. The conditional complexity $K(y|x)$ itself is unsuitable as optimal information distance because it is asymmetric: $K(\varepsilon|x)$ is small for all x, yet intuitively a long random string x is not close to the empty string. The asymmetry of the conditional complexity $K(x|y)$ can be remedied by defining the algorithmic informational distance between x and y to be the sum of the relative complexities, $K(y|x) + K(x|y)$. The resulting metric will overestimate the information required to translate between x and y in case there is some redundancy between the information required to get from x to y and the information required to get from y to x.

For a partial recursive function φ, let

$$E_\varphi(x,y) = \min\{\ell(p) : \varphi(p,x) = y \text{ and } \varphi(p,y) = x\} .$$

Lemma 3.2. *There is a universal partial recursive prefix function* ψ_0 *such that for each partial recursive prefix function* ψ *and all* x, y,

$$E_{\psi_0}(x,y) \leq E_\psi(x,y) + c_\psi ,$$

where c_ψ *is a constant that depends on* ψ *but not on* x *and* y.

By Lemma 3.2, for every two universal prefix functions φ_0 and ψ_0, we have for all x, y that $|E_{\varphi_0}(x,y) - E_{\psi_0}(x,y)| \leq c$, with c a constant depending on φ_0 and ψ_0 but not on x and y. Thus the following definition is machine-independent.

Definition 3.4. Fixing a particular universal prefix function ψ_0, we define *information distance* as

$$E_0(x,y) = \min\{\ell(p) : \psi_0(p,x) = y \text{ and } \psi_0(p,y) = x\} . \tag{3.1}$$

Definition 3.5. The *max distance* between x and y is $E(x,y) = \max\{K(x|y), K(y|x)\}$.

It has been proved that up to an additive logarithmic term, the information distance E_0 is equal to the max distance.

3.2.2.1 Maximal Overlap

To what extent can the information required to compute x from y be made to overlap with that required to compute y from x? In some simple cases, complete overlap can be achieved, so that the same minimal program suffices to compute x from y as to compute y from x.

Example 3.1. If x and y are independent random binary strings of the same length n (up to additive constants $K(x|y) = K(y|x) = n$), then their bitwise exclusive-or $x \oplus y$ serves as a minimal program for both computations. Similarly, if $x = uv$ and $y = vw$ where u, v, and w are independent random strings of the same length, then $u \oplus w$ is a minimal program to compute either string from the other.

Now suppose that more information is required for one of these computations than for the other, say, $K(y|x) > K(x|y)$. Then the minimal programs cannot be made identical because they must be of different sizes. Nevertheless, in simple cases, the overlap can still be made complete, in the sense that the larger program (for y given x) can be made to contain all the information in the smaller program, as well as some additional information. This is so when x and y are independent random strings of unequal length, for example u and vw above. Then $u \oplus v$ serves as a minimal program for u from vw, and $(u \oplus v)w$ serves as one for vw from u.

The following *Conversion Theorem* asserts the existence of a difference string p of length $\ell(p) = \max\{K(x|y), K(y|x)\}$, up to an additive logarithmic term, that converts both ways between x and y and at least one of these conversions is optimal. If $K(x|y) = K(y|x)$, then the conversion is optimal in both directions.

Theorem 3.6. *Let x and y be strings such that $K(y|x) \geq K(x|y)$. There is a string r of length $K(y|x) - K(x|y)$ such that*

$$E_0(rx,y) = K(x|y) + K(K(x|y), K(y|x)) + O(1) .$$

Corollary 3.1. $E_0(x,y) = \max\{K(x|y), K(y|x)\} + O(\log \max\{K(x|y), K(y|x)\})$.

3.2.2.2 Universality

Let us assume we want to quantify how much some given objects differ with respect to a specific feature, for instance, the length of files in bits, the number of beats per second in music pieces, or the number of occurrences of some base in genomes. Every specific feature induces a specific distance measure, and conversely every distance measure can be viewed as a quantification of a feature difference.

Every distance measure should be an effectively approximable positive function of the two objects that satisfies a reasonable density condition and obeys the triangle inequality. It turns out that E is minimal up to an additive constant among all such distances. Hence, it is a universal *information distance* that accounts for any effective resemblance between two objects.

Let us consider an example of measuring *distance* between two pictures. Identify digitized black-and-white pictures with binary strings. There are many distances defined for binary strings, for example, the Hamming distance and the Euclidean distance. Such distances are sometimes appropriate. For instance, if we take a binary picture and change a few bits on that picture, then the changed and unchanged pictures have small Hamming or Euclidean distance, and they do look similar.

However, these measures are not always appropriate. The positive and negative prints of a photo have the largest possible Hamming and Euclidean distance, yet they look similar. Also, if we shift a picture one bit to the right, again the Hamming distance may increase by a lot, but the two pictures remain similar.

Many approaches to pattern recognition define distance measures with respect to pictures, language sentences, vocal utterances, and many more. We have already seen evidence that $E(x,y) = \max\{K(x|y), K(y|x)\}$ is a natural way to formalize a notion of algorithmic informational distance between x and y. Let us now show that the distance E is, in a sense, minimal among all reasonable distance measures.

In general we differentiate between distance functions and metrics. The latter are distance measures that satisfy additional conditions as formalized in the following.

Definition 3.6. A *distance function* is a function $D\colon \{0,1\}^* \times \{0,1\}^* \to \mathscr{R}^+$. It is a *metric* if it satisfies the metric (in)equalities:

- $D(x,y) = 0$ if and only if $x = y$, (identity)
- $D(x,y) = D(y,x)$, (symmetry)
- $D(x,y) \leq D(x,z) + D(z,y)$. (triangle inequality)

The value $D(x,y)$ is called the *distance* between x and y. As a familiar example of a distance function that is also a metric, consider the Euclidean metric, the everyday distance $D_E(a,b)$ between two geographical objects a,b expressed in, say, meters. Clearly, this distance satisfies the properties $D_E(a,a) = 0$, $D_E(a,b) = D_E(b,a)$, and $D_E(a,b) \leq D_E(a,c) + D_E(c,b)$ (for instance, $a =$ Amsterdam, $b =$ Beijing, and $c =$ Chicago). Our goal is to generalize this concept of distance from our physical space to the cyberspace and characterize the set of all reasonable distance functions that would measure informational distances between objects.

For a distance function or metric to be reasonable, it has to satisfy a certain additional condition, referred to as *density condition*. Intuitively this means that for every object x and value $d \in \mathscr{R}^+$ there is at most a certain, finite number of objects y at distance d from x. This requirement excludes degenerate distance measures like $D(x,y) = 1$ for all $x \neq y$. Exactly how fast we want the distances of the strings y from x to go to infinity is not important, it is only a matter of scaling. For convenience, we will require the following *density conditions*:

$$\sum_{y:y \neq x} 2^{-D(x,y)} \leq 1 \qquad \text{and} \qquad \sum_{x:x \neq y} 2^{-D(x,y)} \leq 1 \,. \tag{3.2}$$

Finally we allow only distance measures that are computable in some broad sense, which will not be seen as unduly restrictive. More precisely, only upper semicomputability of D will be required. This is reasonable: as we have more and more time to process x and y we may discover newer and newer similarities among them, and thus may revise our upper bound on their distance. The next definition summarizes the class of distance measures we are concerned with.

Definition 3.7. An *admissible information distance* is a total, possibly asymmetric, nonnegative function on the pairs x, y of binary strings that is 0 if and only if $x = y$, is upper semicomputable, and satisfies the density requirement (3.2).

Example 3.2. The Hamming distance between two strings $x = x_1 \ldots x_n$ and $y = y_1 \ldots y_n$ is defined as $d(x,y) = |\{i : x_i \neq y_i\}|$. This distance does not directly satisfy the density requirements (3.2). With minor modification, we can scale it to satisfy these requirements. In representing the Hamming distance m between x and y, strings of equal length n differing in positions $i_1, \ldots, i_m$, we can use a simple prefix-free encoding of $(n, m, i_1, \ldots, i_m)$ in $H_n(x,y) = 2\log n + 4\log\log n + 2 + m\log n$ bits. We encode n and m prefix-free in $\log n + 2\log\log n + 1$ bits each and then the literal indexes of the actual flipped-bit positions. Thus, $H_n(x,y)$ is the length of a prefix code word specifying the positions where x and y differ. This modified Hamming distance is symmetric, and it is an admissible distance by the Kraft inequality $\sum_{y:y \neq x} 2^{-H_n(x,y)} \leq 1$. It is easy to verify that H_n is a metric in the sense that it satisfies the metric (in)equalities up to $O(\log n)$ additive precision.

The following theorem is the fundamental result about the max distance E. It states that E is an optimal admissible information distance.

Theorem 3.7. *The function E with $E(x,y) = \max\{K(x|y), K(y|x)\}$ is an admissible information distance and a metric. It is minimal in the sense that for every admissible information distance D, we have $E(x,y) \leq D(x,y) + O(1)$.*

The quantitative difference in a certain feature between two objects can be considered as an admissible distance. Theorem 3.7 shows that the information distance E is universal in that among all admissible distances it is always least. That is, it accounts for the dominant feature in which two objects are alike. For that reason E is also called the *universal information distance*.

Many admissible distances are absolute, but if we want to express similarity, then we are more interested in relative ones. For example, if two strings of 1,000,000 bit differ by 1,000 bit, then we are inclined to think that those strings are relatively similar. But if two strings of 1,000 bit differ by 1,000 bit, then we find them very different.

Example 3.3. Consider the problem of comparing genomes. The *E. coli* genome is about 4.8 megabase long, whereas *H. influenza*, a sister species of *E. coli*, has genome length only 1.8 megabase. The information distance E between the two genomes is dominated by their length difference rather than the amount of information they share. Such a measure will trivially classify *H. influenza* as being closer to a more remote species of similar genome length such as *A. fulgidus* (2.18 megabase), rather than with *E. coli*. In order to deal with such problems, we need to normalize.

Our objective now is to normalize the universal information distance $E(x,y) = \max\{K(x|y), K(y|x)\}$ to obtain a universal *similarity* distance. It should give a similarity with distance 0 when objects are maximally similar and distance 1 when they are maximally dissimilar.

3.2.3 Normalized Information Distance

It is paramount that the normalized version of the universal information distance metric is also a metric. Were it not, then the relative relations between the objects in the space would be disrupted and this could lead to anomalies, if, for instance, the triangle inequality would be violated for the normalized version.

In order to obtain a normalized universal information distance function, both versions of information distance discussed so far, E_0 and E, can be normalized. We will only discuss how to normalize the max distance E and call it the *normalized information distance*.

Definition 3.8. The *normalized information distance* (NID) between two strings x and y is defined as

$$e(x,y) = \frac{\max\{K(x|y), K(y|x)\}}{\max\{K(x), K(y)\}}. \tag{3.3}$$

Dividing by $\max\{K(x), K(y)\}$ is not the most obvious idea for normalizing E, but the more obvious ideas do not work:

- Divide by the length. Then no matter whether we divide by the sum or maximum of the length, the triangle inequality is not satisfied.
- Divide by $K(x,y)$. Then the distances will be $1/2$ whenever x and y satisfy $K(x) \approx K(y) \approx K(x|y) \approx K(y|x)$. In this situation, however, x and y are completely dissimilar, and we would expect distance values of about 1.

That the NID is indeed a normalized metric is a remarkable fact [18].

Theorem 3.8. *The normalized information distance* $e(x,y)$ *takes values in the range* $[0,1]$ *and is a metric, up to ignorable discrepancies.*

This concludes our discussion of the theoretical foundations of the NID. We continue with demonstrations of how these insights can be put to use in practical settings.

3.3 Normalized Compression Distance

The normalized information distance is theoretically appealing, but impractical since it cannot be computed. In this section we discuss the normalized *compression* distance, an efficiently computable, and thus practically applicable, form of the normalized information distance.

3.3.1 Introduction

The normalized information distance $e(x,y)$, is called *universal* because it accounts for the dominant difference between two objects. It depends on the uncomputable function K, and is therefore also uncomputable. First we observe that using $K(x,y) = K(xy) + O(\log \min\{K(x), K(y)\}$ and Item 5 of Theorem 3.3 we obtain

$$E(x,y) = \max\{K(x|y), K(y|x)\} = K(xy) - \min\{K(x), K(y)\}\ , \qquad (3.4)$$

up to an additive logarithmic term $O(\log K(xy))$ which we ignore in the sequel.

By rewriting E as in (3.4) we manage to remove all conditional complexity terms and obtain a formula with only the non-conditional terms $K(x), K(y), K(xy)$. This comes in handy if we interpret $K(x)$ as the length of the string x after being maximally compressed. With this in mind, it is an obvious idea to approximate $K(x)$ with the length of the string x under an efficient real-world compressor. Any correct and lossless data compression program can provide an upper-bound approximation to $K(x)$, and most good compressors detect a large number of statistical regularities.

Substituting the numerator of (3.3) with (3.4) and subsequently using a real-world compressor Z (such as `gzip`, `bzip2`, `PPMZ`) to heuristically replace the Kolmogorov complexity, we obtain the distance e_Z, often called the *normalized compression distance* (NCD), defined by

$$e_Z(x,y) = \frac{Z(xy) - \min\{Z(x), Z(y)\}}{\max\{Z(x), Z(y)\}}\ , \qquad (3.5)$$

where $Z(x)$ denotes the binary length of the compressed version of the string x compressed with compressor Z. The distance e_Z is actually a family of distances parametrized with the compressor Z. The better Z is, the closer e_Z approaches the normalized information distance, the better the results are expected to be.

Under mild conditions on compressor Z, the distance e_Z is computable, takes values in $[0,1]$, and is a metric [9]. More formally, a compressor Z is *normal* if it satisfies the axioms

- $Z(xx) = Z(x)$ and $Z(\varepsilon) = 0$, (identity)
- $Z(xy) \geq Z(x)$, (monotonicity)
- $Z(xy) = Z(yx)$, (symmetry)
- $Z(xy) + Z(z) \leq Z(xz) + Z(yz)$, (distributivity)

up to an additive $O(\log n)$ term, with n the maximal binary length of a string involved in the (in)equality concerned.

Then the unnormalized distance $E_Z(x,y) = Z(xy) - \min\{Z(x), Z(y)\}$, with Z a normal compressor, is computable, satisfies the density requirement in (3.2), and satisfies the metric (in)equalities up to additive $O(\log n)$ terms, with n the maximal binary length of a string involved in the (in)equality concerned.

Moreover, the normalized distance e_Z of (3.5), with Z a normal compressor, has values in $[0,1]$ and satisfies the metric (in)equalities up to additive $O((\log n)/n)$ terms, with n the maximal binary length of a string involved in the (in)equality concerned.

Informal experiments [9] have shown that these axioms are in various degrees satisfied by good real-world compressors like `bzip2`, and `PPMZ`, with `PPMZ` being best among the ones tested. The compressor `gzip` performed not so well, and in all cases some compressor-specific window or block size determines the maximum useable length of the arguments x and y (32 KB for `gzip`, 450 KB for `bzip2`, unlimited for `PPMZ`). Cebrián et al. [4] systematically investigated how far the performance of real-world compressors `gzip`, `bzip2`, and `PPMZ` satisfy the identity axiom $Z(xx) = Z(x)$ of a normal compressor.

The normalized information distance e is intended to be universally applicable. In practice, various computable distances, including e_Z, can be viewed as approximations to e. Moreover, many of the measures used in the data mining community (see Tan et al. [24]) may, after normalization, be viewed as various degrees of approximations to e.

The NCD has been put to numerous tests. Keogh et al. [14, 15] have tested a closely related metric as a parameter-free and feature-free data mining tool on a large variety of sequence benchmarks. Comparing the NCD method with 51 major parameter-loaded methods found in the eight major data-mining conferences (SIGKDD, SIGMOD, ICDM, ICDE, SSDB, VLDB, PKDD, and PAKDD) in the last decade, on all data bases of time sequences used, ranging from heart beat signals to stock market curves, they established clear superiority of the NCD method for clustering heterogeneous data, and for anomaly detection, and competitiveness in clustering domain data.

Apart from providing a theoretical justification for these practical distances, the normalized information distance does more in that it embodies all approximations. The broad range of successful applications of e_Z will be demonstrated in the remainder of this section, where we will discuss applications in bioinformatics, linguistics, music, and plagiarism detection.

3.3.2 Phylogenies

DNA sequences seem ideally suited for the compression distance approach. A DNA sequence is a finite string over a four-letter alphabet $\{A,C,G,T\}$. We used the whole mitochondrial genomes of 20 mammals, each of about 18,000 base pairs, to test a hypothesis of Eutherian orders. It has been hotly debated in biology which two of the three main placental mammalian groups, Primates, Ferungulates, and Rodents, are more closely related. One cause of the debate is that the standard maximum likelihood method, which depends on the multiple alignment of sequences corresponding to an individual protein, gives (Rodents, [Ferungulates, Primates]) for half of the proteins in the mitochondrial genome, and (Ferungulates, [Primates, Rodents]) for the other half.

In recent years, when people use more sophisticated methods, together with biological evidences, it is believed that (Rodents, [Ferungulates, Primates]) reflects the true evolutionary history. We confirm this from the whole genome perspective using the distance e_Z. We use the complete mitochondrial genome sequences from following 20 species: rat (*Rattus norvegicus*), house mouse (*Mus musculus*), gray seal (*Halichoerus grypus*), harbor seal (*Phoca vitulina*), cat (*Felis catus*), white rhino (*Ceratotherium simum*), horse (*Equus caballus*), finback whale (*Balaenoptera physalus*), blue whale (*Balaenoptera musculus*), cow (*Bos taurus*), gibbon (*Hylobates lar*), gorilla (*Gorilla gorilla*), human (*Homo sapiens*), chimpanzee (*Pan troglodytes*), pygmy chimpanzee (*Pan paniscus*), orangutan (*Pongo pygmaeus*), Sumatran orangutan (*Pongo pygmaeus abelii*), with opossum (*Didelphis virginiana*), wallaroo (*Macropus robustus*) and platypus (*Ornithorhynchus anatinus*) as the outgroup.

For every pair of mitochondrial genome sequences *x* and *y*, evaluate the formula in (3.5) using a special-purpose DNA sequence compressor `DNACompress` [7], or a good general-purpose compressor like `PPMZ`. The resulting distances are the entries in a 20×20 distance matrix. Constructing a phylogeny tree from the distance matrix, using common tree reconstruction software, gives the tree in Fig. 3.1. This tree confirms the accepted hypothesis of (Rodents, [Primates, Ferungulates]), and every single branch of the tree agrees with the current biological classification.

Similarity of sequences in biology is currently primarily handled using alignments. However, the alignment methods seem inadequate for post-genomic studies since they do not scale well with data set size and they seem to be confined only to genomic and proteomic sequences. Therefore, alignment-free similarity measures are actively pursued. Ferragina et al. [13] experimentally tested the normalized information distance using 25 compressors to obtain the NCD, and six data sets of relevance to molecular biology. They compared the methodology with methods based on alignments and not. They assessed the intrinsic ability of the methodology to discriminate and classify biological sequences and structures. The compression program `PPMd`, based on `PPM` (Prediction by Partial Matching), for generic data and `GenCompress` [17] for DNA, are the best performers among the compression algorithms they used. The quantitative analysis supports the conclusion that the normalized information/compression method is worth using because of its robustness,

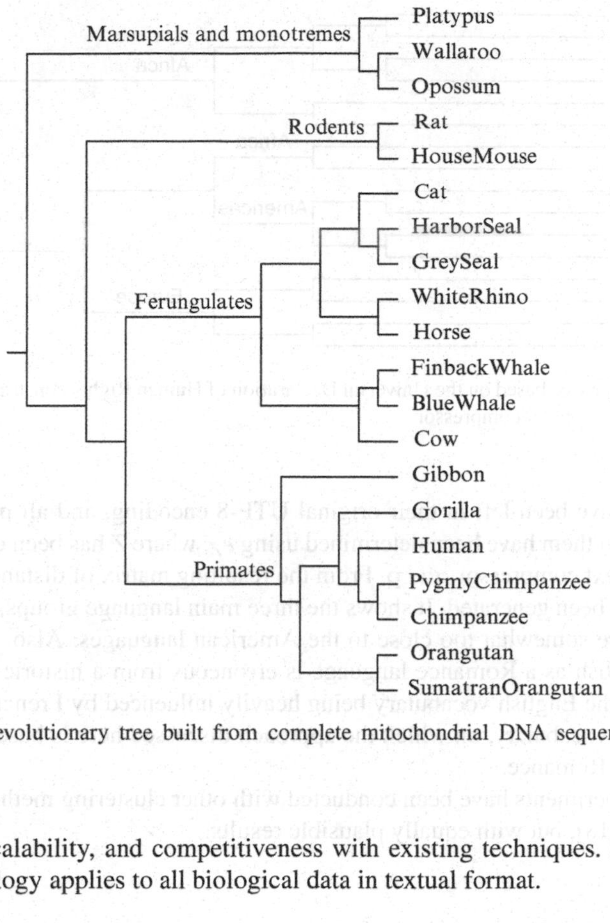

Fig. 3.1 The evolutionary tree built from complete mitochondrial DNA sequences of several mammals

flexibility, scalability, and competitiveness with existing techniques. In particular, the methodology applies to all biological data in textual format.

3.3.3 Language Trees

The similarity between languages can, to some extent, be determined by the similarity of their vocabulary. This means that given two translations of the same text in different languages, one can estimate the similarity of the languages by the similarity of the words occurring in the translations. This has been exploited by Benedetto et al. [2], who use a compression method related to NCD to construct a language tree of 52 Euroasian languages from translations of the Universal Declaration of Human Rights [1].

In this section we present an experiment [9] that uses the NCD method with translations of the Universal Declaration of Human Rights into 16 languages. Among these languages are four European (German, English, Spanish, Dutch), eight African (Pemba, Dendi, Ndebele, Kicongo, Somali, Rundi, Ditammari, Dagaare), and four American (Chikasaw, Purhepecha, Mazahua, Zapoteco).

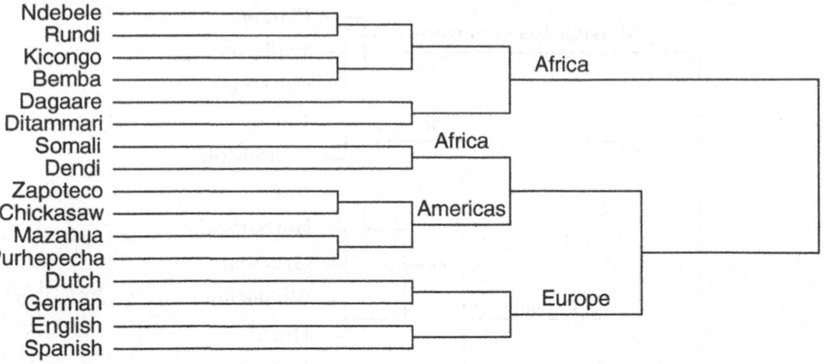

Fig. 3.2 Language tree based on the Universal Declaration of Human Rights constructed using the NCD, based on the `gzip` compressor

The files have been left in their original UTF-8 encoding, and all pairwise distances between them have been determined using e_Z, where Z has been chosen to be the standard text compressor `gzip`. From the resulting matrix of distances, the tree in Fig. 3.2 has been generated. It shows the three main language groups, only Dendi and Somali are somewhat too close to the American languages. Also, the classification of English as a Romance language is erroneous from a historic perspective and is due to the English vocabulary being heavily influenced by French and Latin. Therefore the vocabulary, on which the approach discussed here is based, is indeed to a large part Romance.

Similar experiments have been conducted with other clustering methods or other languages [9, 18], but with equally plausible results.

3.3.4 Plagiarism Detection

It is a common observation in university courses with programming assignments that some programs are plagiarized from others. That means that large portions are copied from other programs. What makes this hard to detect is that it is relatively easy to change a program syntactically without changing its semantics, for example, by renaming variables and functions, inserting dummy statements or comments, or reordering obviously independent statements. Nevertheless a plagiarized program is somehow close to its source and therefore the idea of using a distance measure on programs in order to uncover plagiarism is obvious.

We briefly describe the SID system [6] that uses a variant of the NID to measure the distance between two source code files. This variant is called the *sum distance*, and its Kolmogorov theoretic formulation is

$$e_{\text{sum}}(x,y) = \frac{K(x|y)+K(y|x)}{K(x,y)} .$$

This function takes values in the interval $[0,1]$ and is a metric according to Definition 3.6 [17].

To compute the similarity between two Java source code files, SID first tokenizes the programs and then uses a customized compressor to approximate the sum distance. This compressor is a variant of the Lempel–Ziv compression scheme [26] that has an unbounded buffer size and can thus detect repetitions over the entire file. Moreover it also takes advantage of *approximate* repetitions to increase the compression rate.

Evaluating plagiarism detection systems is difficult, but field experiments indicate that SID performs competitively and is more robust against certain attempts to circumvent detection, such as insertion of irrelevant code.

3.3.5 Clustering Music

The previous examples of NCD applications were based on text, be it source code or the Declaration of Human Rights. But The NCD method can also be applied to multimedia data like music, if it is present in the right format.

Music files in the MIDI format can be transformed into files that can be successfully clustered with the NCD. This transformation involves stripping the files of all instrument indicators, MIDI control signals and meta information such as title and composer. What essentially remains of a file is a list of musical notes of the piece. These preprocessed files can than be treated as text files.

A number of experiments has been performed [9, 11] with such files. We present a single, representative one, in which the set of musical pieces comprises four preludes from Chopin's Opus 28, two preludes and two fugues from Bach's "Das wohltemperierte Klavier," and the four movements from Debussy's "Suite Bergamesque." After preprocessing the MIDI files as described above, the pairwise e_Z values, with `bzip2` as compressor, are computed. To generate the final hierarchical clustering as shown in Fig. 3.3, a special quartet method [9, 11] is used.

Perhaps with the exception of Chopin's Prélude no. 5, which seems somewhat closer to the Bach pieces, the results agree with one's expectations.

3.3.6 Clustering Heterogeneous Data

We test gross classification of files based on heterogeneous data of markedly different file types:

1. Four mitochondrial gene sequences, from a black bear, polar bear, fox, and rat obtained from the GenBank Database [3] on the World Wide Web
2. Four excerpts from the novel *The Zeppelin's Passenger* by E. Phillips Oppenheim, obtained from the Project Gutenberg Edition on the World Wide Web

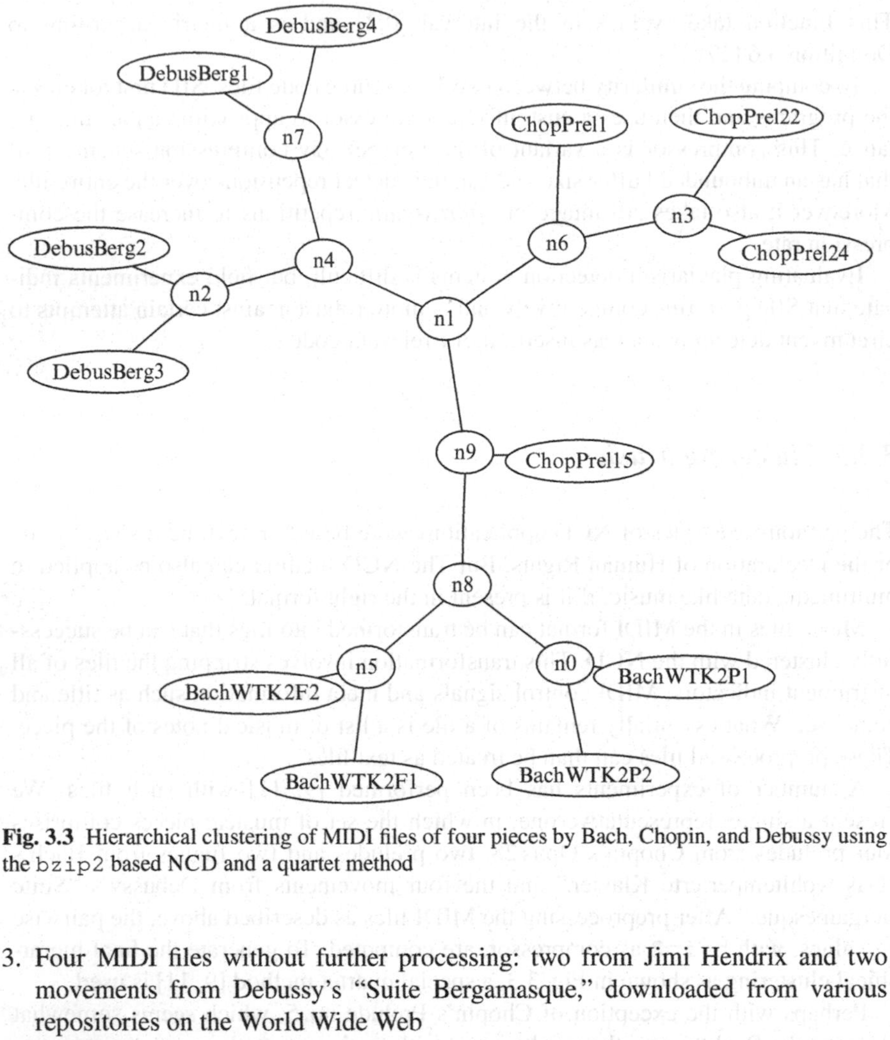

Fig. 3.3 Hierarchical clustering of MIDI files of four pieces by Bach, Chopin, and Debussy using the `bzip2` based NCD and a quartet method

3. Four MIDI files without further processing: two from Jimi Hendrix and two movements from Debussy's "Suite Bergamasque," downloaded from various repositories on the World Wide Web
4. Two Linux x86 ELF executables (the `cp` and `rm` commands), copied directly from the RedHat 9.0 Linux distribution
5. Two compiled Java class files, generated by ourselves

The program correctly classifies each of the different types of files together with like near like. The result is reported in Fig. 3.4. This experiment shows the power and universality of the method: no features of any specific domain of application are used. We believe that there is no other method known that can cluster data that are so heterogeneous this reliably.

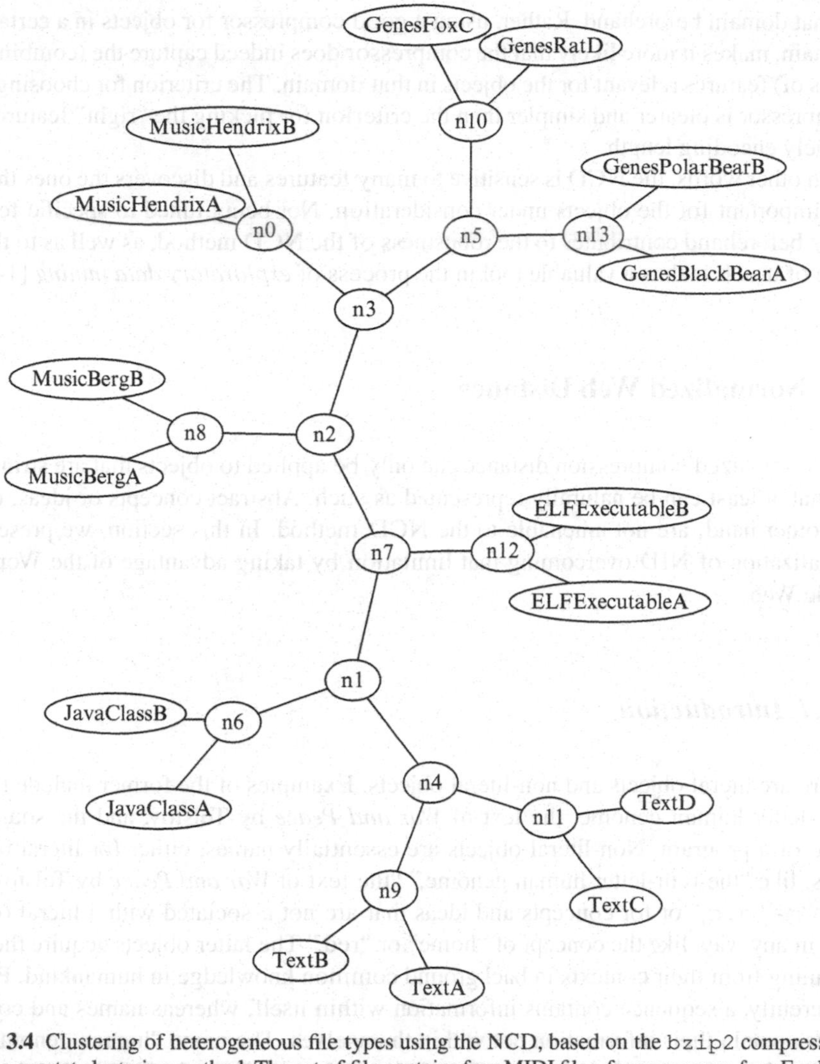

Fig. 3.4 Clustering of heterogeneous file types using the NCD, based on the `bzip2` compressor, and a quartet clustering method. The set of file contains four MIDI files, four genomes, four English texts, and two Java class files and Linux executables

3.3.7 Conclusion

The NCD is universal, in a mathematical sense as approximation of the universal NID, but also in a practical sense, as witnessed by the wide range of successful applications. Nevertheless the practical universality is of a different flavor because the NCD is a family of distance measures parametrized by a compressor. This means that one has to pick a suitable compressor for the application domain at hand. It does, however, not mean that one has to know the relevant features of the objects in that domain beforehand. Rather, using a good compressor for objects in a certain domain, makes it more likely that the compressor does indeed capture the (combinations of) features relevant for the objects in that domain. The criterion for choosing a compressor is clearer and simpler than the criterion for picking the "right" features, namely encoding length.

In other words, the NCD is sensitive to many features and discovers the ones that are important for the objects under consideration. Not being tuned to specific features beforehand contributes to the robustness of the NCD method, as well as to the ease of use. It is thus a valuable tool in the process of *exploratory data mining* [14].

3.4 Normalized Web Distance

The normalized compression distance can only be applied to objects that are strings or that at least can be naturally represented as such. Abstract concepts or ideas, on the other hand, are not amenable to the NCD method. In this section, we present a realization of NID overcoming that limitation by taking advantage of the World Wide Web.

3.4.1 Introduction

There are literal objects and non-literal objects. Examples of the former include the four-letter human genome, the text of *War and Peace* by Tolstoy, and the source code of a program. Non-literal objects are essentially names, either for literal objects, like "the four-letter human genome," "the text of *War and Peace* by Tolstoy," and "`main.c`," or for concepts and ideas that are not associated with a literal object in any way, like the concept of "home" or "red." The latter objects acquire their meaning from their contexts in background common knowledge in humankind. Put differently, a sequence contains information within itself, whereas names and concepts contain their information not within themselves. The name "human genome" implies 3 gigabases of information. The phrase "*War and Peace* by Tolstoy" perhaps carries information even beyond the book.

Let $\mathbf{W}$ be the set of pages of the World Wide Web, and let $\mathbf{x} \subseteq \mathbf{W}$ be the set of pages containing the search term x. By the Conditional Coding Theorem we have $\log 1/\mathbf{m}(\mathbf{x}|\mathbf{x} \subseteq \mathbf{W}) = K(\mathbf{x}|\mathbf{x} \subseteq \mathbf{W}) + O(1)$, where $\mathbf{m}$ is the universal lower semicomputable discrete semimeasure. This equality relates the incompressibility of the set of pages on the Web, containing a given search term, to its universal probability. While we do not know how to compute $\mathbf{m}$, a natural heuristic now is to use the distribution of x in the Web to approximate $\mathbf{m}(\mathbf{x}|\mathbf{x} \subseteq \mathbf{W})$. (We give a simplified approach.) Let us define the probability mass function $g(x)$ to be the probability that the search term x appears in a page indexed by a given internet search engine G, that is, the number of pages returned divided by the overall number of pages indexed. Then the Shannon–Fano code [23] associated with g can be set at

$$G(x) = \log \frac{1}{g(x)}\,.$$

Replacing $Z(x)$ by $G(x)$ in the formula in (3.5), we obtain the distance e_G, often called the *normalized Web distance* (NWD):

$$\begin{aligned} e_G(x,y) &= \frac{G(xy) - \min\{G(x), G(y)\}}{\max\{G(x), G(y)\}} \qquad (3.6)\\ &= \frac{\max\{\log f(x), \log f(y)\} - \log f(x,y)}{\log N - \min\{\log f(x), \log f(y)\}}\,. \end{aligned}$$

where $f(x)$ is the number of pages containing x, the frequency $f(x,y)$ is the number of pages containing both x and y, and N is the total number of indexed pages. We can view the search engine G as a compressor using the Web, and $G(x)$ as the binary length of the compressed version of the set of all pages containing the search term x, given the indexed pages on the Web. The distance e_G is actually a family of distances parametrized with the search engine G. It was originally called normalized Google distance (NGD) and thus featured a particular search engine [10]. The name *normalized Web distance* is more generic and more in line with the name NCD, which also does not mention a concrete compressor.

Example 3.4. We describe an experiment, using a popular search engine, performed in the year 2004, at which time it indexed $N = 8{,}058{,}044{,}651$ pages. A search for "horse" returns a page count of 46,700,000. A search for "rider" returns a page count of 12,200,000. A search for both "horse" and "rider" returns a page count of 2,630,000. Thus $e_G(\text{horse}, \text{rider}) = 0.443$. It is interesting to note that this number stayed relatively fixed as the number of pages indexed by the used search engine increased.

The distance e_G is actually a family of distances parametrized with the search engine G. The better G is, the closer the e_G approaches the normalized information distance, and the better the results are expected to be. The distance e_G is computable, takes values primarily (but not exclusively) in $[0,1]$, and is symmetric, that is, $e_G(x,y) = e_G(y,x)$. It only satisfies "half" of the identity property, namely $e_G(x,x) = 0$ for all x, but $e_G(x,y) = 0$ can hold even if $x \neq y$, for example, if the terms x and y always occur together.

The NWD also does *not* satisfy the triangle inequality $e_G(x,y) \leq e_G(x,z) + e_G(z,y)$ for all x, y, z. To see that, choose x, y, and z such that x and y never occur together, z occurs exactly on those pages on which x or y occurs, and $f(x) = f(y) = \sqrt{N}$. Then $f(x) = f(y) = f(x,z) = f(y,z) = \sqrt{N}$, $f(z) = 2\sqrt{N}$, and $f(x,y) = 0$. This yields $e_G(x,y) = \infty$ and $e_G(x,z) = e_G(z,y) = 2/\log N$, which violates the triangle inequality for all N. It follows that the NWD is not a metric. Indeed, we should view the distance e_G between two concepts as a relative similarity measure between those concepts. Then, while concept x is semantically close to concept y and concept y is semantically close to concept z, concept x can be semantically very different from concept z.

Another important property of the NWD is its *scale-invariance*. This means that, if the number N of pages indexed by the search engine grows sufficiently large, the number of pages containing a given search term goes to a fixed fraction of N, and so does the number of pages containing conjunctions of search terms. This means that if N doubles, then so do the f-frequencies. For the NWD to give us an objective semantic relation between search terms, it needs to become stable when the number N of indexed pages grows. Some evidence that this actually happens was given in Example 3.4.

The NWD can be used as a tool to investigate the meaning of terms and the relations between them as given by common sense. This approach can be compared with the *Cyc* project [16], which tries to create artificial common sense. Cyc's knowledge base consists of hundreds of microtheories and hundreds of thousands of terms, as well as over a million hand-crafted assertions written in a formal language called CycL [21]. CycL is an enhanced variety of first order predicate logic. This knowledge base was created over the course of decades by paid human experts. It is therefore of extremely high quality. The Web, on the other hand, is almost completely unstructured, and offers only a primitive query capability that is not nearly flexible enough to represent formal deduction. But what it lacks in expressiveness the Web makes up for in size; Web search engines have already indexed more than ten billion pages and show no signs of slowing down. Therefore search engine databases represent the largest publicly-available single corpus of aggregate statistical and indexing information so far created, and it seems that even rudimentary analysis thereof yields a variety of intriguing possibilities. It is unlikely, however, that this approach can ever achieve 100% accuracy like in principle deductive logic can, because the Web mirrors humankind's own imperfect and varied nature. But, as we will see below, in practical terms the NWD can offer an easy way to provide results that are good enough for many applications, and which would be far too much work if not impossible to program in a deductive way.

In the following sections we present a number of applications of the NWD: hierarchical clustering and classification of concepts and names in a variety of domains, finding corresponding words in different languages, and a system that answers natural language questions.

3.4.2 Hierarchical Clustering

To perform the experiments in this section, we used the *CompLearn* software tool [8], the same tool that has been used in Sect. 3.3 to construct trees representing hierarchical clusters of objects in an unsupervised way. However, now we use the normalized Web distance (NWD) instead of the normalized compression distance (NCD). Recapitulating, the method works by first calculating a distance matrix using NWD among all pairs of terms in the input list. Then it calculates a best-matching unrooted ternary tree using a novel quartet-method style heuristic based on randomized hill-climbing using a new fitness objective function optimizing the summed costs of all quartet topologies embedded in candidate trees [9].

3.4.2.1 Colors and Numbers

In the first example [10], the objects to be clustered are search terms consisting of the names of colors, numbers, and some tricky words. The program automatically organized the colors towards one side of the tree and the numbers towards the other, Fig. 3.5. It arranges the terms which have as only meaning a color or a number, and nothing else, on the farthest reach of the color side and the number side, respectively. It puts the more general terms black and white, and zero, one, and two, towards the center, thus indicating their more ambiguous interpretation. Also, things which were not exactly colors or numbers are also put towards the center, like the word "small." We may consider this an example of automatic ontology creation.

3.4.2.2 Dutch Seventeenth Century Painters

In the example of Fig. 3.6, the names of 15 paintings by Steen, Rembrandt, and Bol were entered [10]. The names of the associated painters were not included in the input, however they were added to the tree display afterwards to demonstrate the separation according to painters. This type of problem has attracted a great deal of attention [22]. A more classical solution would use a domain-specific database for similar ends. The present automatic oblivious method obtains results that compare favorably with the latter feature-driven method.

3.4.2.3 Chinese Names

In the example of Fig. 3.7, several Chinese names were entered. The tree shows the separation according to concepts like regions, political parties, people, etc. See Fig. 3.8 for English translations of these names. The dotted lines with numbers in-between each adjacent node along the perimeter of the tree represent the NWD values between adjacent nodes in the final ordered tree. The tree is presented in such a way that the sum of these values in the entire ring is minimized. This generally results in trees that makes the most sense upon initial visual inspection, converting an unordered binary tree to an ordered one. This feature allows for a quick visual inspection around the edges to determine the major groupings and divisions among coarse structured problems. It grew out of an idea originally suggested by Rutledge [22].

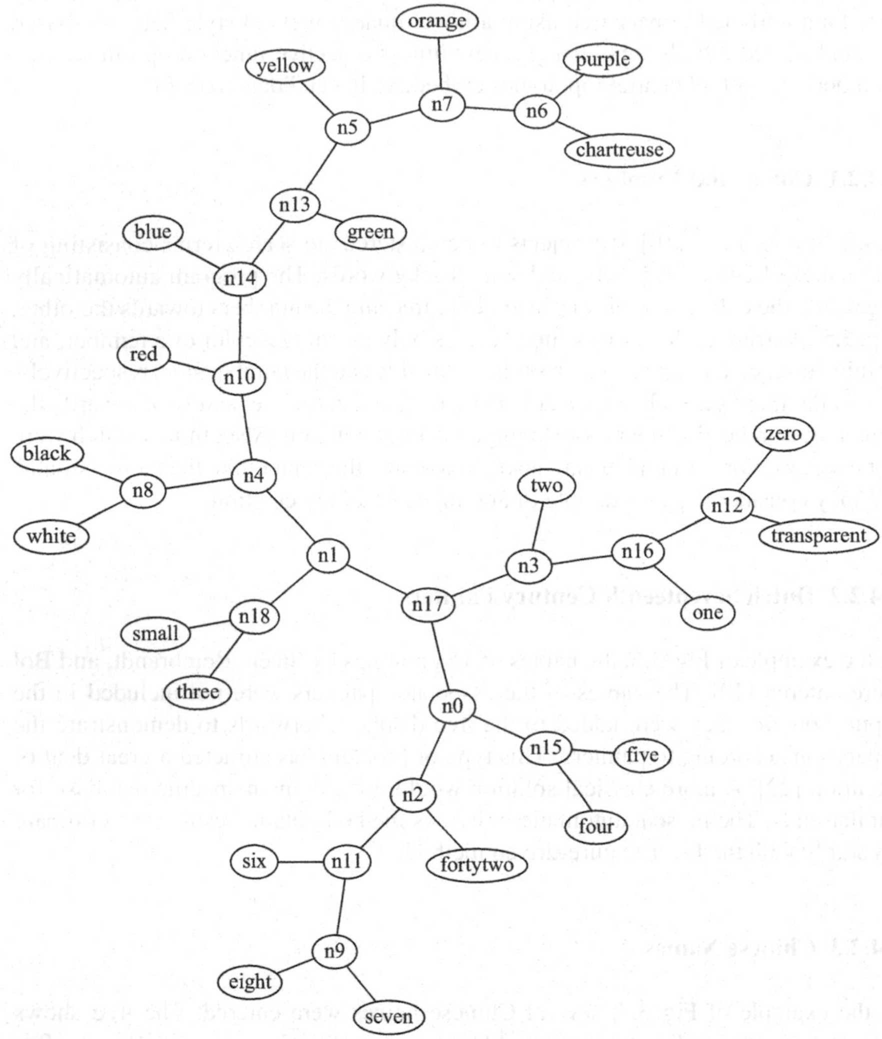

Fig. 3.5 Colors, numbers, and other terms arranged into a tree based on the normalized Web distances between the terms

3.4.3 Support Vector Machine Learning

We augment the NWD method by adding a trainable component of the learning system. This allows us to consider classification rather than clustering problems. Here we use the Support Vector Machine (SVM) as a trainable component. For all SVM experiments, the LIBSVM software [5] has been used.

The setting is a binary classification problem on examples represented by search terms. We require a human expert to provide a list of at least 40 *training words*,

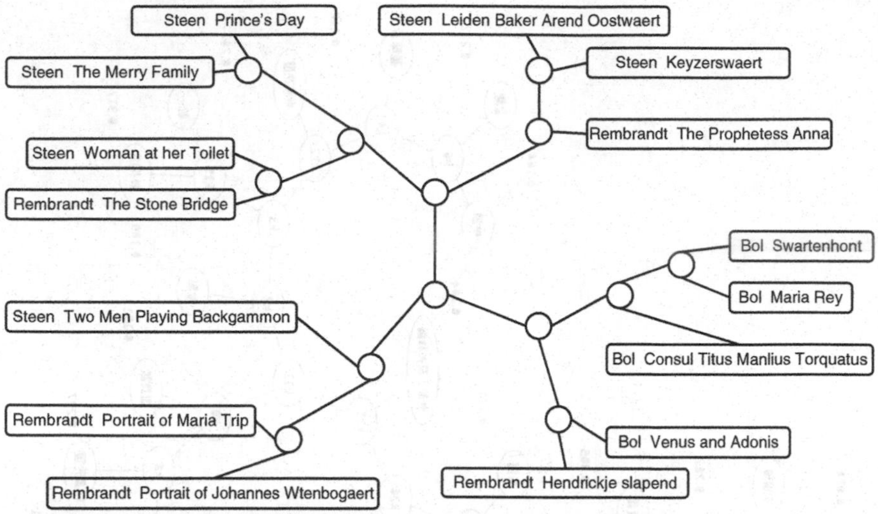

Fig. 3.6 Fifteen paintings by three painters arranged into a tree by hierarchical clustering. To determine the normalized Web distances between the paintings, only the title names were used; the painter prefixes shown in the diagram were added afterwards to assist in interpretation

consisting of at least 20 positive examples and 20 negative examples, to illustrate the contemplated concept class. The expert also provides, say, six *anchor words* $a_1, \ldots, a_6$, of which half are in some way related to the concept under consideration. Then, we use the anchor words to convert each of the 40 training words $w_1, \ldots, w_{40}$ to 6-dimensional *training vectors* $\mathbf{v}_1, \ldots, \mathbf{v}_{40}$. The entry $v_{j,i}$ of $\mathbf{v}_j = (v_{j,1}, \ldots, v_{j,6})$ is defined as $v_{j,i} = e_G(w_i, a_j)$ $(1 \leq i \leq 40,\ 1 \leq j \leq 6)$. The training vectors are then used to train an SVM to learn the concept, and then test words may be classified using the same anchors and trained SVM model. We present all positive examples as x-data (input data), paired with $y = 1$. We present all negative examples as x-data, paired with $y = -1$.

The above method for transforming concepts into real valued vectors is not limited to be used with SVMs, but can be combined with any machine learning algorithm that can handle numeric inputs.

3.4.3.1 Learning Prime Numbers

In Fig. 3.9 the method learns to distinguish prime numbers from non-prime numbers by example [10]. This example illustrates several common features the NWD method that distinguish it from the strictly deductive techniques. It is common for the classifications to be good but imperfect, and this is due to the unpredictability and uncontrolled nature of the Web distribution.

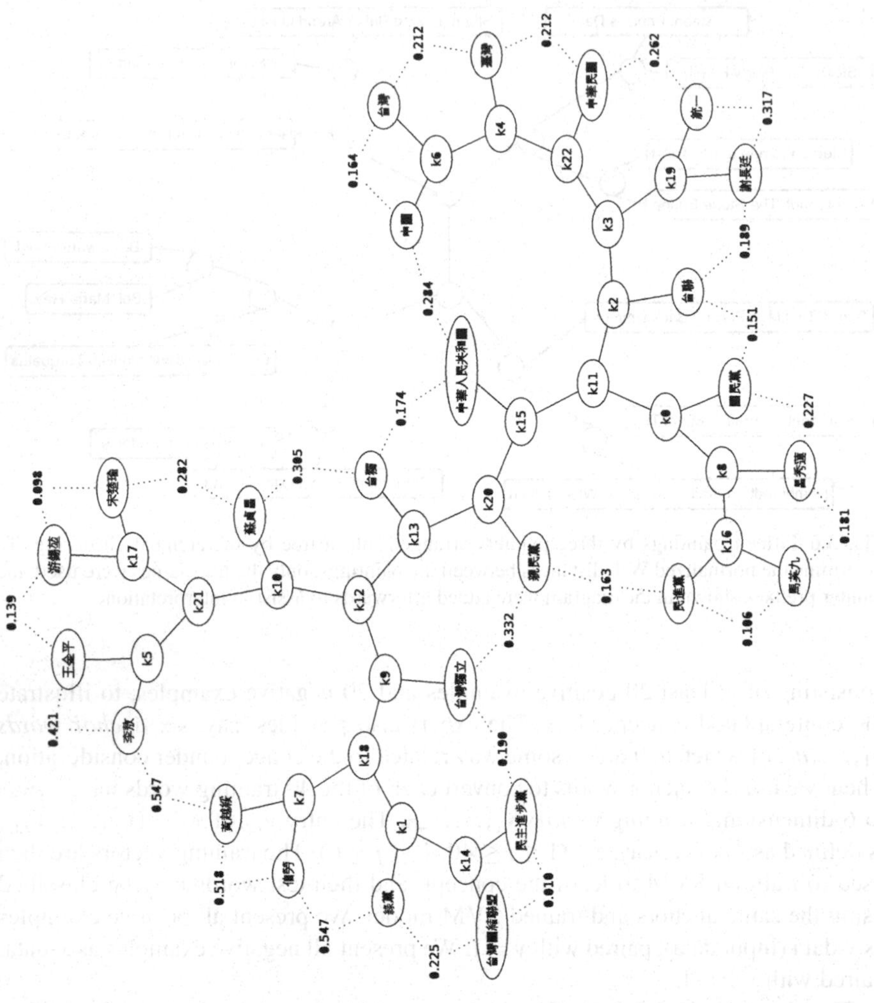

Fig. 3.7 Names of several Chinese people, political parties, regions, and others. The nodes and *solid lines* constitute a tree constructed by a hierarchical clustering method based on the normalized Web distances between all names. The numbers at the perimeter of the tree represent NWD values between the nodes pointed to by the *dotted lines*. For an explanation of the names, refer to Fig. 3.8

3.4.3.2 WordNet Semantics: Learning Religious Terms

The next example (see the preliminary version of [10]) has been created using WordNet [12], which is a semantic concordance of English. It also attempts to focus on the meaning of words instead of the word itself. The category we want to learn here is termed "religious" and represents anything that may pertain to religion. The negative examples are constituted by simply everything else (see Fig. 3.10). Negative examples were chosen randomly and uniformly from a dictionary of English

中國 China
中華人民共和國 People's Republic of China
中華民國 Republic of China
伯勞 shrike (bird) [outgroup]
台灣 Taiwan (with simplified character "tai")
台灣團結聯盟 Taiwan Solidarity Union [Taiwanese political party]
台灣獨立 Taiwan independence
台獨 (abbreviation of the above)
台聯 (abbreviation of Taiwan Solidarity Union)
呂秀蓮 Annette Lu
國民黨 Kuomintang
宋楚瑜 James Soong
李敖 Li Ao
民主進步黨 Democratic Progressive Party
民進黨 (abbreviation of the above)
游錫堃 Yu Shyi-kun
王金平 Wang Jin-pyng
統一 unification [Chinese unification]
綠黨 Green Party
臺灣 Taiwan (with traditional character "tai")
蘇貞昌 Su Tseng-chang
親民黨 People First Party [political party in Taiwan]
謝長廷 Frank Hsieh
馬英九 Ma Ying-jeou
黃越綏 a presidential advisor and 2008 presidential hopeful

Fig. 3.8 Explanations of the Chinese names used in the experiment that produced Fig. 3.7

words. This category represents a typical expansion of a node in the WordNet hierarchy. The accuracy on the test set is 88.89%.

3.4.3.3 WordNet Semantics: 100 Experiments

The previous example shows only one hand-crafted special case. To investigate the more general statistics, a method was devised to estimate how well the NWD-SVM approach agrees with WordNet in a large number of automatically selected semantic categories [10].

Before we explain how each category was automatically generated, we first review the structure of WordNet; the following is paraphrased from the official WordNet documentation available online. WordNet is called a semantic concordance of the English language. It seeks to classify words into many categories and interrelate

Training Data

Positive examples (21 cases)

11	13	17	19	2
23	29	3	31	37
41	43	47	5	53
59	61	67	7	71
73				

Negative examples (22 cases)

10	12	14	15	16
18	20	21	22	24
25	26	27	28	30
32	33	34	4	6
8	9			

Anchors (5 dimensions)

composite, number, orange, prime, record

Testing Results

	Positive tests	**Negative tests**
Positive Predictions	101, 103, 107, 109, 79, 83, 89, 91, 97	110
Negative Predictions		36, 38, 40, 42, 44, 45, 46, 48, 49

Accuracy: 18/19 = 94.74%

Fig. 3.9 NWD-SVM learning of prime numbers. All examples, i.e., numbers, were converted into vectors containing the NWD values between that number and a fixed set of anchor concepts. The classification was then carried out on these vectors using a support vector machine. The only error made is classifying 110 as a prime

the meanings of those words. WordNet contains synsets. A synset is a synonym set; a set of words that are interchangeable in some context, because they share a commonly-agreed upon meaning with little or no variation. Each word in English may have many different senses in which it may be interpreted; each of these distinct senses points to a different synset. Every word in WordNet has a pointer to at least one synset. Each synset, in turn, must point to at least one word. Thus, we have a many-to-many mapping between English words and synsets at the lowest level of WordNet. It is useful to think of synsets as nodes in a graph. At the next level we have lexical and semantic pointers. Lexical pointers are not investigated in this section; only the following semantic pointer types are used in our comparison: A semantic pointer is simply a directed edge in the graph whose nodes are synsets.

Training Data

Positive examples (22 cases)

Allah	Catholic	Christian	Dalai Lama	God
Jerry Falwell	Jesus	John the Baptist	Mother Theresa	Muhammad
Saint Jude	The Pope	Zeus	bible	church
crucifix	devout	holy	prayer	rabbi
religion	sacred			

Negative examples (23 cases)

Abraham Lincoln	Ben Franklin	Bill Clinton	Einstein	George Washington
Jimmy Carter	John Kennedy	Michael Moore	atheist	dictionary
encyclopedia	evolution	helmet	internet	materialistic
minus	money	mouse	science	secular
seven	telephone	walking		

Anchors (6 dimensions)

evil	follower	history	rational	scripture	spirit

Testing Results

	Positive tests	Negative tests
Positive Predictions	altar, blessing, communion, heaven, sacrament, testament, vatican	earth, shepherd
Negative Predictions	angel	Aristotle, Bertrand Russell, Greenspan, John, Newton,Nietzsche, Plato, Socrates, air, bicycle, car, fire, five, man, monitor, water, whistle

Accuracy: 24/27 = 88.89%

Fig. 3.10 NWD-SVM learning of religious terms. All training and test examples were converted into vectors containing the NWD values between that example concept and a fixed set of anchor concepts. The classification was then carried out on these vectors using a support vector machine

The pointer has one end we call a *source* and the other end we call a *destination*. The following relations are used:

1. *Hyponym*: X is a hyponym of Y if X is a (kind of) Y.
2. *Part meronym*: X is a part meronym of Y if X is a part of Y.
3. *Member meronym*: X is a member meronym of Y if X is a member of Y.
4. *Attribute*: A noun synset for which adjectives express values. The noun *weight* is an attribute, for which the adjectives *light* and *heavy* express values.
5. *Similar to*: A synset is similar to another one if the two synsets have meanings that are substantially similar to each other.

Using these semantic pointers we may extract simple categories for testing. First, a random semantic pointer (or edge) of one of the types above is chosen from the WordNet database. Next, the source synset node of this pointer is used as a root. Finally, we traverse outward in a breadth first order starting at this root and following only edges that have an identical semantic pointer type; that is, if the original semantic pointer was a hyponym, then we would only follow hyponym pointers in constructing the category. Thus, if we were to pick a hyponym link initially that says a *tiger* is a *cat*, we may then continue to follow further hyponym relationships in order to continue to get more specific types of cats. See the WordNet homepage [20] documentation for specific definitions of these technical terms.

Once a category is determined, it is expanded in a breadth first way until at least 38 synsets are within the category. 38 was chosen to allow a reasonable amount of training data to be presented with several anchor dimensions, yet also avoiding too many. Here, a rule of thumb is helpful: it states that the number of dimensions in the input data must not exceed one tenth the number of training samples. If the category cannot be expanded this far, then a new one is chosen. Once a suitable category is found, and a set of at least 38 members has been formed, a training set is created using 25 of these cases, randomly chosen. Next, three are chosen randomly as anchors. And finally the remaining ten are saved as positive test cases. To fill in the negative training cases, random words are chosen from the WordNet database. Next, three random words are chosen as unrelated anchors. Finally, 10 random words are chosen as negative test cases.

For each case, the SVM is trained on the training samples, converted to 6-dimensional vectors using NWD. The SVM is trained on a total of 50 samples. The kernel-width and error-cost parameters are automatically determined using five-fold cross validation. Finally testing is performed using 20 examples in a balanced ensemble to yield a final accuracy.

There are several caveats with this analysis. It is necessarily rough, because the problem domain is difficult to define. There is no protection against certain randomly chosen negative words being accidentally members of the category in question, either explicitly in the greater depth transitive closure of the category, or perhaps implicitly in common usage but not indicated in WordNet. Another detail to notice is that WordNet is available through some Web pages, and so undoubtedly contributes something to Web page counts. Further experiments comparing the results when filtering out WordNet images on the Web suggest that this problem does not usually affect the results obtained, except when one of the anchor terms happens to be very rare and thus receives a non-negligible contribution towards its page count from WordNet views. In general, the previous NCD based methods, as in [9], exhibit large-granularity artifacts at the low end of the scale; for small strings we see coarse jumps in the distribution of NCD for different inputs which makes differentiation difficult. With the Web based NWD we see similar problems when page counts are less than a hundred.

We ran 100 experiments. The histogram of agreement accuracies is shown in Fig. 3.11. On average, the NWD method turns out to agree well with the WordNet semantic concordance made by human experts. The mean of the accuracies of agreements is 0.8725. The variance is approximately 0.01367, which gives a standard deviation of approximately 0.1169. Thus, it is rare to find agreement less than 75%.

We conclude this section with a more abstract view of the NWD-SVM method. As we have seen, this method does not use an individual word in isolation, but instead uses an ordered list of its NWD relationships with fixed anchors. Therefore

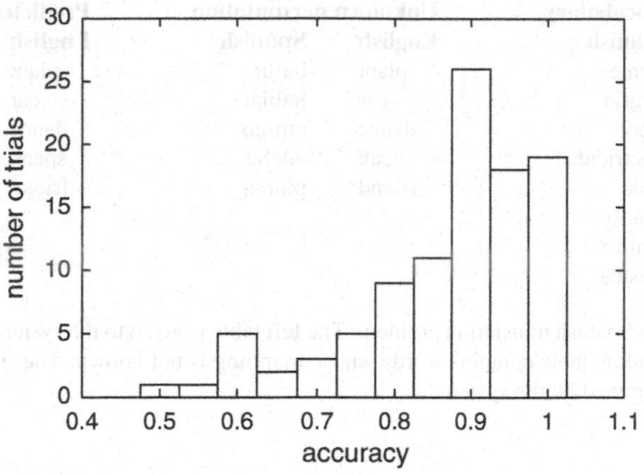

Fig. 3.11 Histogram of accuracies over 100 trials of WordNet experiment. The average accuracy achieved in the experiments is 0.8725

nothing can be attached to the isolated interpretation of a literal term, but only to the ordered list by which it is represented. That is to say, the inputs to our SVM are not directly search terms, but instead an image of the search term through the lens of the Web distribution, and relative to other fixed terms which serve as a grounding for the term. In most schools of ontological thought, and indeed in the WordNet database, there is imagined a two-level structure that characterizes language: a many-to-many relationship between word-forms or utterances and their many possible meanings. Each link in this association will be represented in the Web distribution with strength proportional to how common that usage is found on the Web. The NWD then amplifies and separates the many contributions towards the aggregate page count sum, thereby revealing some components of the latent semantic Web. In almost every informal theory of cognition we have the idea of connectedness of different concepts in a network, and this is precisely the structure that the NWD experiments attempt to explore.

3.4.4 Matching the Meaning

Yet another potential application of the NWD method is in natural language translation. In the experiment below [10] we do not use SVMs to obtain the result, but determine correlations instead. Suppose we are given a system that tries to infer a translation-vocabulary among English and Spanish. Assume that the system has already determined that there are five words that appear in two different matched sentences, but the permutation associating the English and Spanish words is, as yet, undetermined. This setting can arise in real situations, because English and Spanish have different rules for word-ordering. Thus, at the outset we assume a pre-existing vocabulary of eight English words with their matched Spanish translation. Can we

Given starting vocabulary		Unknown permutation		Predicted permutation	
English	**Spanish**	**English**	**Spanish**	**English**	**Spanish**
tooth	diente	plant	bailar	plant	planta
joy	alegria	car	hablar	car	coche
tree	arbol	dance	amigo	dance	bailar
electricity	electricidad	speak	coche	speak	hablar
table	tabla	friend	planta	friend	amigo
money	dinero				
sound	sonido				
music	musica				

Fig. 3.12 English–Spanish translation problem. The left table is given to the system as background knowledge, the middle table contains words whose mapping is not known. The right table shows the mapping determined by the system

infer the correct permutation mapping the unknown English words using the pre-existing vocabulary as a basis?

We start by forming an NWD matrix using additional English words of which the translation is known, Fig. 3.12. We label the columns by the translation-known English words, the rows by the translation-unknown words. The entries of the matrix are the NWDs of the English words labeling the columns and rows. This constitutes the English basis matrix. Next, consider the known Spanish words corresponding to the known English words. Form a new matrix with the known Spanish words labeling the columns in the same order as the known English words. Label the rows of the new matrix by choosing one of the many possible permutations of the unknown Spanish words. For each permutation, form the NWD matrix for the Spanish words, and compute the pairwise correlation of this sequence of values to each of the values in the given English word basis matrix. Choose the permutation with the highest positive correlation. If there is no positive correlation, report a failure to extend the vocabulary. In this example, the computer inferred the correct permutation for the testing words, see the right table in Fig. 3.12.

3.4.5 Question–Answer System

A typical procedure for finding an answer on the World Wide Web consists in entering some terms regarding the question into a Web search engine and then browsing the search results in search for the answer. This is particularly inconvenient when one uses a mobile device with a slow internet connection and small display. Question–answer (QA) systems attempt to solve this problem. They allow the user to enter a question in natural language and generate an answer by searching the Web autonomously.

In this section we describe some parts of the QA system QUANTA [25] that uses a variant of the NID to identify the correct answer to a question out of several candidates for answers. QUANTA is remarkable in that it uses neither NCD nor

NWD introduced so far, but a variation that is nevertheless based on the same theoretical principles. This variation is tuned to the particular needs of a QA system. We begin with some motivation for this particular variant and then describe it formally within the Kolmogorov framework. We shall focus on the new distance measure, not on other, interesting, details of the system, such as parsing and chunking the question, interfacing Web search engines, or generating candidate answers. We thus assume that for a given question a set of possible answers is already available, and the system only has to pick the (or a) right one.

As an example we consider the question "Which city is Lake Washington by?," which allows for many answers, among them Seattle, Bellevue, or Dallas. The first two cities are correct answers, but the preferred answer would be Seattle as the more well-known city. In a straightforward attempt to finding the right answer using the normalized Web distance we could compute e_G(Lake Washington, Bellevue), e_G(Lake Washington, Seattle) and e_G(Lake Washington, Dallas) and pick the city with the least distance. An experiment performed in February 2008 with a popular Web search engine yielded

- e_G(Lake Washington, Bellevue) $= 0.4658$,
- e_G(Lake Washington, Seattle) $= 0.5716$,
- e_G(Lake Washington, Dallas) $= 0.8302$,

so that Bellevue would have been chosen. Without normalization the respective distance values are 6.33, 7.54 and 10.95. Intuitively, the reason for Seattle being relatively far away from Lake Washington (in terms of e_G) is that, due to Seattle's size and popularity, it has many concepts in its neighborhood not all of which can be close. For the less known city of Bellevue, however, Lake Washington is relatively more important. Put differently, the concept "Seattle" contains a lot of information that is irrelevant for its being situated at Lake Washington. Symmetrically, Lake Washington encompasses much information unrelated to Seattle. A variation of (3.1) that accounts for possible irrelevant information is then

$$E_{\min,0}(x,y) = \min\{\ell(p) : U(x,p,r) = y \text{ and } U(y,p,q) = x \text{ and } \ell(p)+\ell(q)+\ell(r) \le E_0(x,y)\}\,. \quad (3.7)$$

Here, r represents the irrelevant information in y and q the irrelevant information in x. The additional restriction $\ell(p)+\ell(q)+\ell(r) \le E_0(x,y)$ ensures that the amount of irrelevant information is limited. Without it, one could set $r = y$ and $q = x$ and always use a program p of constant size that merely outputs one of its arguments.

Similarly as E_0 in (3.1), $E_{\min,0}$ cannot be used practically right away, it must be converted into a formula based on K [25]:

Theorem 3.9. $E_{\min,0}(x,y) = \min\{K(x|y), K(y|x)\} + O(\log \ell(xy))$.

The term $\min\{K(x|y), K(y|x)\}$ is also called the *min distance* and denoted by $E_{\min}$. The min distance is not a metric since it does not satisfy the triangle inequality. But in question–answer systems on the internet, distances are measured with partial information anyway, hence it is unreasonable to require the triangle inequality to hold. Furthermore, $E_{\min}$ satisfies the density conditions in (3.2) only for strings x with $K(x) \geq \ell(x) + O(1)$. It does not hold for objects with a low Kolmogorov complexity, which correspond to concepts with high frequency, such as Seattle. That $E_{\min}$ violates (3.2) for such objects intuitively means that popular concepts are allowed to have a denser neighborhood. This property is therefore rather a feature of $E_{\min}$ than a bug.

In another step paralleling the development of the NID, the min distance can be normalized. Analogously to e, we define the normalized version $e_{\min}$ of $E_{\min}$ as

$$e_{\min}(x,y) = \frac{E_{\min}(x,y)}{\min\{K(x),K(y)\}} = \frac{\min\{K(x|y),K(y|x)\}}{\min\{K(x),K(y)\}}.$$

It follows, though not obviously, that $e_{\min}(x,y) \leq e(x,y)$ for all x and y.

The normalized min distance $e_{\min}$ can be approximated by Web statistics in the same way as the NWD approximates the NID (cf. (3.6)), namely using the formula

$$e_{G,\min}(x,y) = \frac{\min\{\log f(x),\log f(y)\} - \log f(x,y)}{\log N - \max\{\log f(x),\log f(y)\}}.$$

Applying this normalized min Web distance to our above example question and answers, we obtain:

- $e_{G,\min}$(Lake Washington, Bellevue) $= 0.4496$,
- $e_{G,\min}$(Lake Washington, Seattle) $= 0.4281$,
- $e_{G,\min}$(Lake Washington, Dallas) $= 0.7746$,

that is, the answer "Seattle" would now be preferred over "Bellevue," and Dallas is still out of the question.

Regardless of whether we used e_G or $e_{G,\min}$, statistics would be obtained for the (co-)occurrence of the following single words and pairs:

- "Lake Washington"
- "Seattle"
- "Bellevue"
- "Lake Washington" and "Seattle"
- "Lake Washington" and "Bellevue"

But there is nothing hinting to the fact that we are looking for the co-location of a city and a lake. Of course, in this example it is reasonably clear. If, however, the question was represented by "Alan Turing," and candidate answers were "London," "Wilmslow," and "Paris," it would be unclear whether we are looking for Turing's place of birth, place of death, or any other place related to him. Clearly, the veracity of any of these answers depends on the particular question. It is therefore necessary to add some clues as to what the question is to the queries given to the Web search engine. For example adding the phrase "is born in" to all queries would (hopefully) limit the obtained statistics to Web pages that are concerned with Turing's birth and therefore result in "London" being chosen as answer.

The idea of such side information can easily be incorporated into the underlying theory by adding a condition to all terms in $e_{\min}$ (or e for that matter), yielding

$$e_{\min}(x,y|c) = \frac{\min\{K(x|y,c), K(y|x,c)\}}{\min\{K(x|c), K(y|c)\}},$$

where c denotes the conditional information, such as "is born in" in the above example. The extraction of a suitable c requires some sophistication and is beyond the scope of our discussion here.

The conditional version of $e_{\min}$ is at the core of the QUANTA system, whose question answering capabilities compare favorably with other QA systems [25]. The beneficial properties of $e_{\min}$ can perhaps best seen in comparison to other measures such as the normalized max distance e or the unnormalized distances E and $E_{\min}$. Replacing $e_{\min}$ with e results in answers that are still technically correct but often less popular and therefore less "good." We already mentioned Bellevue being preferred over Seattle as a city located at Lake Washington. Another example is the question "When was CERN founded?," which would be answered by e with "52 years ago," correct in 2006, whereas $e_{\min}$ responds more accurately with "1954."

Using the unnormalized E gives overly much weight to popular concepts. For instance, "Who is the greatest scientist of all?" would be answered with "God," whereas $e_{\min}$ would give "Newton," the reason for this discrepancy being that, in terms of Web pages, God is much more popular than Newton. More generally, experiments have shown [25] that $E_{\min}$ and E perform about 8% worse than $e_{\min}$.

The development of $e_{\min}$ to pick the most plausible answer in a QA system demonstrates how distance measures customized to special applications can be derived from first principles of Kolmogorov complexity theory, which in turn shows the power and flexibility of this theoretical approach.

3.5 Conclusions

The approach described in this chapter rests upon the simple idea that an *information distance* between two objects can be measured by the size of the shortest description that transforms each object into the other one. This idea is most naturally expressed mathematically using Kolmogorov complexity. Kolmogorov complexity, moreover, provides mathematical tools to show that such a measure is, in a proper sense, universal among all (upper semi)computable distance measures satisfying a natural density condition. These comprise most, if not all, distances one may be interested in. This information distance happens to be a metric. Since two large, very similar, objects may have the same information distance as two small, very dissimilar, objects, in terms of similarity it is the relative distance we are interested in. Hence we normalize the information metric to a relative similarity (also metric) in between 0 and 1. However, the normalized information metric is uncomputable. We approximate its Kolmogorov complexity parts by off-the-shelf compression programs (in the case of the normalized compression distance) or readily available statistics from the internet (in case of the normalized Web distance). The outcome are two practical distance measures, for literal as well as for non-literal data, that have been proved useful in numerous applications, some of which have been presented in the previous sections.

Just as important as the successes of these practical measures, however, is the underlying process used to derive them. The derivations of NCD and NWD are special instances of this process, which can roughly be broken into three steps: (1) devising an abstract distance notion, (2) transforming it inside the abstract mathematical realm into an equivalent, yet more easily realizable, formula, and (3) using real-world algorithms or data to practically realize the theoretically conceived measure. That this approach does not work by chance just for the information distance, is demonstrated by the derivation of the minimum distance, which employs the same three step process, just with different starting requirements for the distance measure.

Central design principles behind these Kolmogorov-based distance measures are the requirement of universality and the use of absolute measures of information content to achieve this universality. From these principles it follows naturally that the resulting distance measures are independent of fixed feature sets and do not require parameters for tuning. They can thus be used to build feature- and parameter-free methods that are suited for many tasks in exploratory data mining, alignment-free genomics, and elsewhere.

Appendix

List of Symbols

- $\bar{x}$: self-delimiting encoding of string x
- $\ell(\cdot)$: length of a string
- ε: empty string
- $|\ldots|$: cardinality
- $\mathscr{R}^+$: set of all non-negative rational numbers
- $\mathscr{R}$: set of all rational numbers
- $\mathscr{N}$: set of all natural numbers
- $\mathscr{Q}$: set of all rational numbers
- $\lfloor\ldots\rfloor$: floor function
- $\langle\ldots\rangle$: pairing function
- $C(\cdot|\cdot)$: conditional Kolmogorov complexity
- $C(\cdot)$: unconditional Kolmogorov complexity
- $K(\cdot|\cdot)$: conditional Kolmogorov prefix complexity
- $K(\cdot)$: unconditional Kolmogorov prefix complexity
- $\mathbf{m}$: universal upper semicomputable discrete semimeasure
- $\oplus$: bitwise xor
- E: max distance
- E_0: information distance

- e: normalized information distance
- e_Z: normalized compression distance
- e_G: normalized Web distance
- **W**: set of all Web pages
- **x**: set of all Web pages containing term x
- $f(x)$: number of Web pages containing term x
- $g(x)$: probability that a Web page contains term x
- $E_{\min}$: min distance
- $e_{\min}$: normalized min distance
- $e_{G,\min}$: normalized min Web distance

References

1. United Nations General Assembly resolution 217 A (III) of 10 December 1948: Universal Declaration of Human Rights
2. Benedetto, D., Caglioti, E., Loreto, V.: Language trees and zipping. Physical Review Letters **88**(4), 048702 (2002)
3. Benson, D.A., Karsch-Mizrachi, I., Lipman, D.J., Ostell, J., Wheeler, D.L.: GenBank. Nucleic Acids Research **36**(Database-Issue), 25–30 (2008)
4. Cebrián, M., Alfonseca, M., Ortega, A.: Common pitfalls using normalized compression distance: what to watch out for in a compressor. Communications in Information and Systems **5**(4), 367–384 (2005)
5. Chang, C.C., Lin, C.J.: LIBSVM: a library for support vector machines (2001). Software available at http://www.csie.ntu.edu.tw/~cjlin/libsvm
6. Chen, X., Francia, B., Li, M., McKinnon, B., Seker, A.: Shared information and program plagiarism detection. IEEE Transactions on Information Theory **50**(7), 1545–1551 (2004)
7. Chen, X., Li, M., Ma, B., Tromp, J.: DNACompress: fast and effective DNA sequence compression. Bioinformatics **18**, 1696–1698 (2002)
8. Cilibrasi, R.L., Cruz, A.L., de Rooij, S., Keijzer, M.: CompLearn software system, http://www.complearn.org
9. Cilibrasi, R.L., Vitányi, P.M.B.: Clustering by compression. IEEE Transactions on Information Theory **51**(4), 1523–1545 (2005)
10. Cilibrasi, R.L., Vitányi, P.M.B.: The Google similarity distance. IEEE Transactions on Knowledge and Data Engineering **19**(3), 370–383 (2007). Preliminary version: "Automatic Meaning Discovery Using Google", Arxiv preprint cs.CL/0412098, 2004, arxiv.org
11. Cilibrasi, R.L., Vitányi, P.M.B., de Wolf, R.: Algorithmic clustering of music based on string compression. Computer Music Journal **28**(4), 49–67 (2004)
12. Fellbaum, C.: Wordnet: An Electronic Lexical Database. MIT, Cambridge (1998)
13. Ferragina, P., Giancarlo, R., Greco, V., end G. Valiente, G.M.: Compression-based classification of biological sequences and structures via the Universal Similarity Metric: experimental assessment. BMC Bioinformatics **8**(1), 252 (2007)
14. Keogh, E., Lonardi, S., Ratanamahatana, C.: Toward parameter-free data mining. In: Proc. 10th ACM SIGKDD Intn'l Conf. Knowledge Discovery and Data Mining, pp. 206–215. Seattle, Washington, USA (2004). August 22–25, 2004
15. Keogh, E., Lonardi, S., Ratanamahatana, C.A., Wei, L., Lee, S.H., Handley, J.: Compression-based data mining of sequential data. Data Mining and Knowledge Discovery **14**(1), 99–129 (2007)
16. Lenat, D.B.: CYC: A large-scale investment in knowledge infrastructure. Communications of the ACM **38**(11), 33–38 (1995)

17. Li, M., Badger, J.H., Chen, X., Kwong, S., Kearney, P., Zhang, H.: An information-based sequence distance and its application to whole mitochondrial genome phylogeny. Bioinformatics **17**(2), 149–154 (2001)
18. Li, M., Chen, X., Li, X., Ma, B., Vitanyi, P.M.B.: The similarity metric. IEEE Transactions on Information Theory **50**(12), 3250–3264 (2004)
19. Li, M., Vitányi, P.M.B.: An Introduction to Kolmogorov Complexity and Its Applications, second edn. Springer, New York (1997)
20. Miller, G.A., Fellbaum, C., Tengi, R., Wakefield, P., Poddar, R., Langone, H., Haskell, B.: WordNet, A Lexical Database for the English Language. Cognitive Science Lab, Princeton University, http://wordnet.princeton.edu/
21. Reed, S.L., Lenat, D.B.: Mapping ontologies into cyc. In: Proc. AAAI Conference 2002 Workshop on Ontologies for the Semantic Web. Edmonton, Canada
22. Rutledge, L., Alberink, M., Brussee, R., Pokraev, S., van Dieten, W., Veenstra, M.: Finding the story – broader applicability of semantics and discourse for hypermedia generation. In: Proc. 14th ACM Conf. Hypertext and Hypermedia, pp. 67–76. Nottingham, UK (2003). August 23–27, 2003
23. Shannon, C.: The mathematical theory of communication. Bell System Technical Journal **27**, 379–423, 623–656 (1948)
24. Tan, P.N., Kumar, V., Srivastava, J.: Selecting the right interestingness measure for associating patterns. In: Proc. eighth ACM-SIGKDD Conf. Knowledge Discovery and Data Mining, pp. 491–502. ACM (2002)
25. Zhang, X., Hao, Y., Zhu, X., Li, M.: Information distance from a question to an answer. In: Proc. 13th ACM SIGKDD Int. Conf. Knowledge Discovery and Data Mining, pp. 874–883. ACM (2007)
26. Ziv, J., Lempel, A.: A universal algorithm for sequential data compression. IEEE Transactions on Information Theory **23**(3), 337–343 (1977)

Chapter 4
The Application of Data Compression-Based Distances to Biological Sequences

Attila Kertész-Farkas, András Kocsor, and Sándor Pongor

Abstract Text compressor algorithms can be used to construct metric distance measures (CBDs) suitable for character sequences. Here we review the principle of various types of compressor algorithms and describe their general behaviour with respect to the comparison of protein and DNA sequences. We employ reduced and enlarged alphabets, and model biological rearrangements like domain shuffling. In the classification experiments evaluated with ROC analysis, CBDs perform less well than substring-based methods such as the BLAST and the Smith–Waterman algorithms, but perform better than distances based on word composition. CBDs outperformed substring methods with respect to domain shuffling, and in some cases showed an increased performance when the alphabet was reduced.

4.1 Introduction

The term biological sequence denotes the most characteristic data type used in biology today. Data on the structure of genes and proteins is collected and stored in the form of long character sequences, written in a four-letter alphabet for DNA and in a 20-letter alphabet for proteins. The collection, organisation and computational analysis of sequence databanks is one of the main tasks of bioinformatics. The classification of sequence data is at the heart of this work, since the annotation of raw sequence data is primarily based on similarity searches against existing databanks,

A. Kertész-Farkas (✉) and A. Kocsor
Research Group on Artificial Intelligence, Aradi vértanúk tere 1, 6720 Szeged, Hungary
e-mail: kfa@inf.u-szeged.hu, kocsor@inf.u-szeged.hu

S. Pongor
Protein Structure and Bioinformatics Group, International Centre for Genetic Engineering and Biotechnology, Padriciano 99, 34012 Trieste, Italy and Bioinformatics Group, Biological Research Centre, Hungarian Academy of Sciences, Temesvári krt. 62, 6701 Szeged, Hungary
e-mail: pongor@icgeb.org

F. Emmert-Streib, M. Dehmer (eds.), *Information Theory and Statistical Learning*,
DOI: 10.1007/978-0-387-84816-7_4,

whereby a new sequence is assigned to one of the known classes (similarity groups). In order to accomplish this task, one needs sequence similarity measures that reflect the similarity of two sequences a realistic way.

Sequence similarity measures which are currently used fall into two main categories [2]. One of them is based on shared common substrings that are determined by approximate string matching heuristics, such as the well-known BLAST program [3], and dynamic programming methods like the Smith–Waterman algorithm [31]. The other class includes variants of n-gram distances whereby a sequence is first represented as a vector of n-gram word counts (e.g. character doublets, triplets and so on), and then compared in terms of an Euclidean distance or some other vector similarity measure [35]. Even though it is less efficient than a substring similarity search, n-gram methods are frequently included as fast screening tools into genome annotation pipelines.

The interest in compression-based methods was fostered by Vitnyi et al.'s seminal paper in which they proved that a simple index of sequence compressibility is actually better than n-gram techniques. Compression-based methods were soon employed in protein sequences as well [22], and it was also shown that different compressors perform differently on various sequence data [16].

The aim of the present work is to characterise compression algorithms with respect to two salient problems: invariance to biological changes and the resolution of sequence representations (alphabet size). In general, a similarity measure is expected to be invariant with respect to certain changes. For instance, the similarity of 3D shapes should be invariant with respect to the position of the objects; the similarity of melodies should be invariant to the pitch, the similarity of uttered words should be independent of the speaker, and so on. By the same token, we should expect that a similarity measure of biological sequences remain invariant with respect to a certain set of evolutionary events. Two of these are noteworthy: (1) Point mutations – sequences differing in only a few mutations should remain similar in the mathematical sense. This can be characterised by classification experiments followed by a ROC analysis [22]; (2) Major rearrangements can occur in the evolution of protein families. For instance, large segments of proteins can be displaced through a process called "domain shuffling". It is a moot question whether or not, or to what extent, rearranged sequences maintain similarity. This question can be studied by producing artificially rearranged sequences and comparing them with the various similarity measures [22].

The question of alphabet size is not merely one of technical interest. Reduced/enlarged alphabets are often used by bioinformaticians to represent protein and DNA sequences, but there is no systematic data telling us whether these really influence our ability to distinguish or classify biological sequences. Since the speed and memory requirements of compression algorithms depend on the algorithm's size, it would be interesting to see whether or not the classification efficiency of compression algorithms depends on this variable. And of course, all of these effects may depend on how closely the sequences are related.

In the next section we will provide a short introduction to the notion of the Kolmogorov Complexity-induced Information Metric as well as a description of the various compressor types used in our experiments. In Sect. 4.3 we will describe the experiments with our conclusions drawn for them. Finally, in Sect. 4.4 we will summarise our findings and draw some conclusions.

4.2 Definitions and Methods

The notion of an information distance between two discrete objects is the quantity of information in an object in terms of the number of bits needed to reconstruct the other. This notion arises from the theory of the thermodynamics of computation, which was first mentioned in [27, 39] and in [30] in the context of image similarity. Later, an introduction with related definitions and theory was published in [5] in 1998. More formal definitions, theory and related details can be found in [26].

4.2.1 Information Distance

A distance function D is a positive real-valued function defined on the Cartesian product of an arbitrary set X. It is also called a metric on X if it satisfies the following so-called metric properties:

- Non-negativity and identity: $D(x,y) \geq 0$ and $D(x,y) = 0$ iff $x = y$
- Symmetry: $D(x,y) = D(y,x)$
- Triangle inequality: $D(x,z) \leq D(x,y) + D(y,y)$

for every $x,y,z \in X$. These metric properties not only provide a useful characterisation of distance functions that satisfy them, but a metric function is also good as a reliable distance function.

Strings: Any finite object can be represented by binary strings without loss of generality (w.l.o.g). For example any genome sequence, arbitrary number, program represented by its Gödel number, image, structure, and term can be encoded by binary strings. Here, the string x is a finite binary string and its length is defined in the usual way and will be denoted by $l(x)$. An empty string will be denoted by ε and its length $l(\varepsilon) = 0$. The concatenation of x and y will be simply denoted by xy. A set of strings $S \subseteq \{0,1\}^*$ is called a prefix-free set if any string member is not a prefix of another member. A prefix-free set has a useful characterisation, namely it satisfies the Kraft inequality. Formally, for a prefix-free set S, we have

$$\sum_{x \in S} 2^{-l(x)} \leq 1. \tag{4.1}$$

An important consequence of this property is that in a sequence $s_1 s_2 \ldots s_n$ ($s_i \in S$) the end of string s_i is immediately recognizable; that is, the concatenation of strings can be separated by commas into codewords without looking at subsequent symbols. This sort of code is also called self-punctuating, self-delimiting or instantaneous code.

Kolmogorov Complexity: A partial recursive function $F(p,x)$ is called a prefix computer if for each x, the set $\{p \mid F(p,x) < \infty\}$ is a prefix-free set. The argument p is called a prefix program because no punctuation is required to tell F where p ends and the input to the machine can be simply the concatenation px. The conditional Kolmogorov Complexity of the string x with respect to (w.r.t.) y, with prefix computer F, is defined by

$$K_F(x \mid y) = \min\{l(p) \mid F(p,x) = y\}. \tag{4.2}$$

When such a p does not exist, its value is infinite. The invariance theorem states that there is a universal or optimal prefix computer U for every other prefix computer F and for any x,y such that

$$K_U(x \mid y) \leq K_F(x \mid y) + c_F, \tag{4.3}$$

where c_F depends on F but not on x,y. This universal computer U is fixed as the standard reference, and we call $K(x \mid y) = K_U(x \mid y)$ the conditional Kolmogorov Complexity of x w.r.t. y; moreover, the unconditional Kolmogorov Complexity of a string x is defined by $K(x \mid \varepsilon)$. The Kolmogorov Complexity of x w.r.t. y is the shortest binary prefix program that computes x with additional information obtained from y. However, it is non-computable in the Turing sense; that is, no program exists that can compute it in practice since it can be reduced to the halting problem [26].

Information Distance: With the information content, the information distance between strings x and y is defined as the length of the shortest binary program on the reference universal prefix computer with input y, which computes x and vice versa, formally:

$$ID(x,y) = \min\{l(p) \mid U(p,x) = y,\ \ U(p,y) = x\}. \tag{4.4}$$

This is clearly symmetric, and it has been proven that it satisfies the triangle inequality up to an additive constant. ID can be computed by reversible computations up to an additive constant, but there is an easier but weaker approximation of ID. It has been shown that ID, up to an additive logarithmic term, can be computed by the so-called max distance E:

$$E(x,y) = \max\{K(x \mid y), K(y \mid x)\}. \tag{4.5}$$

In general, the "up to an additive logarithmic term" means that the information required to reconstruct x from y is always maximally correlated with the information required to reconstruct y from x that is dependent on the former amount of information. Thus E is also a suitable approximation for the information distance.

Density, Universality and Metric Properties: In a discrete space with a distance function D, the rate of growth of the number of elements in balls of size d, centred at x, denoted by $\#B_D(x,d)$, can be considered as a certain characterisation of the space. For example, the distance function $D(x,y) = 1$ for all $x \neq y$ is not a realistic distance function, but it satisfies the triangle inequality. As for the Hamming distance, H, $\#B_H(x,d) = 2^d$, hence it is finite. For a function D to be a realistic distance function it needs to be satisfy the so-called normalization property:

$$\sum_{y:y \neq x} 2^{-D(x,y)} \leq 1. \tag{4.6}$$

This holds for the information distance $E(x,y)$ and also for ID because they both satisfy the Kraft inequality. Moreover,

$$log(\#B_E(x,d)) = d - K(d \mid x) \tag{4.7}$$

up to an additive constant. This means that the number of elements in the ball $B_E(x,d)$ grows exponentially w.r.t. d up to an additive constant. In addition, a more complex string has fewer neighbours and this has been shown by the theorem "tough guys have few neighbours thermodynamically" [27]. E is universal (up to an additive error term) in the sense that it is smaller than every other upper-semi-computable function f which satisfies the normalization property. That is, we have

$$E(x,y) \leq f(x,y) + O(log(k)/k), \tag{4.8}$$

where $k = \max\{K(x), K(y)\}$. This seems quite reasonable, as we have greater time to process x and y, and we may discover additional similarities between them; then we can revise our upper bound on their distance. As regards semi-computability, $E(x,y)$ is the limit of a computable sequence of upper bounds.

The non-negativity and symmetry properties of the information distance $E(x,y)$ are a consequence of the definition, but $E(x,y)$ satisfies the triangle inequality only up to an additive error term [5].

Compression-Based Distances: The non-normalized information distance is not a proper evolutionary distance measure because of the length factor of strings. For a given pair of strings x and y the normalized information distance is defined by

$$D(x,y) = \frac{\max\{K(x \mid y), K(y \mid x)\}}{\max\{K(x), K(y)\}}. \tag{4.9}$$

In [25] it was shown this satisfies the triangle inequality and vanishes when $x = y$ with a negligible error term. The proof of its universality was given in [24], and the proof that it obeys the normalization property is more technical (for details, see [12, 25]).

The numerator can be rewritten in the form $\max\{K(xy) - K(x), K(yx) - K(y)\}$ within logarithmic additive precision due to the additive property of prefix Kolmogorov complexity [12]. Thus we get

$$D(x,y) = \frac{K(xy) - \min\{K(x), K(y)\}}{\max\{K(x), K(y)\}}. \tag{4.10}$$

Since the Kolmogorov complexity cannot be computed, it has to be approximated, and for this purpose, real file compressors are employed. Let C(x) be the length of the compressed string compressed by a particular compressor like gzip or arj. Then the approximation for the information distance E can be obtained by using the following formula:

$$CBD(x,y) = \frac{C(xy) - \min\{C(x),C(y)\}}{\max\{C(x),C(y)\}}. \tag{4.11}$$

A CBM is a metric up to an additive constant and satisfies the normalization property if C satisfies the following properties up to an additive term:

- Idempotency: $C(xx) = C(x)$
- Symmetry: $C(xy) = C(yx)$
- Monotonicity: $C(xy) \geq C(x)$
- Distributivity: $C(xy) + C(z) \leq C(xy) + C(xy)$

The proof can be found in Theorem 6.2 of [12]. A compressor satisfying these properties is called a normal compressor.

There is no bound for the difference between the information distance and its approximation; that is, $|E - CBD|$ is unbounded. For example, the Kolmogorov complexity of the first few million digits of the number π , denoted by *pi*, is a constant because its digits can be generated by a simple program but $C(pi)$ is proportional to $l(pi)$ for every known text compressor C.

4.2.2 Data Compression Algorithms

In computer science, the purpose of data compression is to store the data more economically without redundancy, and it can be compressed whenever some events are more likely than others. In general, this can be done by assigning a short description code to the more frequent patterns and a longer description code to the less frequent patterns. If the original data can be fully reconstructed, it is called lossless compression. If the original data cannot be exactly reconstructed from those descriptions, it is known as lossy compression. This form is widely used in the area of image and audio compression because the fine details can be removed without being noticed by the listeners. If a set of codes S satisfies the Kraft inequality, it is a lossless compression and it is a prefix-free set; conversely, if it does not satisfy the Kraft inequality, it is a lossy compression and it is not a prefix-free set. None of data or text string can be losslessly compressed to an arbitrary small size by any compression method. The limit of the compression process that is needed to fully describe the original data is related to Shannon entropy or to the information content [13]. A good collection of compressors and their related descriptions are available at http://datacompression.info/. Now, we will briefly describe the compressors used in our experiments.

Adaptive Huffman (Dynamic Huffman Coding): Here, the assumption is that the sequence is generated by a source over a d-ary alphabet D over a probability distribution P; that is, for each $x \in D$, the $p(x)$ is the probability of the letter x and $\sum_{x \in D} p(x) = 1$. Let $C(.)$ be a source code and $C(x)$ be the codeword associated with x, and for ease of notation, $l(x)$ let stand for $l(C(x))$, the length of the codeword. It should be demanded that C is non-singular; that is, $x \neq y \rightarrow C(x) \neq C(y)$ and it gives a prefix-free set. The expected length $L(C)$ of a source code is defined by

$$L(C) = \sum_{x \in D} p(x) l(x). \tag{4.12}$$

The Huffman coding is the optimal source coding; that is

$$L^* = \min_{C}\{L(C)\}, \tag{4.13}$$

subject to C satisfying the Kraft inequality. Solving this constrained optimization problem using Lagrange multipliers yields optimal code lengths $l^* = -\log_d p(x)$. The non-integer choice of codeword length gives $L^* = H(P)$, that is, the length of the optimal compression of data is equal to the Shannon entropy of the distribution P of the codewords. Moreover, for integer codeword length, we have $H(P) \leq L^* \leq H(P) + 1$ and the codes can be built by a method like Shannon–Fano coding [13].

In practice, the key part of the Huffman coding method is to assess the probability of letter occurrence. The adaptive Huffman coding constructs a code when the symbols are being transmitted, having no initial knowledge of the source distribution, which allows one-pass encoding and adaptation to changing conditions in the data. For our experiments we used http://www.xcf.berkeley.edu/~ali/K0D/Algorithms/huff/

Lempel–Ziv–Welch (LZW): It is a one-pass stream parsing algorithm that is widely used because of its simplicity and fast execution speed. It divides a sequence into distinct phrases such that each block is the shortest string which is not a previously parsed phrase. For example, let $x = 01111000110$ be a string of length 11, then the LZW compressor $C(x)$ produces six codes: 0,1,11,10,00,110, thus $l(C(x)) = 6$. Here, the assumption is that sequences are generated by a higher but finite order stationary Markov process P over a finite alphabet. The entropy of P can be estimated by the formula $n^{-1} C(x) \log_2 C(x)$ and the convergence is almost guaranteed as the length of the sequence x tends to infinity. In practice, a better compression ratio can sometimes be achieved by LZW than Huffman coding because of this more realistic assumption. Here we used our own implementation of the original LZW algorithm.

GenCompress (GC): This was developed for the compression of large genomes by exploiting hidden regularities and properties in them, such as repetitions and mutations. The main idea will now be described. First, let us consider a genome sequence $x = uv$ and suppose that its first portion u has already been compressed. Find the largest prefix v' of v with an edit distance method such that it is maximally partial matched by a substring u' in u that has minimal edit operations needed to obtain v' from u'. After, only the position of u' and the edit operations list is encoded by Fibonacci coding. GenCompress is a one-pass algorithm and it is the state-of-the-art compression method for genomic sequences. However, it is not fast in practice because its performance is more important so it is recommended for offline computations [11]. We downloaded this program from http://www.cs.cityu.edu.hk/~cssamk/gencomp/GenCompress1.htm

Prediction by Partial Matching (PPM): PPM is an adaptive statistical data compression technique that uses context tree to record the frequency of characters within their contexts. Then it predicts the next symbol in the stream using arithmetic coding (AC) to encode an event with an interval that is assigned to the event w.r.t. their distribution, and the codeword is the shortest binary number from this interval. A more likely event has a correspondingly longer interval, and thus a shorter codeword can be selected [36]. In our experiments we used the implementation from http://compression.ru/ds/.

Burrows–Wheeler Transform (BWT): While BWT is not a compressor method, it is closely related to data compression and is used as a pre-processing method to achieve a better compression ratio. It is a reversible block-sorting method that produces all n-cyclic shifts of the original sequence then orders them lexicographically and outputs the last column of the sorted table as well as the position of the original sequence in the sorted table [8]. This procedure brings groups of symbols with a similar context closer together and these segments can be used to better estimate the entropy of a Markov process [9]. The implementation that we used was downloaded from http://www.geocities.com/imran_akthar/files/bwt_matlab_code.zip.

Advanced Block-Sorting Compressor (ABC): This compressor contains several advanced compression techniques like BWT, run length encoding, AC, weighted frequency count and sorted inversion frequencies. For details, see [1]. The compression speed is approximately half that of GZIP and BZIP2. This program was downloaded from http://www.data-compression.info/ABC/

4.3 Experiments and Discussion

In this section we will describe several experiments that were carried out to determine how CBDs can exploit the similarity among sequences, from different points of view. Where possible, the Smith–Waterman (SW) and the BLAST alignment-based sequence comparison methods as well as the naive distance were evaluated for comparison. The BLAST version 2.2.4 was used with a cutoff score of 25 and the SW algorithm used was implemented in MATLAB with the BLOSUM 62 matrix [19]. In both cases the gap opening and extension parameters were set to their default values. The naive distance between protein sequences was simply the Euclidean distance of a vector bi-gram counts of sequences.

4.3.1 Experiments on Compressor Properties

In our first experiments, we examined how well each compressor satisfies the properties of a normal compressor. To do this, we chose the SCOP40 (40% identity) database. The sequences were taken from the SCOP database 1.69 [4] and were downloaded from the ASTRAL site (http://astral.berkeley.edu/). Fifty-three non-contiguous domains were discarded and from the remaining 7,237 entries protein families that had at least five members, and at least 10 members outside the family but within the same superfamily were selected. Altogether, we collected 1,357 sequences. This database is available at http://net.icgeb.org/benchmark/ [32]. The average length of the sequences was 194 characters with a relatively high standard deviation of 116. Table 4.1 illustrates how each compressor satisfies the normality properties. In practice, the monotonicity is obeyed for stream-based compressors and only slightly less evident for block-coding types. The distributivity property appears to be satisfied in each case. Symmetry is not precisely obeyed for stream-based compressors and CBDs become slightly asymmetric; however, in practice, the difference $| CBD(x,y) - CBD(y,x) |$ is quite small. A compressor should search exact repetitions and obey idempotency, but in practice this seems to the most difficult condition for compressors and thus CBDs do not vanish when $x = y$.

4.3.2 Invariance Experiments on Rearranged Sequences

Protein evolution often includes rearrangements such as the gain or loss of domains and circular permutations. Therefore the question arises of whether or not CBDs can detect the differences between the products. To answer this, we used the C1S precursor (Swiss-Prot ID: C1S_HUMAN, AC:P09871), a multi-domain protein of

Table 4.1 Results of compressors' properties on SCOP40

Compressor	Length $C(x)$[a]		Idempontecy $\mid C(xx) - C(x) \mid$			Symmetry[b] $\mid C(xy) - C(yx) \mid$		
	avg.	std.	#viol[c]	avg.[d]	std.[e]	#viol[c]	avg.[d]	std.[e]
ABC	142	66.26	1,357	15.98	5.95	845,808	1.07	0.27
AH	121	63.01	1,357	100.84	61.55	1,240,000	1.38	0.62
LZW	154	78.99	1,357	92.31	52.52	1,570,000	2.63	1.70
GC	143	73.79	1,357	4.93	0.80	7,968	1.61	0.76
PPM	160	70.25	1,357	5.46	1.24	1,260,000	1.49	0.79

[a] The average length of the original sequences was 194 characters with a standard deviation of 116
[b] The total number of test cases was 1,841,450
[c] The number of cases when the compressor violated the condition
[d] The average bias
[e] The standard deviation of the bias
With these compressors acting on this dataset, the distributivity and monotonicity properties were not violated.

Table 4.2 The effect of sequence rearrangements on the various similarity/distance measures on C1S

	ABC		AH[c]		GC		LZW		PPM		SW	
	score	%[b]	score	%	score	%	score	%	score	%	score	%
Itself (ABCB′DD′E)[a]	0.088	0	0.939	0	0.013	0	0.646	0	0.020	0	1,895	0
Duplication of C (ABCCB′DD′E)	0.179	11	0.945	−284	0.051	4	0.661	13	0.059	5	1,804	5
Deletion of C (ABB′DD′E)	0.226	17	0.941	−107	0.115	11	0.663	14	0.121	12	1,673	12
Deletion of CB′ (ABDD′E)	0.435	43	0.939	0	0.272	28	0.701	47	0.261	28	1,305	31
Circular permut. (D′EABCB′D)	0.157	9	0.941	−107	0.062	5	0.665	16	0.072	6	966	50
Reverse order (ED′DB′CBA)	0.222	17	0.945	−249	0.082	7	0.665	16	0.075	6	679	65
Duplication (2×ABCB′DD′E)	0.142	7	0.970	−1,294	0.013	0	0.743	82	0.024	0	1,895	0
Random shuffled	0.891	100	0.963	100	0.954	100	0.763	100	0.868	100	19	100

[a]Swiss-Prot AC = P09871, A = Signal peptide (res. 1–15), B, B′ = CUB domains (res. 16–130, 175–290, respectively), C = EGF domain (res. 131–172), D,D′ = Sushi domains (res. 292–356, 357–423, respectively), E = Peptidase S1 (res. 438–688)

[b]Score of C1S with itself = 0%, score of C1S with randomly shuffled C1S = 100%

[c]AH gives abnormal percent values because the distance to itself and to its randomly shuffled sequence are similar.

688 residues consisting of a signal peptide (A), two CUB domains (B,B′), a EGF domain (C), two SUSHI domains (D,D′) and a trypsin-like catalytic domain (E) that is post-translationally cleaved from the precursor. The domain architecture of the native protein can be written as ABCB′DD′E and a hypothetical circular permutant can be written as DD′EABCB′. The results in Table 4.2 show that a reshuffling of the domains does not substantially affect the CBDs. Comparing the C1S precursor with its reshuffled counterparts in which the domain order is reversed, gives a PPM distance of 0.075, while the C1S compared to itself gives a value of 0.020 (The respective SW score values are 679 and 1,895; there is a substantial difference). In addition, the circular permutation of a long sequence has a smaller effect on the compression distance than on the SW score. This property of a CBM may be useful when looking for similarities between rearranged sequences.

4.3.3 The Effect of Alphabet Size

Here we examined the sensitivity of CBDs to sequence manipulation, such as alphabet reduction and expansion. To do this, we chose a set of 131 sequences representing the essentially ubiquitous glycolytic enzyme, 3-phosphoglycerate kinase (3PGK). The sequences are available both as amino acid residues (358–505 residues in length) and nucleotide residues (1,074–1,515) obtained from 15 archaean (consisting of 2 phyla), 83 bacterial (consisting of 3 phyla) and 33 eukaryotic species (consisting of 5 phyla). The dataset is freely available [32].

Experiment: An alphabet reduction was evaluated just on amino acids in such way that they were grouped into different groups and each amino acid residue was replaced by its group identifier. The Dayhoff classes ("AGPST, DENQ, HKR, ILMV, FWY, C") were obtained from [14] and here are referred to as Dayhoff6. Other classes were taken from Table 1 of [33] and were denoted by SB4, SB6, SB8, SB12, and SB16, respectively. The number in the class name denotes the number of amino acid clusters. The alphabet expansion was the residue composition; that is, each bi-gram and tri-gram was considered as a single letter yielding an alphabet with number of elements n^2 and n^3, respectively, where n is the number of a certain alphabet. In DNA sequences it provided an alphabet of 16 or 64 letters, respectively. In amino acid sequences, with a reduced alphabet, we applied this on alphabets SB4, SB6, SB8 and Dayhoff6 classes. The BWT transformation was also evaluated on the original amino acid and nucleotide sequences to see how well it supports compression performance. However, the GenCompress (GC) algorithm was not evaluated on sequences obtained by alphabet expansion because it was originally developed for genome sequences and not for strings of arbitrary characters and alphabets.

Evaluation: To evaluate how well a distance matrix reflects the groups, the ROC analysis [17] was used in the following way. The similarity of two proteins was used as the score of a binary classifier, putting them into the same group such as kingdom and phylum, denoted by AUC-kingdom and AUC-phylum, respectively. Next, the ROC analysis was performed by plotting sensitivity versus 1-specificity at various threshold values, and the resulting curve was integrated to give an "area under the curve" (AUC) value. For a perfect ranking AUC = 1.0, while for a random ranking AUC = 0.5. The calculated AUC value can be interpreted as the probability of a similarity score for a pair from the same group being greater than the score for a sequence pair from a different group [18]. Here, the ROC analysis was evaluated on each distance matrix calculated by a CBD on each dataset constructed from the 3PGK by alphabet reduction, expansion and BWT, respectively. Figure 4.1a shows high correlation between the AUC-kingdom and AUC-phylum, hence in the following just the AUC-phylum is shown. In Fig. 4.1b the correlation between the compression ratio and the AUC-phylum is plotted, where the compression ratio cr is defined by

$$cr = \frac{avg.uncompr.size}{avg.compr.size} \cdot \frac{\lceil \log_2(alphabetsize) \rceil}{\log_2(256)}. \tag{4.14}$$

Here $\lceil . \rceil$ stands for the larger integer. The second constant term is a scaling factor intended to provide a fair comparison for compressors on different size of alphabet of 3PGK. After computing it, we found no apparent connection between the compression performance and the quality of CBDs (measured by AUC), but a non-linear relationship was found between the compression ratio and the size of the alphabet (see Fig. 4.1c). This suggests that there is no connection between the alphabet size and AUC, but in some cases an improvement can be attained. For example, the

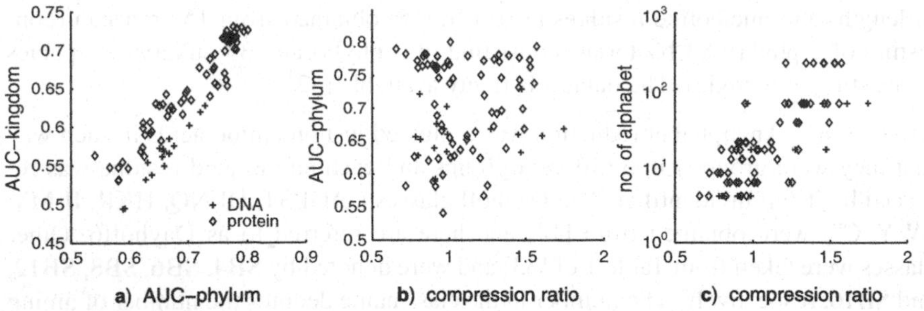

Fig. 4.1 The correlation between methods on nucleotide (*plus*) and amino acid (*diamond*) sequences from 3PGK. (**a**) The relationship between the AUC-phylum and AUC-kingdom; (**b**) the correlation between compression ratio and AUC; (**c**) the connection ratio is related to the alphabet size non-linearly

Table 4.3 AUC results for amino acid sequences with an alphabet reduction and BWT acting on 3PGK

	Original	SB16	SB12	SB8	SB6	SB4	Dayhoff6	BWT
ABC	0.738	0.766	0.771	0.783	<u>0.801</u>	0.764	0.745	0.655
AH	0.658	0.628	0.636	0.625	0.567	0.541	0.583	0.652
GC	0.763	0.759	0.746	0.773	0.762	0.779	0.767	0.689
LZW	0.666	0.620	0.623	0.626	0.589	0.582	0.581	0.599
PPM	0.764	0.762	0.766	0.783	0.779	0.790	0.782	0.706

The pairwise distance/similarity matrix also was calculated on the original amino acids sequences via naive distance and SW, and the AUC values we got were 0.685 and 0.797, respectively. The largest score is underlined.

alphabet reduced to 4–8 can improve the AUC values with some compressors, as Table 4.3 shows. Tables 4.4 and 4.5 list results of expanded alphabets on reduced amino acids and of the original nucleotide sequences, with varying results. At this point it should be mentioned that the compressor of a reduced alphabet can be considered as a lossy compressor. The BWT method was also evaluated both on amino acids and nucleotide sequences, but it did not improve the AUC values perhaps because the sequence lengths were short.

4.3.4 Experiments on Protein Classification

To evaluate these CBDs for protein classification, we chose the SCOP40 database. For the train and test set division, the supervised cross-validation technique was applied, which is a selection of test and train sets that are based on known subtypes within a database [21]. A family with over five members was selected as the positive test and the rest of the superfamily was the positive train set. The negative sets were selected in a similar way. This approach provides a reliable estimation of

Table 4.4 AUC results for an alphabet expansion on 3PGK of a reduced alphabet

	SB8		SB6			SB4			Dayhoff		
	A[a]	AA[b]	A[a]	AA[b]	AAA[c]	A[a]	AA[b]	AAA[c]	A[a]	AA[b]	AAA[c]
ABC	0.783	0.746	<u>0.801</u>	0.668	0.794	0.764	0.672	0.699	0.745	0.738	0.774
AH	0.625	0.639	0.567	0.664	0.746	0.541	0.646	0.675	0.583	0.624	0.750
LZW	0.626	0.696	0.589	0.693	0.763	0.582	0.637	0.720	0.581	0.691	0.732
PPM	0.783	0.771	0.779	0.775	0.777	0.790	0.780	0.773	0.782	0.782	0.780

[a]The values were taken from Table 4.3
[b]Bi-gram residue compositions were used
[c]Tri-gram residue compositions were used
Here, GC compressor was not used. The largest score is underlined.

Table 4.5 AUC results for an nucleotide sequences with an alphabet expansion on 3PGK

	Original	AA[a]	AAA[b]	BWT
ABC	0.676	0.633	0.668	0.639
AH	0.589	0.660	0.669	0.593
GC	0.702	n.a.	n.a.	0.634
LZW	0.652	0.668	0.660	0.628
PPM	0.722	<u>0.729</u>	0.658	0.657

[a]Bi-gram residue compositions were used
[b]Tri-gram residue compositions were used
The pairwise distance/similarity matrix was also calculated on the original amino acids sequences via naive distance and SW, and the AUC values we got were 0.662 and 0.807, respectively. The largest score is underlined.

how an algorithm will generalise to a novel, distantly-related subtype of the known protein classes. Altogether, we obtained 55 classification tasks in this manner. The sequences were represented by the so-called Empirical Feature Map method, where a sequence X is represented by a feature vector $F_X = f_{X_1}, f_{X_2}, \ldots, f_{X_n}$. Here n is the total number of proteins in the training set and f_{X_i} is a similarity/distance score between sequence X and the ith sequence in the training set. For the underlying similarity measure, CBDs along with the BLAST and Smith–Waterman algorithms were applied.

Classification Methods: Nearest Neighbour (1NN) classification [15] is a technique whereby a query sequences is assigned to the priori known class of the database entry that was found most similar to it in terms of a log-likelihood ratio of the nearest positive and negative score via a distance/similarity measure [20]. Artificial Neural Networks (ANNs) were then employed, whose structure consisted of one hidden layer with 40 neurons and the output layer consisted of one neuron. For each neuron, the log-sigmoid function was used as the transfer function and the Scaled Conjugate Gradient (SCG) algorithm was used for training [6]. The package we applied was the Neural Network toolbox version 5.0 of Matlab. The Support Vector Machines (SVMs) [34] method employed was LibSVM [10]. In our experiments, the Radial

Table 4.6 The AUC results on protein classification with several classifiers and featuring method applied on SCOP40

Method name[a] (AUC[b])	1NN	SVM	RF	LogReg	ANN	Avg
ABC (0.699)	0.726	0.843	0.779	0.806	0.834	0.798
AH (0.690)	0.711	0.877	0.824	0.751	0.800	0.793
GC (0.674)	0.644	0.775	0.691	0.753	0.769	0.726
LZW (0.769)	0.751	0.856	0.821	0.718	0.794	0.788
PPM (0.700)	0.798	0.851	0.800	0.813	0.583	0.823
naive (0.597)	0.653	0.882	0.848	0.786	0.837	0.801
BLAST (0.684)	0.775	0.905	0.697	0.836	0.902	0.823
SW (0.823)	0.956	0.946	0.823	0.913	0.897	0.907

[a]Method used in the vectorization step
[b]The AUC values in parentheses were obtained via distance matrices in the way described in Sect. 4.3.3.

Basis Function kernel was used, where the width parameter σ was the median Euclidean distance from any positive training example to the nearest negative example. The Logistic Regression (LogReg) [29] method was part of Weka versions 3–4 [37]. Finally, the Random Forest (RF) technique is a combination of decision trees such that each tree is grown on a bootstrap sample of the training set. For each node, the split is chosen from $m \ll M$ variables (M being the number of dimensions) on which to base the decision [7]. In our experiments, 50 trees were used and the number of features m was set to $\log(l+1)$, where l is the number of training patterns. The RF was part of Weka versions 3–4. For an evaluation of the classifier, ROC analysis was performed by ranking the test object using their score obtained by a learned model. The results are shown in Table 4.6.

4.3.5 Sequence Length and Sequence Complexity

It is known that short sequence similarities are more difficult to locate than long ones. This tendency also appears to be valid for CBDs analysed here. For example, when we plot the classification performance measure AUC for the 55 SCOP40 families as a function of the average sequence length, the low values are predominantly found in the shorter superfamilies (Fig. 4.2). This trend is quite similar for the BLAST and Smith–Waterman algorithms as well (http://www.inf.u-szeged.hu/~kocsor/CBM05).

A dataset of high and low complexity sequences was produced by taking human proteins from the KOG database [23], processing them with the SEG program written by J. Wootton [38], and applying the parameter values of window length = 45, trigger complexity = 3.25, and extension complexity = 3.55, as recommended by the author. Segments with length in the range of 20–1,000 amino acids were chosen for further analysis. From a total of 163,473 high-complexity and 53,849

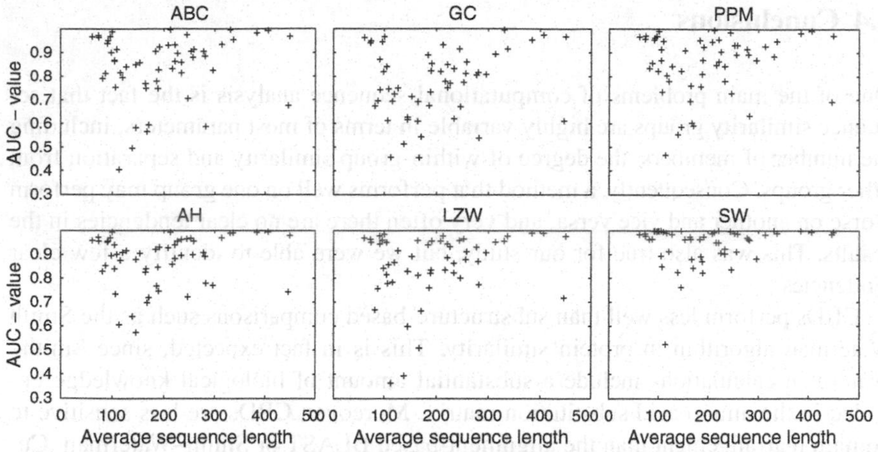

Fig. 4.2 The dependence of classification performance on the average sequence length in the SCOP40 families

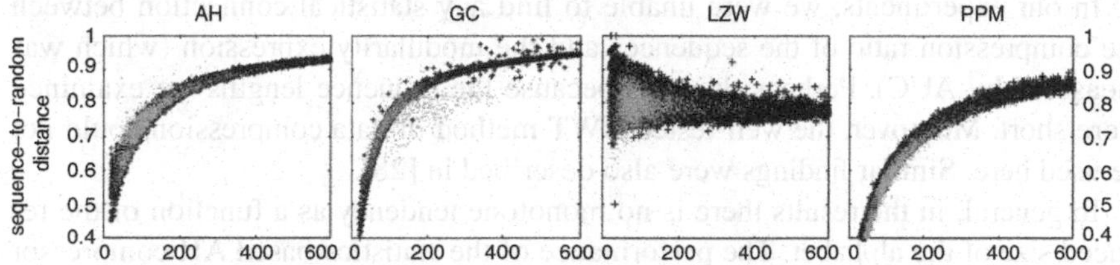

Fig. 4.3 A comparison of native sequences with their random shuffled counterparts in KOG and high- (*grey*) and low-complexity (*black*) segments of human proteins as a function of their length. The CBD distances of a sequence from its random-shuffled counterparts should give a value close to 1.00

low-complexity regions, we randomly selected 8,859 and 3,772 sequences, respectively, for an analysis, for which the results are shown in Fig. 4.3. Low-complexity segments correspond to non-globular regions in proteins, characterised by a biased amino acid composition and/or a repetitive amino acid sequence. Such regions are well known for obscuring the detection of biologically important similarities [38]. We had two particular reasons for testing CBDs on low-complexity regions: (1) repetitive character-sequences can be better compressed than average sequences; (2) our datasets of natural sequences are composed of proteins that are predominantly globular, so they are expected to contain few low complexity regions. We found that the distributions of CBDs values for low- and high-complexity proteins were not markedly different (see Fig. 4.3); however, this seems to be analogous to our earlier finding that the compression ratio is independent of the AUC measure (Fig. 4.1b).

4.4 Conclusions

One of the main problems of computational sequence analysis is the fact that sequence similarity groups are highly variable in terms of most parameters, including the number of members, the degree of within-group similarity and separation from other groups. Consequently, a method that performs well on one group may perform worse on another and vice versa, and very often there are no clear tendencies in the results. This was also true for our study, but we were able to identify a few clear tendencies.

CBDs perform less well than substructure-based comparisons such as the Smith Waterman algorithm in protein similarity. This is in fact expected, since Smith–Waterman calculations include a substantial amount of biological knowledge encoded in the amino acid substitution matrix. Moreover, CBDs are less sensitive to domain rearrangement than the alignment-based BLAST or Smith–Waterman. Curiously, some combinations of CBDs approach with BLAST comparison exceed the performance of Smith–Waterman in protein classification [22]. Unfortunately CBDs and alignment-based measures depend on the length of the sequences.

In our experiments, we were unable to find any statistical connection between the compression ratio of the sequences and the modularity expression (which was measured by AUC). Perhaps this was because the sequence lengths we examined were short. Moreover, the well-tested BWT method of data compression could not be used here. Similar findings were also described in [28].

In general, in the results there is no monotone tendency as a function of the reduced size of the alphabet. The performance of the statistics-based AH compressor decreased when the alphabet was reduced, while the partial matching-based algorithms like PPM and GC improved their performance. This could mean that mutations of amino acids found in the same cluster do not really change their structure and function(s). Furthermore, a compressor applied to reduced alphabet sequences can be viewed as a lossy compressor. An alphabet expansion with bi-gram and tri-gram composition usually increases the performance of the statistical AH compressor, as in the expanded letter distribution the neighbours of the original letters are taken into account.

Overall, partial matching-based compressors (PPM, GC) seem to outperform the various types of compressors available, both in ROC analysis and protein classification.

Acknowledgements A. Kocsor was supported by the Jnos Bolyai fellowship of the Hungarian Academy of Sciences. Work at ICGEB was supported in part by grants from the Ministero dell.Universita' e della Ricerca (D.D. 2187, FIRB 2003 (art. 8), "Laboratorio Internazionale di Bioinformatica").

References

1. Abel, J.: Improvements to the burrows-wheeler compression algorithm: After bwt stages (2003)
2. Ágoston, V., Kájan, L., Carugo, O., Hegedűs, Z., Vlahovicek, K., Pongor, S.: Concepts of similarity in bioinformatics. In: D. Moss, S. Jelaska, S. Pongor (eds.) Essays in Bioinformatics. IOS, Amsterdam (2005)
3. Altschul, S.F., Gish, W., Miller, W., Myers, E.W., Lipman, D.J.: Basic local alignment search tool. J Mol Biol **215**(3), 403–410 (1990)
4. Andreeva, A., Howorth, D., Brenner, C.: Scop database in 2004: refinements integrate structure and sequence family data (2004)
5. Bennett, C.H., Gács, P., Li, M., Vitanyi, P.M.B., Zurek, W.H.: Information distance. IEEETIT: IEEE Trans Inform Theory **44**, 1407–1423 (1998)
6. Bishop, C.M.: Neural networks for pattern recognition. Oxford University Press, Oxford (1996)
7. Breiman, L.: Random forests. Machine Learning **45**(1), 5–32 (2001)
8. Burrows, M., Wheeler, D.J.: A block-sorting lossless data compression algorithm. Tech. Rep. 124, 130 Lytton Avenuve, Palo Alto, CA, 94301 (1994)
9. Cai, H., Kulkarni, S.R., Verdú, S.: Universal entropy estimation via block sorting. IEEE Trans Inform Theory **50**(7), 1551–1561 (2004)
10. Chang, C.C., Lin, C.J.: LIBSVM: a library for support vector machines (2001)
11. Chen, X., Kwong, S., Li, M.: A compression algorithm for DNA sequences and its applications in genome comparison. In: RECOMB, p 107 (2000)
12. Cilibrasi, R., Vitanyi, P.M.B.: Clustering by compression. IEEE Trans Inform Theory **51**(4), 1523–1545 (2005)
13. Cover, T.M., Thomas, J.A.: Elements of information theory. Wiley, New York (1991)
14. Dayhoff, M.O., Schwartz, R.M., Orcutt, B.C.: A model of evolutionary change in proteins. In: M.O. Dayhoff (ed.) Atlas of protein sequence and structure, vol. 5, 345–358. National Biomedical Research Foundation, Washington, D.C., (1978)
15. Duda, R.O., Hart, P.E., Stork, D.G.: Pattern Classification, 2nd edn. Wiley Interscience, New York (2000)
16. Ferragina, P., Giancarlo, R., Greco, V., Manzini, G., Valiente, G.: Compression-based classification of biological sequences and structures via the universal similarity metric: experimental assessment. BMC Bioinformatics **8**, 252 (2007)
17. Gribskov, M., Robinson, N.: Use of receiver operating characteristic (roc) analysis to evaluate sequence matching. Comput Chem **20**, 25–33 (1996)
18. Hanley, J.A., Mcneil, B.J.: The meaning and use of the area under a receiver operating characteristic (roc) curve. Radiology **143**(1), 29–36 (1982)
19. Henikoff, S., Henikoff, J.G., Pietrokovski, S.: Blocks: a non-redundant database of protein alignment blocks derived from multiple compilations. Bioinformatics **15**(6), 471–479 (1999)
20. Kaján, L., Kertész-Farkas, A., Franklin, D., Ivanova, N., Kocsor, A., Pongor, S.: Application of a simple likelihood ratio approximant to protein sequence classification. Bioinformatics **22**(23), 2865–2869 (2006)
21. Kertész-Farkas, A., Dhir, S., Sonego, P., Pacurar M., Netoteia, S., Nijveen, H., Kuzinar, A., Leunissen, J., Kocsor, A., Pongor, S.: Benchmarking protein classification algorithms via supervised cross-validation. J Biochem Biophys Methods **35**, 1215–1223 (2007)
22. Kocsor, A., Kertész-Farkas, A., Kaján, L., Pongor, S.: Application of compression-based distance measures to protein sequence classification: a methodological study. Bioinformatics **22**(4), 407–412 (2006)
23. Koonin, E.V., Fedorova, N.D., Jackson, J.D., Jacobs, A.R., Krylov, D.M., Makarova, K.S., Mazumder, R., Mekhedov, S.L., Nikolskaya, A.N., Rao, B.S., Rogozin, I.B., Smirnov, S., Sorokin, A.V., Sverdlov, A.V., Vasudevan, S., Wolf, Y.I., Yin, J.J., Natale, D.A.: A comprehen-

sive evolutionary classification of proteins encoded in complete eukaryotic genomes. Genome Biol **5**(2) (2004)
24. Li, M.: Information distance and its applications. In: O.H. Ibarra, H.C. Yen (eds.) CIAA, *Lecture Notes in Computer Science*, vol. 4094, 1–9. Springer, Berlin (2006)
25. Li, M., Chen, X., Li, X., Ma, B., Vitányi, P.: The similarity metric. In: SODA '03: Proceedings of the fourteenth annual ACM-SIAM symposium on Discrete algorithms, 863–872. Society for Industrial and Applied Mathematics, Philadelphia (2003)
26. Li, M., Vitanyi, P.: An introduction to kolmogorov complexity and its applications, 2nd edn. Springer, Berlin (1997)
27. Li, M., Vitányi, P.M.: Mathematical theory of thermodynamics of computation. Tech. rep., Centre for Mathematics and Computer Science, Amsterdam, The Netherlands (1992)
28. Nevill-Manning, C.G., Witten, I.H.: Protein is incompressible. In: DCC '99: Proceedings of the Conference on Data Compression, p. 257. IEEE Computer Society, Washington, DC, USA (1999)
29. Rice, J.C.: Logistic regression: An introduction. In: B. Rhompson (ed.) Advances in social science methodology, vol. 3, 191–245. JAI, Greenwich (1994)
30. Schweizer, D., Abu-Mostafa, Y.: Kolmogorov metric spaces. Manuscript, Computer Sciences, 256–80, California Institute of Technology, Pasadena, CA 91125 (1998)
31. Smith, T.F., Waterman, M.S.: Identification of common molecular subsequences. J Mol Biol **147**, 195–197 (1981)
32. Sonego, P., Pacurar, M., Dhir, S., Kertész-Farkas, A., Kocsor, A., Gáspári, Z., Leunissen, J.A.M., Pongor, S.: A protein classification benchmark collection for machine learning. Nucleic Acids Res **35**(Database-Issue), 232–236 (2007)
33. Susko, E., Roger, A.J.: On reduced amino acid alphabets for phylogenetic inference. Mol Biol Evol **24**, 2139–2150 (2007)
34. Vapnik, V.N.: The nature of statistical learning theory, 2nd edn. Springer, Berlin (1999)
35. Vinga, S., Almeida, J.: Alignment-free sequence comparison-a review. Bioinformatics **19**(4), 513–523 (2003)
36. Willems, F.M.J., Shtarkov, Y.M., Tjalkens, T.J.: The context-tree weighting method: basic properties. IEEE Trans Inform Theory, 653–664 (1995)
37. Witten, I.H., Frank, E.: Data mining: practical machine learning tools and techniques with Java implementations. Morgan Kaufmann, San Francisco (1999)
38. Wootton, J.C.: Non-globular domains in protein sequences: automated segmentation using complexity measures. Comput Chem **18**(3), 269–285 (1994)
39. Zurek, W.H.: Thermodynamic cost of computation, algorithmic complexity and the information metric. Nature **341**(6238), 119–124 (1989)

Chapter 5
MIC: Mutual Information Based Hierarchical Clustering

Alexander Kraskov and Peter Grassberger

Abstract Clustering is a concept used in a huge variety of applications. We review a conceptually very simple algorithm for hierarchical clustering called in the following the *mutual information clustering* (MIC) algorithm. It uses mutual information (MI) as a similarity measure and exploits its grouping property: The MI between three objects X, Y, and Z is equal to the sum of the MI between X and Y, plus the MI between Z and the combined object (XY). We use MIC both in the Shannon (probabilistic) version of information theory, where the "objects" are probability distributions represented by random samples, and in the Kolmogorov (algorithmic) version, where the "objects" are symbol sequences. We apply our method to the construction of phylogenetic trees from mitochondrial DNA sequences and we reconstruct the fetal ECG from the output of independent components analysis (ICA) applied to the ECG of a pregnant woman.

5.1 Introduction

Classification or organizing of data is a crucial task in all scientific disciplines. It is one of the most fundamental mechanism of understanding and learning [19]. Depending on the problem, classification can be exclusive or overlapping, supervised or unsupervised. In the following we will be interested only in exclusive unsupervised classification. This type of classification is usually called clustering or cluster analysis.

A. Kraskov (✉)
UCL Institute of Neurology, Queen Square, London WC1N 3BG, UK
e-mail: akraskov@ion.ucl.ac.uk

P. Grassberger
Department of Physics and Astronomy and Institute for Biocomplexity and Informatics, University of Calgary, 2500 University Drive NW, Calgary AB, Canada T2N 1N4
e-mail: pgrassbe@ucalgary.ca

F. Emmert-Streib, M. Dehmer (eds.), *Information Theory and Statistical Learning*,
DOI: 10.1007/978-0-387-84816-7_5,

An instance of a clustering problem consists of a set of objects and a set of properties (called characteristic vector) for each object. The goal of clustering is the separation of objects into groups using only the characteristic vectors. Indeed, in general only certain aspects of the characteristic vectors will be relevant, and extracting these relevant features is one field where mutual information (MI) plays a major role [36], but we shall not deal with this here. Cluster analysis organizes data either as a single grouping of individuals into non-overlapping clusters or as a hierarchy of nested partitions. The former approach is called partitional clustering (PC), the latter one is hierarchical clustering (HC). One of the main features of HC methods is the visual impact of the tree or *dendrogram* which enables one to see how objects are being merged into clusters. From any HC one can obtain a PC by restricting oneself to a "horizontal" cut through the dendrogram, while one cannot go in the other direction and obtain a full hierarchy from a single PC. Because of their wide spread of applications, there are a large variety of different clustering methods in use [19]. In the following we shall only deal with *agglomerative* hierarchical clustering, where clusters are built by joining first the most obvious objects into pairs, and then continues to join build up larger and larger objects. Thus the tree is built by starting at the leaves, and is finished when the main branches are finally joined at the root. This is opposed to algorithms where one starts at the root and splits clusters up recursively. In either case, the tree obtained in this way can be refined later by restructuring it, e.g. using so-called *quartet methods* [7, 34].

The crucial point of all clustering algorithms is the choice of a *proximity measure*. This is obtained from the characteristic vectors and can be either an indicator for similarity (i.e. large for similar and small for dissimilar objects), or dissimilarity. In the latter case it is convenient (but not obligatory) if it satisfies the standard axioms of a *metric* (positivity, symmetry, and triangle inequality). A matrix of all pairwise proximities is called proximity matrix. Among agglomerative HC methods one should distinguish between those where one uses the characteristic vectors only at the first level of the hierarchy and derives the proximities between clusters from the proximities of their constituents, and methods where the proximities are calculated each time from their characteristic vectors. The latter strategy (which is used also in the present paper) allows of course for more flexibility but might also be computationally more costly. There exist a large number of different strategies [19, 30], and the choice of the optimal strategy depends on the characteristics of the similarities: for ultrametric distances, e.g. the "natural" method is UPGMA [30], while *neighbor joining* is the natural choice when the distance matrix is a metric satisfying the four-point condition $d_{ij} + d_{kl} \leq \max(d_{ik} + d_{jl}, d_{il} + d_{jk})$ [30].

In the present chapter we shall use proximities resp. distances derived from mutual information [9]. In that case the distances neither form an ultrametric, nor do they satisfy the above four-point condition. Thus neither of the two most popular agglomerative clustering methods are favored. But we shall see that the distances have another special feature which suggests a different clustering strategy discussed first in [21, 23].

Quite generally, the "objects" to be clustered can be either single (finite) patterns (e.g. DNA sequences) or random variables, i.e. *probability distributions*. In the latter case the data are usually supplied in form of a statistical sample, and one of the simplest and most widely used similarity measures is the linear (Pearson) correlation coefficient. But this is not sensitive to nonlinear dependencies which do not manifest themselves in the covariance and can thus miss important features. This is in contrast to mutual information (MI) which is also singled out by its information theoretic background [9]. Indeed, MI is zero only if the two random variables are strictly independent.

Another important feature of MI is that it has also an "algorithmic" cousin, defined within algorithmic (Kolmogorov) information theory [26] which measures the similarity between individual objects. For a comparison between probabilistic and algorithmic information theories, see [17]. For a thorough discussion of distance measures based on algorithmic MI and for their application to clustering, see [6, 24, 25].

Another feature of MI which is essential for the present application is its *grouping property*: The MI between three objects (distributions) X,Y, and Z is equal to the sum of the MI between X and Y, plus the MI between Z and the combined object (joint distribution) (XY),

$$I(X,Y,Z) = I(X,Y) + I((X,Y),Z). \tag{5.1}$$

Within Shannon information theory this is an exact theorem (see below), while it is true in the algorithmic version up to the usual logarithmic correction terms [26]. Since X,Y, and Z can be themselves composite, (5.1) can be used recursively for a cluster decomposition of MI. This motivates the main idea of our clustering method: instead of using, e.g. centers of masses in order to treat clusters like individual objects in an approximative way, we treat them exactly like individual objects when using MI as proximity measure.

More precisely, we propose the following scheme for clustering n objects with MIC:

(1) Compute a proximity matrix based on pairwise mutual informations; assign n clusters such that each cluster contains exactly one object.
(2) Find the two closest clusters i and j.
(3) Create a new cluster (ij) by combining i and j.
(4) Delete the lines/columns with indices i and j from the proximity matrix, and add one line/column containing the proximities between cluster (ij) and all other clusters. These new proximities are calculated by either treating (X_i,X_j) as a single random variable Shannon version), or by concatenating X_i and X_j (algorithmic version).
(5) If the number of clusters is still >2, go to (2); else join the two clusters and stop.

In the next section we shall review the pertinent properties of MI, both in the Shannon and in the algorithmic version. This is applied in Sect. 5.3 to construct a phylogenetic tree using mitochondrial DNA and in Sect. 5.4 to cluster the output channels of an independent component analysis (ICA) of an electrocardiogram (ECG) of a pregnant woman, and to reconstruct from this the maternal and fetal ECGs. We finish with our conclusions in Sect. 5.5.

5.2 Mutual Information

5.2.1 Shannon Theory

Assume that one has two random variables X and Y. If they are discrete, we write $p_i(X) = \text{prob}(X = x_i)$, $p_i(Y) = \text{prob}(Y = x_i)$, and $p_{ij} = \text{prob}(X = x_i, Y = y_i)$ for the marginal and joint distribution. Otherwise (and if they have finite densities) we denote the densities by $\mu_X(x), \mu_Y(y)$, and $\mu(x,y)$. Entropies are defined for the discrete case as usual by $H(X) = -\sum_i p_i(X)\log p_i(X)$, $H(Y) = -\sum_i p_i(Y)\log p_i(Y)$, and $H(X,Y) = -\sum_{i,j} p_{ij}\log p_{ij}$. Conditional entropies are defined as $H(X|Y) = H(X,Y) - H(Y) = -\sum_{i,j} p_{ij}\log p_{i|j}$. The base of the logarithm determines the units in which information is measured. In particular, taking base two leads to information measured in bits. In the following, we always will use natural logarithms. The MI between X and Y is finally defined as

$$
\begin{aligned}
I(X,Y) &= H(X) + H(Y) - H(X,Y) \\
&= \sum_{i,j} p_{ij} \log \frac{p_{ij}}{p_i(X)p_j(Y)}.
\end{aligned} \tag{5.2}
$$

It can be shown to be non-negative, and is zero only when X and Y are strictly independent. For n random variables $X_1, X_2 \ldots X_n$, the MI is defined as

$$
I(X_1,\ldots,X_n) = \sum_{k=1}^{n} H(X_k) - H(X_1,\ldots,X_n). \tag{5.3}
$$

This quantity is often referred to as (generalized) redundancy, in order to distinguish it from different "mutual informations" which are constructed analogously to higher order cumulants [9], but we shall not follow this usage. Equation (5.1) can be checked easily,

$$
\begin{aligned}
I(X,Y,Z) &= H(X) + H(Y) + H(Z) - H(X,Y,Z) \\
&= \sum_{i,j,k} p_{ijk} \log \frac{p_{ijk}}{p_i(X)p_j(Y)p_k(Z)} \\
&= \sum_{i,j,k} p_{ijk} \left[\log \frac{p_{ij}(XY)}{p_i(X)p_j(Y)} + \log \frac{p_{ijk}}{p_{ij}(XY)p_k(Z)}\right] \\
&= I(X,Y) + I((X,Y),Z),
\end{aligned} \tag{5.4}
$$

together with its generalization to arbitrary groupings. It means that MI can be *decomposed into hierarchical levels*. By iterating it, one can decompose $I(X_1 \ldots X_n)$ for any $n > 2$ and for any partitioning of the set $(X_1 \ldots X_n)$ into the MIs between elements within one cluster and MIs between clusters.

For continuous variables one first introduces some binning ("coarse-graining"), and applies the above to the binned variables. If x is a vector with dimension m and each bin has Lebesgue measure Δ, then $p_i(X) \approx \mu_X(x)\Delta^m$ with x chosen suitably in bin i, and[1]

$$H_{\text{bin}}(X) \approx \tilde{H}(X) - m \log \Delta, \tag{5.5}$$

where the *differential entropy* is given by

$$\tilde{H}(X) = -\int dx\, \mu_X(x) \log \mu_X(x). \tag{5.6}$$

Notice that $H_{\text{bin}}(X)$ is a true (average) information and is thus non-negative, but $\tilde{H}(X)$ is not an information and can be negative. Also, $\tilde{H}(X)$ is not invariant under homeomorphisms $x \to \varphi(x)$.

Joint entropies, conditional entropies, and MI are defined as above, with sums replaced by integrals. Like $\tilde{H}(X)$, joint and conditional entropies are neither positive (semi-)definite nor invariant. But MI, defined as

$$I(X,Y) = \iint dxdy\, \mu_{XY}(x,y) \log \frac{\mu_{XY}(x,y)}{\mu_X(x)\mu_Y(y)}, \tag{5.7}$$

is non-negative and invariant under $x \to \varphi(x)$ and $y \to \psi(y)$. It is (the limit of) a true information,

$$I(X,Y) = \lim_{\Delta \to 0} [H_{\text{bin}}(X) + H_{\text{bin}}(Y) - H_{\text{bin}}(X,Y)]. \tag{5.8}$$

5.2.2 Estimating Mutual Shannon Information

In applications, one usually has the data available in form of a statistical sample. To estimate $I(X,Y)$ one starts from N bivariate measurements (x_i, y_i), $i = 1, \ldots N$ which are assumed to be iid (independent identically distributed) realizations. For discrete variables, estimating the probabilities p_i, p_{ij}, etc., is straightforward: p_i is just approximated by the ratio n_i/N, where n_i is the number of outcomes $X = x_i$. This approximation gives rise both to a bias in the estimate of entropies, and to statistical fluctuations. The bias can be largely avoided by more sophisticated methods (see, e.g. [16]), but we shall not go into details.

For continuous variables, the situation is worse. There exist numerous strategies to estimate $I(X,Y)$. The most popular include discretization by partitioning the ranges of X and Y into finite intervals [10], "soft"or "fuzzy" partitioning using B-splines [11], and kernel density estimators [28]. We shall use in the following the MI estimators based on k-nearest neighbors statistics proposed in [22], and we refer to this paper for a comparison with alternative methods.

---

[1] Notice that we have here assumed that densities really exists. If not, e.g. if X lives on a fractal set), then m is to be replaced by the Hausdorff dimension of the measure μ.

5.2.3 k-Nearest Neighbors Estimators

There exists an extensive literature on nearest neighbors based estimators for the simple Shannon entropy

$$H(X) = -\int dx\mu(x)\log\mu(x), \tag{5.9}$$

dating back at least to [12, 38]. But it seems that these methods have never been used for estimating MI. In [8, 13–15, 37–39] it is assumed that x is one-dimensional, so that the x_i can be ordered by magnitude and $x_{i+1} - x_i \to 0$ for $N \to \infty$. In the simplest case, the estimator based only on these distances is

$$H(X) \approx -\frac{1}{N-1}\sum_{i=1}^{N-1} \log(x_{i+1} - x_i) - \psi(1) + \psi(N) \,. \tag{5.10}$$

Here, $\psi(x)$ is the digamma function, $\psi(x) = \Gamma(x)^{-1} d\Gamma(x)/dx$. It satisfies the recursion $\psi(x+1) = \psi(x) + 1/x$ and $\psi(1) = -C$ where $C = 0.5772156\ldots$ is the Euler–Mascheroni constant. For large x, $\psi(x) \approx \log x - 1/2x$. Similar formulas exist which use $x_{i+k} - x_i$ instead of $x_{i+1} - x_i$, for any integer $k < N$.

Although (5.10) and its generalizations to $k > 1$ seem to give the best estimators of $H(X)$, they cannot be used for MI because it is not obvious how to generalize them to higher dimensions. Here we have to use a slightly different approach, due to [20].

Assume some metrics to be given on the spaces spanned by X, Y and $Z = (X,Y)$. In the following we shall use always the maximum norm in the joint space, i.e.

$$||z - z'|| = \max\{||x - x'||, ||y - y'||\}, \tag{5.11}$$

independently of the norms used for $||x - x'||$ and $||y - y'||$ (they need not be the same, as these spaces could be completely different). We can then rank, for each point $z_i = (x_i, y_i)$, its neighbors by distance $d_{i,j} = ||z_i - z_j||$: $d_{i,j_1} \le d_{i,j_2} \le d_{i,j_3} \le \ldots$. Similar rankings can be done in the subspaces X and Y. The basic idea of [20] is to estimate $H(X)$ from the average distance to the k-nearest neighbor, averaged over all x_i. Mutual information could be obtained by estimating in this way $H(X)$, $H(Y)$ and $H(X,Y)$ separately and using

$$I(X,Y) = H(X) + H(Y) - H(X,Y) \,. \tag{5.12}$$

But using the same k in both the marginal and joint spaces would mean that the typical distances to the k–th neighbors are much larger in the joint (Z) space than in the marginal spaces. This would mean that any errors made in the individual estimates would presumably not cancel, since they depend primarily on these distances.

Therefore we proceed differently in [22]. We first choose a value of k, which gives the number of neighbors in the joint space. From this we obtain for each point $z_i = (x_i, y_i)$ a length scale ε_i, and then we count the number of neighbors within this distance for each of the marginal points x_i and y_i.

Indeed, for each k two different versions of this algorithm were given in [22]. In the first, neighborhoods in the joint space are chosen as (hyper-)squares, so that the length scales ε_i are the same in x and in y. In the second version, the size of the neighborhood is further optimized by taking them to be (hyper-)rectangles, so that $\varepsilon_{i,x} \neq \varepsilon_{i,y}$. Also, the analogous estimators for the generalized redundancies $I(X_1, X_2, \ldots X_m)$ were given there. Both variants give very similar results. For details see [22].

Compared to other estimators, these estimators are of similar speed (they are faster than methods based on kernel density estimators, slower than the B-spline estimator of [11]) and of comparable speed to the sophisticated adaptive partitioning method of [10]. They are rather robust (i.e. they are insensitive to outliers). Their superiority becomes most evident in higher dimensions, where any method based on partitioning fails. Any estimator has statistical errors (due to sample-to-sample fluctuations) and some bias. Statistical errors decrease with k, while the bias increases in general with k. Thus it is advised to take large k (up to $k/N \approx 0.1$) when the bias is not expected be a problem, and to use small k ($k = 1$, in the extreme), if a small bias is more important than small statistical sample-to-sample fluctuations.

A systematic study of the performance of these estimators and comparison with previous algorithms is given in [22]. Here we will discuss just one further feature of the estimators proposed in [22]: They seem to be strictly unbiased whenever the true mutual information is zero. This makes them particularly useful for a test for independence. We have no analytic proof for this, but very good numerical evidence. As an example, we show in Fig. 5.1 results for Gaussian distributions. More precisely, we drew a large number (typically 10^6 and more) of N-tuples of vectors (x, y) from a bivariate Gaussian with fixed covariance r, and estimated $I(X,Y)$ with $k = 1$ by means of the second variant $\widehat{I}^{(2)}(X,Y)$ of our estimator. The averages over all tuples of $\widehat{I}^{(2)}(X,Y) - I_{\mathrm{Gauss}}(X,Y)$ is plotted in Fig. 5.1 against $1/N$. Here,

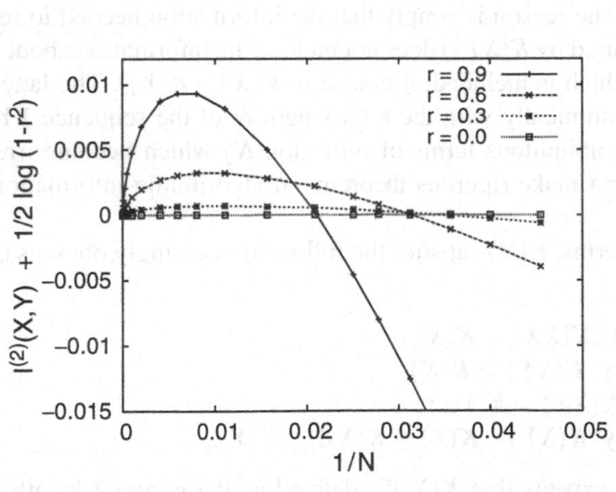

Fig. 5.1 Averages of the estimates of $\widehat{I}^{(2)}(X,Y) - I_{\mathrm{exact}}(X,Y)$ for Gaussians with unit variance and covariances $r = 0.9, 0.6, 0.3$, and 0.0 (*from top to bottom*), plotted against $1/N$. In all cases $k = 1$. The number of realizations over which this is averaged is $>2 \times 10^6$ for $N \leq 1{,}000$, and decreases to $\approx 10^5$ for $N = 40{,}000$. Error bars are smaller than the sizes of the symbols

$$I_{\rm Gauss}(X,Y) = -\frac{1}{2}\log(1-r^2)\,. \tag{5.13}$$

is the exact MI for Gaussians with covariance r [10].

The most conspicuous feature seen in Fig. 5.1, apart from the fact that indeed $I^{(2)}(X,Y) - I_{\rm Gauss}(X,Y) \to 0$ for $N \to \infty$, is that the systematic error is compatible with zero for $r = 0$, i.e. when the two Gaussians are uncorrelated. We checked this with high statistics runs for many different values of k and N (a priori one should expect that systematic errors become large for very small N), and for many more distributions (exponential, uniform, ...). In all cases we found that both variants $\widehat{I}^{(1)}(X,Y)$ and $\widehat{I}^{(2)}(X,Y)$ become exact for independent variables.

5.2.4 *Algorithmic Information Theory*

In contrast to Shannon theory where the basic objects are random variables and entropies are *average* informations, algorithmic information theory deals with individual symbol strings and with the actual information needed to specify them. To "specify" a sequence X means here to give the necessary input to a universal computer U, such that U prints X on its output and stops. The analogon to entropy, called here usually the *complexity* $K(X)$ of X, is the minimal length of any input which leads to the output X, for fixed U. It depends on U, but it can be shown that this dependence is weak and can be neglected in the limit when $K(X)$ is large [9, 26].

Let us denote the concatenation of two strings X and Y as XY. Its complexity is $K(XY)$. It is intuitively clear that $K(XY)$ should be larger than $K(X)$ but cannot be larger than the sum $K(X)+K(Y)$. Even if X and Y are completely unrelated, so that one cannot learn anything about Y by knowing X, $K(XY)$ is slightly smaller that $K(X)+K(Y)$. The reason is simply that the information needed to reconstruct XY (which is measured by $K(XY)$) does not include the information about where X ends and Y starts (which is included of course in $K(X)+K(Y)$). The latter information increases logarithmically with the total length N of the sequence XY. It is one of the sources for ubiquitous terms of order $\log(N)$ which become irrelevant in the limit $N \to \infty$, but make rigorous theorems in algorithmic information theory look unpleasant.

Up to such terms, $K(X)$ satisfies the following seemingly obvious but non-trivial properties [6]:

1. Idempotency: $K(XX) = K(X)$
2. Monotonicity: $K(XY) \geq K(X)$
3. Symmetry: $K(XY) = K(YX)$
4. Distributivity: $K(XY) + K(Z) \leq K(XZ) + K(YZ)$

Finally, one expects that $K(X|Y)$, defined as the minimal length of a program printing X when Y is furnished as auxiliary input, is related to $K(XY) - K(Y)$. Indeed, one can show [26] (again within correction terms which become irrelevant asymptotically) that

$$0 \leq K(X|Y) \simeq K(XY) - K(Y) \leq K(X). \tag{5.14}$$

Notice the close similarity with Shannon entropy.

The algorithmic information in Y about X is finally defined as

$$I_{\text{alg}}(X,Y) = K(X) - K(X|Y) \simeq K(X) + K(Y) - K(XY). \tag{5.15}$$

Within the same additive correction terms, one shows that it is symmetric, $I_{\text{alg}}(X,Y) = I_{\text{alg}}(Y,X)$, and can thus serve as an analogon to mutual information.

From Turing's halting theorem it follows that $K(X)$ is in general not computable. But one can easily give upper bounds. Indeed, the length of any input which produces X (e.g. by spelling it out verbatim) is an upper bound. Improved upper bounds are provided by any file compression algorithm such as gnuzip or UNIX "compress". Good compression algorithms will give good approximations to $K(X)$, and algorithms whose performance does not depend on the input file length (in particular since they do not segment the file during compression) will be crucial for the following. As argued in [6], it is not necessary that the compression algorithm gives estimates of $K(X)$ which are close to the true values, as long as it satisfies points 1–4 above. Such a compression algorithm is called *normal* in [6]. While the old UNIX "compress" algorithm is not normal (idempotency is badly violated), most modern compression algorithms (see [1] for an exhaustive overview) are close to normal.

Before leaving this subsection, we should mention that $K(X|Y)$ can also be estimated in a completely different way, by aligning X and Y [29]. If X and Y are sufficiently close so that a global alignment makes sense, one can form from such an alignment a *translation string* $T_{Y \to X}$ which is such that Y and $T_{Y \to X}$ together determine X uniquely. Then $K(T_{Y \to X})$ is an upper bound to $K(X|Y)$, and can be estimated by compressing $T_{Y \to X}$. In [29] this was applied among others to complete mitochondrial genomes of vertebrates. The estimates obtained with state of the art global alignment and text compression algorithms (MAVID [27] and lpaq1 [1]) were surprisingly close to those obtained by the compression-and-concatenation method with the gene compression algorithm XM [40]. The latter seems at present by far the best algorithm for compressing DNA sequences. Estimates for $K(X|Y)$ obtained with other algorithms such as gencompress [2] were typically smaller by nearly an order of magnitude.

5.2.5 MI-Based Distance Measures

Mutual information itself is a similarity measure in the sense that small values imply large "distances" in a loose sense. But it would be useful to modify it such that the resulting quantity is a metric in the strict sense, i.e. satisfies the triangle inequality. Indeed, the first such metric is well known within Shannon theory [9]: The quantity

$$d(X,Y) = H(X|Y) + H(Y|X) = H(X,Y) - I(X,Y) \tag{5.16}$$

satisfies the triangle inequality, in addition to being non-negative and symmetric and to satisfying $d(X,X) = 0$. The analogous statement in algorithmic information theory, with $H(X,Y)$ and $I(X,Y)$ replaced by $K(XY)$ and $I_{\text{alg}}(X,Y)$, was proven in [24, 25].

But $d(X,Y)$ is not appropriate for our purposes. Since we want to compare the proximity between two single objects and that between two clusters containing maybe many objects, we would like the distance measure to be unbiased by the sizes of the clusters. As argued forcefully in [24, 25], this is not true for $I_{\text{alg}}(X,Y)$, and for the same reasons it is not true for $I(X,Y)$ or $d(X,Y)$ either: A mutual information of 1,000 bit should be considered as large, if X and Y themselves are just 1,000 bit long, but it should be considered as very small, if X and Y would each be huge, say one billion bits.

As shown in [24, 25] within the algorithmic framework, one can form two different distance measures from MI which define metrics and which are normalized. As shown in [21] (see also [41]), the proofs of [24, 25] can be transposed almost verbatim to the Shannon case. In the following we quote only the latter.

Theorem 5.1. *The quantity*

$$D(X,Y) = 1 - \frac{I(X,Y)}{H(X,Y)} = \frac{d(X,Y)}{H(X,Y)} \tag{5.17}$$

is a metric, with $D(X,X) = 0$ and $D(X,Y) \leq 1$ for all pairs (X,Y).

Theorem 5.2. *The quantity*

$$\begin{aligned} D'(X,Y) &= 1 - \frac{I(X,Y)}{\max\{H(X),H(Y)\}} \\ &= \frac{\max\{H(X|Y),H(Y|X)\}}{\max\{H(X),H(Y)\}} \end{aligned} \tag{5.18}$$

is also a metric, also with $D'(X,X) = 0$ and $D'(X,Y) \leq 1$ for all pairs (X,Y). It is sharper than D, i.e. $D'(X,Y) \leq D(X,Y)$.

Apart from scaling correctly with the total information, in contrast to $d(X,Y)$, the algorithmic analogs to $D(X,Y)$ and $D'(X,Y)$ are also *universal* [25]. Essentially this means that if $X \approx Y$ according to any non-trivial distance measure, then $X \approx Y$ also according to D, and even more so (by a factor up to 2) according to D'. In contrast to the other properties of D and D', this is not easy to carry over from algorithmic to Shannon theory. The proof in [25] depends on X and Y being discrete, which is obviously not true for probability distributions. Based on the universality argument, it was argued in [25] that D' should be superior to D, but the numerical studies shown in that reference did not show a clear difference between them. In addition, D is singled out by a further property:

Theorem 5.3. *Let X and Y be two strings, and let W be the concatenation $W = XY$. Then W is a weighted "mid point" in the sense that*

$$D(X,W) + D(W,Y) = D(X,Y), \qquad D(X,W) : D(Y,W) = H(Y|X) : H(X|Y). \tag{5.19}$$

Proof. We present the proof in its Shannon version. The algorithmic version is basically the same, if one neglects the standard logarithmic correction terms.

Since $H(X|W) = 0$, one has $I(X,W) = H(X)$. Similarly, $H(X,W) = H(X,Y)$. Thus

$$D(X,W) + D(W,Y) = 2 - \frac{H(X) + H(Y)}{H(X,Y)} = 1 - \frac{I(X,Y)}{H(X,Y)} = D(X,Y), \tag{5.20}$$

which proofs the first part. The second part is proven similarly by straightforward calculation.

For D' one has only the inequalities $D'(X,Y) \le D'(X,W) + D'(W,Y) \le D(X,Y)$.

Theorem 3 provides a strong argument for using D in MIC, instead of D'. This does not mean that D is *always* preferable to D'. Indeed, we will see in the next section that MIC is not always the most natural clustering algorithm, but that depends very much on the application one has in mind. Anyhow, we found numerically that in all cases D gave at least comparable results as D'.

A major difficulty appears in the Shannon framework, if we deal with continuous random variables. As we mentioned above, Shannon informations are only finite for coarse-grained variables, while they diverge if the resolution tends to zero. This means that dividing MI by the entropy as in the definitions of D and D' becomes problematic. One has essentially two alternative possibilities. The first is to actually introduce some coarse-graining, although it would not have been necessary for the definition of $I(X,Y)$, and divide by the coarse-grained entropies. This introduces an arbitrariness, since the scale Δ is completely ad hoc, unless it can be fixed by some independent arguments. We have found no such arguments, and thus we propose the second alternative. There we take $\Delta \to 0$. In this case $H(X) \sim m_x \log \Delta$, with m_x being the dimension of X. In this limit D and D' would tend to 1. But using similarity measures

$$S(X,Y) = (1 - D(X,Y)) \log(1/\Delta), \tag{5.21}$$
$$S'(X,Y) = (1 - D'(X,Y)) \log(1/\Delta) \tag{5.22}$$

instead of D and D' gives *exactly* the same results in MIC, and

$$S(X,Y) = \frac{I(X,Y)}{m_x + m_y}, \qquad S'(X,Y) = \frac{I(X,Y)}{\max\{m_x, m_y\}}. \tag{5.23}$$

Thus, when dealing with continuous variables, we divide the MI either by the sum or by the maximum of the dimensions. When starting with scalar variables and when X is a cluster variable obtained by joining m elementary variables, then its dimension is just $m_x = m$.

5.2.6 Properties and Presentation of MIC Trees

MIC gives rooted trees: The leaves are original sequences/variables $X,\dots,Z$, internal nodes correspond to subsets of the set of all leaves, and the root represents the total set of all leaves, i.e. the joint variable $(X \dots Z)$. A bad choice of the metric and/or of the clustering algorithm will in general manifest itself in long "caterpillar-like" branches, while good choices will tend towards more balanced branchings.

When presenting the tree, it is useful to put all leaves on the x-axis, and to use information about the distances/similarities to control the height. We have essentially two natural choices, illustrated in Figs. 5.2 and 5.5, respectively. In Fig. 5.5, the height of a node is given by the mutual information between all leaves in the subtree below it. If the node is a leave, its height is zero. If it is an internal node (including the root) corresponding to a subset $\mathscr{S}$ of leaves, then its height is given by the mutual information between all members of $\mathscr{S}$,

$$height(\mathscr{S}) = I(\mathscr{S}) \qquad \text{method 1.} \tag{5.24}$$

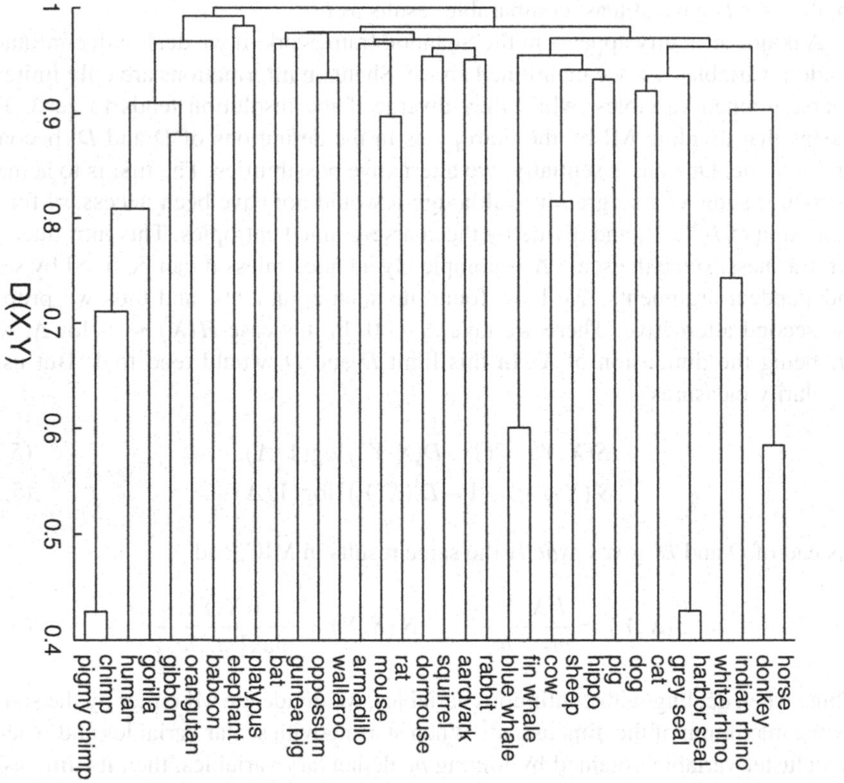

Fig. 5.2 Phylogenetic tree for 34 mammals (31 eutherians plus 3 non-placenta mammals), with mutual informations estimated by means of GenCompress. In contrast to Fig. 5.6, the heights of nodes are here and in the following tree equal to the distances between the joining daughter clusters

Let us assume that $\mathscr{S}$ has the two daughters X and Y, i.e. $\mathscr{S} = (XY)$. X and Y themselves might be either leaves or also internal nodes. Then the grouping property implies that $height(\mathscr{S}) - height(X) = I(\mathscr{S}) - I(X)$ is the MI between X and all the other sequences/variables in $\mathscr{S}$ which are not contained in X. This is non-negative by the positivity of MI, and the same is true when X is exchanged with Y. Thus the trees drawn in this way are always *well formed* in the sense that a mother node is located above its daughters. Any violation of this rule must be due to imprecise estimation of MIs.

A drawback of this method of representing the tree is that very close pairs have long branches, while relatively distant pairs are joined by short branches. This is the opposite of what is usually done in phylogenetic trees, where the branch lengths are representative of the distance between a mother and its daughter. This can be achieved by using the alternative representation employed in Fig. 5.2. There, the height of a mother node W which has two daughters X and Y is given by

$$height(W) = D(X,Y) \qquad \text{method 2.} \tag{5.25}$$

Although it gives rise to more intuitive trees, it also has one disadvantage: It is no longer guaranteed that the tree is well formed, but it may happen that $height(W) < height(X)$. To see this, consider a tree formed by three variables X, Y, and Z which are pairwise independent but globally dependent: $I(X,Y) = I(X,Z) = I(Y,Z) = 0$ but $I(X,Y,Z) > 0$.[2] In this case, all pairwise distances are maximal, thus also the first pair to be joined has distance 1. But the height of the root is less than 1. In our numerical applications we indeed found occasionally such "glitches", but they were rare and were usually related to imprecise estimates of MI or to improper choices of the metric.

5.3 Mitochondrial DNA and a Phylogenetic Tree for Mammals

As a first application, we study the mitochondrial DNA of a group of 34 mammals (see Fig. 5.2). Exactly the same species [3] had previously been analyzed in [21, 24, 31]. This group includes non-eutherians,[3] rodents and lagomorphs,[4] ferungulates,[5]

---

[2] An example is provided by three binary random variables with $p_{000} = p_{011} = p_{101} = p_{110} = 1/2 + \varepsilon$ and $p_{001} = p_{010} = p_{100} = p_{111} = 1/2 - \varepsilon$.

[3] Opossum (*Didelphis virginiana*), wallaroo (*Macropus robustus*), and platypus (*Ornithorhyncus anatinus*).

[4] Rabbit (*Oryctolagus cuniculus*), guinea pig (*Cavia porcellus*), fat dormouse (*Myoxus glis*), rat (*Rattus norvegicus*), squirrel (Sciurus vulgaris), and mouse (*Mus musculus*).

[5] Horse (*Equus caballus*), donkey (*Equus asinus*), Indian rhinoceros (*Rhinoceros unicornis*), white rhinoceros (*Ceratotherium simum*), harbor seal (*Phoca vitulina*), grey seal (*Halichoerus grypus*), cat (*Felis catus*), dog (*Canis lupus familiaris*), fin whale (*Balaenoptera physalus*), blue whale (*Balenoptera musculus*), cow (*Bos taurus*), sheep (*Ovis aries*), pig (*Sus scrofa*) and hippopotamus (*Hippopotamus amphibius*).

primates,[6] members of the African clade,[7] and others.[8] It had been chosen in [24] because of doubts about the relative closeness among these three groups [5, 31].

Obviously, we are here dealing with the algorithmic version of information theory, and informations are estimated by lossless data compression. For constructing the proximity matrix between individual taxa, we proceed essentially a in [24]. But in addition to using the special compression program GenCompress [2], we also tested several general purpose compression programs such as BWTzip, the UNIX tool bzip2, and lpaq1 [1], and durilca [1]. Finally, we also tested XM [40] (the abbreviation stands for "expert model"), which is according to its authors the most efficient compressor for biological (DNA, proteins) sequences. Indeed, we found that XM was even better than expected. While the advantage over GenCompress and other compressors was a few per cent when applied to single sequences [4], the estimates for MI between not too closely related species were higher by up to an order of magnitude. This is possible mainly because the MIs estimated by means of GenCompress and similar methods are reliably positive (unless the species are from different phyla) but extremely small. Thus even a small improvement on the compression rate can make a big change in MI.

In [24], the proximity matrix derived from MI estimates was then used as the input to a standard HC algorithm (neighbor-joining and hypercleaning) to produce an evolutionary tree. It is here where our treatment deviates crucially. We used the MIC algorithm described in Sect. 5.1, with distance $D(X,Y)$. The joining of two clusters (the third step in the MIC algorithm) is obtained by simply concatenating the DNA sequences. There is of course an arbitrariness in the order of concatenation sequences: XY and YX give in general compressed sequences of different lengths. But we found this to have negligible effect on the evolutionary tree. The resulting evolutionary tree obtained with Gencompress is shown in Fig. 5.2, while the tree obtained with XM is shown in Fig. 5.3.

Both trees are quite similar, and they are also similar to the most widely accepted phylogeny found, e.g. in [3]. All primates are, e.g. correctly clustered and the ferungulates are joined together. There are however a number connections (in both trees) which obviously do not reflect the true evolutionary tree. As shown in Fig. 5.2 the overall structure of this tree closely resembles the one shown in [31]. All primates are correctly clustered and also the relative order of the ferungulates is in accordance with [31]. On the other hand, there are a number of connections which obviously do not reflect the true evolutionary tree, see for example the guinea pig with bat and elephant with platypus in Fig. 5.2, and the mixing of rodents with African clade and armadillo in Fig. 5.3. But all these wrong associations are between species which have a very large distance from each other and from any other species within this sample. All in all, the results in Figs. 5.2 and 5.3 are in surprisingly good agreement (showing that neither compression scheme has obvious faults)

[6] Human (*Homo sapiens*), common chimpanzee (*Pan troglodytes*), pigmy chimpanzee (*Pan paniscus*), gorilla (*Gorilla gorilla*), orangutan (*Pongo pygmaeus*), gibbon (*Hylobates lar*), and baboon (*Papio hamadryas*).

[7] African elephant (*Loxodonta africana*), aardvark (*Orycteropus afer*).

[8] Jamaican fruit bat (*Artibeus jamaicensis*), armadillo (*Dasypus novemcintus*).

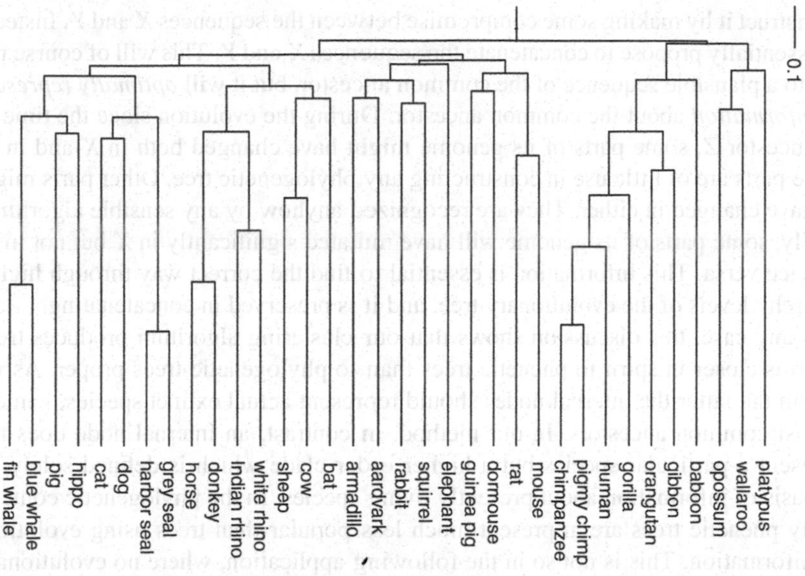

Fig. 5.3 Same as in Fig. 5.2, but with mutual informations estimated by means of XM. Notice that the *x*-axis covers here, in contrast to Fig. 5.2, the entire interval from 0 to 1

and capture surprisingly well the known relationships between mammals. Dividing MI by the total information is essential for this success. If we had used the non-normalized $I_{\rm alg}(X,Y)$ itself, results obtained in [24] would not change much, since all 34 DNA sequences have roughly the same length. But our MIC algorithm would be completely screwed up: After the first cluster formation, we have DNA sequences of very different lengths, and longer sequences tend also to have larger MI, even if they are not closely related.

One recurrent theme in the discussion of mammalian phylogenetic trees is the placement of the rodents [31, 32]. Are they closer to ferungulates, or closer to primates? Our results are inconclusive. On the one hand, the average distances between all 14 rodents and all 104 ferungulates in the Genebank data (Feb. 2008), estimated with XM, is 0.860 – which is clearly smaller than the average distance 0.881 to all 28 primates. On the other hand, the distances within the group of rodents are very large, suggesting that this group is either not monophylic, or that its mtDNA has evolved much faster than, say, that of ferungulates. Either possibility makes a statement about the classification of rodents with respect to ferungulates and primates based on these data very uncertain.

A heuristic reasoning for the use of MIC for the reconstruction of an evolutionary tree might be given as follows: Suppose that a proximity matrix has been calculated for a set of DNA sequences and the smallest distance is found for the pair (X,Y). Ideally, one would remove the sequences X and Y, replace them by the sequence of the common ancestor (say Z) of the two species, update the proximity matrix to find the smallest entry in the reduced set of species, and so on. But the DNA sequence of the common ancestor is not available. One solution might be that one tries to reconstruct it by making some compromise between the sequences X and Y. Instead, we essentially propose to concatenate the sequences X and Y. This will of course not lead to a plausible sequence of the common ancestor, but it will *optimally represent the information* about the common ancestor. During the evolution since the time of the ancestor Z, some parts of its genome might have changed both in X and in Y. These parts are of little use in constructing any phylogenetic tree. Other parts might not have changed in either. They are recognized anyhow by any sensible algorithm. Finally, some parts of its genome will have mutated significantly in X but not in Y, and vice versa. This information is essential to find the correct way through higher hierarchy levels of the evolutionary tree, and it is preserved in concatenating.

In any case, this discussion shows that our clustering algorithm produces trees which is closer in spirit to phenetic trees than to phylogenetic trees proper. As we said, in the latter the internal nodes should represent actual extinct species, namely the last common ancestors. In our method, in contrast, an internal node does not represent a particular species but a higher order clade which is defined solely on the basis of information about presently living species. In the phylogenetic context, purely phenetic trees are at present much less popular than trees using evolutionary information. This is not so in the following application, where no evolutionary aspect exists and the above discussion is irrelevant.

5.4 Clustering of Minimally Dependent Components in an Electrocardiogram

As our second application we choose a case where Shannon theory is the proper setting. We show in Fig. 5.4 an ECG recorded from the abdomen and thorax of a pregnant woman (8 channels, sampling rate 500 Hz, 5 s total). It is already seen from this graph that there are at least two important components in this ECG: the heartbeat of the mother, with a frequency of $\approx$3 beat s^{-1}, and the heartbeat of the fetus with roughly twice this frequency. Both are not synchronized. In addition there is noise from various sources (muscle activity, measurement noise, etc.). While it is easy to detect anomalies in the mother's ECG from such a recording, it would be difficult to detect them in the fetal ECG.

As a first approximation we can assume that the total ECG is a linear superposition of several independent sources (mother, child, noise$_1$, noise$_2$,...). A standard method to disentangle such superpositions is *independent component analysis* (ICA) [18]. In the simplest case one has n independent sources $s_i(t)$, $i = 1 \dots n$ and n measured channels $x_i(t)$ obtained by instantaneous superpositions with a time independent non-singular matrix $\mathbf{A}$,

$$x_i(t) = \sum_{j=1}^{n} A_{ij} s_j(t) \,. \tag{5.26}$$

In this case the sources can be reconstructed by applying the inverse transformation $\mathbf{W} = \mathbf{A}^{-1}$ which is obtained by minimizing the (estimated) mutual informations

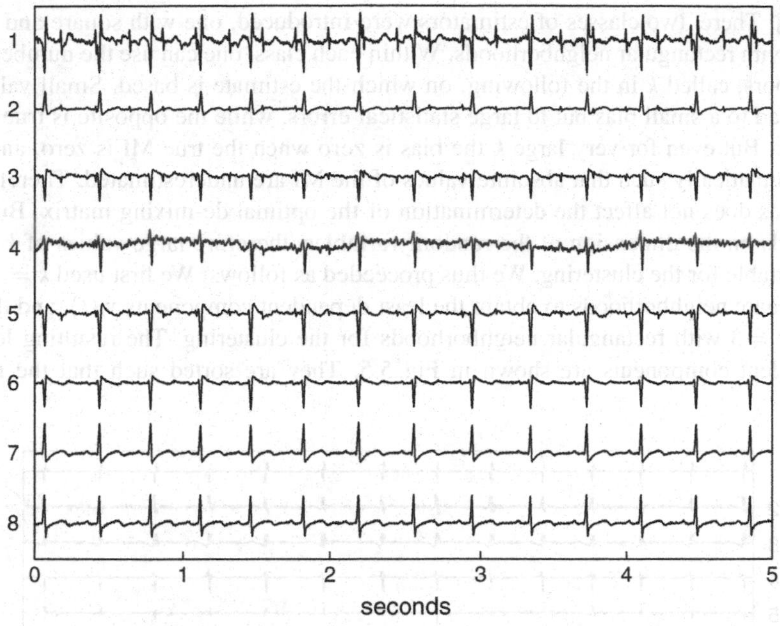

Fig. 5.4 ECG of a pregnant woman

between the transformed components $y_i(t) = \sum_{j=1}^{n} W_{ij}x_j(t)$. If some of the sources are Gaussian, this leads to ambiguities [18], but it gives a unique solution if the sources have more structure.

In reality things are not so simple. For instance, the sources might not be independent, the number of sources (including noise sources!) might be different from the number of channels, and the mixing might involve delays. For the present case this implies that the heartbeat of the mother is seen in several reconstructed components y_i, and that the supposedly "independent" components are not independent at all. In particular, all components y_i which have large contributions from the mother form a cluster with large intra-cluster MIs and small inter-cluster MIs. The same is true for the fetal ECG, albeit less pronounced. It is thus our aim to:

1) Optimally decompose the signals into least dependent components
2) Cluster these components hierarchically such that the most dependent ones are grouped together
3) Decide on an optimal level of the hierarchy, such that the clusters make most sense physiologically
4) Project onto these clusters and apply the inverse transformations to obtain cleaned signals for the sources of interest

Technically we proceeded as follows [33]: Since we expect different delays in the different channels, we first used Takens delay embedding [35] with time delay 0.002 s and embedding dimension 3, resulting in 24 channels. We then formed 24 linear combinations $y_i(t)$ and determined the de-mixing coefficients W_{ij} by minimizing the overall mutual information between them, using the MI estimator proposed in [22]. There, two classes of estimators were introduced, one with square and the other with rectangular neighborhoods. Within each class, one can use the number of neighbors, called k in the following, on which the estimate is based. Small values of k lead to a small bias but to large statistical errors, while the opposite is true for large k. But even for very large k the bias is zero when the true MI is zero, and it is systematically such that absolute values of the MI are underestimated. Therefore this bias does not affect the determination of the optimal de-mixing matrix. But it depends on the dimension of the random variables, therefore large values of k are not suitable for the clustering. We thus proceeded as follows: We first used $k = 100$ and square neighborhoods to obtain the least dependent components $y_i(t)$, and then used $k = 3$ with rectangular neighborhoods for the clustering. The resulting least dependent components are shown in Fig. 5.5. They are sorted such that the first

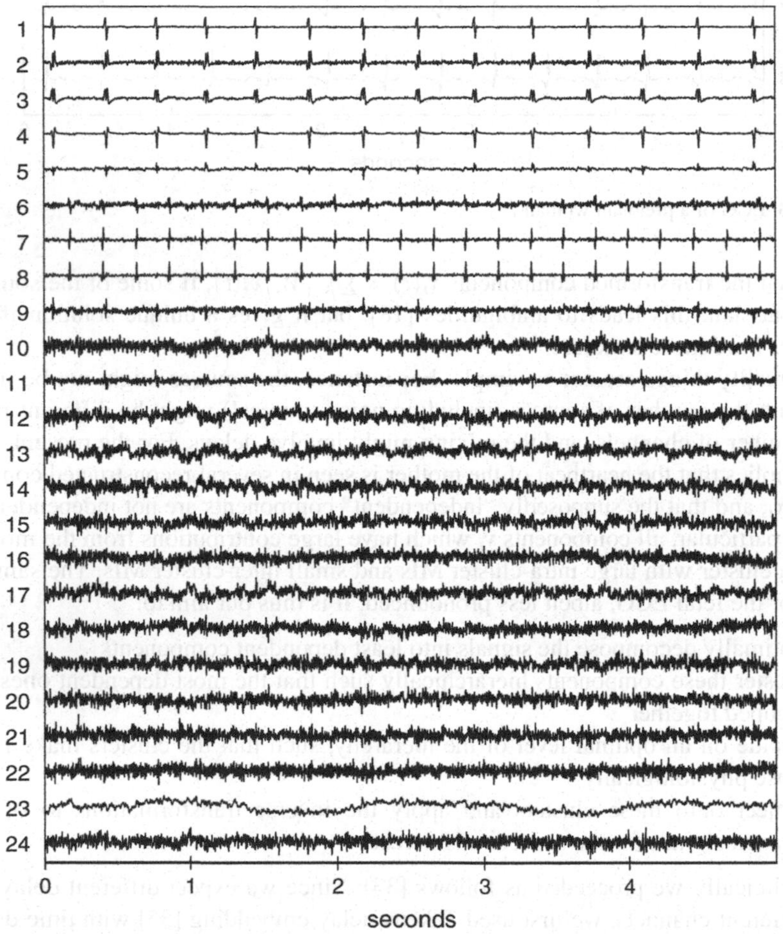

Fig. 5.5 Least dependent components of the ECG shown in Fig. 5.4, after increasing the number of channels by delay embedding

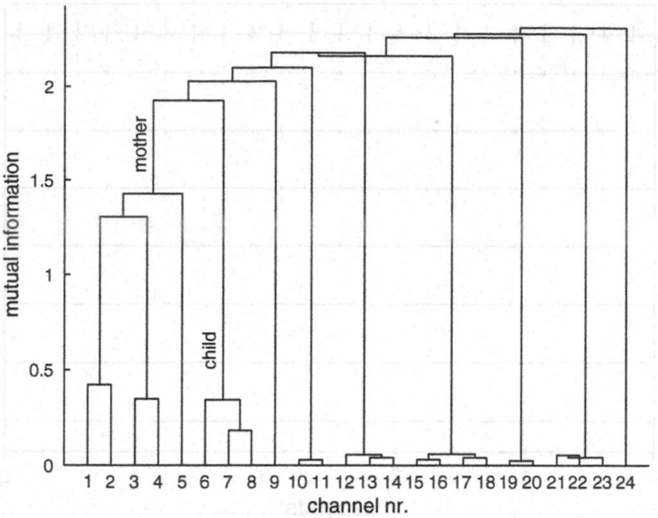

Fig. 5.6 Dendrogram for least dependent components. The height where the two branches of a cluster join corresponds to the MI of the cluster

components (1–5) are dominated by the maternal ECG, while the next three contain large contributions from the fetus. The rest contains mostly noise, although some seem to be still mixed.

These results obtained by visual inspection are fully supported by the cluster analysis. The dendrogram is shown in Fig. 5.6. In constructing it we used $S(X,Y)$ (5.23) as similarity measure to find the correct topology. Again we would have obtained much worse results if we had not normalized it by dividing MI by $m_X + m_Y$. In plotting the actual dendrogram, however, we used the MI of the cluster to determine the height at which the two daughters join. The MI of the first five channels, e.g. is ≈ 1.43, while that of channels 6–8 is ≈ 0.34. For any two clusters (tuples) $X = X_1 \dots X_n$ and $Y = Y_1 \dots Y_m$ one has $I(X,Y) \geq I(X) + I(Y)$. This guarantees, if the MI is estimated correctly, that the tree is drawn properly. The two slight glitches (when clusters (1–14) and (15–18) join, and when (21–22) is joined with 23) result from small errors in estimating MI. They do in no way effect our conclusions.

In Fig. 5.6 one can clearly see two big clusters corresponding to the mother and to the child. There are also some small clusters which should be considered as noise. For reconstructing the mother and child contributions to Fig. 5.4, we have to decide on one specific clustering from the entire hierarchy. We decided to make the cut such that mother and child are separated. The resulting clusters are indicated in Fig. 5.6 and were already anticipated in sorting the channels. Reconstructing the original ECG from the child components only, we obtain Fig. 5.7.

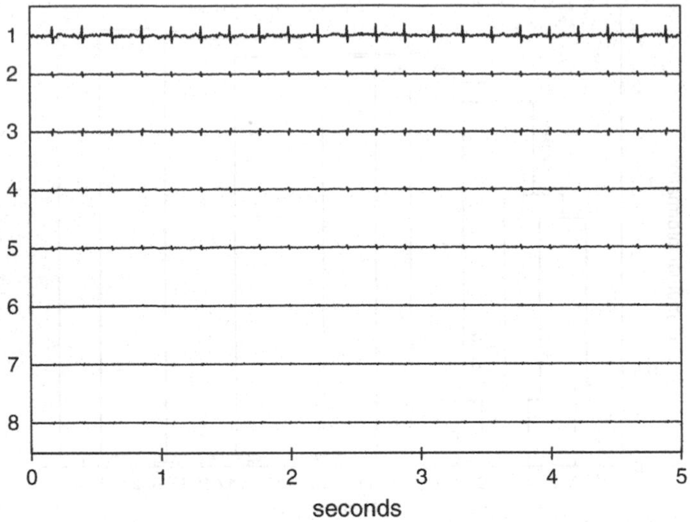

Fig. 5.7 Original ECG where all contributions except those of the child cluster have been removed

5.5 Conclusions

We have shown that MI can not only be used as a proximity measure in clustering, but that it also suggests a conceptually very simple and natural hierarchical clustering algorithm. We do not claim that this algorithm, called *mutual information clustering* (MIC), is always superior to other algorithms. Indeed, MI is in general not easy to estimate. Obviously, when only crude estimates are possible, also MIC will not give optimal results. But as MI estimates are becoming better, also the results of MIC should improve. The present paper was partly triggered by the development of a new class of MI estimators for continuous random variables which have very small bias and also rather small variances [22].

We have illustrated our method with two applications, one from genetics and one from cardiology. For neither application MIC might give the very best clustering, but it seems promising and indicative of the inherit simplicity of our method that one common method gives decent results in two very different applications.

If better data become available, e.g. in the form of longer time sequences in the application to ECG or of more complete genomes (so that mutual information can be estimated more reliably), then the results of MIC should improve. It is not obvious what to expect when one wants to include more data in order to estimate larger trees. On the one hand, more species within one taxonomic clade would describe this clade more precisely, so results should improve. On the other hand, as clusters become bigger and bigger, also the disparities of the sequences lengths which describe these clusters increase. It is not clear whether in this case a strict normalization of the distances as in (5.17, 5.18) is still appropriate, and whether the available compression algorithms will still be able to catch the very long resulting correlations within the concatenated sequences. Experiments with phylogenetic trees of animals with up to 360 species (unpublished results) had mixed success.

As we said in the introduction, after a construction of a first tree one can try to improve on it. One possibility is to change the topology of the tree, using, e.g. quartet moves and accepting them based on some heuristic cost function. One such cost function could be the sum of all distances between linked nodes in the tree. Alternatively, one could try to keep the topology fixed and change the sequences representing the internal nodes, i.e. deviate from simple concatenation. We have not tried either.

There are two versions of information theory, algorithmic and probabilistic, and therefore there are also two variants of MI and of MIC. We discussed in detail one application of each, and showed that indeed common concepts were involved in both. In particular it was crucial to normalize MI properly, so that it is essentially the *relative* MI which is used as proximity measure. For conventional clustering algorithms using algorithmic MI as proximity measure this had already been stressed in [24, 25], but it is even more important in MIC, both in the algorithmic and in the probabilistic versions.

In the probabilistic version, one studies the clustering of probability distributions. But usually distributions are not provided as such, but are given implicitly by finite random samples drawn (more or less) independently from them. On the other hand, the full power of algorithmic information theory is only reached for infinitely long sequences, and in this limit any individual sequence defines a sequence of probability measures on finite subsequences. Thus the strict distinction between the two theories is somewhat blurred in practice. Nevertheless, one should not confuse the similarity between two sequences (two English books, say) and that between their subsequence statistics. Two sequences are maximally different if they are completely random, but their statistics for short subsequences is then identical (all subsequences appear in both with equal probabilities). Thus one should always be aware of what similarities or independencies one is looking for. The fact that MI can be used in similar ways for all these problems is not trivial.

Acknowledgements This work was done originally in collaboration with H. Stögbauer and R.G. Andrzejak. We are very much indebted to them for their contributions. We also would like to thank Arndt von Haeseler, Walter Nadler, Volker Roth, Orion Penner and Maya Pczuski for many useful and critical discussions.

References

1. http://cs.fit.edu/ mmahoney/compression/text.html
2. Gencompress. http://monod.uwaterloo.ca/downloads/gencompress/
3. Genebank. http://www.ncbi.nlm.nih.gov
4. XM-results. ftp://ftp.infotech.monash.edu.au/software/DNAcompress-XM/2007/results.html
5. Cao, Y., Janke, A., Waddell, P.J., Westerman, M., Takenaka, O., Murata, S., Okada, N., Paabo, S., Hasegawa, M.: Conflict among individual mitochondrial proteins in resolving the phylogeny of eutherian orders. Journal of Molecular Evolution **47**(3), 307–322 (1998)

6. Cilibrasi, R., Vitanyi, P.M.B.: Clustering by compression. IEEE Transaction on Information Theory **51**, 1523–1545 (2005)
7. Cilibrasi, R., Vitanyi, P.M.B.: A new quartet tree heuristic for hierarchical clustering. arXiv:cs/0606048 (2006)
8. Correa, J.C.: A new estimator of entropy. Communications in Statistics-Theory and Methods **24**(10), 2439–2449 (1995)
9. Cover, T., Thomas, J.: Elements of Information Theory. Wiley, New York (1991)
10. Darbellay, G.A., Vajda, I.: Estimation of the information by an adaptive partitioning of the observation space. IEEE Transactions on Information Theory **45**(4), 1315–1321 (1999)
11. Daub, C.O., Steuer, R., Selbig, J., Kloska, S.: Estimating mutual information using B-spline functions an improved similarity measure for analysing gene expression data. BMC Bioinformatics **5**, 118 (2004)
12. Dobrushin, R.: A simplified method for experimental estimate of the entropy of a stationary sequence. Theory of Probability and Its Applications **3**, 462 (1958)
13. Dudewicz, E.J., Van der Meulen, E.C.: Entropy-based tests of uniformity. Journal of the American Statistical Association **76**(376), 967–974 (1981)
14. Ebrahimi, N., Pflughoeft, K., Soofi, E.S.: Two measures of sample entropy. Statistics and Probability Letters **20**(3), 225–234 (1994)
15. van Es, B.: Estimating functionals related to a density by a class of statistics based on spacings. Scandinavian Journal of Statistics **19**(1), 61–72 (1992)
16. Grassberger, P.: Entropy estimates from insufficient samplings. arXiv:physics/0307138 (2003)
17. Grunwald, P., Vitanyi, P.: Shannon information and Kolmogorov complexity. arXiv:cs/0410002 (2004)
18. Hyvärinen, A., Karhunen, J., Oja, E.: Independent component analysis. Wiley, New York (2001)
19. Jain, A., Dubes, R.: Algorithms for clustering data. Prentice Hall, Englewood Cliffs (1988)
20. Kozachenko, L.F., Leonenko, L.: Sample estimate of the entropy of a random vector. Problems of Information Transmission **23**, 95–101 (1987)
21. Kraskov, A., Stogbauer, H., Andrzejak, R.G., Grassberger, P.: Hierarchical clustering based on mutual information. arXive:q-bio/0311039 (2003)
22. Kraskov, A., Stogbauer, H., Grassberger, P.: Estimating mutual information. Physical Review E **69**(6), 066,138 (2004)
23. Kraskov, A., Stogbauer, H., Andrzejak, R.G., Grassberger, P.: Hierarchical clustering using mutual information. Europhysics Letters, **70**(2), 278–284 (2005)
24. Li, M., Badger, J.H., Chen, X., Kwong, S., Kearney, P., Zhang, H.Y.: An information-based sequence distance and its application to whole mitochondrial genome phylogeny. Bioinformatics **17**(2), 149–154 (2001)
25. Li, M., Chen, X., Li, X., Ma, B., Vitanyi, P.M.B.: The similarity metric. IEEE Transactions on Information Theory **50**(12), 3250–3264 (2004)
26. Li, M., Vitanyi, P.: An introduction to Kolmogorov complexity and its applications. Springer, New York (1997)
27. Bray, N., Pachter, L.: MAVID: Constrained ancestral alignment of multiple sequences. Genome Research **14**, 693–699 (2004)
28. Moon, Y.-I., Rajagopalan, B., Lall, U.: Estimation of mutual information using kernel density estimators Physical Review E **52**, 2318–2321 (1995)
29. Penner, O., Grassberger, P., Paczuski, M.: to be published (2008)
30. Press, W.H., Teukolski, S.A., Vetterling, W.T., Flannery, B.P.: Numerical Recipes in C++, 3rd edn, chap. 16.4. Cambridge University Press, New York (2007)
31. Reyes, A., Gissi, C., Pesole, G., Catzeflis, F.M., Saccone, C.: Where do rodents fit? evidence from the complete mitochondrial genome of sciurus vulgaris. Molecular Biology and Evolution **17**(6), 979–983 (2000)
32. Reyes, A., Gissi, C., Catzeflis, F., Nevo, E., Pesole, G.: Congruent Mammalian trees from mitochondrial and nuclear genes using Bayesian methods. Molecular Biology and Evolution **21**(2), 397–403 (2004)

33. Stogbauer, H., Kraskov, A., Astakhov, S.A., Grassberger, P.: Least-dependent-component analysis based on mutual information. Physical Review E **70**(6), 066,123 (2004)
34. Strimmer, K., von Haeseler, A.: Quartet puzzling: a quartet maximum likelihood method for reconstructing tree topologies. Molecular Biology and Evolution **13**, 964–969 (1996)
35. Takens, F.: Detecting strange attractors in turbulence. In: D. Rand, L. Young (eds.) Dynamical Systems and Turbulence, vol. 898, p. 366. Springer, Berlin (1980)
36. Tishby, N., Pereira, F., Bialek, W.: The information bottleneck method. In: 37-th Annual Allerton Conference on Communication, Control and Computing, p. 368 (1997)
37. Tsybakov, A.B., Van der Meulen, E.C.: Root-n consistent estimators of entropy for densities with unbounded support. Scandinavian Journal of Statistics **23**(1), 75–83 (1996)
38. Vasicek, O.: Test for normality based on sample entropy. Journal of the Royal Statistical Society Series B-Methodological **38**(1), 54–59 (1976)
39. Wieczorkowski, R., Grzegorzewski, P.: Entropy estimators – improvements and comparisons. Communications in Statistics-Simulation and Computation **28**(2), 541–567 (1999)
40. Cao, M.D., Dix, T.I., Allison, L., Mears, C.: A simple statistical algorithm for biological sequence compression. IEEE Data Compression Conference (DCC), pp. 43–52 (2007)
41. Yu, Z.G., Mao, Z., Zhou, L.Q., Anh, V.V.: A mutual information based sequence distance for vertebrate phylogeny using complete mitochondrial genomes. Third International Conference on Natural Computation (ICNC 2007), pp. 253–257 (2007)

Chapter 6
A Hybrid Genetic Algorithm for Feature Selection Based on Mutual Information

Jinjie Huang and Panxiang Rong

Abstract Feature selection aims to reduce the dimensionality of patterns for classificatory analysis by selecting the most informative rather than irrelevant and/or redundant features. In this study, a hybrid genetic algorithm for feature selection is presented to combine the advantages of both wrappers and filters. Two stages of optimization are involved. The outer optimization stage completes the global search for the best subset of features in a wrapper way, in which the mutual information between the predictive labels of a trained classifier and the true classes serves as the fitness function for the genetic algorithm. The inner optimization performs the local search in a filter manner, in which an improved estimation of the conditional mutual information acts as an independent measure of feature ranking. This measure takes into account not only the relevance of the candidate feature to the output classes but also the redundancy to the features already selected. The inner and outer optimizations cooperate with each other and achieve the high global predictive accuracy as well as the high local search efficiency. Experimental results demonstrate both parsimonious feature selection and excellent classification accuracy of the method on a range of benchmark data sets.

6.1 Introduction

Coming with the rapid growth of high dimensional data collected in many areas such as text categorization and gene selection there is an increasing demand for the feature selection in classificatory analysis [1, 18]. To describe the domain of applications as good as possible, real-world data sets are often characterized by many irrelevant and/or redundant features due to the lack of *prior* knowledge about specific problems [10]. If these features are not properly excluded, they may significantly

J. Huang (✉) and P. Rong,
Department of Automation, Harbin University of Science and Technology, Xuefu Road 52, Harbin 150080, China
e-mail: jinjiehyh@yahoo.com.cn, pxrong@hrbust.edu.cn

F. Emmert-Streib, M. Dehmer (eds.), *Information Theory and Statistical Learning*, 125
DOI: 10.1007/978-0-387-84816-7_6,

hamper the model accuracy and the learning speed. Because the primary task of classificatory analysis is to extract knowledge (e.g., in the form of classification rules) from the training data the presence of a large number of irrelevant or redundant features can make it difficult to extract the core regularities of the data. Conversely, if the learned rules are based on a small number of relevant features, they are more concise and hence easier to understand and use [18, 30]. Therefore, it is very important to reduce the dimensionality of the raw input feature space in classificatory analysis to ensure the practical feasibility of the classifier.

Feature selection is to select a subset of original features that is good enough regarding its ability to describe the training data set and to predict for future cases. Broadly, methods for feature selection fall into three categories: the filter approach, the wrapper approach and the embedded method. In the first category, the filter approach is first utilized to select the subsets of features before the actual model learning algorithm is applied. The best subset of features is selected in one pass by evaluating some predefined criteria independent of the actual generalization performance of the learning machine. So a faster speed can usually be obtained. The filter approach is argued to be computational less expensive and more general. However, it might fail to select the right subset of features if the used criterion deviates from the one used for training the learning machine. Another drawback involved in the filter approach is that it may also fail to find a feature subset that would jointly maximize the criterion, since most filters estimate the significance of each feature just by means of evaluating one feature a time [25]. Thus, the performance of the learning models is degraded.

Methods from the second category, on the other hand, utilize the learning machine as a fitness function and search for the best subset of features in the space of all feature subsets. This formulation of the problem allows the use of standard optimization techniques with the learning machine of interest as a black box to score subsets of features according to their predictive power. Therefore, the wrapper approach generally outperforms the filter approach in the aspect of the final predictive accuracy of a learning machine. The wrapper methodology is greatly popularized by Kohavi and John [24], and offers a simple but powerful way to address the problem of feature selection, despite the fact that it involves some more computational complexity and requires more execution time than that of the filter methodology.

Besides wrappers and filters, the embedded methods are another category of feature selection algorithms, which perform feature selection in the process of training and are usually specific to given learning machines [18]. Some examples of the embedded methods are decision tree learners, such as ID3 [33] and C4.5 [34], or the recursive feature elimination (RFE) approach, which is a recently proposed feature selection algorithm derived based on support vector machine (SVM) theory and has been shown good performance on the problems of gene selection for microarray data [19, 35, 43]. The embedded methods are argued to be more efficient because they avoid retraining a predictor from the scratch for every subset of features investigated. However, they are much intricate and limited to a specific learning machine.

Recently, research on feature selection is mainly focused on two aspects: criteria and search strategies. As we known, an optimal subset is always optimal relative to a certain criterion. In general, different criteria may not lead to the same optimal feature subset. Typically, a criterion tries to measure the discriminating ability of a feature or a subset to distinguish the different class labels. M. Dash called these criteria the evaluation functions and grouped them into five categories [10]: *distance*, *information* (or *uncertainty*), *dependence*, *consistency* and *classifier error*. The *distance* measure, e.g., the Euclidean distance measure, is a very traditional discrimination or divergence measure. The dependence measure, also called the correlation measure, is mainly utilized to find the correlation between two features or a feature and a class. The *consistency* measure relies heavily on the training data set and is discussed for feature selection in [11]. These three measures are all sensitive to the concrete values of the training data; hence they are easily affected by noise or outlier data. In contrast, the *information* measures, such as the entropy or mutual information, investigate the amount of information or the uncertainty of a feature for the classification. The data classification process is aimed at reducing the amount of uncertainty or gaining information about the classification. In Shannon's information theory [9, 38], information is defined as something that removes or reduces uncertainty. For a classification task, the more information we get, the higher the accuracy of a classification model becomes, because the predicted classes of new instances are more likely to correspond to their true classes. A model that does not increase the amount of information is useless and its prediction accuracy is not expected to be better than just a random guess [28]. Thus, the *Information* measure is different from the above three measures by its metric-free nature: it depends only on the probability distribution of a random variable rather than on its concrete values. The *Information* measures have been widely used in feature selection [4, 17, 22, 29], including many famous learning algorithms such as ID3 [33] and C4.5 [34].

Searching for the best *m* features out of the *n* available for the classification task is known to be a NP-hard problem and the number of local minima can be quite large [3]. Exhaustive evaluation of possible feature subsets is usually unfeasible in practice due to the large amount of computational effort required. A wide range of heuristic search strategies have been used including forward selection [4], backward elimination [6], hill-climbing [7], branch and bound algorithms [40], and the stochastic algorithms like simulated annealing [12] and genetic algorithms (GAs) [5, 36, 44, 45]. Kudo and Sklansky [26] made a comparison among many of the feature selection algorithms and explicitly recommended that GAs should be used for large-scale problems with more than 50 candidate variables. They also described a practical implementation of GAs for feature selection. The advantages of GAs for feature selection are often summarized as follows: First, compared with those deterministic algorithms, they are more capable of avoiding getting stuck in local optima often encountered in feature selection problems. Second, they may be classified into a kind of *anytime* algorithms [46], which can generate currently best subsets constantly and keep improving the quality of selected features as time goes on. However, the limitations of a simple GA algorithm have been uncovered in many applications, such as premature convergence, poor ability of fine-tuning near local optimum points. A practical and effective way to overcome these limitations is to incorporate domain-specific knowledge into the GA. In fact, some hybrid GAs have been developed in diverse applications and successful performance has been obtained [23, 37].

In this chapter we present a hybrid GA designed to solve the feature selection problem. This approach combines the global search ability of GA with some efficient local search heuristic strategies on the basis of mutual information, achieving both the high accuracy of wrappers and the efficiency of filters to some extent. Moreover, an improved formula is derived to estimate the conditional mutual information between the candidate feature and the output classes given a subset of selected features. The formula is utilized to rank features in the local search process, taking account both the relevance of the candidate feature to the classes and the redundancy between the candidate feature and the features already selected. Hence, a good subset of features can be obtained for machine learning.

The rest of the chapter is organized as follows. In Sect. 6.2, the basics of information theory, such as entropy and mutual information, are briefly introduced. In Sect. 6.3, an improved formula to compute the conditional mutual information between the candidate feature and the classes given a subset of features selected is derived based on the results in [27], which is utilized to rank all the candidate features in the local search process. In Sect. 6.4, the framework of the hybrid genetic algorithm for feature selection is proposed, and the local search schemes are discussed in detail. The results of various experiments performed on a range of data sets are reported in Sect. 6.5. Finally, Sect. 6.6 closes the chapter with some concluding remarks.

6.2 Evaluation of Mutual Information

In feature selection problems, the relevant features contain important information about output, whereas the irrelevant features contain little information regarding output. The task for solving the problems is to find those input features that contain as much information about the output as possible. For this purpose, Shannon's information theory provides us with a way to measure the information of random variables by entropy and mutual information [9, 38].

Evaluating the mutual information between two discrete random variables is feasible and convenient through histograms. Given a data set of two discrete variables X and Y, and let $|X| = m, |Y| = n$ be the number of the discrete values of X and Y respectively. Denote G as the joint histogram matrix, where g_{ij} is the number of times over the data set, an input pattern with X equal to the ith discrete value and Y equal to its jth discrete value, (for simplicity, just denote as X=i, Y=j, $1 \leq i \leq m, ; 1 \leq j \leq n$. Notice that in what follows a summation over i is a sum over values of X, i.e., over various rows of a given column of the joint histogram matrix. Similarly, a summation over j is a sum over values of Y, i.e., across various columns of a given row of the joint histogram matrix. Both variables are summed from 1 to m and from 1 to n respectively. The sum $N = \sum_{i=1}^{m} \sum_{j=1}^{n} g_{ij}$ is the total number of input samples. Using this matrix $G_{m \times n}$, it is possible to estimate all the relevant probabilities and the mutual information $I(X;Y)$ as follows:

$$P(X=i) = \frac{1}{N}\sum_{j=1}^{n} g_{ij}\,, \tag{6.1}$$

$$P(Y=j) = \frac{1}{N}\sum_{i=1}^{m} g_{ij}\,, \tag{6.2}$$

$$P(X=i|Y=j) = \frac{g_{ij}}{\sum_{i=1}^{m} g_{ij}}\,, \tag{6.3}$$

where $P(X=i)$ is the empirical prior probability of $X=i$; $P(Y=j)$ is the frequency with $Y=j$, and $P(X=i|Y=j)$ (more conveniently written as P_{ij}) is the empirical probability of $X=i$ given $Y=j$. The relevant empirical entropies are now given by

$$H(X) = -\sum_{i=1}^{m} P(X=i)\log(P(X=i))\,, \tag{6.4}$$

$$H(X|Y=j) = -\sum_{i=1}^{m} P_{ij}\log P_{ij}\,, \tag{6.5}$$

$$H(X|Y) = \sum_{j=1}^{n} P(Y=j)H(X|Y=j). \tag{6.6}$$

Then the estimated value of the mutual information between X and Y is given in terms of the above entropies, simply by

$$I(X;Y) = H(X) - H(X|Y). \tag{6.7}$$

Note that the computation of mutual information involves only discrete variables that typically assume a small number of values. For continuous random variables, we usually divide them into several discrete partitions and calculate the entropy and mutual information using the definitions for discrete cases. There are mainly two approaches for this purpose: equal distance partitioning [4] and equiprobable partitioning [16]. In our study, we use the equal distance partitioning method as follows. If the distribution of the values in a variable f is not known a priori, we compute its mean μ and standard deviation σ, and then cut the interval $[\mu-2\sigma, \mu+2\sigma]$ into p equally spaced segments. Here, the value of p corresponds to the number of discrete values of f. The points falling outside are assigned to the extreme left (right) segments. Each segment corresponds to a discrete value.

The mutual information may serve as a statistics that summarizes the degree of dependence of X on Y. When random variables X and Y are independent, i.e., $p(x,y) = p(x)p(y)$, (the usual definition of independence), the mutual information of X and Y goes to zero ($I(X;Y) = 0$). This also means that Y can tell us nothing about X.

6.3 Information-Theoretic Criteria for Feature Selection

6.3.1 The Feature Selection Problem and Related Studies

In classificatory analysis, the input to a classifier is a training data set of instances. The instances are labeled by a class variable C. An unlabeled instance is an element of the n dimensional feature space $f_1 \times f_2 \times \cdots \times f_n$. The feature selection problem is to find the subset with $k < n$ features that is "maximally informative" about the class by excluding irrelevant and/or redundant features among the ones extracted from the raw data. In the framework of Shannon's Information Theory [9, 38], a natural measure of "informative" is the mutual information. So the problem of selecting input features can be solved by computing the mutual information between input features and output classes. This was formulated by Battiti as a "feature reduction" problem as [FRn-k].

[FRn-k] Given an initial set F with n features and C set of all output classes, find the subset $S \subset F$ with k features that minimizes $H(C|S)$, i.e., that maximizes the mutual information $I(C;S)$ [4].

However, there are some difficulties in computing $I(C;f_i,S)$ directly. First, since the probability distribution functions of input features are actually hard to know exactly, the best way to estimate the mutual information may be using a histogram of the data. But though the histogram-based mutual information estimation works with two or even three variables, it fails in higher dimensions due to the sparsity of data in high-dimensional spaces. Second, computing $I(C;f_i,S)$ requires a large amount of storage. For example, assuming to select m features, if the output classes are composed of K_c categories, and the jth input feature is partitioned into P_j (usually $P_j \leq 10$) intervals to get the histogram, then there must be $K_c \times \prod_{j=1}^{m} P_j$ cells for storage to compute $I(C;f_i,S)$ [27]. In this case, even to select only 10 input features ($m = 10$), $K_c \times 10^{10}$ memories would be needed. Obviously, this is difficult in practice. To overcome the above difficulties, some alternative methods to compute the mutual information $I(C;S)$ have been devised [4, 27].

Assume S is the subset of features already selected, F is the subset of unselected features, $S \cap F = \emptyset$, C is the output classes. For a feature $f_i \in F$ to be selected, the mutual information $I(C;\{S,f_i\})$ should be the largest one among those $I(C;\{S,f_i\})$'s, $f_i \in F$.

Notice that, the mutual information $I(C;\{S,f_i\})$ can be represented as

$$I(C;\{S,f_i\}) = I(C;S) + I(C;f_i|S). \tag{6.8}$$

For a given feature subset S, since $I(C;S)$ is a constant, to maximize $I(C;\{S,f_i\})$, the conditional mutual information $I(C;f_i|S)$ should be maximized. Furthermore, the conditional mutual information $I(C;f_i|S)$ can be represented as

$$I(C;f_i|S) = I(C;f_i) - I(C;f_i;S). \tag{6.9}$$

Here, the mutual information $I(C;f_i;S)$ is the common information found among the output classes C, the candidate feature f_i and the features already selected in the current subset S. It is also the redundant information between the candidate feature f_i and the already-selected features in S with respect to classes C. Denote $I(C;f_i;S)$ by I_r, Battiti used only the mutual information between the candidate feature f_i and each of the already-selected features in S to estimate I_r with a parameter β predefined by the user according to the degree of redundancy,

$$I_r = \beta \sum_{f_s \in S} I(f_i;f_s), \qquad \beta \in [0,1]. \tag{6.10}$$

Thus, the conditional mutual information can be estimated as

$$I(C;f_i|S) = I(C;f_i) - \beta \sum_{f_s \in S} I(f_i;f_s). \tag{6.11}$$

Based on (6.11), Battiti formulated his greedy selection algorithm, MIFS [4]. As one can see, parameter β has a great effect on the right feature selection in MIFS. Unfortunately, we are actually unable to know what an appropriate value the parameter β should take in (6.11). As a result, the performance of MIFS is greatly degraded. To decrease the influence of parameter β on Battiti's MIFS, we consider a more accurate estimation of the redundant information I_r.

Proposition 1. For any $f_s \in S$, $f_i \in F$, suppose that the information is distributed uniformly throughout the regions of $H(f_s)$, $H(f_i)$ and $H(C)$, and that the classes C do not change the ratio of entropy of f_s for the mutual information between f_s and f_i, if all the selected features in S are completely independent to each other, the total redundant information of the candidate feature f_i to all the selected features in subset S with respect to output classes C, denoted by I_r, can be calculated by the simple summation as defined in (6.12).

$$I_r = \sum_{f_s \in S} \frac{I(f_i;f_s)}{H(f_s)} I(C;f_s). \tag{6.12}$$

Proof. Consider any features $f_s \in S$, $f_i \in F$, the relations between the entropies of f_s, f_i and C are illustrated in Fig. 6.1. The common information among features f_s, f_i and output classes C is represented by the area 4 in Fig. 6.1, and can be computed as

$$I(C;f_i;f_s) = I(f_s;f_i) - I(f_s;f_i|C), \tag{6.13}$$

where the conditional mutual information $I(f_s;f_i|C)$ corresponds to the area 1 in Fig. 6.1. Under the assumption that the information is distributed uniformly throughout the regions of $H(f_s)$, $H(f_i)$ and $H(C)$ in Fig. 6.1, and that the classes C does not change the ratio of entropy of f_s to the mutual information between f_s and f_i, then the following relation holds

$$\frac{I(f_i;f_s|C)}{H(f_s|C)} = \frac{I(f_i;f_s)}{H(f_s)}. \tag{6.14}$$

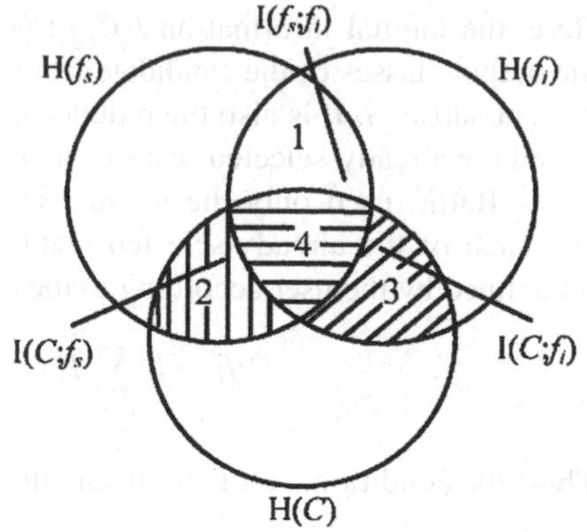

Fig. 6.1 The relations between input features and output classes

Consequently, $I(f_i; f_s|C)$ can be represented as

$$I(f_i; f_s|C) = \frac{I(f_i; f_s)}{H(f_s)} H(f_s|C). \tag{6.15}$$

Substituting for (6.13), we get

$$I(C; f_i; f_s) = (1 - \frac{H(f_s|C)}{H(f_s)}) I(f_i; f_s) = \frac{I(C; f_s)}{H(f_s)} I(f_i; f_s). \tag{6.16}$$

When all the selected features in S are completely independent to each other, the mutual information of any couple of features in subset S is zero. Hence, the total redundant information of the candidate feature f_i to all the selected features in subset S with respect to output classes C, denoted by I_r for convenience, can be calculated by the simple summation as follows

$$I_r = \sum_{f_s \in S} I(C; f_i; f_s) = \sum_{f_s \in S} \frac{I(f_i; f_s)}{H(f_s)} I(C; f_s). \tag{6.17}$$

□

From (6.12) we can further deduce that the coefficient of $I(C; f_s)$ is the ratio of the mutual information between the candidate feature f_i and the selected feature f_s for the entropy of f_s. Obviously, $0 \le I(f_i; f_s)/H(f_s) \le 1$. This ratio also provides the proportion for the redundant information between feature f_i and feature f_s taking up in the mutual information between the selected feature f_s and output classes C.

Based on (6.12), Kwak and Choi gave the following estimation of the conditional mutual information $I(C; f_i|S)$ in their MIFS-U algorithm [27],

$$I(C; f_i|S) = I(C; f_i) - \beta \sum_{f_s \in S} \frac{I(C; f_s)}{H(f_s)} I(f_i; f_s). \tag{6.18}$$

The second term of the right hand in (6.18) is used to estimate the redundant information between the candidate feature f_i and the already-selected features in S with respect to classes C. Here again, the parameter β is used as a factor for controlling the redundancy penalization. As β grows, it penalizes the redundancy strongly. Therefore, different values of β can still lead to different orders of feature selection, though the influence of β has been reduced to a certain extent by the weights $I(C;f_s)/H(f_s)$. However, how to choose an appropriate value for β remains an open issue.

6.3.2 An Improved Formula to Estimate the Conditional Mutual Information

Let's consider the estimation of the conditional mutual information $I(C;f_i|S)$ without any predefined parameter β under the assumption that the information of all features is distributed uniformly. From (6.9) and Proposition 1, when all the selected features in S are completely independent to each other, $I(C;f_i|S)$ can be represented as

$$I(C;f_i|S) = I(C;f_i) - I_r = I(C;f_i) - \sum_{f_s \in S} \frac{I(f_i;f_s)}{H(f_s)} I(C;f_s). \tag{6.19}$$

While there is some redundancy among the selected features in S, the estimation of $I(C;f_i|S)$ ought to take into account not only the redundancy of feature f_i to each feature in S, but also the redundancy among features in S. The latter redundancy will affect the calculation of the whole redundancy of feature f_i to the features in S. Taking any two redundant selected features f_{s1} and f_{s2} from S as an illustration, we now derive a formula for directly computing the redundant information of f_{s1} and f_{s2} to the candidate feature f_i with respect to output classes C, which has been denoted by I_r.

Firstly, as viewed from f_{s1}, under the assumption that the information is distributed uniformly, we get the common information of f_i and f_{s2} given f_{s1} as

$$I(f_i;f_{s2}|f_{s1}) = I(f_i;f_{s2}) - \frac{I(f_i;f_{s1})}{H(f_{s1})} I(f_{s2};f_{s1}). \tag{6.20}$$

Since the entropies $H(f_{s1})$ and $H(f_{s2}|f_{s1})$ are completely independent, based on Proposition 1, the redundant information I_r can be computed as

$$\begin{aligned} I_r &= \frac{I(f_i;f_{s1})}{H(f_{s1})} I(C;f_{s1}) + \frac{I(f_i;f_{s2}|f_{s1})}{H(f_{s2})} I(C;f_{s2}) \\ &= \frac{I(f_i;f_{s1})}{H(f_{s1})} I(C;f_{s1}) + \frac{I(f_i;f_{s2})}{H(f_{s2})} I(C;f_{s2}) - \frac{I(f_i;f_{s1})}{H(f_{s1})} \frac{I(f_{s2};f_{s1})}{H(f_{s2})} I(C;f_{s2}). \end{aligned} \tag{6.21}$$

Let's further interpret (6.21). Since the information of f_{s1}, f_{s2} and f_i is distributed uniformly, the following relation holds

$$\frac{I(f_i;f_{s1})}{H(f_{s1})} = \frac{I(f_i;f_{s1};f_{s2})}{I(f_{s1};f_{s2})}. \tag{6.22}$$

By substituting (6.22) for (6.21), we have

$$I_r = \frac{I(f_i;f_{s1})}{H(f_{s1})} I(C;f_{s1}) + \frac{I(f_i;f_{s2})}{H(f_{s2})} I(C;f_{s2}) - \frac{I(f_i;f_{s1};f_{s2})}{H(f_{s2})} I(C;f_{s2}). \tag{6.23}$$

Now, return to (6.21). Since the common information of feature f_{s1}, f_{s2} and f_i has been consisted in the right first term, it should not be considered repeatedly when computing the redundant information of f_i and f_{s2} with respect to C in the right second term. Therefore, this common information is ruled out in terms of its proportion to the entropy of f_{s2} as done in the right third term.

Similarly, as viewed from f_{s2}, I_r can be written as

$$\begin{aligned} I_r = {} & \frac{I(f_i;f_{s2})}{H(f_{s2})} I(C;f_{s2}) + \frac{I(f_i;f_{s1})}{H(f_{s1})} I(C;f_{s1}) \\ & - \frac{I(f_i;f_{s2})}{H(f_{s2})} \frac{I(f_{s2};f_{s1})}{H(f_{s1})} I(C;f_{s1}). \end{aligned} \tag{6.24}$$

Add up both sides of (6.21) and (6.24) respectively. After arranging, we can derive

$$\begin{aligned} I_r = {} & \left[\frac{I(f_i;f_{s1})}{H(f_{s1})} - \frac{1}{2} \frac{I(f_i;f_{s2})}{H(f_{s2})} \frac{I(f_{s2};f_{s1})}{H(f_{s1})} \right] I(C;f_{s1}) \\ & + \left[\frac{I(f_i;f_{s2})}{H(f_{s2})} - \frac{1}{2} \frac{I(f_i;f_{s1})}{H(f_{s1})} \frac{I(f_{s1};f_{s2})}{H(f_{s2})} \right] I(C;f_{s2}). \end{aligned} \tag{6.25}$$

Hence, for all features in S, if we neglect the mutual information among three or more selected features (usually these values are small enough to be neglected), the conditional mutual information for feature ranking, $I(C;f_i|S)$, can be computed as follows,

$$\begin{aligned} I(C;f_i|S) = {} & I(C;f_i) \\ & - \sum_{f_k \in S} \left[\frac{I(f_i;f_k)}{H(f_k)} - \frac{1}{2} \sum_{\substack{f_j \in S \\ f_j \neq f_k}} \frac{I(f_i;f_j)}{H(f_j)} \frac{I(f_j;f_k)}{H(f_k)} \right] I(C;f_k). \end{aligned} \tag{6.26}$$

Thus, we have the following proposition.

Proposition 2. For any $f_s \in S$, $f_i \in F$, suppose that the information is distributed uniformly throughout the regions of $H(f_s)$, $H(f_i)$ and $H(C)$ that the classes C does not change the ratio of the entropy of f_s for the mutual information between f_s and f_i, the conditional mutual information $I(C;f_i|S)$, can be computed as (6.26).

For simplicity, we just denote $f_j \in S$, $f_j \neq f_k$ as $j \in S$, $j \neq k$, and denote

$$\varphi_{ik} = \frac{I(f_i;f_k)}{H(f_k)} \tag{6.27}$$

then (6.26) can be written as

$$I(C;f_i|S) = I(C;f_i) - \sum_{k \in S} (\varphi_{ik} - \frac{1}{2} \sum_{\substack{j \in S \\ j \neq k}} \varphi_{ij}\varphi_{jk}) I(C;f_k). \tag{6.28}$$

Thus, with formula (6.26) or (6.28), it avoids successfully the guess for the value of parameter β in MIFS and MIFS-U methods.

6.4 Hybrid Algorithm for Feature Selection Based on Mutual Information

6.4.1 Framework of the Hybrid GA for Feature Selection

Let's consider designing a hybrid genetic algorithm to solve the feature selection problem in the context of pattern classification based on mutual information. In general, a pattern classification problem can be described as follows: assume that an input feature space $\mathscr{X}$ is constructed from m features $X_i, i = 1, \ldots, m$, i.e., $\mathscr{X} = span\{X_1, X_2, \ldots, X_m\}$, and that patterns drawn from $\mathscr{X}$ are associated with c categories, whose labels constitute the set $\mathscr{Y} = \{1, 2, \ldots, c\}$. Given a training data set (x_i, y_i), $i = 1, \ldots, N$, $x_i \in \mathscr{X}$, $y \in \mathscr{Y}$, find a classifier $f : \mathscr{X} \to \mathscr{Y}$ that exhibits good generalization ability on unseen patterns.

As known, there are two main factors that affect the generalization ability of a classifier. One is to choose an optimal feature subset; the other is to find the proper model that is learnt from the selected feature subset. These two problems can be solved within a wrapper framework through cooperations of the inner and outer optimization based on maximum information.

Let S be a subset of features on X, i.e., $S \subseteq A = \{X_1, X_2, \cdots, X_m\}$, and φ_S be the class of functions that map S to Y. Let Y and Y_f $(= f(x), x \in \mathscr{X})$ be the discrete random variables over $\mathscr{Y}$ describing the unknown true labels and the labels predicted by the classifier respectively. The optimization problem we would ideally like to solve is the following:

$$J = \max_{S \subseteq A} \max_{f \subseteq \varphi_S} I(Y; Y_f), \tag{6.29}$$

where $I(Y;Y_f)$ is the mutual information between Y and Y_f. Since this is the average rate of information delivered by the classifier via its output, the quantity is referred to as the classifier output information [39].

The main task of inner optimization includes two aspects: pruning the feature subset and training a classifier on the pruned feature subset. Feature subset pruning is based on mutual information and considered as a local search, which will be detailed in the next section. Training a classifier on a given set of input features can employ most popular learning machines such as SVM, MLP and Naïve Bayes methods, whose training objective functions are usually related to the error rate of classifiers. In our study, SVM is used as the training procedure, which is a state-of-the-art classification algorithm that is shown to be particularly successful in pattern recognition [42]. For SVM, a quadratic optimization problem is solved in order to maximize the margin of separation between examples of two classes either in the original input space or in an implicitly mapped higher dimensional space by the use of kernel functions. To construct the multiclass SVMs, the *one-against-rest* strategy is adopted, where a binary SVM is trained to separate each class from the other classes. The performance of the multiclass SVMs is the prediction accuracy of the induction algorithm, and estimated by the *k-fold* cross-validation method.

The outer maximization procedure deals with the problem of choosing the optimal feature subset based on the output information of classifiers trained on those candidate feature subsets. This is performed through a genetic algorithm framework, within which the fitness function is evaluated with the mutual information between the true label and the label predicted by the trained classifier, i.e., $I(Y;Y_f)$. The overall scheme of this hybrid GA for feature selection, HGFS for short or called the HGA wrapper, is outlined in Fig. 6.2. It can be observed from Fig. 6.2 that both the global and the local search of feature selection are guided by maximizing the mutual information criterion, while the training objective function of the classifier calls for the classification error rate minimization principle. In fact, there exists a relationship between the probability of the error and the output information of a classifier.

In classification, an optimal criterion should reflect the Bayes risk in the selected variable space. The Bayes risk is usually defined in terms of a problem specific loss function. For the simplest case of *0/1-loss*, the Bayes error can be represented as a mathematic expectation as

$$p_e(X) = E_x[P_r(Y \neq Y_f)] = \int_x p(\boldsymbol{x})(1 - \max_i(p(y_i|\boldsymbol{x})))d\boldsymbol{x}, \tag{6.30}$$

where Y_f denotes the estimated class variable and Y the true class variable value, $\boldsymbol{x}$ is the input variable vector in the selected feature space X and y_i the class label. Note that the direct use of this formula would require the full knowledge of posterior probability density functions of classes $p(y_i|\boldsymbol{x})$. To estimate the Bayes error, estimating the posterior probability density functions of classes and calculating the numerical integration of a nonlinear function would thus be inevitable, and they are difficult in practice when only a training data set is given. On the other hand, in our approach, feature selection is achieved by maximizing the mutual information between the true label and the label predicted by the trained classifier, i.e., $I(Y;Y_f)$.

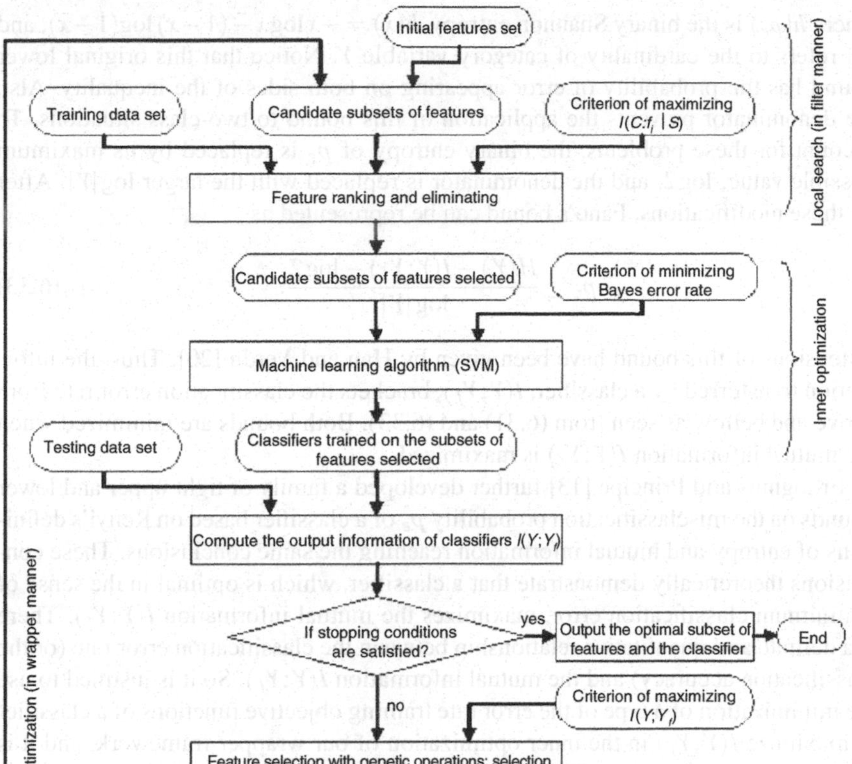

Fig. 6.2 The overall scheme of the hybrid GA for feature selection

Then, what's the relationship between $I(Y;Y_f)$ and classification accuracy in theory? Or how is the classification performance influenced by the amount of information transferred through the classifier? In fact, many researchers have done good works with insights in this question. An upper bound for the probability of error in classification

$$p_e \leq \frac{1}{2}(H(Y) - I(Y;Y_f)) \tag{6.31}$$

was obtained by Hellman and Raviv [21] for the binary case and Feder and Merhav [15] for the general case. A lower bound on the error involving mutual information is also given by Fano's inequality [14]

$$p_e \geq \frac{H(Y) - I(Y;Y_f) - h(p_e)}{\log(|Y|-1)}, \tag{6.32}$$

where $h(p_e)$ is the binary Shannon entropy, $h(x) = -x\log x - (1-x)\log(1-x)$, and $|Y|$ refers to the cardinality of category variable Y. Notice that this original lower bound has the probability of error appearing on both sides of the inequality. Also the denominator prevents the application of this bound to two-class situations. To account for these problems, the binary entropy of p_e is replaced by its maximum possible value, $\log 2$, and the denominator is replaced with the larger $\log|Y|$. After all these modifications, Fano's bound can be represented as

$$p_e \geq \frac{H(Y) - I(Y;Y_f) - \log 2}{\log|Y|}. \tag{6.33}$$

Extensions of this bound have been given by Han and Verdu [20]. Thus, the information transferred by a classifier, $I(Y;Y_f)$, brackets the classification error rate from above and bellow as seen from (6.31) and (6.33). Both bounds are minimized when the mutual information $I(Y;Y_f)$ is maximized.

Erdogmus and Principe [13] further developed a family of tight upper and lower bounds on the misclassification probability p_e of a classifier based on Renyi's definitions of entropy and mutual information reaching the same conclusions. These conclusions theoretically demonstrate that a classifier, which is optimal in the sense of a minimum classification error, maximizes the mutual information $I(Y;Y_f)$. There is a definite and quantitative relationship between the classification error rate (or the classification accuracy) and the mutual information $I(Y;Y_f)$. So it is justified to use the minimization of a type of the error rate training objective functions of a classifier to maximize $I(Y;Y_f)$ in the inner optimization of our wrapper framework, and it is naturally justifiable to use $I(Y;Y_f)$ as a proxy to the classification error rate in the outer optimization of the proposed wrapper framework.

There is another advantage when taking $I(Y;Y_f)$ as the measure for model performance. In many real world applications, it is quite likely that there are fewer data points for certain classes. For such uneven class distribution problems, classifiers tend to improve their overall classification accuracy by learning to ignore the smaller classes. So the overall accuracy criterion is not a good choice to compare feature-driven classifiers in this sense. But $I(Y;Y_f)$, as an evaluation criterion for model performance, can take into account the distribution of errors across various classes, and it is immune to biased input samples.

The *kappa* statistic is another common measure to model performance that can allow for input sampling bias. It was first introduced by Cohen [8] and has been used in ensemble feature selection methods [32, 41]. The *kappa* statistic is commonly used to measure the agreement of two classifiers. Let N_{ij} be the number of instances in a data set, which is recognized as class i by the first classifier and as class j by the second one, N_{i*} be the number of instances recognized as class i by the first classifier, and N_{*i} be the number of instances recognized as class i by the second classifier. Then define $P(A)$ and $P(E)$ as

$$P(A) = \frac{\sum_{i=1}^{c} N_{ii}}{N}, \qquad P(E) = \sum_{i=1}^{c} \left(\frac{N_{i*}}{N} \cdot \frac{N_{*i}}{N} \right), \tag{6.34}$$

where c is the number of classes and N is the total number of instances. $P(A)$ estimates the probability that the two classifiers agree, and $P(E)$ is a correction term for $P(A)$, which estimates the probability that the two classifiers agree simply by chance (in the case where each classifier chooses to assign a class label randomly). The *kappa* statistic of agreement is then defined as

$$k = \frac{P(A) - P(E)}{1 - P(E)}. \tag{6.35}$$

Kappa values can also be used to assess the agreement of the predictive output of a classifier to the true classes of inputs. $k = 0$ means that the agreement is not different from chance, and $k = 1$ means perfect agreement on every example.

Consider the confusion matrices of a three-class problem for case (a), (b) and (c) given in Table 6.1. There are 20, 5 and 35 true instances in class 1, 2 and 3, respectively, for the three cases. The classification accuracy, *kappa* statistic and output information of the classifiers are also shown in Table 6.1. Obviously, the classification accuracy for all cases is 83.33%. The observer can demonstrate good accuracy with the classifier in (c) by labeling all instances in class 2 as class 3. However, the *kappa* value in case (c) is less than that in both case (b) and case (a). The same phenomenon occurs for the output information measure, $I(Y;Y_f)$ in the three cases. This means that, as a measure of model performance, both *kappa* statistic and output information $I(Y;Y_f)$ can avoid choosing the classifier in (c). For the *kappa* statistic, it is much addressed to the balance agreement among classes, so the classifier in (b) with the largest *kappa* value is chosen. While for the output information, $I(Y;Y_f)$, it tends to put a high premium on certainty. Consequently, the classifier in (a) with the largest output information is preferred. To some extent, this classifier performs better on (a) because with the information that the classifier has output class 1 or 2, the observer can be confident about the true class of the input. In (b), when the classifier outputs class 1 or class 2, it maintains slightly greater uncertainty than (a) by sometimes also claiming for patterns of other classes. But one should notice that even in (b), the classifier output class 3 is more reliable than that in (a). In our approach, the scheme of case (a) is adopted.

Table 6.1 Confusion matrices of a three-class problem and the output information, kappa statistic and classification accuracy measures of classifiers for case (a), (b) and (c)

Y_f	(a)			(b)			(c)		
Y	1	2	3	1	2	3	1	2	3
1	12	0	8	14	2	4	16	0	4
2	0	3	2	1	3	1	0	0	5
3	0	0	35	1	1	33	1	0	34
$I(Y;Y_f)$	0.5872			0.5405			0.5101		
Kappa	0.661			0.6875			0.6581		
Acc	0.8333			0.8333			0.8333		

6.4.2 Local Search for Feature Selection

It has been proved in theory that genetic algorithms can find the optimal solution to a problem in the sense of probability in a random manner. However, simple genetic algorithms have some weaknesses like premature convergence, poor ability of fine-tuning near local optimum points in applications.

On the other hand, some other optimization methods, such as steepest descent method, hill-climbing and simulated annealing, usually have powerful local search ability. In addition, some heuristic algorithms with the problem-specific knowledge are also of high efficiency. Therefore, to improve the fine-tuning capability and efficiency of simple GAs, some hybrid GAs for feature selection have been developed by incorporating the above optimization methods or heuristic algorithms [23, 37]. Here, a novel hybrid GA for the feature selection problem is proposed, in which the feature's conditional mutual information is used as a measure to rank the candidate features, and local search operations are performed in a filter manner.

As discussed in Sect. 6.3, the conditional mutual information $I(C;f_i|S)$ measures the new information to the output class C contributed by feature f_i given the subset S of features selected. The bigger the value of $I(C;f_i|S)$, is, the more new information is provided by the candidate feature f_i. In order to apply this conditional mutual information measure for the feature selection task, a numerical threshold value *dam* is set to $I(C;f_i|S)$. This can help the algorithm to be immunized against noise in the data and also to overcome the over-fitting problem to a certain extent. Let A be the set of all input features, and $F = A - S$ be the subset of unselected features. When applying $I(C;f_i|S)$ to feature selection, depending on feature f_i is an element of the selected feature subset S or not, the local search algorithm handles two different cases:

(a) In the case of $f_i \in S$

- If $I(C;f_i|S-\{f_i\}) \leq dam$, then remove f_i from S to F
- If $I(C;f_i|S-\{f_i\}) > dam$, then let f_i remain in S

(b) In the case of $f_i \in F = A - S$

- If $I(C;f_i|S) > dam$, then add f_i to S from F
- If $I(C;f_i|S) \leq dam$, then let f_i remain in F

In a generation of the hybrid GA, each chromosome of the population corresponds to a scheme of feature selection. The first operation (a) aims to find features in the selected subset S that are less informative to classification and remove them from S; whereas the second operation (b) tries to find features in the unselected subset F that are most informative to classification and add them to the subset S. These two operations constitute the local search of feature selection. However, it should be pointed out that, to execute both the operations for every chromosome in each generation, the hybrid GA would require considerable processing time, and seems to be more likely to lead to local optima. In addition, the search function of operation (b) may be repeated by the GA operations and result in a waste of computational resources. Therefore, the final implementation of our algorithm executes only the first operation (a), whereas the second operation (b) is not executed, whose search function is just delivered to the GA to be accomplished. These strategies can help to search the global optimal subset of features and improve the efficiency of the algorithm.

The above local search operations are carried out in a filter manner. The local search measure $I(C; f_i | S)$ is independent of the learning machine, but it is consistent with the training objective function of the learning machine. These local search operations inherit all the merits of filters such as high efficiency, fast computation and simplicity. They first remove the "bad genes" from every chromosome of a population, corresponding to removing the insignificant features out of every subset generated by GA in each generation. Then the learning machines are trained on the preprocessed feature subsets, and finally the optimal subset of features is found by the outer GA optimization through generations according to the output information of the series of learning machines trained. Thus, a perfect cooperation of the inner and out optimization is achieved.

Another advantage of this algorithm is that the threshold *dam* can be use to adaptively control the number of selected features. This manipulation seems to be more reasonable than predefining a fixed number of the selected features. If the number of feature selected is expected to be no more than k, we may just lower the threshold value and modify the local search operations as follows:

(c) Let $S_q = \{f_i | I(C; f_i | S - \{f_i\}) > dam\}$

- If $|S_q| > k$, then remove the last $|S_q| - k$ features with the least values of $I(C; f_i | S - \{f_i\})$ out of S
- If $|S_q| < k$, then add the first $k - |S_q|$ features with values of $I(C; f_i | S) > dam$ in the unselected feature subset F to S

6.4.3 Implementation of the Hybrid GA for Feature Selection

The implementation of the hybrid genetic algorithm for feature selection mainly includes the encoding schemes of chromosomes, evaluating fitness function, local searching operations, designing for the selection, crossover and mutation genetic operations, and stopping criterion. In the proposed algorithm, each individual in the population of chromosomes represents a candidate solution to the feature subset selection problem. A chromosome is encoded by a binary digit series where "1" means "selected" and "0" means "unselected". Each digit (or gene) corresponds to a feature, so the gene length of a chromosome is equal to the total number of input features available. As discussed in Sect. 6.4.1, the output information $I(Y; Y_f)$ of the trained SVM classifier is used as the fitness function of the hybrid genetic algorithm. It is evaluated by the *k-fold cross-validation* method and k takes the value from 2 to 10 depending on the data set scale. The local search operations have been described in Sect. 6.4.2.

As for genetic operations, the strategies are detailed as follows. First, our design adopts the *rank-based roulette wheel* selection scheme. To guarantee the fast convergence ability, an *elitism* strategy is also used so that the best 10% of chromosomes in current population can enter the next generation directly without undergoing the crossover and mutation operations. Next, an adaptive crossover strategy is employed. When the total number of features is less than 20, the *single-point* crossover operator is used; while the total number of features is more than 20, the *double-point* crossover operator is used. The crossover probability p_c is assigned as 0.7. Last, a simple mutation operator with the probability 0.1 is used.

The hybrid GA stops when the number of generations reaches the preset maximum generation T. The overall procedure of the hybrid GA for feature selection (HGFS) is outlined below.

Algorithm 6.4.1 (Procedure of HGFS)

```
M is the size of population P(t)
Begin
    Initialize P(0);
    t = 0;
    While (t ≤ T) do
        Local improvement of each chromosome of P(t) by I(C; f_i|S);
        for i = 1 to M do
            Evaluate fitness I(Y;Y_f) of individuals of P(t);
        end for
        if stopping conditions are satisfied, break;
        Selection operation to P(t);
        Crossover operation to P(t);
        Mutation operation to P(t);
        for i = 1 to M do
            P(t+1) = P(t);
        End for
        t = t+1;
    End while
End
```

6.5 Experiments and Discussions

Some aspects of the proposed HGFS approach are studied and discussed via various experiments in this section. First, we investigate the performance of HGFS on a specific data set, the wine dataset and two synthetic data sets. The wine dataset is chosen from the UCI Machine Learning Repository [31]. Some characteristics of the HGFS are well studied by experiments including time analysis, efficiency of the local search operations, and affects of different number of samples, dimensions as well as bins used in the histogram computation. Then, more real-world benchmark datasets with a diverse mixture of attribute types (continuous, nominal), a large number of instances and a wide range of input feature dimensions are chosen for experiments to verify the effectiveness and applicability of the HGFS approach. All the results are compared with other feature selection methods such as Battiti's MIFS, Il-seok Oh's HGA and the recursive feature selection method (RFE).

6.5.1 Performance of the HGFS Approach

Time Analysis and Comparisons with Some Methods

In this subsection, the proposed HGFS method will be briefly compared with Battiti's MIFS, recursive feature elimination (RFE) and Il-seok Oh's HGA methods. A simple time analysis and comparison is performed through both designing of algorithms and experimental results.

In fact, the above four approaches are very typical feature selection methods. First, as viewed from the evaluation measures for feature selection, Battiti's MIFS is a classical filter method. It evaluates each candidate feature with the mutual information between the selected features and output classes without involving any learning algorithms. Whereas Il-seok Oh's HGA and the proposed HGFS algorithms are typical wrapper methods in which a specific learning machine is utilized to assess the goodness of candidate feature subsets. RFE may be categorized into one type of embedded feature selection methods, for it chooses features according to the weights given by the classifier itself. Second, as seen from the search strategies, MIFS is a forward selection approach that starts with an empty subset and iteratively adds features according to a mutual information measure until a stopping criterion is met. Whereas RFE is a backward elimination method that starts with all features and iteratively removes one feature or bunches of features according to the weight vector. Unlike MIFS and RFE, Il-seok Oh's HGA and the proposed HGFS algorithms are stochastic search methods that generate the next feature subset in a heuristic but non-deterministic fashion by means of amounts of calculations. These distinctions among the four algorithms result in different computational complexity.

In MIFS, the feature selection procedure doesn't need to train any learning machines such as SVMs. If n_0 out of n features are required to be selected, it needs only to compute the mutual information criterion (6.11) n_0 times. When histogram with bins for each feature no more than ten is employed (this number of bins is usually enough for feature selection), the time to compute the mutual information criterion (6.11) is a tiny fraction of that to train an SVM. Consequently, MIFS is a very fast feature selection method.

In RFE, it operates by trying to choose the n_0 features which lead to the largest margin of class separation, using an SVM classifier. This combinatorial problem is solved in a greedy fashion with iterations of training by removing the feature that decreases the margin the least until only n_0 features remain. Hence, $n - n_0$ SVMs have to be trained in the procedure. The algorithm can be accelerated by removing more than one feature at a time, for example, half of the features remain.

Whereas in HGFS, each chromosome represents a candidate subset of features. The local search operations mainly involve computation of the conditional mutual information criterion (6.26) to evaluate each feature in the chromosome. If an initial chromosome has n' features, the times of computing the criterion for this chromosome, denoted by t_P, will be

$$t_P = \begin{cases} n' - n_0, & n' \geq n_0, \\ n_0 - n', & n' < n_0. \end{cases} \tag{6.36}$$

Meantime, an SVM classifier needs to be trained to assess the improved chromosome. If the population of GA has N_{ind} chromosomes, the sum of computational cost in one generation includes training N_{ind} SVM classifiers and calculating the criterion (6.26) $N_{ind} \cdot t_P$ times.

Finally, in Il-seok Oh's HGA (denoted by HGA-o), a local search operation *ripple*(r) was presented. The basic idea of *ripple*(r) procedure can be stated as follows: first remove the least significant feature from the subset of already-selected features r times, and then add the most significant unselected feature $r-1$ times again, thus one feature is added; contrariwise, one feature is removed. If the number of features in current subset is equal to the required number, then just do both removing and adding operations r times respectively. The parameter r is called the ripple factor taking value of integers like 1, 2. Each evaluation of significance of features needs to train and test a classifier. If an initial chromosome has n' features, the number of classifiers to be trained for a chromosome in the rippling local operations, denoted by t_{oh}, will be

$$t_{oh} = \begin{cases} (n' - n_0) \cdot (2r-1), & n' > n_0, \\ 2 \cdot (2r-1), & n' = n_0, \\ (n_0 - n') \cdot (2r-1), & n' < n_0. \end{cases} \tag{6.37}$$

If a population of GA has N_{ind} chromosomes and SVM is assumed the type of classifiers to be trained, the sum of computational cost in one generation of Il-seok Oh's HGA includes training $N_{ind} \cdot t_{oh}$ SVM classifiers. Obviously, this is a very time-consuming methodology.

The *Wine* dataset has 13 numerical input features and one label attribute with three classes. It consists of 178 instances in which 57, 59 and 61 instances belong to the three classes respectively. Perform the above four feature selection methods on Wine dataset with 2-*fold* cross-validation and find the best subset of 3, 5, 8 and 10 features required in turn. All the computer hardware conditions are the same. In MIFS, the parameter A takes the recommended value 0.7. In both Il-seok Oh's HGA and the HGFS, the size of population is set to 10, and other parameters are set as described in Sect. 6.4.3. The time to find the best subset of required numbers of features on Wine dataset with the four methods is recorded in Table 6.2, in which for Il-seok Oh's HGA and HGFS, the time is accumulated by the iterative generation time till the two methods converge to the optimal subsets of required number features.

Table 6.2 Time to find the best subset of required numbers of features with MIFS, RFE, Oh's HGA and HGFS methods for the Wine dataset

Methods	Time (s)			
	3 features	5 features	8 features	10 features
MIFS	0.06	0.10	0.16	0.21
RFE	5.81	4.94	3.33	2.39
HGFS	29.01	16.42	21.49	17.17
HGA-o	217.73	275.22	332.56	274.94

From Table 6.2 one can see that the time of MIFS is the least, while the time of the Il-seok Oh's HGA is the most. As an improved approach, time of the HGFS method is much less than that of the Il-seok Oh's HGA, but still more than that of RFE approach. In addition, one can find that the time of MIFS increases with the number of required features increasing as it is a forward selection method, while the time of RFE decreases with the number of required features increasing due to being a backward elimination approach. However, this property doesn't appear in Il-seok Oh's HGA and HGFS methods since they are the stochastic search algorithms.

Study on Local Search Operations in HGFS

Rippling Operation and Stepwise Eliminating Operation

Different from the rippling local search operations employed in Il-seok Oh's HGA, in the HGFS approach, the local search involves no classifiers training; only the feature whose measure $I(C;f_s,S)$ is the least and less than the threshold *dam* is eliminated stepwise as described in Sect. 6.4.2. The search process of HGFS with rippling or stepwise-eliminating local operation is illustrated in Fig. 6.3. After 100 generations of evaluation, the HGFS with stepwise-eliminating local operation finds the best feature subset $\{1,3,5,7,11,13\}$ with the maximal output information of the trained classifier $I(Y;Y_f) = 1.46$ and the predictive accuracy 98.31%, while the HGFS with *ripple*(r) ($r = 2$) local search operation finds the "best" feature subset $\{1,2,6,7,10,11,12,13\}$ with the maximal output information of the trained classifier $I(Y;Y_f) = 0.84$ and the predictive accuracy 84.83%. Obviously, the latter is trapped into the local optimum. Moreover, the HGFS with the ripple local operations consumes much more runtime in each generation than the stepwise-eliminating local operation. This shows that the ripple local search operations may weaken the global search ability of the hybrid GA for feature selection and do little improvement on predictive accuracy in our approach.

Fitness Evaluation on Redundant Chromosomes

When running the HGFS procedures, the most time-consuming computation is to evaluate the fitness of chromosomes in a population. To compute the fitness, a series of classifiers are trained and tested for each chromosome of the population in a

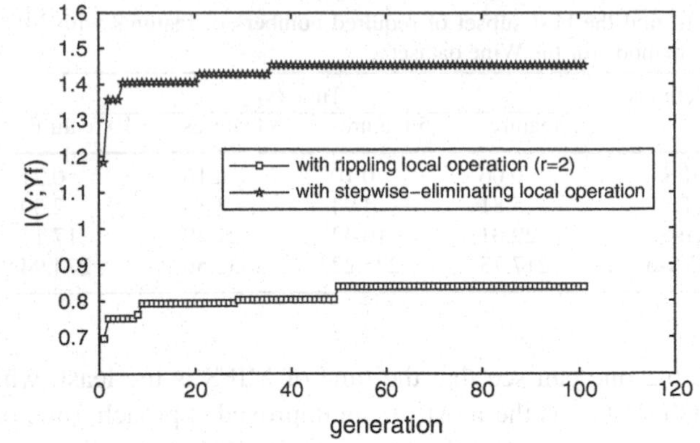

Fig. 6.3 The searching process of the HGFS approach with rippling or stepwise-eliminating local operations

generation. However, as the generations of the hybrid GA wrapper are evaluated, the chromosomes represented by a string of binary digits of zeros and ones would converge to the optimal subset of features gradually, and more and more identical chromosomes would appear in the same population. Since the main objective of HGFS is to find the best subset of features, not to search the best value of fitness on the best subset of feature, and the predictive accuracy of a classifier is most likely determined by the subset of features selected rather than by the randomness of the training data, so it is no need to compute the fitness values repeatedly for those identical chromosomes in the same population, and only the evaluation of fitness of different chromosomes in a population is necessary. In the hybrid GA wrapper approach, the strategy of fitness evaluation for unique chromosomes is adopted. The running time of every generation with 20 chromosomes in a population by 10-*fold* cross-validation is shown in Fig. 6.4. As one can see from Fig. 6.4, when this strategy is employed, the running time of one generation is decreased with the number of generations increasing, and the average running time of one generation is 103.5 s. Whereas this strategy is not used, the average running time of one generation is 135.1 s.

Effectiveness of the Local Search Operations

Finally, we inspect the effectiveness of the local search operations for the HGFS approach. Figure 6.5 shows the searching processes of HGFS with and without the local search operations. Corresponding to the two cases, the best subsets of features $\{1,3,5,7,11,13\}$ and $\{1,4,9,10,11,13\}$ are found after 50 generations, the maximal output information of the classifiers trained are 1.46 and 1.37, and the predictive accuracy is 98.31% and 96.63%, respectively.

From the experimental results one can see obviously that the stepwise-eliminating local search operations can greatly enforce the local searching ability and make the algorithm fast reaching its optimum. As a result, the performance of the HGFS approach is improved.

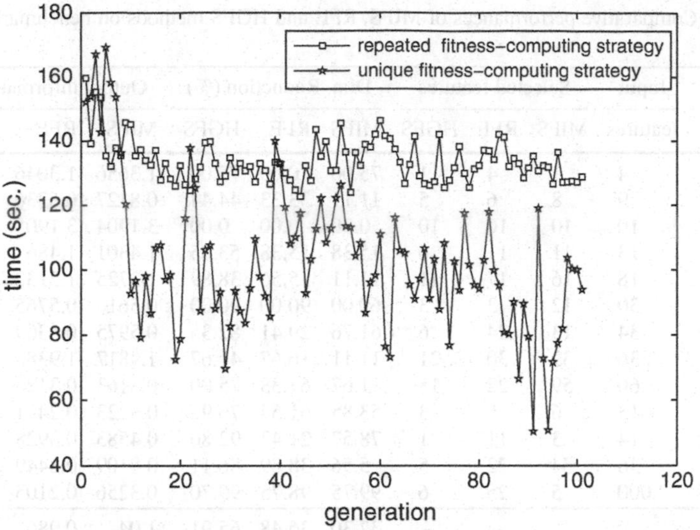

Fig. 6.4 The searching time of generations in HGFS with repeated or unique fitness-computing strategies

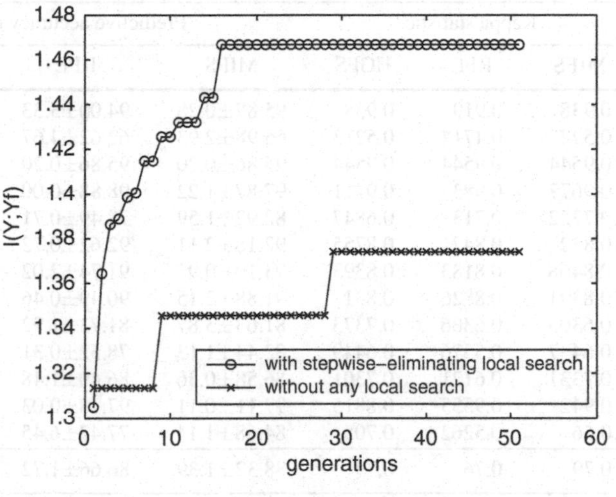

Fig. 6.5 The searching processes of the HGFS approach with or without the local search operations

6.5.2 Experimental Results on Benchmark Datasets

To study the applicability of the hybrid GA wrapper for feature selection (HGFS), 13 benchmark datasets are chosen here for experiments (see Tables 6.3 and 6.4). The first 12 datasets are posted at the UCI Machine Learning Repository [31], and

Table 6.3 Comparative performances of MIFS, RFE and HGFS methods on benchmark datasets (part 1)

Datasets	Input	Selected features			Dim. Reduction (%)			Output information (bit)		
	features	MIFS	RFE	HGFS	MIFS	RFE	HGFS	MIFS	RFE	HGFS
Iris	4	1	4	1	75.00	0.00	75.00	1.3656	1.3046	1.3656
Glass	9	8	6	5	11.11	33.33	44.44	0.8127	0.6736	0.8208
Vowel	10	10	10	10	0.00	0.00	0.00	3.1904	3.1905	3.1904
Wine	13	11	11	6	15.38	15.38	53.85	1.4301	1.486	1.4628
Vehicle	18	16	17	11	11.11	5.56	38.89	1.3725	1.2132	1.23
WDBC	30	12	3	3	60.00	90.00	90.00	0.561	0.5765	0.6353
Ionosphere	34	13	24	6	61.76	29.41	82.35	0.5975	0.5309	0.5693
Satellite	36	32	30	21	11.11	16.67	41.67	1.9817	1.9388	1.9755
Sonar	60	59	22	15	1.67	63.33	75.00	0.3163	0.3125	0.4461
Heart	13	6	5	3	53.85	61.54	76.92	0.3223	0.2421	0.3266
Credit	14	3	11	1	78.57	21.43	92.86	0.4583	0.3928	0.4527
Chess	36	34	22	5	5.56	38.89	86.11	0.8109	0.8449	0.6811
Colon	2,000	5	25	6	99.75	98.75	99.70	0.3256	0.2105	0.3835
Average	–	–	–	–	37.30	36.48	65.91	1.04	0.986	1.04

Table 6.4 Comparative performances of MIFS, RFE and HGFS methods on benchmark datasets (part 2)

Datasets	Kappa statistic			Predictive accuracy (%)		
	MIFS	RFE	HGFS	MIFS	RFE	HGFS
Iris	0.938	0.919	0.938	95.87±0.28	94.00±3.33	95.87±0.50
Glass	0.5227	0.4714	0.5213	65.98±2.93	62.62±4.67	65.51±2.44
Vowel	0.9544	0.9544	0.9544	95.86±0.20	95.86±0.20	95.86±0.20
Wine	0.9677	0.983	0.9711	97.87±1.22	98.88±0.00	98.31±0.57
Vehicle	0.7722	0.713	0.6847	82.92±1.59	78.49±0.71	76.36±1.95
WDBC	0.832	0.8421	0.8755	92.18±2.11	92.62±0.72	94.24±0.39
Ionosphere	0.8468	0.8183	0.8393	93.19±0.97	91.74±2.02	92.76±1.09
Satellite	0.8391	0.8826	0.831	91.88±2.15	90.49±0.46	91.47±1.89
Sonar	0.6303	0.6306	0.7373	81.63±3.87	81.73±1.92	87.02±0.00
Heart	0.6437	0.5576	0.6449	82.44±1.13	78.52±0.31	82.59±2.11
Credit	0.7331	0.6121	0.7301	86.58±0.36	86.40±1.48	86.43±0.00
Chess	0.942	0.9555	0.8815	97.11±0.11	97.78±0.03	94.10±0.02
Colon	0.66	0.5262	0.705	84.68±1.14	77.42±6.45	86.77±2.58
Average	0.79	0.76	0.79	88.32±1.39	86.66±1.72	88.25±1.06

the last one, the prostate cancer data set, is available from [2]. All these datasets are widely used by the data mining community for evaluating learning algorithms. The datasets selected here comprise a diverse mixture of feature types, ranging from purely continuous to purely nominal attribute domains. The size of the datasets varies from 150 to 6,435 cases, and the original feature dimension of the datasets is up to 2,000.

The performance of the HGFS approach is compared with that of other two feature selection algorithms: Battiti's MIFS and the recursive feature selection method (RFE) algorithm. The results are mainly compared in terms of dimensionality reduction, output information ($I(Y;Y_f)$), *kappa* statistic and the predictive accuracy.

The parameters are set as follows: all the three feature selection methods take SVM learning algorithm as the classifier with parameter $C = 100$ and the *rbf* kernel function is used with parameter $\sigma = 2$. For multiclass problems, the *one-against-rest* strategy is employed. In MIFS, parameter takes the recommended value 0.7. In RFE, the strategy that half of the features left are removed at a time is adopted. In MIFS and HGFS, by the equiprobable partitioning approach described in Sect. 6.2, the continuous features are discretised into four bins for data sets that the number of samples is less than 200, six bins for data sets that the number of samples is within 200 and 600, and 10 bins for data sets that the number of samples is above 600. In addition, for the HGFS approach, the size of chromosome population is set to 10, and other GA parameters are set as Sect. 4.3 described. Finally, all the results are evaluated by 2-*fold* cross-validation method for each experiment and presented in Tables 6.3 and 6.4.

Tables 6.3 and 6.4 show the initial number of candidate input features in each data set, the number of input features selected by the evaluated algorithms, and the reduction in data dimensionality (the portion of candidate input features that were excluded from the model). The way to determine the number of features in the optimal subset is as follows. Let r be the required number of features to be selected, m be the total number of features of a data set or the most number of features expected to select when the initial features of a dataset is too many. Repeat the following operations from $r = 1$ to m : carrying out the feature selection method to find the best feature subset of size r, training and testing an SVM classifier with the best feature subset, and computing the output information $I(Y;Y_f)$ with *k-fold* cross validation approach. The r corresponding to the maximal $I(Y;Y_f)$ is regarded as the number of features in the optimal subset. When m is very large, the r, after which a successive number of $I(Y;Y_f)$'s are not improved saliently, e.g., from $r+1$ to $r+5$, is then considered as the number of features in optimal subset. Usually, the *kappa* value and classification accuracy have the same trend as $I(Y;Y_f)$.

The results in Tables 6.3 and 6.4 show that the models produced by the HGAFS method are significantly smaller than that built by the MIFS and RFE methods in almost all the data sets. The average difference between them in dimensionality reduction is about 28% of the number of available features. This means that the HGFS approach is a much "aggressive" dimensionality reducer than RFE and Battiti's MIFS methods.

Tables 6.3 and 6.4 also show, for each data set, the estimated model performance such as output information, kappa statistic and predictive accuracy of the HGFS approach versus the other two methods. As one can see from Tables 6.3 and 6.4, these three performance indices of HGFS are very similar to that of both RFE and Battiti's MIFS in overall cases. This means that the HGFS approach is able to achieve the highest average model performance using the least number of features, which is less than 40% of the number of available features on average. To sum up, the HGFS method is prominent compared to both Battiti's MIFS and RFE methods.

6.6 Conclusions

A hybrid genetic algorithm for feature selection (HGFS) has been presented based on mutual information. Both the performance and the applicability of the method have been well studied by experiments on various types of datasets. The results demonstrate clearly the effectiveness of the HGFS approach. The results can be summarized as follows:

(a) The mutual information between the predictive labels of a trained classifier and the true classes $I(Y;Y_f)$, called the output information of the classifier, can be used as the fitness function of the hybrid GA for feature selection. It has a close relation to the predictive accuracy of the trained classifier. When a classifier is trained by the objective function of minimal classification error rate, its output information achieves its maximum.

(b) The improved formula to calculate the conditional mutual information between the candidate feature and the classes given a subset of already-selected features, $I(C;f_i|S)$, works well in feature ranking in the local search operations of the HGFS approach. It takes carefully into account both the relevance of the candidate feature to the output classes and the redundancy between the candidate feature and the features already selected.

(c) The wrapper and filter approaches are well incorporated within the two-layer optimization framework of the hybrid GA for feature selection. The outer optimization completes the global search for the best subset of features in a wrapper way, which uses the output mutual information as the fitness function of the genetic algorithm. The inner optimization performs the local search in a filter manner, which uses the conditional mutual information $I(C;f_i|S)$ as an independent measure to rank every feature in a subset and removes those features with inferior values of $I(C;f_i|S)$. The inner and outer optimization cooperate with each other and achieve good performances of both the high global predictive accuracy and the high local search efficiency.

The overall framework of HGFS adopts the wrapper formulation, so it inevitably inherits the weakness like a long run time needed. The value of the HGFS approach lies in the fact that once the optimal subset of features is found with this method, the subsequent learning machines or data representations may be trained or processed directly with the selected features with much less time.

Acknowledgements This work was partly supported by the National Natural Science Foundation of China under Grant No.60575036, by the Science and Technology Foundation for Innovative

Talents of Harbin City of P. R. China (2007RFXXG023), and by the Science Foundation for Top Talents with the Spirit of Innovation of Harbin University of Science and Technology.

References

1. Ahmad, A., Dey, L.: A feature selection technique for classificatory analysis. Pattern Recognit Lett 26(1), 43–56, (2005).
2. Alon U., Barkai N., et al.: The colon microarray data set. http://microarray.princeton.edu/oncology/affydata/, 1999.
3. Amaldi, E., Kann, V.: On the approximation of minimizing non zero variables or unsatisfied relations in linear systems. Theor Comput Sci 209(1C2), 237–260, (1998).
4. Battiti, R.: Using mutual information for selecting features in supervised neural net learning. IEEE Trans Neural Netw 5(4), 537–550, (1994).
5. Bhanu, B., Lin, Y.: Genetic algorithm based feature selection for target detection in SAR images. Image Vision Comput 21(7), 591– 608, (2003).
6. Bishop, C.M.: Neural Networks for Pattern Recognition. Oxford University Press, Oxford (1995).
7. Caruana, R., Freitag, D.: Greedy attribute selection. In: Proc. of the 11th Internat. Conf. on Machine Learn., New Brunswick, NJ, USA, pp. 28–36, (1994).
8. Cohen, J.: A coefficient of agreement for nominal scales. Educ Psychol Meas 20(1), 37–46, (1960).
9. Cover, T.M., Thomas, J.A.: Elements of Information Theory. Wiley, New York, (1991).
10. Dash, M., Liu, H.: Feature selection for classification. Intell Data Anal 1(3), 131–156, (1997).
11. Dash, M., Liu, H.: Consistency-based search in feature selection. Artif Intell 151(1–2), 155–176, (2003).
12. Doak, J.: An evaluation of feature selection methods and their application to computer security. CSE Technical Report 92-18, University of California at Davis, (1992).
13. Erdogmus, D., Principe, J.: Lower and upper bounds for misclassification probability based on Renyis information. J VLSI Signal Process Systems 37(2–3), 305–317, (2004).
14. Fano, R.M.: Transmission of Information: A Statistical Theory of Communications. Wiley, New York, (1961).
15. Feder M and Merhav N.: Relations between entropy and error probability. IEEE Trans Inform Theory, 40(1), 259–266, (1994)
16. Fraser, A.M., Swinney, H.L.: Independent coordinates for strange attractors from mutual information. Phys Rev A 33(2), 1134–1140, (1986).
17. Grall-Maes, E., Beauseroy, P.: Mutual information-based feature extraction on the time-frequency plane. IEEE Trans Signal Process 50(4), 779–790, (2002).
18. Guyon, I., Elisseeff, A.: An introduction to variable and feature selection. J Machine Learn Res 2003(3), 1157–1182, (2003).
19. Guyon, I., Weston, J., Barnhill, S., Vapnik, V.: Gene selection for cancer classification using support vector machines. Machine Learn 46(1–3), 389–422, (2002).
20. Han, TS. and Verdu, S.: Generalizing the fano inequality. IEEE Trans Inform Theory 40(4), 1147–1157, (1994)
21. Hellman, M.E., Raviv, J.: Probability of error, equivocation and the chernoff bound. IEEE Trans Inform Theory 16(4), 368–372, (1970).
22. Huang, D., Chow, T.W.S.: Effective feature selection scheme using mutual information. Neurocomputing 63(1), 325–343, (2005).
23. Oh, Il-Seok, Lee, Jin-Seon, Moon, Byung-Ro: Hybrid genetic algorithms for feature selection. IEEE Trans Pattern Anal Machine Intell 26(11), 1424–1437, (2004).
24. Kohavi, R., John, G.H.: Wrappers for feature subset selection. Artif Intell 97(1–2), 273–324, (1997).

25. Koller, D., Sahami, M.: Toward optimal feature selection. In: Proc. Internat. Conf. Machine Learn., Bari, Italy, pp. 284–292, (1996)
26. Kudo, M., Sklansky, J.: Comparison of algorithms that select features for pattern classifiers. Pattern Recognit 33(1), 25–41, (2000).
27. Kwak, N., Choi, C.H.: Input feature selection for classification problems. IEEE Trans Neural Netw 13(1), 143–159, (2002).
28. Last, M., Maimon, O.: A compact and accurate model for classification. IEEE Trans Knowledge Data Eng 16(2), 203–215, (2004).
29. Last, M., Kandel, A., Maimon, O.: Information-theoretic algorithm for feature selection. Pattern Recogni Lett 22(6–7), 799–811, (2001).
30. Liu H., Motoda H.: Feature Selection for Knowledge Discovery and Data Mining. Kluwer, Boston (1998)
31. Merz, C.J., Murphy, P.M.: UCI Repository of machine learning databases. University of California, Department of Information and Computer Science, Irvine, CA. http:// www.ics.uci.edu/ mlearn/ MLRepository.html. Accessed 15 Jan 2006.
32. Opitz, D.: Feature selection for ensembles. In: Proc. 16th National Conf. on Artificial Intelligence. AAAI Press, pp. 379–384, (1999).
33. Quinlan, J.R.: Induction of decision trees. Machine Learn 1(1), 81–106, (1986).
34. Quinlan, J.R.: Improved use of continuous attributes in C4.5. J Artif Intell Res 4(1), 77–90, (1996).
35. Rakotomamonjy, A.: Variable selection using SVM-based criteria. J. Machine Learn Res 2003(3), 1357–1370, (2003).
36. Raymer, M.L., Punch, W.F., Goodman, E.D., Kuhn, L.A., Jain, A.K.: Dimensionality reduction using genetic algorithms. IEEE Trans Evolut Comput 4(2), 164–171, (2000).
37. Salcedo-Sanz, S., Camps-Valls, G., Perez-Cruz, F., Sepulveda-Sanchis, J., Bousono-Calzon, C.: Enhancing genetic feature selection through restricted search and Walsh analysis. IEEE Trans Syst Man Cybern C 34(4), 398–406, (2004).
38. Shannon, C.E., Weaver, W.: The Mathematical Theory of Communication. University of Illinois Press, Urbana, IL, (1949).
39. Sindhwani, V., Rakshit, S., Deodhare, D., Erdogmus, D., Principe, J.C., Niyogi, P.: Feature selection in MLPs and SVMs based on maximum output information. IEEE Trans Neural Netw 15(4), 937–948, (2004).
40. Somol, P., Pudil, P., Kittler, J.: Fast branch and bound algorithms for optimal feature selection. IEEE Trans Pattern Anal Machine Intell 26(7), 900–912, (2004).
41. Tsymbal, A., Pechenizkiy, M., Cunningham, P.: Diversity in search strategies for ensemble feature selection. Inform Fusion 6(1), 83–98, 2005).
42. Vapnik, V.: Statistical Learning Theory. Wiley, New York, (1998).
43. Weston, J., Elisseff, A., Schoelkopf, A.B., Tipping, M.: Use of the zero norm with linear models and kernel methods. J Machine Learn Res 2003(3), 1439–1461, (2003).
44. Yang, J.H., Honavar, V.: Feature subset selection using a genetic algorithm. IEEE Intell Systems 13(2), 44–49, (1998).
45. Zhu, F., Guan, S.: Feature selection for modular GA-based classification. Appl Soft Comput J 4(4), 381–393, (2004).
46. Zilberstein, S.: Using anytime algorithms in intelligent systems. AI Mag 17(3), 73–83, (1996).

Chapter 7
Information Approach to Blind Source Separation and Deconvolution

Pham Dinh-Tuan

7.1 Introduction

Blind separation of sources aims to recover the sources from a their mixtures without relying on any specific knowledge of the sources and/or on the mixing mechanism [1]. (That is why the separation is called blind). Instead, it relies on the basic assumption that the sources are mutually independent.[1] A popular measure of dependence is the mutual information. This chapter attempts to provide a systematic approach to blind source separation based on the mutual information.

We shall focus on noiseless mixtures. There are few methods which deal explicitly with noises. Often a preprocessing step is done to reduce noises, or a method designed for a noiseless model is found to be rather insensitive to noise and can thus be applied when some noises are present. A general noiseless mixture model can be written as $\mathbf{x}(\cdot) = \mathbf{A}\{\mathbf{s}(\cdot)\}$, where $\mathbf{s}(n)$ et $\mathbf{x}(n)$ represent the observation and the source vector at time n et $\mathbf{A}$ is some transformation, which can be instantaneous, i.e. operating on each $\mathbf{s}(n)$ to produce $\mathbf{x}(n)$, or global (i.e. operating on the whole sequence $\mathbf{s}(\cdot)$ of the source vectors. The transformation $\mathbf{A}$ is not completely arbitrary, one often assumes it belongs to a certain class $\mathscr{A}$, the most popular ones are the class of linear (or affine) instantaneous transformation and the class of linear convolutions. More complex non linear transformations have been considered, but for simplicity we shall limit ourselves to the above two linear classes. Separation may be realized by applying an inverse transformation $\mathbf{A}^{-1}$ to $\mathbf{x}(\cdot)$. However, $\mathbf{A}$ is unknown, so is its inverse. The natural idea is to apply a transformation $\mathbf{B} \in \mathscr{A}^{-1}$, the set of all transformations which are inverses of a transformation in $\mathscr{A}$, and is chosen

P. Dinh-Tuan
Laboratory Jean Kuntzmann, CNRS-INPG-UJF BP 53, 38041 Grenoble Cedex, France
e-mail: Dinh-Tuan.Pham@imag.fr

[1] In telecommunication however, the discrete nature of the sources provides a powerful means for their separation even without the independence assumption. Also, for non stationary and/or non white source, a weak form of independence – second order independence – can be enough.

F. Emmert-Streib, M. Dehmer (eds.), *Information Theory and Statistical Learning*,
DOI: 10.1007/978-0-387-84816-7_7,

to minimize some criterion. We consider here the independence criterion based on the mutual information measure.

This chapter contains to parts: the first one concerns instantaneous (linear) mixtures and the second one convolutive (linear) mixtures. In the last case, if the observed and source vectors are both scalar, the problem reduces to that of blind deconvolution, since the observation sequence is the convolution of the source sequence with some filter and the aim is to deconvolve the observation sequence to recover the source. Here, temporal independence of the source is essential for blind deconvolution. In general, for convolutive mixture, the sources can be recovered only up to a filtering (as it will be seen later) so the temporal independence assumption of the source may be introduced to eliminate this ambiguity. Then the problem may be called multichannel blind deconvolution or more appropriately blind separation deconvolution as one tries to both separate the source and deconvolve them.

7.2 Blind Separation of Linear Instantaneous Mixtures

We begin with the definition and some properties of the mutual information

7.2.1 Mutual Information Between Random Vectors

Let $\mathbf{y}_1,\ldots,\mathbf{y}_K$ be K random vectors with joint density $p_{\mathbf{y}_1,\ldots,\mathbf{y}_K}$ and marginal densities $p_{\mathbf{y}_1},\ldots,p_{\mathbf{y}_K}$, the *mutual information* between them is defined as the Kullback–Leibler divergence between the product density of the marginal densities $\prod_{k=1}^K p_{\mathbf{y}_k}$ and the joint density $p_{\mathbf{y}_1,\ldots,\mathbf{y}_K}$:

$$I(\mathbf{y}_1\ldots,\mathbf{y}_K) = -\mathbf{E}\left[\log\frac{p_{\mathbf{y}_1}(\mathbf{y}_1)\cdots p_{\mathbf{y}_K}(\mathbf{y}_K)}{p_{\mathbf{y}_1,\ldots,\mathbf{y}_K}(\mathbf{y}_1,\ldots,\mathbf{y}_K)}\right]. \tag{7.1}$$

This is a dependence measure as it is non negative (but can be infinity) and can vanish if and only if the random vectors are independent [4]. One can also write the mutual information in term of the (differential) entropy as [4]

$$I(\mathbf{y}_1\ldots,\mathbf{y}_K) = \sum_{i=1}^{K} H(\mathbf{y}_k) - H(\mathbf{y}_1\ldots,\mathbf{y}_K), \tag{7.2}$$

where $H(\mathbf{y}_1\ldots,\mathbf{y}_K)$ et $H(\mathbf{y}_1),\ldots,H(\mathbf{y}_K)$ are joint and marginal entropies of $\mathbf{y}_1,\ldots,$ $\mathbf{y}_K$:

$$H(\mathbf{y}_k) = -\int \log[p_{\mathbf{y}_k}(\mathbf{y}_k)]p_{\mathbf{y}_k}(\mathbf{y}_k)d\mathbf{y}_k = -\mathbf{E}\{\log[p_{\mathbf{y}_k}(\mathbf{y}_k)]\} \tag{7.3}$$

and $H(\mathbf{y}_1\ldots,\mathbf{y}_K)$ is the same as $H([\mathbf{y}_1^{\mathrm{T}}\ \cdots\ \mathbf{y}_K^{\mathrm{T}}]^{\mathrm{T}})$, ^{T} denoting la transposition, and is defined similarly as $H(\mathbf{y}_k)$ with $p_{\mathbf{y}_k}$ replaced by $p_{\mathbf{y}_1,\ldots,\mathbf{y}_K}$.

The entropy possesses the following interesting property with respect to invertible transformation:

Lemma 7.1. *Let* $\mathbf{x}$ *be a random vector and* $\mathbf{y} = g(\mathbf{x})$ *where* g *is an invertible differentiable transformation with Jacobian (matrix of derivatives)* g'. *Then*

$$H(\mathbf{y}) = H(\mathbf{x}) + \mathbf{E}\log|\det g'(\mathbf{x})|.$$

The proof of this result follows easily from the definition (7.3) of the entropy and the equality $p_{\mathbf{x}}(x) = p_{\mathbf{y}}[g(x)]|\det g'(x)|$.

The above result shows that the mutual information between a set of random vectors is unchanged when each vector is undergone an invertible transformation: $I(\mathbf{y}_1, \ldots, \mathbf{y}_K) = I[g_1(\mathbf{y}_1), \ldots, g_K(\mathbf{y}_K)]$ where $g_1, \ldots, g_K$ are invertible differentiable maps. This is compatible with the fact that independent random vectors remain independent under such transformation.

A particular and most interesting application of Lemma 7.1 is where g is linear. It shows that the entropy is translation invariant and scale equi-variant: $H(ax+b) = H(x) + \log|a|$ for any real random variable x and real number a, b. More generally $H(\mathbf{A}\mathbf{x}+\mathbf{b}) = H(\mathbf{x}) + \log|\det \mathbf{A}|$ for any random vector $\mathbf{x}$, matrix $\mathbf{A}$ and vector $\mathbf{b}$.

7.2.2 The Mutual Information Criterion

Consider the linear instantaneous mixture model $\mathbf{x}(n) = \mathbf{A}\mathbf{s}(n)$ where $\mathbf{x}(n)$ and $\mathbf{s}(n)$ are the vectors of observed mixtures and of sources (at time n) and $\mathbf{A}$ is a matrix. We assume that there are a same number K of mixtures as the number of sources and the $K \times K$ matrix $\mathbf{A}$ is invertible. The sources may be recovered as the components of $\mathbf{y}(n) = \mathbf{B}\mathbf{x}(n)$, $\mathbf{B}$ being a matrix chosen such that these components are as independent as is possible. Adopting the mutual information as a measure of dependence, one is led to the following criterion

$$C(\mathbf{B}) = I(y_1, \ldots, y_k) = \sum_{k=1}^{K} H(y_k) - \log|\det \mathbf{B}| - H(\mathbf{x}), \tag{7.4}$$

where y_k is the kth component of $\mathbf{y}$. Here the time index n is dropped as the entropies does not depend on time, because of the (assumed) stationarity of the sequence $\{\mathbf{s}(n)\}$. Note that the term $H(\mathbf{x})$ does not depend on $\mathbf{B}$, hence will be dropped. Thus the above criterion only involves marginal entropies and not joint entropy.

Scale and Permutation Ambiguities

From the scale equi-variance of the entropy, it is easily seen that the criterion is unchanged when $\mathbf{B}$ is left multiplied by a diagonal matrix. It is also clearly unchanged when the rows of $\mathbf{B}$ are permuted. Thus the sources can be recovered only up to a scaling and a permutation. These well known ambiguities actually are a consequence of the fact that separation is based only on the independence assumption of the sources.

7.2.3 Estimation of the Criterion

The criterion (7.4) is only theoretical since it involves the unknown entropies $H(y_k)$. In practice, it has to be replaced by an estimator, obtained by replacing these unknown entropies by their estimates. Thus the main problem is to estimate the entropy.

There exist several entropy estimators in the literature. As we will need many entropy estimations in the course of minimizing the estimated criterion, it is of interest to use an estimator which can be computed quickly. With this consideration in mind, we introduce an estimator based on the kernel density estimator and the discretization of the formula (7.3) for the entropy. Indeed, by the *binning* technique and the use of *cardinal spline* kernel (described below), the kernel density estimate *over a regular grid* can be computed rather quickly.

The kernel density estimator of a random variable y based on a sample $y(1), \dots, y(N)$ is given by [16]

$$\widehat{p}_y(u) = \frac{1}{Nh\widehat{\sigma}_y} \sum_{n=1}^{N} \kappa\Big[\frac{u - y(n)}{h\widehat{\sigma}_y}\Big], \tag{7.5}$$

where κ is a density function called kernel, h is a bandwidth parameter and $\widehat{\sigma}_y^2 = N^{-1}\sum_{n=1}^{N}[y(n)-\bar{y}]^2$ is the sample variance of y, $\bar{y} = N^{-1}\sum_{n=1}^{N} y(n)$ being the sample mean. A natural estimate of $H(y)$ could be $-\sum_{m=-\infty}^{\infty} \widehat{p}_y(m\delta)\delta \log \widehat{p}_y(m\delta)$ where δ is the discretization step. This estimator however has the defect that it is neither translation invariant nor scale equi-variant, while the entropy is. For this reason, we will apply this estimator to the normalized variable $y' = (y-\bar{y})/\widehat{\sigma}_y$ instead and add the term $\log \widehat{\sigma}_y$, which yields the entropy estimator of y:

$$\widehat{H}(y) = -\sum_{m=-\infty}^{\infty} \widehat{p}_{y'}(m\delta)\delta \log[\widehat{p}_{y'}(m\delta)] + \log \widehat{\sigma}_y\,,$$

where $\widehat{p}_{y'}$ is the kernel density estimate of $y' = (y-\bar{y})/\widehat{\sigma}_y$. From (7.5), one gets

$$\widehat{p}_{y'}(m\delta) = \frac{1}{Nh} \sum_{n=1}^{N} \kappa\Big[\frac{m\delta\widehat{\sigma}_y + \bar{y} - y(n)}{h\widehat{\sigma}_y}\Big] = p_y(m\delta\widehat{\sigma}_y + \bar{y})\widehat{\sigma}_y\,. \tag{7.6}$$

Therefore

$$\widehat{H}(y) = -\sum_{m=-\infty}^{\infty} \widehat{p}_y(m\delta\widehat{\sigma}_y + \bar{y})\delta\widehat{\sigma}_y \log[\widehat{p}_y(m\delta\widehat{\sigma}_y + \bar{y})] + \log\widehat{\sigma}_y\Big[1 - \sum_{m=-\infty}^{\infty} \widehat{p}_{y'}(m\delta)\delta\Big].$$

Thus $\widehat{H}(y)$ is almost the discretization of $-\int \widehat{p}_y(u)\log[\widehat{p}_y(u)]du$ with a grid spacing $\delta\widehat{\sigma}_y$ and origin $\bar{y}$. The last term in the above formula is small as $\sum_{m=-\infty}^{\infty} \widehat{p}_{y'}(m\delta)\delta \approx \int p_{y'}(y)du = 1$. In fact, we shall limit ourselves to kernel κ having *the partition of unity property relative to* δ/h, in the sense that $\sum_{m=-\infty}^{\infty} \kappa(u+m\delta/h)\delta/h = 1$, $\forall u$, then $\widehat{H}(y)$ is exactly the discretization of $-\int \widehat{p}_y(u)\log[\widehat{p}_y(u)]du$ since by (7.6)

$$\sum_{m=-\infty}^{\infty} p_{y'}(m\delta)\delta = \frac{1}{N}\sum_{n=1}^{N}\sum_{m=-\infty}^{\infty} \kappa\Big[\frac{\bar{y}-y(n)}{h\widehat{\sigma}_y} + m\frac{\delta}{h}\Big]\frac{\delta}{h} = 1.$$

Put $\widehat{\pi}_y(m) = \widehat{p}_y(m\delta\widehat{\sigma}_y + \bar{y})\delta\widehat{\sigma}_y$, the $\widehat{\pi}_y(m)$ sum to 1 and can be interpreted as probabilities. Then

$$\widehat{H}(y) = -\sum_{m=-\infty}^{\infty} \widehat{\pi}_y(m)\log[\widehat{\pi}_y(m)] + \log(\delta\widehat{\sigma}_y), \tag{7.7}$$

which is the sum of a discrete entropy plus the term $\log(\delta\widehat{\sigma}_y)$ to take into account of the discretization.

7.2.3.1 The Cardinal Spline Kernel

We shall use the cardinal splines as kernel. The cardinal spline of order r is defined as the indicator function of the interval $[0,1)$ convolved with itself r times and denoted by $\mathbb{1}_{[0,1)}^{\star r}$. This function is symmetric around $r/2$, so we shift it by $r/2$, that is we take as kernel the function $\mathbb{1}_{[0,1)}^{\star r}(\cdot + r/2) = \mathbb{1}_{[-1/2,1/2)}^{\star r}$, the indicator function of the interval $[1/2,1/2)$ convolved with itself r times. This family of kernels is not unusual. For $r = 1$, one gets the rectangular kernel which yields the histogram. For $r = 2$, one gets triangular kernel which is also common. For $r \to \infty$, the kernel tends to a Gaussian kernel (since sum of independent uniform $[0,1]$ variables tends to be Gaussian). Even with r as low as 3, the function $\mathbb{1}_{[-1/2,1/2)}^{\star r}$ already has a very similar shape to the Gaussian density. Further, such kernel has many advantages:

1. It is very easy to compute (for small r) and has compact support, therefore the density estimate has finite support and the summation in (7.7) is a finite.
2. It possesses the partition of unity relative to 1, hence to $1/p$ for all integers p:

$$\frac{1}{p}\sum_{m=-\infty}^{\infty} \mathbb{1}_{[0,1)}^{\star r}\Big(u + \frac{m}{p}\Big) = \frac{1}{p}\sum_{k=0}^{p-1}\sum_{m=-\infty}^{\infty} \mathbb{1}_{[0,1)}^{\star r}\Big(u + \frac{k}{p} + m\Big) = 1.$$

3. The scaled kernel with an integer scale p, can be expressed as a convex combination of the same but integer shifted kernel:

$$\frac{1}{p}\mathbb{1}_{[0,1)}^{\star r}\Big(\frac{u}{p}\Big) = \sum_{l=0}^{pr-r} c_l^{p,r}\mathbb{1}_{[0,1)}^{\star r}(u-l),$$

where $c_l^{p,r}, l = 0, \ldots, pr - r$ are the coefficients of the polynomial $(p^{-1}\sum_{l=0}^{p-1} z^l)^r$. This result can be obtained by noting that $\mathbb{1}_{(0,1]}(\cdot/p) = (\sum_{l=0}^{p-1} B^l)\mathbb{1}_{(0,1]}$ where B is the backward shift operator: $Bf = f(\cdot - 1)$ and that this operator commutes with the convolution.

The last property is very interesting as it permits the use of binning technique.

7.2.3.2 Binning Technique

The binning technique [7, 14] has been introduced to speed up the computation of the kernel density estimate *over a regular grid*. It has been viewed as an approximation to such estimate. However, we have shown in [10] that there is no approximation if the kernel can be expressed as combination of easy to compute shifted kernels and the grid size is right. Although the idea is very general, it is best applied to the case of the kernel $\kappa = \mathbb{1}^{\star r}_{[1/2,1/2)}$, which from the property 3 above can be expressed as

$$\kappa(u) = p \sum_{l=0}^{pr-r} c_l^{p,r} \mathbb{1}^{\star r}_{(0,1]}\Big(pu + \frac{pr}{2} - l\Big).$$

Taking $\delta = h/p$, a *sub multiple of h*, then the formula (7.5) for $\widehat{p}_y$ becomes

$$\widehat{p}_y(u) = \frac{1}{\delta\widehat{\sigma}_y} \sum_{l=0}^{L} c_l^{p,r} \frac{1}{N} \sum_{n=1}^{N} \mathbb{1}^{\star r}_{[0,1)}\Big[\frac{u - y(n)}{\delta\widehat{\sigma}_y} + \frac{pr}{2} - l\Big].$$

We are actually interested in computing not the density estimate but the probabilities:

$$\tilde{\pi}_y\Big(m - \frac{pr}{2}\Big) = \widehat{p}_y\Big[\Big(m - \frac{pr}{2}\Big)\delta\widehat{\sigma}_y + \bar{y}\Big]\delta\widehat{\sigma}_y = \sum_{l=0}^{pr-r} c_l^{p,r} \frac{1}{N} \sum_{n=1}^{N} \mathbb{1}^{\star r}_{[0,1)}\Big[m - \frac{y(n) - \bar{y}}{\delta\widehat{\sigma}_y} - l\Big].$$

This computation can be done in two steps:

(1) Compute $\tilde{\pi}_y(m - r/2) = (1/N)\sum_{n=1}^{N} \mathbb{1}^{\star r}_{[0,1)}\{m - [y(n) - \bar{y}]/\delta\widehat{\sigma}_y\}, \forall m$

(2) Compute $\tilde{\pi}_y(m - pr/2) = \sum_{l=0}^{pr-r} c_l^{p,r} \tilde{\pi}_y(m - r/2 - l)$

Step (2) can be implemented as the output of a (smoothing) filter with impulse response $\{c_l^{p,r}\}$ and input $\{\tilde{\pi}_Y(m - r/2)\}$. It is fast for small $pr - r$. Note that larger p correspond to smaller discretization step, but there is no significant gain in accuracy by taking a large p. In fact our experience shows that one can even take $p = 1$ (i.e. $\delta = h$) thus eliminate step (2) altogether.

Step (1) is fast for small r. For $r = 1$ it consists simply in counting the fraction of time $\tilde{y}(n) = [y(n) - \bar{y}]/\delta\widehat{\sigma}_y$ falls into the interval $(m - 1, m]$. For general $r > 1$, this step can be implemented as follows. We note that $\mathbb{1}^{\star r}_{[0,1)}$ vanishes outside $(0, r)$, hence the term $\mathbb{1}^{\star r}_{[0,1)}[m - \tilde{y}(n)]$ is nonzero only for $m = \lfloor\tilde{y}(n)\rfloor + 1, \ldots, \lfloor\tilde{y}(n)\rfloor + r$ where $\lfloor x\rfloor$ denotes the largest signed integer not exceeding x. Thus the computation of $\tilde{\pi}_Y(m - r/2)$, $m \ldots, -1, 0, 1, \ldots,$ can be done as follows:

(a) Initialize $\tilde{\pi}_y(m - r/2) = 0, \forall m$
(b) For $n = 1, \ldots, N$ add $\mathbb{1}^{\star r}_{(0,1]}[\lfloor \tilde{y}(n) \rfloor + j - \tilde{y}(n)]/N$ to $\tilde{\pi}_Y(\lfloor \tilde{y}(n) \rfloor + j - r/2)$, $j = 1, \ldots, r$

We recommend taking $r = 3$. For $r = 2$, the density estimate has jumps in its derivative so that $\widehat{H}(\mathbf{bx})$, where $\mathbf{b}$ is a row vector, is not differentiable with respect to $\mathbf{b}$. Since the computation cost increases with r with no proven gain in performance, $r = 3$ is a good choice. The function $\mathbb{1}^{\star 3}_{[0,1)}$ is given explicitly below:

$$\mathbb{1}^{\star 3}_{(0,1]}(x) = \begin{cases} x^2/2, & 0 \le x \le 1, \\ 1/2 + (x-1)(2-x), & 1 \le x \le 2, \\ (3-x)^2/2, & 2 \le x \le 3, \\ 0, & \text{otherwise}. \end{cases}$$

Note that for r odd, the probabilities $\tilde{\pi}_y$ are computed only at the center-grid points $m + 1/2$, $m = \ldots, -1, 0, 1, \ldots$ and the same is true for $\widehat{\pi}_y$ if p is odd. In this case, the definition (7.7) of the entropy estimator should be slightly changed by changing m to $m + 1/2$.

7.2.4 *Minimization of the Criterion*

To minimize the criterion $\widehat{C}$ defined as in (7.4) but with $\widehat{H}$ in place of H, a gradient descent, or better a quasi Newton, iteration may be used. In both cases, one will need the gradient of the criterion. Instead of the gradient, we will work with the relative gradient $\widehat{C}'$ of $\widehat{C}$, defined via the first order Taylor expansion: $\widehat{C}(\mathbf{B} + \mathscr{E}\mathbf{B}) = \widehat{C}(\mathbf{B}) + \mathrm{tr}[\widehat{C}'(\mathbf{B})\mathscr{E}^{\mathrm{T}}] +$ higher order terms in $\mathscr{E}$, where tr denotes the trace. The concept of relative gradient has been introduced in [2].

7.2.4.1 Relative Gradient of the Criterion

Introducing the estimated score function $\widehat{\psi}_y$ of y defined by

$$\widehat{\psi}_y[y(n)] = N \frac{\partial \widehat{H}(y)}{\partial y(n)}. \tag{7.8}$$

Strictly speaking, this defines $\widehat{\psi}_y$ only at the data points but the definition can be naturally extended to any point and in any case only the values of $\widehat{\psi}_y$ at the data points will be needed. Then, denoting by $\mathscr{E}_{k\cdot}$ the kth row of $\mathscr{E}$:

$$\widehat{H}(y_k + \mathscr{E}_{k\cdot}\mathbf{y}) = \widehat{H}(y_k) + \mathscr{E}_{k\cdot}\Big\{\frac{1}{N}\sum_{n=1}^{N} \widehat{\psi}_{y_k}[y_k(n)]\mathbf{y}(n)\Big\} + \cdots.$$

Therefore, from (7.4) and noting that $\log\det(\mathbf{I} + \mathscr{E}) = \mathrm{tr}(\mathscr{E}) + \cdots$, one gets

$$\widehat{C}'(\mathbf{B}) = \frac{1}{N}\sum_{n=1}^{N} \widehat{\psi}_{\mathbf{y}}[\mathbf{y}(n)]\mathbf{y}^{\mathrm{T}} - \mathbf{I}, \tag{7.9}$$

where

$$\widehat{\psi}_{\mathbf{y}}[\mathbf{y}(n)] = [\widehat{\psi}_{y_1}[y_1(n)] \ \cdots \ \widehat{\psi}_{y_K}[y_K(n)]]^{\mathrm{T}}. \tag{7.10}$$

The above estimated score function $\widehat{\psi}_y$ has been introduced in [10]. We will see that this is indeed an estimator of the score function of y, defined as the derivative ψ_y of $-\log p_y$. We first derive explicitly an expression for $\widehat{\psi}_y$. By (7.7), we have

$$\widehat{\psi}_y[y(n)] = -N \sum_{m=-\infty}^{\infty} \log[\widehat{\pi}_y(m)] \frac{\partial \widehat{\pi}_y(m)}{\partial y(n)} + \frac{N}{\widehat{\sigma}_y}\frac{\partial \widehat{\sigma}_y}{\partial y(n)}$$

since $\sum_{m=-\infty}^{\infty} \widehat{\pi}_y(m) = 1$ and hence its partial derivatives vanish. Since $\widehat{\sigma}_y^2$ can be written as $(1/N)\sum_{n=1}^{N} y^2(n) - \bar{y}^2$ and $\partial \bar{y}/\partial y(n) = 1/N$:

$$\frac{N}{\widehat{\sigma}_y}\frac{\partial \widehat{\sigma}_y}{\partial y(n)} = \frac{N}{2\widehat{\sigma}_y^2}\frac{\partial \widehat{\sigma}_y^2}{\partial y(n)} = \frac{y(n) - \bar{y}}{\widehat{\sigma}_y^2}.$$

Further, from (7.6) and the above formula and putting $y' = (y - \bar{y})/\widehat{\sigma}_y$ for short,

$$N\frac{\partial \widehat{\pi}_y(m)}{\partial y(n)} = -\frac{\delta}{h^2\sigma_y}\left\{\kappa'\left[\frac{m\delta - y'(n)}{h}\right] - \frac{1}{N}\sum_{l=1}^{N} \kappa'\left[\frac{m\delta - y'(l)}{h}\right]\left[1 + y'(l)\frac{y(n) - \bar{y}}{\sigma_y}\right]\right\},$$

where κ' is the derivative of κ. Therefore

$$\widehat{\psi}_y(u) = \tilde{\psi}_y(u) - \frac{1}{N}\sum_{l=1}^{N} \tilde{\psi}_y[y(l)] + \left\{1 - \frac{1}{N}\sum_{l=1}^{N} \tilde{\psi}_y[y(l)][y(l) - \bar{y}]\right\}\frac{u - \bar{y}}{\widehat{\sigma}_y^2}, \tag{7.11}$$

where

$$\tilde{\psi}_y(u) = \frac{\delta}{h^2\widehat{\sigma}_y}\sum_{m=-\infty}^{\infty} \log[\widehat{\pi}(m)]\kappa'\left(\frac{m\delta\widehat{\sigma}_y + \bar{y} - u}{h\sigma_y}\right).$$

Interpretation: One may write $\tilde{\psi}_y$ as

$$\tilde{\psi}_y(u) = -\frac{d}{du}\sum_{m=-\infty}^{\infty} \log[\widehat{p}_y(m\delta\widehat{\sigma}_y + \bar{y})]\frac{\delta}{h}\kappa\left(\frac{m\delta\widehat{\sigma}_y + \bar{y} - u}{h\sigma_y}\right)$$

assuming that κ possesses the partition of unity property relative to δ/h so that $\sum_{m=-\infty}^{\infty} \kappa'(m\delta/h + u) = 0$, $\forall u$. Thus $\tilde{\psi}_y$ appears as the derivative of a doubly smoothed estimator of $-\log p_y$, hence it is an estimate of the score function ψ_y of y. Our estimated score function $\widehat{\psi}_y$ is a corrected form of $\tilde{\psi}_y$, obtained by a centering and adding a linear function, so that it satisfies

$$\frac{1}{N}\sum_{n=1}^{N} \widehat{\psi}_y[y(n)] = 0, \qquad \frac{1}{N}\sum_{n=1}^{N} \widehat{\psi}_y[y(n)]y(n) = 1. \tag{7.12}$$

These equalities mimic that of the score function: $\mathbf{E}[\psi_y(y)] = 0$ and $\mathbf{E}[\psi_y(y)y] = 1$ (which can be obtained by integration by parts). Further, the second equality entails that $\widehat{C}'$ given in (7.9) has zero diagonal.

In the particular case where $\kappa = \mathbb{1}^{\star 3}_{[-1/2,1/2)}$ and $\delta = h$, a direct calculation yields

$$\tilde{\psi}_y[\bar{y} + (m+u)\delta\widehat{\sigma}_y] = \{(1-u)\log[\tilde{\pi}_y(m-\tfrac{1}{2})] + (2u-1)\log[\tilde{\pi}_y(m+\tfrac{1}{2})] - u\log[\tilde{\pi}_y(m+\tfrac{3}{2})]\}/(\delta\widehat{\sigma}_y),$$

for integer m and real $u \in [0,1)$. One can see that $\tilde{\psi}_y$ is the linear interpolation of the function taking the value $\{\log[\widehat{\pi}_y(m-\frac{1}{2})] - \log[\widehat{\pi}_y(m+\frac{1}{2})]\}/(\delta\widehat{\sigma}_y)$ at $\bar{y} + m\delta\widehat{\sigma}_y$.

7.2.4.2 Approximate Relative Hessian of the Criterion

The concept of relative Hessian is similar to that of relative gradient. Specifically, the relative Hessian of the criterion $\widehat{C}(\mathbf{B})$ is composed of the second derivatives of $\widehat{C}(\mathbf{B} + \mathscr{E}\mathbf{B})$ with respect to the elements of $\mathscr{E}$ at $\mathscr{E} = \mathbf{0}$. From the definition of relative gradient, the ordinary derivarive of $\widehat{C}(\mathbf{B} + \mathscr{E}\mathbf{B})$ with respect to $\mathscr{E}$ is $\widehat{C}'(\mathbf{B} + \mathscr{E}\mathbf{B})(\mathbf{I} + \mathscr{E})^{-1\mathrm{T}}$. Thus one has to compute its derivative with respect to $\mathscr{E}$, at $\mathscr{E} = \mathbf{0}$.

To proceed, we introduce three kinds of approximation: (1) sample average is the same as expectation, (2) the estimated score function $\widehat{\psi}_{y_i}$ is the same score function ψ_{y_i} (or a weaker assumption that it is a non random function) and (3) the variables $y_1, \ldots, y_K$ are independent. The point (3) is justified if $\mathbf{B}$ is close to a matrix which separate the sources. Under (1)–(3), $\widehat{C}'(\mathbf{B})$ would be negligible hence the derivative of $\widehat{C}'(\mathbf{B} + \mathscr{E}\mathbf{B})(\mathbf{I} + \mathscr{E})^{-1\mathrm{T}}$ at $\mathscr{E} = \mathbf{0}$ reduces to that of $\widehat{C}'(\mathbf{B} + \mathscr{E}\mathbf{B})$ at $\mathscr{E} = \mathbf{0}$. Since the diagonal elements of $\widehat{C}'(\mathbf{B} + \mathscr{E}\mathbf{B})$ vanish for all $\mathscr{E}$, we conclude that the second derivative of $\widehat{C}(\mathbf{B} + \mathscr{E}\mathbf{B})$ with respect to a diagonal element of $\mathscr{E}$ and *any* other element of $\mathscr{E}$ at $\mathscr{E} = \mathbf{0}$ also vanishes (approximately). Thus, we need only to compute the second derivatives of $\widehat{C}(\mathbf{B} + \mathscr{E}\mathbf{B})$ with respect to the off diagonal elements of $\mathscr{E}$ at $\mathscr{E} = \mathbf{0}$, which are (approximately) the derivatives of the off diagonal elements of $\widehat{C}'(\mathbf{B} + \mathscr{E}\mathbf{B})$ with respect to those of $\mathscr{E}$ at $\mathscr{E} = \mathbf{0}$.

The ij element of $\widehat{C}'(\mathbf{B} + \mathscr{E}\mathbf{B})$ for $i \neq j$, is

$$\frac{1}{N}\sum_{n=1}^{N} \widehat{\psi}_{y_i + \mathscr{E}_{i\cdot}\mathbf{y}}[y_i(n) + \mathscr{E}_{i\cdot}\mathbf{y}(n)][y_j(n) + \mathscr{E}_{j\cdot}\mathbf{y}(n)].$$

This expression does not depend on $\mathscr{E}_{k\cdot}$ for $k \notin \{i,j\}$, its derivative with respect to $\mathscr{E}_{jk}$, $k \neq j$ at $\mathscr{E} = 0$ is $N^{-1}\sum_{n=1}^{N} \widehat{\psi}_{y_i}[y_i(n)]y_k(n)$, and with respect to $\mathscr{E}_{ik}$, $k \neq i$ at $\mathscr{E} = 0$ is

$$\frac{1}{N}\sum_{n=1}^{N} \widehat{\psi}'_{y_i}[y_i(n)]y_k(n)y_j(n) + \frac{1}{N}\sum_{n=1}^{N} \frac{\partial \widehat{\psi}_{y_i + \mathscr{E}_{ik}y_k}}{\partial \mathscr{E}_{ik}}[y_i(n)]\bigg|_{\mathscr{E}_{ik}=0} y_j(n),$$

where $\widehat{\psi}'_{y_i}$ is the derivative of $\widehat{\psi}_{y_i}$.

Under (1)–(3) above, the second term in the above expression can be neglected since y_i is independent with y_j. Likewise, the first term can be neglected unless $k = j$ in which case it is approximately $\{N^{-1}\sum_{n=1}^{N}\widehat{\psi}'_{y_i}[y_i(n)]\}\widehat{\sigma}^2_{y_j}$. The term $N^{-1}\sum_{n=1}^{N}\widehat{\psi}_{y_i}[y_i(n)]y_k(n)$ is also negligible unless $k = i$ in which case it equals 1 (by (7.12)). With these approximations, the Hessian matrix is block diagonal with 2×2 diagonal blocks

$$\begin{bmatrix} \{N^{-1}\sum_{n=1}^{N}\widehat{\psi}'_{y_i}[y_i(n)]\}\widehat{\sigma}^2_{y_j} & 1 \\ 1 & \{N^{-1}\sum_{n=1}^{N}\widehat{\psi}'_{y_j}[y_j(n)]\widehat{\sigma}^2_{y_i}\} \end{bmatrix}$$

corresponding to the derivatives with respect to the pair $\mathscr{E}_{ij}, \mathscr{E}_{ji}$.

The above matrix however may not be positive. A positive definite approximation to the Hessian is desirable. For this purpose, note that $N^{-1}\sum_{n=1}^{N}\widehat{\psi}'_{y_i}[y_i(n)]$ is an estimator of $\mathbf{E}[\psi'_{y_i}(y_i)]$ but we know (by integration by parts) that this expectation equals $\mathbf{E}[\psi^2_{y_i}(y_i)] = J(y_i)$ the Fisher information of y_i. Therefore, we may approximate the Hessian by a block diagonal matrix with diagonal blocks

$$\begin{bmatrix} \widehat{J}(y_i)\widehat{\sigma}^2_{y_j} & 1 \\ 1 & \widehat{J}(y_j)\widehat{\sigma}^2_{y_i} \end{bmatrix}, \quad \text{where} \quad \widehat{J}(y) = \frac{1}{N}\sum_{n=1}^{N}\widehat{\psi}^2_y[y(n)].$$

By the Schwartz inequality and the equalities (7.12), $\widehat{J}(y)\widehat{\sigma}^2_y > 1$ (unless $\widehat{\psi}_y$ is a linear function). Hence the above approximate Hessian is positive definite.

7.2.4.3 The Quasi Newton Algorithm

The Newton algorithm consists in replacing the function to be minimized by its second order expansion around the current point, then minimizing this expansion to obtain the new point. In the quasi Newton algorithm , the second order terms are replaced by some approximations. In our case, as we work with relative gradient and Hessian, we consider the (approximate) expansion

$$C(\mathbf{B}+\mathscr{E}\mathbf{B}) \approx C(\mathbf{B}) + \sum_{i \neq j}\left\{\mathscr{E}_{ij}\frac{1}{N}\sum_{n=1}^{N}\widehat{\psi}_{y_i}[y_i(n)]y_j(n) + \frac{1}{2}[\mathscr{E}^2_{ij}\widehat{J}(y_i)\widehat{\sigma}^2_{y_j} + \mathscr{E}_{ij}\mathscr{E}_{ji}]\right\}.$$

The minimization of the above right-hand side yields $\mathscr{E}_{ij}$, $j \neq i$. Then $\mathbf{B}$ is changed to $\mathbf{B}+\mathscr{E}\mathbf{B}$ for the next step of the algorithm (the diagonal of $\mathscr{E}$ may be put to zero since it essentially affects the scale of the extracted sources). Explicitly

$$\begin{bmatrix} \mathscr{E}_{ij} \\ \mathscr{E}_{ji} \end{bmatrix} = -\begin{bmatrix} \widehat{J}(y_i)\widehat{\sigma}^2_{y_j} & 1 \\ 1 & \widehat{J}(y_j)\widehat{\sigma}^2_{y_i} \end{bmatrix}^{-1} \begin{bmatrix} N^{-1}\sum_{n=1}^{N}\widehat{\psi}_{y_i}[y_i(n)]y_j(n) \\ N^{-1}\sum_{n=1}^{N}\widehat{\psi}_{y_j}[y_j(n)]y_i(n) \end{bmatrix}$$

There is no guarantee that the criterion is decreased at each step. In practice, if this is not the case, one reduces the step size sufficiently to obtain a decrease of the criterion. This is always possible as long as the approximate Hessian matrix is positive definite. A method for reducing the step size is described in [13, p. 384].

7.2.5 Statistical Performance

A first question concerning the performance of the above method is that whether it correctly extracts the sources *up to a permutation and a scaling* under ideal conditions where an infinite (very large) number of observations obeying exactly the mixture model (7.13) is available.[2] In this ideal situation the estimated criterion $\widehat{C}(\mathbf{B})$ may be regarded as identical to the theoretical criterion $C(\mathbf{B})$ defined in (7.4). Therefore the question is whether $-C(\cdot)$ is a contrast in the sense of Comon [3], that is it attains it global maximum if and only if $\mathbf{BA}$ is a product of a nonsingular diagonal and a permutation matrices. The answer is yes if there is no more than one Gaussian sources, since by the Darmois Theorem [3], no two set of linear combinations of independent variables can be independent unless at least two of these variables are Gaussian.

The above discussion requires implicitly that $h \to 0$ tends to 0 as $N \to \infty$ (hence the same holds for δ as the ratio δ/h is kept fixed) so that $\widehat{C}(\mathbf{B}) \to C(\mathbf{B})$. In fact we shall show below that h can be kept fixed (or tends to a non zero limit) provided that it (or its limit) is small enough. In practice, as only a finite but large sample is available, this means that one can choose h moderately small. Then $\widehat{C}(\mathbf{B})$ may be considered as the same as $C_{h,\delta}(\mathbf{B})$, its limit as $N \to \infty$ and h, δ fixed. We now show that $-C_{h,\delta}(\cdot)$ is still a contrast provided that h, δ are small enough. By the scale and permutation invariance of $C_{h,\delta}(\cdot)$, it suffices to shows that $C_{h,\delta}(\mathbf{A}^{-1}) < C_{h,\delta}(\mathbf{GA}^{-1})$ for any matrix $\mathbf{G}$ with rows of unit norm which is not a permutation matrix.

The crucial point is that $\mathbf{A}^{-1}$ *is a stationary point of* $-C_{h,\delta}(\cdot)$. It may be expected and indeed may shown that the relative gradient of $C_{h,\delta}(\mathbf{B})$ is the limit (as $N \to \infty$) of the relative gradient $\widehat{C}'(\mathbf{B})$ of $C(\mathbf{B})$, which from the result of Sect. 7.2.4.1 is $\mathbf{E}[\psi_{h,\delta,\mathbf{y}}(\mathbf{y})\mathbf{y}^{\mathrm{T}}] - \mathbf{I}$ where $\psi_{h,\delta,\mathbf{y}}(\mathbf{y})$ is the limit of $\widehat{\psi}_{\mathbf{y}}(\mathbf{y})$. The important point is that $\widehat{\psi}_{h,\delta,y_k}(y_k)$, the kth component of $\psi_{h,\delta,\mathbf{y}}(\mathbf{y})$, is a non random function of y_k satisfying $\mathbf{E}[\psi_{h,\delta,y_k}(y_k)] = 0$, this equality being obtained from the first equality of (7.12). For $\mathbf{B} = \mathbf{A}^{-1}$, $\mathbf{y} = \mathbf{s}$, hence the off diagonal elements of $\mathbf{E}[\psi_{h,\delta,\mathbf{y}}(\mathbf{y})\mathbf{y}^{\mathrm{T}}]$ vanish. The diagonal elements of $\mathbf{E}[\psi_{h,\delta,\mathbf{y}}(\mathbf{y})\mathbf{y}^{\mathrm{T}}] - \mathbf{I}$ also vanish since those of $\widehat{C}'(\mathbf{B})$ vanish for all N.

Consider now the Taylor expansion of $C_{h,\delta}(\mathbf{GA}^{-1})$ with respect to $\mathbf{G} \approx \mathbf{I}$:

$$C_{h,\delta}(\mathbf{GA}^{-1}) = C_{h,\delta}(\mathbf{A}^{-1}) + \frac{1}{2} \sum_{i \neq j, k \neq l} C''_{h,\delta,ij,kl} G_{ij} G_{kl} + \cdots,$$

where $C''_{h,\delta,ij,kl}$ are the elements of the relative Hessian of $C_{h,\delta}(\cdot)$ at $\mathbf{A}^{-1}$ and G_{ij} the elements of $\mathbf{G}$. It may be shown that as $(h,\delta) \to (0,0)$, $C''_{h,\delta,ij,kl} \to C''_{ij,kl}$,

[2] We already know that these ambiguities cannot be avoided in a blind context.

the elements of the relative Hessian of $C(\cdot)$ at $\mathbf{A}^{-1}$. Since $-C(\cdot)$ is a contrast, $\sum_{i\neq j,k\neq l} C''_{ij,kl} G_{ij} G_{kl} > 0$ for any non diagonal matrix $\mathbf{G}$, hence so is $\sum_{i\neq j,k\neq l} C''_{h,\delta,ij,kl} G_{ij} G_{kl}$ if h,δ are small enough. Thus for such h,δ, $C_{h,\delta}(\mathbf{A}^{-1}) < C_{h,\delta}(\mathbf{G}\mathbf{A}^{-1})$ for $\mathbf{G}$ close enough to $\mathbf{I}$ but distinct from $\mathbf{I}$. Since the criterion is permutation invariant, this inequality also holds for $\mathbf{G}$ is an open neighborhood $\mathscr{P}$ of all permutation matrices but is not a permutation matrix. Further, since $-C(\cdot)$ is a contrast $C(\mathbf{A}^{-1}) < C(\mathbf{G}\mathbf{A}^{-1})$ for all $\mathbf{G} \in \mathscr{G}$, the set of matrices not in $\mathscr{P}$ with rows of unit norm. Since this set is compact, the infimum of $C(\mathbf{G}\mathbf{A}^{-1})$ over it is attained, implying that there is a $\varepsilon > 0$ such that $C(\mathbf{G}\mathbf{A}^{-1}) > C(\mathbf{A}^{-1}) - \varepsilon$ for all $\mathbf{G} \in \mathscr{G}$. Finally it may be shown that $C_{h,\delta}(\cdot)$ converge to $C(\cdot)$ *uniformly on any compact set* as $(h,\delta) \to (0,0)$, so that for h,δ small enough $C_{h,\delta}(\mathbf{A}) < C_{h,\delta}(\mathbf{G}\mathbf{A}^{-1})$ for all $\mathbf{G} \in \mathscr{G}$. □

It is of interest to point out the similarity of the present approach with the quasi maximum likelihood approach [12]. The last approach estimates the separation matrix by maximum likelihood using hypothetical (*a priori*) densities for the sources, which leads to the estimating equation

$$\frac{1}{N}\sum_{n=1}^{N} \psi_i[y_j(n)]y_j(n) = 0, \qquad 1 \le i \neq j \le K,$$

where ψ_i is the score function of the hypothetical density of the ith source. From the result of Sect. 7.2.4.1, the mutual information approach also leads to a similar estimating equation, only that ψ_i is replaced with $\widehat{\psi}_{y_i}$, an estimator of the true score function ψ_{y_i}. Assuming without loss of generality that the sources $s_i(n)$ have unit variance and the separation matrix has been permuted and rescaled so that $y_i(n)$ has unit variance and estimates $s_i(n)$, the paper [12] provides formula for the asymptotic distribution of the off diagonal elements $G_{ij}, i \neq j$ of the global matrix $\mathbf{G} = \mathbf{BA}$. (These elements may be viewed as the contamination coefficients, as $y_i(n) = (1 - \sum_{j\neq i} G_{ij}^2)^{1/2} s_i(n) + \sum_{j\neq i} G_{ij} s_j(n)$ and $(1 - \sum_{j\neq i} G_{ij}^2)^{1/2} \approx 1$.) Specifically, the asymptotic covariance matrix of the $G_{ij}, i \neq j$ is block diagonal with 2×2 diagonal block

$$\frac{1}{N}\begin{bmatrix} \lambda_i & 1 \\ 1 & \lambda_j \end{bmatrix}^{-1} \begin{bmatrix} \rho_i^{-2} & 1 \\ 1 & \rho_j^{-2} \end{bmatrix} \begin{bmatrix} \lambda_i & 1 \\ 1 & \lambda_j \end{bmatrix}^{-1},$$

where $\lambda_i = \mathbf{E}[\psi_i'(s_i)]\sigma_{s_i}^2 / \mathbf{E}[\psi_i(s_i)s_i]$ and $\rho_i = \mathbf{E}[\psi_i(s_i)s_i]/\{\mathbf{E}[\psi_i^2(s_i)]\sigma_{s_i}^2\}^{1/2}$, $\sigma_{s_i}^2$ being three variance of y_i (which is 1 but is included here to show that the above parameters are dimensionless). It was also shown in [12] that this covariance matrix is smallest when the ψ_i are chosen (miraculously) equal to the (unknown) true score functions of the sources. In the mutual information approach, the above result should still holds with ψ_i replaced by ψ_{h,δ,s_i} if h,δ are kept fixed as $N \to \infty$ or ψ_{s_i} if $(h,\delta) \to 0$ as $N \to \infty$ *with a slow enough rate*. Thus the mutual information approach is (assymtotically) optimal. This is not an coincidence as there is a strong link between this approach and the maximum likelihood method [1].

The fact that h,δ can be fixed show that the algorithm is quite robust with respect to their choice: a bad choice simply result in a small loss of performance. To chose

h one may follow the simple approach in [16], which minimizes the (asymptotic) integrated mean square errors of the density estimate *under Gaussian assumption.* This assumption is of course false, but it provides a convenient and easy way to choose h. As the true density is often less smooth than the Gaussian density, it tends to over smooth the estimated density by providing a larger h than it should be. But this is fine with us, as we have seen that h can even be fixed. For the third cardinal kernel, this approach yields $h = 2.107683N^{-1/5}$. (The formula $h = 1.06N^{-1/5}$ in [16] corresponds to the Gaussian kernel.)

For $\psi_i = \psi_{s_i}$, $\lambda_i = \rho_i^{-2}$ hence the asymptotic covariance matrix of $G_{ij}, i \neq j$ is $1/N$ times the inverse of the block diagonal matrix with 2×2 diagonal blocks having ρ_i^{-2} and ρ_j^{-2} on the diagonal and 1 elsewhere. This matrix is also approximately the inverse of the Hessian matrix of the criterion (see Sect. 7.2.4.2). Note that ρ_i is the correlation between $\psi_i(s_i)$ and s_i and hence always less than 1, unless ψ_i is linear. For $\psi_i = \psi_{s_i}$, the parameter $\lambda_i = \rho_i^{-2}$ measures the nonlinearity of the score function of s_i as it takes the smallest value 1 if and only if this function is linear. The non linearity of the score function translates into the non Gaussianity of the variable since only Gaussian variables have linear score function. The more non Gaussian the variables $s_i(n)$ and $s_j(n)$, the higher ρ_i^{-2} and ρ_j^{-2} and from the above results, the smaller the optimal asymptotic covariance matrix of G_{ij}, G_{ji}.

7.2.6 An Example of Simulation

As an example, we have generate three sources of length 512. the first is a sum of two sine waves of different frequencies and phase, the second is a triangular signal and the third is a Gaussian autoregressive moving average (ARMA) process of AR coefficients 1.4, −0.9 and MA coefficient 0.2. The two sine waves have amplitude 1 and the triangular signal has amplitude $\sqrt{3}$, so that the first two sources have average power 1. The innovation variance of the ARMA source is also determined so that this source has unit variance. The three sources are plotted in the upper left part of Fig. 7.1. They are mixed by the mixing matrix

$$\mathbf{A} = \begin{bmatrix} 1 & 1 & 1 \\ 1 & -1 & 1 \\ 0.5 & 1 & 1 \end{bmatrix},$$

yielding three mixtures which are plotted in the upper right part of Fig. 7.1.

We apply the algorithm to these mixtures starting with the identity matrix as the initial separating matrix **B**. The result is displayed in the lower right part of Fig. 7.1, in which the nine elements of the global matrix **BA** is plotted against the iteration number. It can be seen that the algorithm converges well, to a product of a permutation and a diagonal matrix. All except three elements of the global matrix converge to nearly zero. Those which don't are the (1,2), (2,1) and (3,3) elements, as indicated in the graph. This shows that the first two extracted sources are permuted.

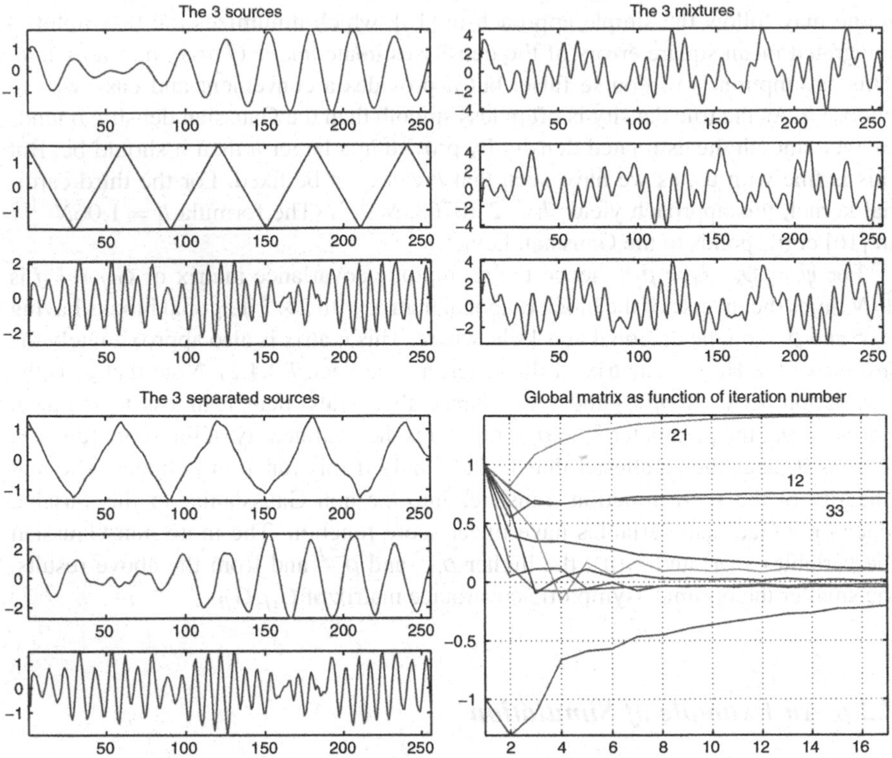

Fig. 7.1 Results of application of the algorithm: the three sources (*upper left*), the three mixtures (*upper right*), the three separated sources (*lower left*) and the nine elements of the global matrix as function of the iteration number (*lower right*)

This is also clearly visible in the lower part of Fig. 7.1 where the three separated sources are plotted. One can see that the graph of them are very similar to that of the original sources.

For entropy estimation, we choose a discretization step δ equal to the smoothing parameter h and estimate the density with the third cardinal spline as kernel and the parameter h chosen as described in Sect. 7.2.5. Even with such simple choice of h and a rather coarse discretization step the algorithm works quite well. Note that the ARMA source is Gaussian, but the algorithm still works since one Gaussian source is allowed.

7.3 Multichannel Blind Deconvolution

In the case of convolutive mixture the independence assumption of the sources alone can only permit to separate them up to a convolution, since the outputs of the separation remain independent if each of them is filtered. Thus further assumption may be introduced to eliminate this ambiguity. A common assumption is the temporal independence of the source signals. In this case the problem may be called blind separation-deconvolution or multichannel blind deconvolution, since it reduces to the well known blind deconvolution problem when there are only one source and one sensor. Specifically, we consider the model

$$\mathbf{x}(n) = \sum_{j=-\infty}^{\infty} \mathbf{A}(j)\mathbf{s}(n-j) = (\mathbf{A}\star\mathbf{s})(n), \tag{7.13}$$

where again the observed vector $\mathbf{x}(n)$ and source vector $\mathbf{s}(n)$ have the same dimension K and $\{\mathbf{A}(j)\}$ is a sequence of $K\times K$ matrices and $\star$ denotes the convolution. The goal is to recover the sources from the observations, using only the independence of the source sequences (blind separation) and their temporal independence (blind deconvolution).

Naturally, the recovered source sequences $\{y_k(n)\}, k = 1,\dots,K$ are obtained through a deconvolution matrix filter

$$\mathbf{y}(n) = \sum_{j=-\infty}^{\infty} \mathbf{B}(j)\mathbf{x}(n-j) = (\mathbf{B}\star\mathbf{y})(n), \tag{7.14}$$

where $\mathbf{y}(n) = [y_1(n) \;\; \cdots \;\; y_K(n)]^{\mathrm{T}}$ and $\{\mathbf{B}(j)\}$ is a sequence of matrices to be determined.

Note 7.1. In order that the sources can be recovered at all, the matrix sequence $\{\mathbf{A}(j)\}$ is assumed to be invertible in the convolutive sense, that is there exists an inverse sequence $\{\mathbf{A}^{\dagger}(j)\}$ such that $\sum_{k=-\infty}^{\infty}\mathbf{A}^{\dagger}(j-k)\mathbf{A}(k) = \mathbf{I}$ if $k = 0$, $= \mathbf{0}$ otherwise. Naturally, the sequence $\{\mathbf{B}(j)\}$ will also be restricted to be invertible.

7.3.1 The Mutual Information Criterion

Our objective is to make the recovered sources as independent as it is possible, both temporally and among themselves. Therefore, we adopt as criterion the average mutual information between the variables $y_k(n)$, $n = 1,2,\dots,k = 1,\dots,K$:

$$\lim_{L\to\infty}\frac{1}{L}I[y_1(1),\dots,y_1(L),\dots,y_K(1),\dots,y_K(L)],$$

From the relation between the mutual information and entropy and noting that $H[y_k(n)]$ does not depend on n by stationarity and hence will be denoted simply by $H(y_k)$, the above criterion can be written as $\sum_{k=1}^{K} H(y_k) - H[\mathbf{y}(\cdot)]$ where

$$H[\mathbf{y}(\cdot)] = \lim_{L\to\infty}\frac{1}{L}H[\mathbf{y}(1),\dots,\mathbf{y}_k(L)]$$

is (by definition) the entropy rate of the $\{\mathbf{y}(n)\}$ sequence [4]. The limit in this definition is known to exists for stationary the sequences [4], hence the criterion is well defined.

It has been proved in [9] that if the sequence $\{\mathbf{y}(n)\}$ is related to $\{\mathbf{x}(n)\}$ via a filter as in (7.14), then

$$H[\mathbf{y}(\cdot)] = H[\mathbf{x}(\cdot)] + \int_0^{2\pi} \log\det|\mathbf{B}(\omega)| \frac{d\omega}{2\pi},$$

where $\mathbf{B}(\omega) = \sum_{j=-\infty}^{\infty} \mathbf{B}(j)e^{-ij\omega}$ is the Fourier transform (or Fourier series) of the sequence $\{\mathbf{B}(j)\}$. *For simplicity of notation, a same symbol will be used for a sequence in the time domain and its Fourier transform in the frequency domain, the ambiguity is lifted by using a Greek letter (such as ω) or a non integer for the frequency variable and a Roman letter (such as j) for the time index.* Therefore, one is led to the mutual information criterion

$$C[\mathbf{B}(\cdot)] = \sum_{k=1}^{K} H(y_k) - \int_0^{2\pi} \log\det|\mathbf{B}(\omega)| \frac{d\omega}{2\pi}. \tag{7.15}$$

Note that in this formula, $|\mathbf{B}(\omega)|$ may be replaced by $\mathbf{B}(\omega)$ since it equals the complex conjugate of $\mathbf{B}(2\pi-\omega)$ and hence the imaginary part of $\log\det\mathbf{B}(\omega)$ integrates to zero.

7.3.2 The Practical Criterion

As before, the theoretical criterion (7.15) should in practice be replaced an estimated criterion, obtained by replacing the unknown entropies by their estimators. The resulting criterion is denoted by $\widehat{C}(\mathbf{B})$.

We have already discussed entropy estimation in Sect. 7.2.3. There is however an issue here with such estimation: the sequence $\{\mathbf{y}(n)\}$ as defined in (7.14) may require the knowledge of the whole sequence $\{\mathbf{x}(n)\}$ but actually only a finite section, $\mathbf{x}(0),\ldots,\mathbf{x}(N-1)$ say, of it is observed. To solve this problem, we adopt the simple approach which consists in extending the observed sequence periodically, and thus redefine

$$\mathbf{y}(n) = \sum_{j=-\infty}^{\infty} \mathbf{B}(j)\mathbf{x}(n-j \bmod N). \tag{7.16}$$

Other approaches are possible but the above is most convenient as it is well adapted to the use of the discrete Fourier transform (DFT). Indeed, recall that the DFT of the sequence $\mathbf{x}(0),\ldots,\mathbf{x}(N-1)$ is defined as $d_{\mathbf{x}}(2\pi k/N) = \sum_{n=0}^{N-1} \mathbf{x}(n)e^{-i2\pi kn/N}$ and similarly for other sequences, it can be checked that the DFT of the redefined $\mathbf{y}(0),\ldots,\mathbf{y}(N-1)$ is related to that of $\mathbf{x}(0),\ldots,\mathbf{x}(N-1)$ by $d_{\mathbf{y}}(2\pi k/N) = \mathbf{B}(2\pi k/N)d_{\mathbf{x}}(2\pi k/N)$. Thus, one may compute the $\mathbf{y}(n)$ by computing the DFT of $\mathbf{x}(0),\ldots,\mathbf{x}(N-1)$, multiplying it by $\mathbf{B}(2\pi k/N)$ then doing an inverse DFT.

In the case where the sequence $\{\mathbf{B}(j)\}$ is of finite support, $[-m_1, m_2]$ say, a better approach would be to limit oneself to those $\mathbf{y}$ which can be computed from the observations. We assume in this case, referred to as the FIR case, that $\mathbf{x}(-m_2), \ldots, \mathbf{x}(N+m_1-1)$ are observed so that $\mathbf{y}(n)$ can be computed according to (7.14) for $n = 0, \ldots, N-1$. Thus in all cases, $\widehat{H}(y_k)$ will be based on the observations $y_k(0), \ldots, y_k(N-1)$.

7.3.3 Minimization of the Criterion

Since the sequence $\{\mathbf{B}(j)\}$ is potentially of infinite support, in a parametric setup we need to assume that it can be parameterized by some vector parameter θ. In the FIR case, this vector could be composed of all elements of the matrices $\mathbf{B}(j)$ for j in the support of sequence $\{\mathbf{B}(j)\}$.

We begin by deriving the gradient and approximate Hessian of the criterion $\widehat{C}(\mathbf{B})$. For simplicity, we assume that sources have zero means. Otherwise one would subtract the data its sample mean, which does not change the criterion since it is translation invariant.

7.3.3.1 Gradient of the Criterion

From the definition (7.8) of the estimated score function:

$$\frac{\partial}{\partial \theta_\mu} \sum_{k=1}^{K} \widehat{H}(y_k) = \operatorname{tr}\left\{ \frac{1}{N} \sum_{n=0}^{N-1} \widehat{\psi}_{\mathbf{y}}^{\mathrm{T}}[\mathbf{y}(n)] \frac{\partial \mathbf{y}(n)}{\partial \theta_\mu} \right\}, \tag{7.17}$$

where $\widehat{\psi}_{\mathbf{y}}[\mathbf{y}(n)]$ is defined in (7.10). From (7.16)

$$\frac{\partial \mathbf{y}(n)}{\partial \theta_\mu} = \sum_{j=-\infty}^{\infty} \frac{\partial \mathbf{B}(j)}{\partial \theta_\mu} \mathbf{x}(n - j \bmod N) \tag{7.18}$$

and noting that $\partial \log \det \mathbf{B}(\omega)/\partial \theta_\mu = \operatorname{tr}\{[\partial \mathbf{B}(\omega)/\partial \theta_\mu] \mathbf{B}(\omega)^{-1}\}$, one gets

$$\begin{aligned} \frac{\partial \widehat{C}[\mathbf{B}(\cdot)]}{\partial \theta_\mu} &= \operatorname{tr}\left\{ \sum_j \frac{\partial \mathbf{B}(j)}{\partial \theta_\mu} \frac{1}{N} \sum_{n=0}^{N-1} \mathbf{x}(n - j \bmod N)\, \widehat{\psi}_{\mathbf{y}}^{\mathrm{T}}[\mathbf{y}(n)] \right. \\ &\quad \left. - \int_0^{2\pi} \frac{\partial \mathbf{B}(\omega)}{\partial \theta_\mu} \mathbf{B}(\omega)^{-1} \frac{d\omega}{2\pi} \right\}. \end{aligned} \tag{7.19}$$

The last term in this formula involves an integral, which for numerical calculation may be replaced by a Riemann sum, based on the points $0, 2\pi/N, \ldots, 2\pi(N-1)/N$, for example.

The above formula is most useful in the FIR case where $\mathbf{y}$ is computed by (7.14) and hence the modulo N disappears and the summation over j is finite. In the case where $\mathbf{y}$ is computed by (7.16), it is not convenient since it involves infinite summation. In this case, a more useful alternative formula in the Fourier domain may be used. We note that the right-hand side of (7.17) involves scalar products of sequences and by the (discrete) Parseval equality, the scalar product of two sequences equals that of their DFT divided by their length. By a similar calculation as that near the end of Sect. 7.3.2, the DFT of the sequence $\partial\mathbf{y}(0)/\partial\theta_\mu,\ldots,\partial\mathbf{y}(N-1)/\partial\theta_\mu$ can be obtained as $[\partial\mathbf{B}(2\pi k/N)/\partial\theta_\mu]d_\mathbf{x}(2\pi k/N)$ where $d_\mathbf{x}(2\pi k/N)$ is the DFT of $\mathbf{x}(0),\ldots,\mathbf{x}(N-1)$. Therefore

$$\frac{\partial}{\partial\theta_\mu}\sum_{k=1}^{K}\widehat{H}(y_k) = \operatorname{tr}\left\{\frac{1}{N}\sum_{k=0}^{N-1}\frac{\partial\mathbf{B}(2\pi k/N)}{\partial\theta_\mu}S_{\mathbf{x}\widehat{\psi}_\mathbf{y}}\left(\frac{2\pi k}{N}\right)\right\},$$

where

$$S_{\mathbf{x}\widehat{\psi}_\mathbf{y}}\left(\frac{2\pi k}{N}\right) = \frac{1}{N}d_\mathbf{x}\left(\frac{2\pi k}{N}\right)d^T_{\widehat{\psi}_\mathbf{y}}\left(2\pi\frac{N-k}{N}\right)$$

is the cross periodogram between the sequences $\mathbf{x}(0),\ldots,\mathbf{x}(N-1)$ and $\widehat{\psi}_\mathbf{y}(0),\ldots,$ $\widehat{\psi}_\mathbf{y}(N-1)$. Since one also has $d_\mathbf{y}(2\pi k/N) = \mathbf{B}(2\pi k/N)d_\mathbf{x}(2\pi k/N)$, one can write $S_{\mathbf{x}\widehat{\psi}_\mathbf{y}}(2\pi k/N) = \mathbf{B}(2\pi k/N)^{-1}S_{\mathbf{y}\widehat{\psi}_\mathbf{y}}(2\pi k/N)$, $S_{\mathbf{y}\widehat{\psi}_\mathbf{y}}$ being defined similarly as $S_{\mathbf{x}\widehat{\psi}_\mathbf{y}}$. Finally, replacing the integral in the formula (7.19) by the Riemann sum based on $0, 2\pi/N,\ldots,2\pi-2\pi/N$, one gets

$$\frac{\partial\widehat{C}[\mathbf{B}(\cdot)]}{\partial\theta_\mu} = \operatorname{tr}\left\{\frac{1}{N}\sum_{k=0}^{N-1}\frac{\partial\mathbf{B}(2\pi k/N)}{\partial\theta_\mu}\mathbf{B}\left(\frac{2\pi k}{N}\right)^{-1}\left[S_{\mathbf{y}\widehat{\psi}_\mathbf{y}}\left(\frac{2\pi k}{N}\right)-\mathbf{I}\right]\right\}. \quad (7.20)$$

7.3.3.2 Approximate Hessian of the Criterion

For simplicity, we limit ourselves to the FIR case. The case where the sequence $\{\mathbf{B}(n)\}$ has infinite support may be handled by truncating this sequence to have support length a small fraction of the sample length N (we are interested in approximating the Hessian for large N). Thus we have to compute the derivative of $\partial\widehat{C}[\mathbf{B}(\cdot)]/\partial\theta_\mu$ given in (7.19), with respect to θ_ν.

The derivative of the first sum in (7.19) can be split into $S_1+S_2+S_3$ where

$$S_1 = \sum_j \frac{\partial^2\mathbf{B}(j)}{\partial\theta_\mu\partial\theta_\nu}\frac{1}{N}\sum_{n=1}^{N}\mathbf{x}(n-j)\widehat{\psi}_\mathbf{y}[\mathbf{y}(n)], \quad S_2 = \frac{1}{N}\sum_{n=1}^{N}\frac{\partial\mathbf{y}(n)}{\partial\theta_\mu}\frac{\partial\mathbf{y}^\mathrm{T}(n)}{\partial\theta_\nu}\operatorname{diag}\widehat{\psi}'_\mathbf{y}[\mathbf{y}(n)],$$

$'$ denoting the derivative and diag the operator which builds a diagonal matrix from its vector argument, and $S_3 = N^{-1}\sum_{n=1}^{N}[\partial\mathbf{y}(n)/\partial\theta_\mu](\partial\widehat{\psi}^\mathrm{T}_\mathbf{y}/\partial\theta_\nu)[\mathbf{y}(n)]$.

The derivation of an approximation to the Hessian is done in several steps.

Step 1: Replacing sample average by expectation and $\widehat{\psi}_\mathbf{y}$ *by the theoretical score* $\psi_\mathbf{y}$. This leads to

$$S_1 \approx \sum_j \frac{\partial^2 \mathbf{B}(j)}{\partial\theta_\mu \partial\theta_\nu} \mathbf{E}\{\mathbf{x}(-j)\psi_\mathbf{y}[\mathbf{y}(0)]\}, \qquad S_2 \approx \mathbf{E}\left\{\frac{\partial \mathbf{y}(0)}{\partial\theta_\mu}\frac{\partial \mathbf{y}^\mathrm{T}(0)}{\partial\theta_\nu} \mathrm{diag}\,\psi'_\mathbf{y}[\mathbf{y}(0)]\right\}$$

and $S_3 \approx \mathbf{E}\{[\partial\mathbf{y}(0)/\partial\theta_\mu](\partial\widehat{\psi}_\mathbf{y}^\mathrm{T}/\partial\theta_\nu)[\mathbf{y}(0)]\}$.

Step 2: Simplifying the expectations by treating the variables $\{y_k(n),\ n = \ldots, -1, 0, 1, \ldots, k = 1, \ldots, K\}$ *as independent.* For this purpose, we first express other random variables in terms of the $\mathbf{y}(n)$. Let $\{\mathbf{B}^\dagger(j)\}$ be the inverse sequence of $\{\mathbf{B}(j)\}$ in the convolutive sense, that is $\sum_k \mathbf{B}^\dagger(k)\mathbf{B}(j-k) = \mathbf{I}$ if $j = 0$, $= \mathbf{I}$ otherwise. Then

$$\mathbf{x}(-j) = \sum_k \mathbf{B}^\dagger(k)\mathbf{y}(-j-k).$$

Further, from (7.18), $\partial\mathbf{y}(0)/\partial\theta_\mu = \sum_j [\partial\mathbf{B}(j)/\partial\theta_\mu]\mathbf{B}^\dagger(k)\mathbf{y}(-j-k)$, therefore, putting

$$\mathbf{C}_\mu(l) = \sum_j [\partial\mathbf{B}(j)/\partial\theta_\mu]\mathbf{B}^\dagger(l-j)$$

gives

$$\partial\mathbf{y}(0)/\partial\theta_\mu = \sum_l \mathbf{C}(l)\mathbf{y}(-l).$$

From the above expression for $\mathbf{x}(-j)$:

$$S_1 \approx \sum_j \sum_k \frac{\partial^2 \mathbf{B}(j)}{\partial\theta_\mu \partial\theta_\nu} \mathbf{B}^\dagger(k)\mathbf{E}\{\mathbf{y}(-j-k)\psi_\mathbf{y}[\mathbf{y}(0)]\} = \sum_j \frac{\partial^2 \mathbf{B}(j)}{\partial\theta_\mu \partial\theta_\nu}\mathbf{B}^\dagger(-j),$$

the last equality follows from the independence of the $\{y_k(n)\}$ and the property $\mathbf{E}[y_k\psi_{y_k}(y_k)] = 1$ of the score function.

From the above expression for $\partial\mathbf{y}(0)/\partial\theta_\mu$:

$$S_2 \approx \sum_l \sum_m \mathbf{E}\{\mathbf{C}_\mu(l)\mathbf{y}(-l)\mathbf{y}^\mathrm{T}(-m)\mathbf{C}_\mu^T(m)\mathrm{diag}\,\psi'_\mathbf{y}[\mathbf{y}(0)]\}.$$

By the temporal independence of the sequence $\{\mathbf{y}(n)\}$, the above sum reduces to

$$\sum_{l\neq 0} \mathbf{C}_\mu(l)\mathbf{E}(\mathbf{y}\mathbf{y}^\mathrm{T})\mathbf{C}_\mu^T(l)\mathbf{E}[\mathrm{diag}\,\psi'_\mathbf{y}(\mathbf{y})] + \mathbf{E}[\mathbf{C}_\mu(0)\mathbf{y}\mathbf{y}^\mathrm{T}\mathbf{C}_\mu^T(0)\mathrm{diag}\,\psi'_\mathbf{y}(\mathbf{y})],$$

the time index of the variable $\mathbf{y}$ being omitted because it is the same everywhere and the expectation does not depend on it. The components of $\mathbf{y}$ being independent, the trace of the above sum, after a somewhat tedious algebra, reduces to

$$\sum_{j,k=1}^K \sum_l \mathbf{E}(y_j^2)\mathbf{E}[\psi'_{Y_k}(y_k)]C_{\mu,kj}(l)C_{\nu,kj}(l) + \sum_{k=1}^K \mathrm{cov}[y_k^2, \psi'_{y_k}(y_k)]C_{\mu,kk}(0)C_{\nu,kk}(0),$$

where $C_{\mu,kj}$ denotes the k,j element of the matrix $\mathbf{C}_\mu$ and $\mathrm{cov}[y_k^2, \psi'_{y_k}(y_k)] = \mathbf{E}[y_k^2\psi'_{y_k}(y_k)] - \mathbf{E}(y_k^2)\mathbf{E}[\psi'_{y_k}(y_k)]$.

Again from the above expression for $\partial \mathbf{y}(0)/\partial \theta_\mu$:

$$S_3 \approx \sum_l \mathbf{C}_\mu(l)\mathbf{E}\{\mathbf{y}(-l)(\partial \psi_\mathbf{y}^\mathrm{T}/\partial \theta_\nu)[\mathbf{y}(0)]\}.$$

The expectation term the above sum, by the independence of the $y_k(n)$, vanishes if $l \neq 0$ and is a diagonal matrix if $l = 0$. Thus the trace of the above sum reduces to $\sum_{k=1}^K C_{\mu,kk}(0)\mathbf{E}[y_k(\partial \psi_{y_k}/\partial \theta_\nu)(y_k)]$. But the score function ψ_{y_k} satisfies $\mathbf{E}[y_k \psi_{y_k}(y_k)] = 1$, hence by taking derivative:

$$\mathbf{E}[y_k(\partial \psi_{y_k}/\partial \theta_\nu)(y_k)] = -\mathbf{E}\{[\psi_{y_k}(y_k) + y_k \psi'_{y_k}(y_k)]\partial y_k/\partial \theta_\nu\}.$$

Using again the expression for $\partial y_k/\partial \theta_\nu$ and the independence of the variables $\{y_k(n)\}$, the above right hand reduces to $-C_{\nu,kk}(0)\mathbf{E}[y_k \psi_{y_k}(y_k) + y_k^2 \psi'_{y_k}(y_k)]$. Thus

$$\mathrm{tr}(S_3) \approx -\sum_{k=1}^K C_{\mu,kk}(0)C_{\nu,kk}(0)\{1 + \mathbf{E}[y_k^2 \psi'_{y_k}(y_k)]\}.$$

Step 3: Derivative of the second sum in (7.19): It is

$$\int_0^{2\pi} \frac{\partial^2 \mathbf{B}(\omega)}{\partial \theta_\mu \partial \theta_\nu} \mathbf{B}(\omega)^{-1} \frac{d\omega}{2\pi} - \int_0^{2\pi} \frac{\partial \mathbf{B}(\omega)}{\partial \theta_\mu} \mathbf{B}(\omega)^{-1} \frac{\partial \mathbf{B}(\omega)}{\partial \theta_\nu} \mathbf{B}(\omega)^{-1} \frac{d\omega}{2\pi},$$

The first integral, by the Parseval equality is precisely $\sum_j [\partial^2 \mathbf{B}(j)/\partial \theta_\mu \partial \theta_\nu]\mathbf{B}^\dagger(-j)$, which is no other than the approximation to S_1 obtained before.

Step 4: Approximation to the Hessian. From the above results,

$$\frac{\partial^2 \widehat{C}[\mathbf{B}(\cdot)]}{\partial \theta_\mu \partial \theta_\nu} \approx \sum_{j,k=1}^K \left\{ \int_0^{2\pi} \left[\frac{\partial \mathbf{B}(\omega)}{\partial \theta_\mu} \mathbf{B}(\omega)^{-1} \right]_{kj} \left[\frac{\partial \mathbf{B}(\omega)}{\partial \theta_\nu} \mathbf{B}(\omega)^{-1} \right]_{jk} \frac{d\omega}{2\pi} \right.$$

$$\left. \sum_l \sigma_{y_j}^2 \mathbf{E}[\psi'_{Y_k}(y_k)] C_{\mu,kj}(l) C_{\nu,kj}(l) \right\} - \sum_{k=1}^K \{\sigma_{y_k}^2 \mathbf{E}[\psi'_{y_k}(y_k)] + 1\} C_{\mu,kk}(0) C_{\nu,kk}(0),$$

$[\mathbf{M}]_{kj}$ denoting the k,j element of the matrix $\mathbf{M}$.

The above formula involves the sequence $\{\mathbf{C}_\mu(l)\}$ which is no other than the convolution of the sequence $\{\partial \mathbf{B}/\partial \partial_\mu\}$ with the sequence $\{\mathbf{B}^\dagger(j)\}$, hence its Fourier transforms equal $[\partial \mathbf{B}(\omega)/\partial \theta_\mu]\mathbf{B}^\dagger(\omega)$. But $\mathbf{B}^\dagger(\omega) = \mathbf{B}(\omega)^{-1}$, therefore,

$$\mathbf{C}_\mu(0) = \int_0^{2\pi} \frac{\partial \mathbf{B}(\omega)}{\partial \theta_\mu} \mathbf{B}(\omega)^{-1} \frac{d\omega}{2\pi} \stackrel{\text{def}}{=} \overline{\frac{\partial \mathbf{B}}{\partial \theta_\mu} \mathbf{B}^{-1}}$$

(the over bar means average). Further, by the Parseval equality

$$\sum_l C_{\mu,kj}(l) C_{\nu,kj}(l) = \int_0^{2\pi} \left[\frac{\partial \mathbf{B}(\omega)}{\partial \theta_\mu} \mathbf{B}(\omega)^{-1} \right]_{kj} \left[\frac{\partial \mathbf{B}(-\omega)}{\partial \theta_\mu} \mathbf{B}(-\omega)^{-1} \right]_{kj} \frac{d\omega}{2\pi}.$$

Therefore one may rewrite the approximate Hessian as

$$\frac{\partial^2 \widehat{C}[\mathbf{B}(\cdot)]}{\partial\theta_\mu \partial\theta_\nu} \approx \sum_{j,k=1}^{K} \int_0^{2\pi} \left\{ \sigma_{y_k}^2 \mathbf{E}[\psi'_{y_k}(y_k)] \left[\frac{\partial \mathbf{B}(\omega)}{\partial\theta_\mu}\mathbf{B}(\omega)^{-1}\right]_{kj} \left[\frac{\partial \mathbf{B}(-\omega)}{\partial\theta_\nu}\mathbf{B}(-\omega)^{-1}\right]_{jk} + \left[\frac{\partial \mathbf{B}(\omega)}{\partial\theta_\mu}\mathbf{B}(\omega)^{-1}\right]_{kj} \left[\frac{\partial \mathbf{B}(\omega)}{\partial\theta_\nu}\mathbf{B}(\omega)^{-1}\right]_{jk} \right\} \frac{d\omega}{2\pi}$$

$$- \sum_{k=1}^{K} \{\sigma_{y_k}^2 \mathbf{E}[\psi'_{y_k}(y_k)] + 1\} \left[\overline{\frac{\partial \mathbf{B}}{\partial\theta_\mu}\mathbf{B}^{-1}}\right]_{kk} \left[\overline{\frac{\partial \mathbf{B}}{\partial\theta_\mu}\mathbf{B}^{-1}}\right]_{kk}. \tag{7.21}$$

7.3.3.3 The Quasi Newton Algorithm (in the Pure Deconvolution Case)

One can construct a quasi Newton algorithm from the formulas (7.19) or (7.20) for the gradient and (7.21) for the approximate Hessian. However, in the multichannel case formula (7.21) is too complex to yield an useful algorithm. It is in fact much simpler and no more expensive to compute numerically an approximate Hessian from the computed gradients at previous and current iterations. This is actually done in the BFGS (Broyden–Fletcher–Goldfarb–Shanno) algorithm [5, 13], which behaves like a quasi Newton algorithm and yet requires only a formula for the gradient.

In the unichannel case however, through a clever parameterization, formula (7.21) yields an approximate Hessian matrix of block diagonal or diagonal form. This results in a particularly simple and fast quasi Newton algorithm. In this case the matrix **B** is a scalar, now denoted by B, and thus $[\partial B(\omega)/\partial\theta_\mu]B(\omega)^{-1}$ is no other than $\partial \log B(\omega)/\partial\theta_\mu$. Therefore, formula (7.21) for the approximate Hessian becomes:

$$\frac{\partial 2\widehat{C}(B)}{\partial\theta_\mu \partial\theta_\nu} \approx \int_0^{2\pi} \left[\lambda \frac{\partial \log B(\omega)}{\partial\theta_\mu}\frac{\partial \log B(-\omega)}{\partial\theta_\nu} + \frac{\partial \log B(\omega)}{\partial\theta_\mu}\frac{\partial \log B(\omega)}{\partial\theta_\nu}\right]\frac{d\omega}{2\pi} - (\lambda+1)\overline{\frac{\partial \log B}{\partial\theta_\mu}}\,\overline{\frac{\partial \log B}{\partial\theta_\mu}}$$

where $\lambda = \sigma_y^2 \mathbf{E}[\psi'_y(y)]$ (since **y** is now a scalar, it is written as y).

Further, by separating the real and imaginary parts of $\log B(\omega)$ and noting that $\overline{\partial \log B/\partial\theta_\mu}$ is real hence equal its real part:

$$\frac{\partial^2 \widehat{C}(B)}{\partial\theta_\mu \partial\theta_\nu} \approx \int_0^{2\pi} \left\{ (\lambda+1)\left[\Re\frac{\partial \log B(\omega)}{\partial\theta_\mu} - \overline{\frac{\partial \log B}{\partial\theta_\mu}}\right]\left[\Re\frac{\partial \log B(\omega)}{\partial\theta_\nu} - \overline{\frac{\partial \log B}{\partial\theta_\nu}}\right] + (\lambda-1)\left[\Im\frac{\partial \log B(\omega)}{\partial\theta_\mu}\Im\frac{\partial \log B(\omega)}{\partial\theta_\nu}\right] \right\}\frac{d\omega}{2\pi},$$

where $\Re$ and $\Im$ denote the real and imaginary parts. This formula shows that it is of interest to parameterize the real and imaginary parts of $\log B(\omega)$ (or equivalently the modulus and phase $\mathbf{B}(\omega)$) by two different sets of parameters, since then there is a decoupling between these sets in the quasi Newton algorithm. Recall that the imaginary part of the logarithm of a complex number z is its phase (or its argument) which we shall denote by $\arg(z)$. Let $|B(\omega)]$ and $\arg[B(\omega)]$ be parameterized by two independent vector parameter θ_R and θ_I, then the Hessian of $\widehat{C}(B)$ is approximately block diagonal with diagonal blocks $(\lambda+1)\mathbf{H}_R$ and $(\lambda-1)\mathbf{H}_I$ where

$$\mathbf{H}_R = \int_0^{2\pi} \left[\frac{\partial \log|B(\omega)|}{\partial \theta_R} - \overline{\frac{\partial|\log B|}{\partial \theta_R}}\right]\left[\frac{\partial|\log B(\omega)|}{\partial \theta_R} - \overline{\frac{\partial|\log B|}{\partial \theta_R}}\right]^{\mathrm{T}} \frac{d\omega}{2\pi}, \tag{7.22}$$

$$\mathbf{H}_I = \int_0^{2\pi} \frac{\partial \arg[B(\omega)]}{\partial \theta_I}\left\{\frac{\partial \arg[B(\omega)]}{\partial \theta_I}\right\}^{\mathrm{T}} \frac{d\omega}{2\pi}, \tag{7.23}$$

$\partial/\partial\theta_R$ and $\partial/\partial\theta_R$ denoting the vectors of partial derivatives with respect to the components of θ_R and to the components of θ_I, respectively.

A simple interesting parameterization for $\log B(\omega)$ is

$$\log B(\omega) = \theta_0 + \sum_{\mu=1}^{m} [\theta_\mu \cos(\mu\omega) + i\theta_{m+\mu}\sin(\mu\omega)]. \tag{7.24}$$

The parameters $\theta_0, \ldots, \theta_m$ and $\theta_{m+1}, \ldots, \theta_{2m}$ specify the real and imaginary parts of $\log B(\omega)$ respectively. These parts are even and odd functions respectively, hence the use of the cosine and sine functions to represent them. The parameter θ_0 corresponds to the scale of the source and cannot be estimated. We may and will put it to 0. The mutual information criterion then reduces to $\widehat{C}(B) = \widehat{H}(By)$ since $\int \log B(\omega) d\omega/(2\pi) = 0$.

For the above parameterization, the formula (7.20) for the gradient becomes

$$\frac{\partial \widehat{C}(B)}{\partial \theta_\mu} = \begin{cases} (r_\mu + r_{-\mu})/2, & \mu = 1, \ldots, m\,, \\ r_{\mu-m} - r_{m-\mu})/2, & \mu = m+1, \ldots, 2m\,, \end{cases}$$

where

$$r_\mu = \frac{1}{N}\sum_{k=1}^{N-1} e^{2\pi\mu k/N} S_{y\widehat{\psi}_y}\left(\frac{2\pi k}{N}\right) = \frac{1}{N}\sum_{n=0}^{N-1} y(n+\mu)\widehat{\psi}_y(n). \tag{7.25}$$

The last equality follows from the fact that the DFT transforms (circular) convolution into products. (Note that $y(n)$ here is computed by (7.16) and hence is periodic of period N.) The above formula shows that r_μ are the circular cross covariances between $y(0), \ldots, y(n-1)$ and $\widehat{\psi}_y|y(0)], \ldots, \widehat{\psi}_y[y(N-1)]$. Of more interest is the fact that the matrices $\mathbf{H}_R$ and $\mathbf{H}_I$ corresponding to the vector parameters $\theta_R = [\theta_1 \ \cdots \theta_m]^{\mathrm{T}}$ and $\theta_I = [\theta_{m+1} \ \cdots \theta_{2m}]^{\mathrm{T}}$ and given by (7.22) and (7.23), reduce to one half the identity matrix. Thus, the quasi Newton algorithm reduces to the fixed point iteration

$$\theta_\mu \leftarrow \theta_\mu - \begin{cases} (r_\mu + r_{-\mu})/(\widehat{\lambda}+1), & \mu = 1, \ldots, m\,, \\ (r_{\mu-m} - r_{m-\mu})/(\widehat{\lambda}-1), & \mu = m+1, \ldots, 2m\,, \end{cases}$$

where $\widehat{\lambda}$ is the current estimate of λ. One may take $\widehat{\lambda} = \widehat{\sigma}_y^2 N^{-1} \sum_{n=0}^{N-1} \psi_y'[y(n)]$ but there is no guarantee that this estimator is always greater than 1. Therefore, as in Sect. 7.2.4.2, we take $\widehat{\lambda} = \widehat{\sigma}_y^2 \widehat{J}(y)$ which is always greater than 1.

One can also work with the parameters $\vartheta_\mu = (\theta_\mu - \theta_{\mu+m})/2$ and $\vartheta_{-\mu} = (\theta_\mu + \theta_{\mu+m})/2$, $\mu = 1,\ldots,m$, so that one has $\log B(\omega) = \theta_0 + \sum_{0<|\mu|\le m} \vartheta_\mu e^{-i\mu\omega}$. The quasi Newton algorithm for these parameters is then

$$\vartheta_\mu \mapsto \vartheta_\mu - \frac{\widehat{\lambda} r_{-\mu} - r_\mu}{\widehat{\lambda}^2 - 1}, \qquad 0 < |\mu| \le m. \tag{7.26}$$

Note that since $d_Y(2\pi k/N) = B(2\pi k/N) d_X(2\pi k/N)$, one can update it directly as

$$d_Y\Big(\frac{2\pi k}{N}\Big) \mapsto d_Y\Big(\frac{2\pi k}{N}\Big) \exp\Big[-\sum_{0<|\mu|\le m} \frac{\widehat{\lambda} r_{-\mu} - r_\mu}{\widehat{\lambda}^2 - 1} e^{-i2\pi\mu k/N}\Big]. \tag{7.27}$$

The criterion $\widehat{H}(y)$ may be computed after each quasi Newton step to check if it decreases. If it does not, the step size should be reduced so that it is so (by the method described in [13, p. 384], for example). The estimation algorithm is summarized in Table 7.1.

It is of interest to note that the parameter $\theta_1,\ldots,\theta_m$ specifying the real part of $\log B(\omega)$ may be estimated directly. Indeed, since the source is white the spectral density of the observed process $\{\mathbf{x}(n)\}$ is proportional to $|A(\omega)|^2$ where $A(\omega)$ is the Fourier transform of the convolution filter. Thus its logarithms equals $2\log|A(\omega)|$ plus a constant, which equals $-2\log|B(\omega)|$ plus another constant since $B(\omega)$ should be inversely proportional to $A(\omega)$. As the periodogram $S_{xx}(2\pi k/N) = |d_x(2\pi k/N)|^2$ is an (asymptotically) unbiased estimator of this density at $\omega = 2\pi k/N$, one can write $\frac{1}{2}\log S_{xx}(2\pi k/N) = -\log|B(2\pi k/N)| +$ a constant $+$ an error term. Thus by adopting the parameterization (7.24), one gets the linear model

Table 7.1 Summary of the Algorithm for the parameterization $B(\omega) = \exp[\sum_{0<|\mu|\le m} \vartheta_\mu e^{i\mu\omega}]$

Initialization: Compute the DFT of the data $x(0,\ldots,x(N-1)$, then the DFT of the initial source estimates using the initial deconvolution filter in the frequency domain. Next compute the initial source estimates $y(0),\ldots,y(N-1)$ by inverse DFT and the corresponding criterion $\widehat{H}(y)$.
Iteration:
1. Compute the score estimates $\widehat{\psi}_y[y(n)]$, $n = 0,\ldots,N-1$, then the circular cross covariance r_μ, $0 < \mu \le m$, between $y(0),\ldots y(N-1)$ and $\widehat{\psi}_y[y(0)],\ldots,\widehat{\psi}_y[y(N-1)]$ according to either one of the right-hand sides of (7.25). Next compute $\widehat{\lambda} = \widehat{\sigma}_y^2 \widehat{J}(y)$.
2. Update $\vartheta_\mu, 0 < |\mu| \le m$ and $d_Y(2\pi k/N)$, $k = 0,\ldots,N-1$, by (7.26) and (7.27), then compute $y(0),\ldots,y(N-1)$ by inverse DFT.
3. Compute the new criterion $\widehat{H}(y)$. If it does not decrease then repeat (2), with decreased step size in (7.26) and (7.27) and recompute the criterion.
Repeat the iteration until convergence.

$$\log \left| d_x \left(\frac{2\pi k}{N} \right) \right| = -\tilde{\theta}_0 - \sum_{\mu=1}^{m} \theta_\mu \cos\left(\frac{2\pi\mu k}{N} \right) + \text{an error term},$$

where we have absorbed the constant into θ_0 to yield $\tilde{\theta}_0$. This suggests estimating $\tilde{\theta}_0, \theta_1, \ldots, \theta_m$ by least squares (we can estimate $\tilde{\theta}_0$ but not θ_0 since we don't know the constant). Specifically, we minimize

$$\sum_{k=0}^{N-1} \left[\tilde{\theta}_0 + \sum_{\mu=1}^{m} \theta_\mu \cos\left(\frac{2\pi\mu k}{N} \right) + \log \left| d_X \left(\frac{2\pi k}{N} \right) \right| \right]^2,$$

The resulting estimators

$$\theta_\mu = -\frac{2}{N} \sum_{k=0}^{N-1} \log \left| d_X \left(\frac{2\pi k}{N} \right) \right| \cos\left(\frac{2\pi\mu k}{N} \right) = -\frac{1}{N} \sum_{k=0}^{N-1} \log \left| d_X \left(\frac{2\pi k}{N} \right) \right| e^{2\pi\mu k/N},$$

$\mu = 1, \ldots, m$, are sub optimal but can be used to initialize the algorithm. In our limited experience, they can vastly improve the convergence. There is however no simple method to initialize $\theta_{m+1}, \ldots, \theta_{2m}$. One may simply initialize them by zero, which amounts to forcing the initial deconvolution filter to have zero phase.

7.3.4 Discussion

The above approach has been described in more details in [11]. This paper also considers an generalization of the criterion (7.15), obtained by replacing the entropy functional $H(y_k)$ by $\log[Q(y_k)]$ where $Q(\cdot)$ is some class II functional in the sense of Huber [6]: A functional $Q(y)$ (of the distribution of the random variable y) is said to be of class II is it is translation invariant[3] and scale equi-variant, that is $Q(ay+b) = |a|Q(y)$ for any real numbers a, b. A simple example of such functional is $Q(y) = \mathbf{E}(|y|^\alpha)^{1/\alpha}$, $\alpha > 0$. The main motivation for considering $\log[Q(\cdot)]$ instead of $H(\cdot)$ is that the functional $Q(\cdot)$ may be much easier to estimate than the entropy functional.

The negative of the generalized criterion is called a contrast if it attains its maximum at and only at the sequences $\{\mathbf{B}(j)\}$ for which the component sequences $\{y_k(n)\}$ of the sequence $\{\mathbf{y}(n)\}$ defined in (7.14) coincide with the source sequences *up to a permutation, a scaling and a delay.* It was shown in [8] that if the functional $Q(\cdot)$ is super-additive in the sense of Huber [6], that is $Q^2(y_1 + y_2) \geq Q^2(y_1) + Q^2(y_2)$ for any pair of independent random variables y_1, y_2, with equality if and only if they are Gaussian, then the negative of the corresponding generalized criterion is a contrast, unless there is more than one Gaussian source. This result contains as a special case the mutual information criterion (7.15), since the functional $Q(\cdot) = e^{H(\cdot)}$, by the entropy power inequality [4], verify the defining inequality of a

[3] As we deal with zero mean variables, the translation invariance requirement may be dropped.

class II functional. If in the estimated mutual information criterion $\widehat{C}[\mathbf{B}(\cdot)]$, the parameters h, δ are kept fixed as $N \to \infty$, then by the same argument as in Sect. 7.2.5 the negative of the limiting criterion is still a contrast if h, δ is small enough.

The paper [11] also provides results for the performance of the algorithm. Specifically in the case where $(h, \delta) \to (0,0)$ as $N \to \infty$, the estimator of θ admits the asymptotic covariance matrix $1/N$ times the inverse of the approximate Hessian matrix (7.21). (Note that $\widehat{\psi}_{y_k}$ converges here to the true score function ψ_{y_k} and $\mathbf{E}[\psi'_{y_k}(y_k)] = \mathbf{E}|\psi^2_{y_k}(y_k)]$.) In the unichannel case, if one parameterizes $|B(\omega)|$ and $\arg[B(\omega)]$ independently by two vectors parameters θ^R and θ^I, the above result together with that of previous section show that the estimators of θ_R and θ_I are asymptotically independent with covariance matrices $N^{-1}(\lambda+1)^{-1}\mathbf{H}_R^{-1}$ and $N^{-1}(\lambda-1)^{-1}\mathbf{H}_I^{-1}$. Note that λ is also the inverse of the squared correlation ρ^2 between $\psi_y(y)$ and y. In particular, for the parameterization (7.24), the matrices $\mathbf{H}_R$ and $\mathbf{H}_I$ both reduce to one half of the identity matrix.

The above result shows that it is often much harder to estimated the phase of the deconvolution filter than its amplitude. Indeed, the covariance matrix of θ_I contains the factor $1/(\lambda-1)$ which is greater than the factor $1/(\lambda+1)$ found in the formula for covariance matrix of θ_R. The factor $1/(\lambda-1)$ can be very large for nearly Gaussian source and it becomes infinity for Gaussian source ($\lambda = 1$). In the last case, one cannot estimate the phase of the deconvolution filter, as is well known. But its amplitude can still be estimated, since its square is inversely proportional to the spectral density of the observation sequence.

If the source is strongly non Gaussian however, λ is large and there is little difference between $1/(\lambda-1)$ and $1/(\lambda+1)$ and both amplitude and phase of the deconvolution filter can be well estimated. In this case, one can deconvolve the sources based only on its non Gaussianity. Many earlier blind deconvolution method are based only on a non Gaussianinty criterion such as the kurtosis or higher order cumulants (see [15, 17], ... for example).

7.3.5 Simulation

We first consider the unichannel (pure deconvolution) case. We generate 1,000 observation records of length $N = 1,024$ according to the convolution model: $x(n) = \sum_{j=-2}^{2} A(j)s(n-j)$ where $\{s(n)\}$ is a sequence of independent bilateral exponential (or Laplace) variables.[4] The function $\log[A(\omega)]$ has infinite Fourier series expansion, but as it can be seen from Fig. 7.2, this series may be truncated into a sum of 21 terms of index from -10 to 10. We have computed the relative error due to this truncation which is 0.0112; the relative error is defined as the square root of the ratio of the sum of the squared Fourier coefficients which are dropped to the sum of square of all Fourier coefficients except the one of index 0 (which is irrelevant since it concerns the scale of the filter). Since the true $B(\omega)$ must be inversely

[4] Such variable has density $\exp(-[x|)/2$.

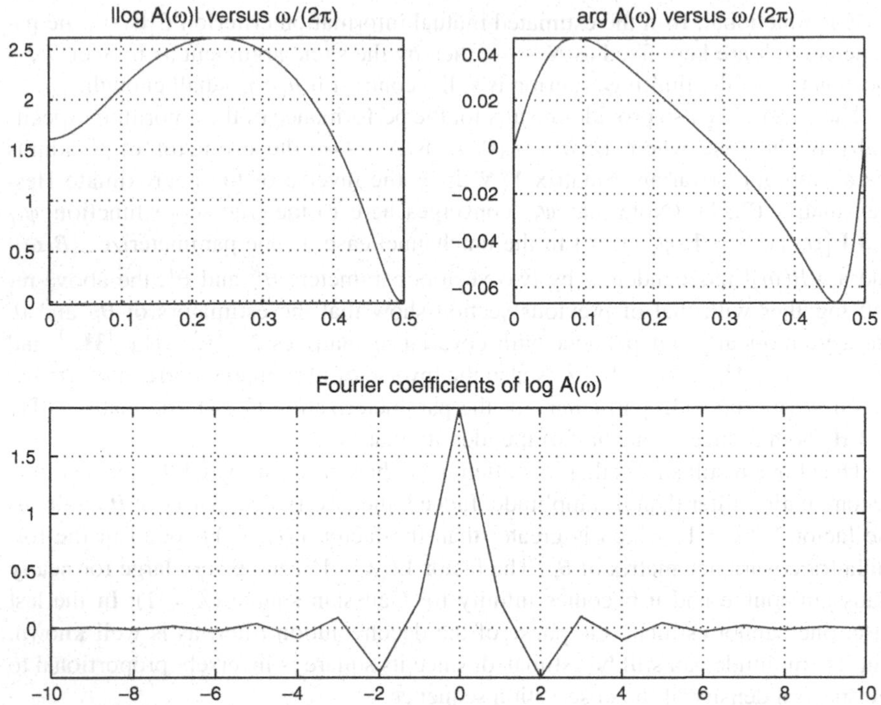

Fig. 7.2 The logarithm of $A(\omega)$

Table 7.2 Values of θ, the mean of the initial estimator $\tilde{\theta}$ and the mutual information estimator $\widehat{\theta}$ and their standard deviation (std.) computed from 1,000 simulations

j	θ_j	$\tilde{\theta}_j$	std. $\tilde{\theta}_j$	$\widehat{\theta}_j$	std. $\widehat{\theta}_j$	θ_{10+j}	$\widehat{\theta}_{10+j}$	std. $\widehat{\theta}_{10+j}$
1	−0.4749	−0.4698	0.0413	−0.4695	0.0301	−0.0082	−0.0089	0.0703
2	0.8265	0.8216	0.0430	0.8235	0.0295	−0.0436	−0.0412	0.0693
3	−0.1916	−0.1855	0.0400	−0.1856	0.0299	0.0072	0.0072	0.0670
4	0.1937	0.1915	0.0413	0.1926	0.0288	−0.0185	−0.0193	0.0688
5	−0.0811	−0.0772	0.0405	−0.0770	0.0291	0.0072	0.0078	0.0668
6	0.0660	0.0655	0.0395	0.0651	0.0295	−0.0088	−0.0096	0.0684
7	−0.0358	−0.0327	0.0407	−0.0330	0.0302	0.0049	0.0084	0.0692
7	0.0266	0.0245	0.0403	0.0244	0.0303	−0.0046	−0.0084	0.0714
8	−0.0164	−0.0143	0.0404	−0.0134	0.0308	0.0030	0.0089	0.0713
9	0.0118	0.0121	0.0394	0.0124	0.0293	−0.0025	−0.0046	0.0692

proportional to $A(\omega)$, we will consider as it is parameterized by (7.24) with $m = 10$ and parameters $\theta_\mu = \vartheta_\mu + \vartheta_{-\mu}$, $\mu = 1, \ldots, 10$, $= \vartheta_{\mu-10} - \vartheta_{10-\mu}$, $\mu = 11, \ldots, 20$, ϑ_μ being minus the Fourier coefficients of $\log A(\omega)$. The values of $\theta_1, \ldots, \theta_{20}$ are displayed in Table 7.2.

The algorithm in Sect. 7.3.3.3 is then applied to obtained the estimates $\widehat{\theta}_j$, $j = 1, \ldots, 20$ of the parameters θ of the deconvolution filter. These estimator are referred to as the mutual information estimator. The first 10 parameters are initialized by the method described at the end of Sect. 7.3.3.3 and the corresponding initial estimators are denoted by $\tilde{\theta}_1,\ldots,\tilde{\theta}_{10}$.

The simulation results are reported in Table 7.2. One can see from Table 3 that the mean values of the estimators are very close to the true values. Concerning the standard deviation, the value λ corresponding to the bilateral exponential variable is 2 hence, the asymptotic standard deviation of the $\widehat{\theta}_j$ is $\sqrt{2/(3\cdot 1{,}024)} = 0.0255$ for $j = 1,\ldots,10$ and $\sqrt{2/1{,}024} = 0.0442$ for $j = 11,\ldots,20$. The empirical standard deviations obtained from the simulation are significantly higher (although they are nearly the same for $\widehat{\theta}_1,\ldots,\widehat{\theta}_{10}$ and for $\widehat{\theta}_{11},\ldots,\widehat{\theta}_{20}$ as predicted by the theory). A fist reason may be that there are so many parameters to be estimated so that the asymptotic results are not yet attained. Another reason may be that the deconvolution filter is not exactly parameterized by (7.24) with $m = 10$ (to be exact m should be infinite) hence such filter cannot recover exactly the sources. This not only induces some bias but also increases the variance, since the extracted source contain some contamination and hence is more Gaussian than the true sources, resulting in a parameter $\lambda < 2$ instead of 2. Actually, the mean value of the λ estimated by the algorithm is 1.9292. In this regard, it is of interest to choose the parameter h smaller than the standard choice described near the end of Sect. 7.2.5. This standard choice tend to over-smooth the density, thus reducing the parameter λ. In the first stage of the algorithm where the extract sources is still very much a mixture, this may result in a value of λ too close to one, which may render the algorithm unstable, since $\lambda - 1$ appears in the denominator of the updating equation. It appears to us that using a smoothing parameter h much smaller than what would be used in density estimation can be quite beneficial here, as it reduces the smoothing effect (at the expense of an increase in variance). In this simulation, we choose h to be half the standard choice.

We next consider the multichannel case. As discussed in Sect. 7.3.3.3, in this case there is no simple quasi Newton algorithm based on the formula (7.21) for the approximated Hessian and therefore we shall use the BFGS algorithm to minimize the criterion.

We consider the observation model $\sum_{j=-2}^{2}\mathbf{B}(j)\mathbf{x}(n-j) = \mathbf{s}(n)$ where $\mathbf{B}(-2),\ldots,$ $\mathbf{B}(2)$ are 2×2 matrices whose entries are listed in Table 7.3, and $\mathbf{s}(n)$ are independent random variables with a uniform distribution in $[-\sqrt{3},\sqrt{3}]$. The process $\{\mathbf{x}(n)\}$ is by definition given by $\mathbf{x}(n) = \sum_{j=-\infty}^{\infty}\mathbf{A}(j)\mathbf{s}(n-j)$ where the (infinite) sequence $\{\mathbf{A}(j)\}$ is the inverse in the convolutive sense of the sequence $\{\mathbf{B}(j)\}$ (by convention $\mathbf{B}(j) = \mathbf{0}$ for $|j| > 2$). However, for computational purpose, we shall use a sequence of support length 32 instead, which is adjusted so that the convolution of $\{\mathbf{B}(j)\}$ with it best approaches in Frobinus norm[5] the identity sequence (with the identity matrix at index 0 and zero elsewhere). It turns out that the best adjustment is reached for a sequence of support $[-20,11]$, again denoted by $\{\mathbf{A}(j)\}$, whose en-

[5] The Frobinus norm of a matrix is the square root of the sum of squares of its elements. As we deal with matrix sequence, we also sum over all matrices as well.

Table 7.3 True and estimate deconvolution filter (std. = standard deviation)

	True value		Mean of estimate		Std. of estimate	
$B(-2)^T$	−0.1647	0.1575	−0.1646	0.1593	0.0168	0.0163
	0.1555	0.2657	0.1561	0.2623	0.0201	0.0233
$B(-2)^T$	0.0601	−0.4636	0.0598	−0.4615	0.0180	0.0153
	−0.3628	−0.3376	−0.3613	−0.3394	0.0195	0.0242
$B(-2)^T$	−0.1838	1.3913	−0.1803	1.3909	0.0189	0.0047
	1.4801	−0.5285	1.4806	−0.5298	0.0082	0.0228
$B(-2)^T$	0.1864	−0.4404	0.1844	−0.4429	0.0163	0.0165
	−0.4553	−0.4320	−0.4547	−0.4283	0.0198	0.0242
$B(-2)^T$	−0.0251	0.1321	−0.0230	0.1356	0.0126	0.0154
	0.1762	0.2364	0.1740	0.2317	0.0210	0.0226

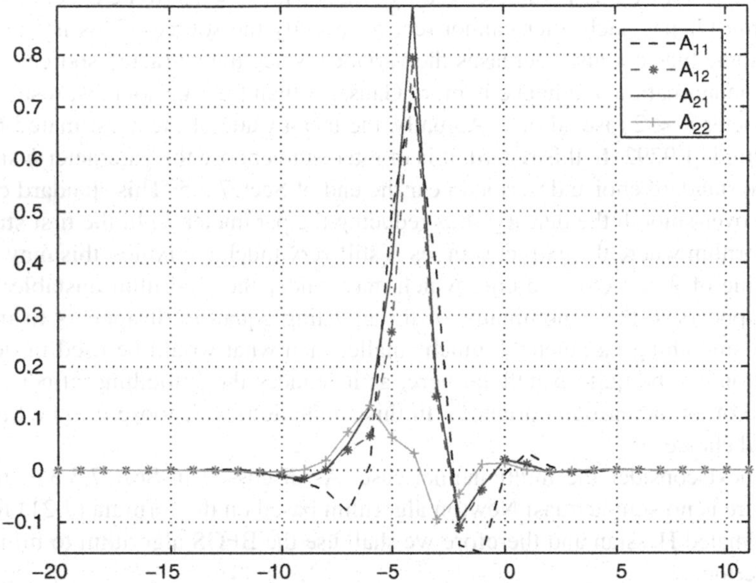

Fig. 7.3 Elements of the matrix $\mathbf{A}(j)$ as function of j

tries are plotted in Fig. 7.3. The difference between the convolution of $\{\mathbf{B}(j)\}$ with this sequence and the identity sequence is very small: 1.0925×10^{-5} in Frobinus norm.

One hundred observation sequences of length 1,028 are generated and for each sequence the BFGS algorithm is applied to minimize the estimated mutual information criterion. Since the separation deconvolution filter is FIR, the data is not periodically extended but instead the output of the deconvolution is made shorter (length 1,024 exactly). Accordingly, the gradient is computed by formula (7.19) (without the "mod N"). Mean and standard deviation of the estimated separation deconvolution filter are listed in Table 7.3. As in the pure deconvolution case, we find that better results are obtained by using a lower value of the smoothing parameter h than the standard choice: it is taken to be half of the standard choice in this simulation. One can see from Table 7.3 that the separation deconvolution filter is well estimated. Note that the above results do not come directly from the output of the BFGS algorithm, since there is a permutation and scale ambiguity. In order to eliminate the permutation ambiguity, we have permuted the rows of $\widehat{\mathbf{B}} = [\widehat{\mathbf{B}}(-1) \;\cdots\; \widehat{\mathbf{B}}(2)]$ ($\widehat{\mathbf{B}}(j)$ being the estimate of $\mathbf{B}(j)$) whenever the product of the diagonal elements of $\widehat{\mathbf{B}}\mathbf{B}^{\mathrm{T}}$, where $\mathbf{B} = [\mathbf{B}(-1) \;\cdots\; \mathbf{B}(2)]$, is less in absolute value than that of the off diagonal elements. To eliminate the scale ambiguity, we have normalized the rows of $\widehat{\mathbf{B}}$ (eventually permuted) so that they have the same Frobinus norm as that of $\mathbf{B}$.

7.4 Conclusion

We have provided a unified approach to blind sources separation and deconvolution based on the mutual information criterion. Practical algorithms have been derived in some details and their theoretical performance discussed.

References

1. Cardoso, J.F.: Blind signal separation: Statistical principles. Proc IEEE, special issue on Blind Estimation and Identification **86**(10), 2009–2025 (1998)
2. Cardoso, J.F., Laheld, B.: Equivariant adaptive sousrce separation. IEEE Trans Signal Process **44**(12), 3017–3030 (1996)
3. Comon, P.: Independence components analysis, a new concept. Signal Process **36**(3), 287–314 (1994)
4. Cover, T., Thomas, J.: Elements of Information Theory. New-York: Wiley (1991)
5. Dai, Y.H.: Convergence properties of the BFGS algorithm. SIAM J Optim **31**(3), 693–701 (2002)
6. Huber, P.J.: Projection pursuit. Ann Statist **13**(2), 435–475 (1985)
7. Jones, M.C.: Discretized and interpolated kernel density estimates. J Am Statist Assoc **84**, 733–741 (1989)
8. Pham, D.T.: Contrast functions for blind seperation and deconvolution of sources. In: Proceeding of ICA 2001 Conference, pp. 37–42. San Diego, USA (2001)
9. Pham, D.T.: Mutual information approach to blind separation of stationary sourcs. IEEE Trans Inform Theory **48**, 1935–1946 (2002)
10. Pham, D.T.: Fast algorithms for mutual information based independent component analysis. IEEE Trans Signal Process **52**(10), 2690–2700 (2004)
11. Pham, D.T.: Generalized mutual information approach to multichannel blind deconvolution. Signal Process **87**(9), 2045–2060 (2007)
12. Pham, D.T., Garat, P.: Blind separation of mixtures of independent sources through a quasi maximum likelihood approach. IEEE Trans Signal Process **45**(7), 1712–1725 (1997)
13. Press, W.H., Flannery, B.P., Teukolsky, S.A., Vetterling, W.T.: Numerical Recipes in C. The Art of Scientific Computing, Second Edition. Cambridge: Cambridge University Press (1993)
14. Scott, D.W., Sheather, S.J.: Kernel density estimation with binned data. Commun Statist – Theory Meth **14**, 1353–1359 (1985)
15. Shalvi, O., Weinstein, E.: Super-exponential methods for blind deconvolution. IEEE Trans Inform Theory **39**(2), 504–519 (1993)

16. Silverman, B.W.: Density Estimation for Statistics and Data Analysis. London: Chapman and Hall (1982)
17. Tugnait, J.K.: Identification and deconvolution of multichannel linear non-gaussian processes using higher order statistics and inverse filter criteria. IEEE Trans Signal Process **45**(3), 658–672 (1997)

Chapter 8
Causality in Time Series: Its Detection and Quantification by Means of Information Theory

Kateřina Hlaváčková-Schindler

Abstract While studying complex systems, one of the fundamental questions is to identify causal relationships (i.e., which system drives which) between relevant subsystems. In this paper, we focus on information-theoretic approaches for causality detection by means of directionality index based on mutual information estimation. We briefly review the current methods for mutual information estimation from the point of view of their consistency. We also present some arguments from recent literature, supporting the usefulness of the information-theoretic tools for causality detection.

8.1 Introduction

During the history of most natural and social sciences, detection and clarification of cause–effect relationships among variables, events or objects have been the fundamental questions. Despite some philosophers of mathematics like B. Russel [76] (1872–1970) tried to deny the existence of the phenomenon "causality" in mathematics and physics, saying that causal relationships and physical equations are incompatible, calling causality to be "a word relic" (see, i.e., [67]), the language of all sciences, including mathematics and physics, has been using this term actively until now. To advocate the Russell's view, any exact and sufficiently comprehensive formulation of what is causality is difficult. Causality can be understood in terms of a "flow" among processes and expressed in mathematical language and mathematically analysed.

K. Hlaváčková-Schindler
Commission for Scientific Visualization, Austrian Academy of Sciences, Donau-City Str. 1, 1220 Vienna, Austria and Institute of Information Theory and Automation of the Academy of Sciences of the Czech Republic, Pod Vodárenskou věží 4, 18208 Praha 8, Czech Republic
e-mail: katerina.schindler@assoc.oeaw.ac.at

F. Emmert-Streib, M. Dehmer (eds.), *Information Theory and Statistical Learning*,
DOI: 10.1007/978-0-387-84816-7_8,

The general philosophical definition of causality from the Wikipedia Encyclopedia [92] states: "The philosophical concept of causality or causation refers to the set of all particular 'causal' or 'cause-and-effect' relations. Most generally, causation is a relationship that holds between events, objects, variables, or states of affairs. It is usually presumed that the cause chronologically precedes the effect." Causality expresses a kind of a "law" necessity, while probabilities express uncertainty, a lack of regularity. Probability theory seems to be the most used "mathematical language" of most scientific disciplines using causal modeling, but it seems not to be able to grasp all related questions. In most disciplines, adopting the above definition, the aim is not only to detect a causal relationship but also to measure or quantify the relative strengths of these relationships. This can be done by information theory tools. In [43] we provided a detailed overview of the information-theoretic approaches for measuring of a causal influence in multi-variate time series. Here we mainly focus on the methods using mutual information for the computation of the causal directional index and entropy estimation methods. The methods are discussed from the point of view of their consistency properties and for their detailed description we refer the reader to [43]. The outline of the paper is the following. Section 8.1.1 presents measures for causality detection. In Sect. 8.2 we define the basic information-theoretic functionals and from them derived causality measurements. Sections 8.3 and 8.4 briefly present the non-parametric and parametric methods including their consistency properties. Granger causality is discussed in Sect. 8.5 and conclusion is in Sect. 8.6.

8.1.1 Causality and Causal Measures

Most of the earlier research literature attempts to discuss unique causes in deterministic situations, and two conditions are important for deterministic causation [34]: (1) necessity: if X occurs, then Y must occur, and (2) sufficiency: if Y occurs, then X must have occurred. However, deterministic formulation, albeit appealing and analytically tractable, is not in accordance with reality, as no real-life system is strictly deterministic (i.e., its outcomes cannot be predicted with complete certainty). So, it is more realistic if one modifies the earlier formulation in terms of likelihood (i.e., if X occurs, then the likelihood of Y occurring increases). This can be illustrated by a simple statement such as if the oil price increases, the carbon emission does not necessarily decrease, but there is a good likelihood that it will decrease. The probabilistic notion of causality is nicely described by Suppes (1970) [86] as follows: An event X is a cause to the event Y if (1) X occurs before Y, (2) likelihood of X is non zero, and (3) likelihood of occurring Y given X is more than the likelihood of Y occurring alone. Although this formulation is logically appealing, there are some arbitrariness in practice in categorizing an event [34]. Till 1970, the causal modeling was mostly used in social sciences. This was primarily due to a pioneering work by Selltiz et al. [82] who specified three conditions for the existence of causality:

1. There must be a concomitant covariation between X and Y.

2. There should be a temporal asymmetry or time ordering between the two observed sequences.
3. The covariance between X and Y should not disappear when the effects of any confounding variables (i.e., those variables which are causally prior to both X and Y) are removed.

The first condition implies a correlation between a cause and its effect, though one should explicitly remember that a perfect correlation between two observed variables in no way implies a causal relationship. The second condition is intuitively based on the arrow of time. The third condition is problematic since it requires that one should rule out all other possible causal factors. Theoretically, there are potentially an infinite number of unobserved confounding variables available, yet the set of measured variables is finite, thus leading to indeterminacy in the causal modeling approach. In order to avoid this, some structure is imposed on the adopted modeling scheme which should help to define the considered model. The way in which the structure is imposed is crucial in defining as well as in quantifying causality.

The first definition of causality which could be quantified and measured computationally, yet very general, was given in 1956 by N. Wiener [91]: "...For two simultaneously measured signals, if we can predict the first signal better by using the past information from the second one than by using the information without it, then we call the second signal causal to the first one."

The introduction of the concept of causality into the experimental practice, namely into analyses of data observed in consecutive time instants, time series, is due to Clive W. J. Granger, the 2003 Nobel prize winner in economy. In his Nobel lecture [35] he recalled the inspiration by the Wiener's work and identified two components of the statement about causality: (1) The cause occurs before the effect; (2) The cause contains information about the effect that is unique, and is in no other variable.

As Granger put it, a consequence of these statements is that the causal variable can help to forecast the effect variable after other data has been first used [35]. This restricted sense of causality, referred to as *Granger causality*, GC thereafter, characterizes the extent to which a process X_t is leading another process Y_t, and builds upon the notion of incremental predictability. It is said that the *process* X_t *Granger causes another process* Y_t if future values of Y_t can be better predicted using the past values of X_t and Y_t rather then only past values of Y_t. The standard test of GC developed by Granger [32] is based on a linear regression model

$$Y_t = a_o + \sum_{k=1}^{L} b_{1k} Y_{t-k} + \sum_{k=1}^{L} b_{2k} X_{t-k} + \xi_t, \tag{8.1}$$

ξ_t are uncorrelated random variables with zero mean and variance σ^2, L is the specified number of time lags, and $t = L+1, \ldots, N$. The null hypothesis that X_t does not Granger cause Y_t is supported when $b_{2k} = 0$ for $k = 1, \ldots, L$, reducing (8.1) to

$$Y_t = a_o + \sum_{k=1}^{L} b_{1k} Y_{t-k} + \tilde{\xi}_t. \tag{8.2}$$

This model leads to two well-known alternative test statistics, the Granger–Sargent and the Granger–Wald test. The Granger–Sargent test is defined as

$$GS = \frac{(R_2 - R_1)/L}{R_1/(N-2L)}, \tag{8.3}$$

where R_1 is the residual sum of squares in (8.1) and R_2 is the residual sum of squares in (8.2). The GS test statistic has an F-distribution with L and $N-2L$ degrees of freedom. On the other hand, the Granger–Wald test is defined as

$$GW = N\frac{\widehat{\sigma}^2_{\tilde{\xi}_t} - \widehat{\sigma}^2_{\xi_t}}{\widehat{\sigma}^2_{\xi_t}}, \tag{8.4}$$

where $\widehat{\sigma}^2_{\tilde{\xi}_t}$ is the estimate of the variance of $\tilde{\xi}_t$ from model (8.2) and $\widehat{\sigma}^2_{\xi_t}$ is the estimate of the variance of ξ_t from model (8.1). The GW statistic follows the χ^2_L distribution under the null hypothesis of no causality.

This linear framework for measuring and testing causality has been widely applied not only in economy and finance (see Geweke [31] for a comprehensive survey of the literature), but also in diverse fields of natural sciences such, where specific problems of multichannel electroencephalogram recordings were solved by generalizing the Granger causality concept to multivariate case [14]. Nevertheless, the limitation of the present concept to linear relations required further generalizations.

Recent development in nonlinear dynamics [1] evoked lively interactions between statistics and econometrics on one side, and physics and other natural sciences on the other side. In the field of economy, Baek and Brock [9] and Hiemstra and Jones [40] proposed a nonlinear extension of the Granger causality concept. Their non-parametric dependence estimator is based on so-called correlation integral, a probability distribution and entropy estimator, developed by physicists Grassberger and Procaccia in the field of nonlinear dynamics and deterministic chaos as a characterization tool of chaotic attractors [36]. A non-parametric approach to non-linear causality testing, based on non-parametric regression, was proposed by Bell et al. [12]. Following Hiemstra and Jones [40], Aparicio and Escribano [6] succinctly suggested an information-theoretic definition of causality which include both linear and nonlinear dependence.

In physics and nonlinear dynamics, a considerable interest recently emerged in studying cooperative behavior of coupled complex systems [15, 69]. Synchronization and related phenomena were observed not only in physical, but also in many biological systems. Examples include the cardio-respiratory interaction [63, 77] and the synchronization of neural signals [58, 73]. In such physiological systems it is not only important to detect synchronized states, but also to identify drive–response relationships and thus the causality in evolution of the interacting (sub)systems. Schiff et al. [78] and Quyen et al. [73] used ideas similar to those of Granger, however, their cross-prediction models utilize zero-order nonlinear predictors based on mutual nearest neighbors. A careful comparison of these two papers [73, 78] reveals how complex is the problem of inferring causality in nonlinear systems. The authors of the two papers use contradictory assumptions for interpreting the differences in prediction errors of mutual predictions, however, both teams were able to present numerical examples in which their approaches apparently worked.

While the latter two papers use the method of mutual nearest neighbors for mutual prediction, Arnhold et al. [8] proposed asymmetric dependence measures based on averaged relative distances of the (mutual) nearest neighbors. As pointed out by Quian Quiroga et al. [74] and by Schmitz [79], these measures, however, might be influenced by different dynamics of individual signals and different dimensionality of the underlying processes, rather than by asymmetry in coupling.

Another nonlinear extension of the Granger causality approach was proposed by Chen et al. [19] using local linear predictors. An important class of nonlinear predictors are based on so-called radial basis functions [17] which were used for nonlinear parametric extension of the Granger causality concept [4]. A nonparametric method for measuring causal information transfer between systems was proposed by Schreiber [81]. His *transfer entropy* is designed as a Kullback–Leibler distance ((8.14) in Sect. 8.2.1) of transition probabilities. This measure is in fact an information-theoretic functional of probability distribution functions.

Paluš et al. [58] proposed to study synchronization phenomena in experimental time series by using the tools of information theory. Mutual information, an information-theoretic functional of probability distribution functions, is a measure of general statistical dependence. For inferring causal relation, conditional mutual information can be used. It was shown that, with proper conditioning, the Schreiber's transfer entropy [81] is equivalent to the conditional mutual information [58]. The latter, however, is a standard measure of information theory.

Turning our attention back to econometrics, we can follow further development due to Diks and DeGoede [23]. They again applied a nonparametric approach to nonlinear Granger causality using the concept of correlation integrals [36] and pointed out the connection between the correlation integrals and information theory. Diks and Panchenko [24] critically discussed the previous tests of Hiemstra and Jones [40]. As the most recent development in economics, Baghli [10] proposes information-theoretic statistics for a model-free characterization of causality, based on an evaluation of conditional entropy. The nonlinear extension of the Granger causality based the information-theoretic formulation has found numerous applications in various fields of natural and social sciences. Let us mention just a few examples. Schreiber's transfer entropy was used in physiology and neurophysiology [46]. Paluš et al. [58] applied their conditional mutual information based measures in analyses of electroencephalograms of patients suffering from epilepsy. Other applications of the conditional mutual information in neurophysiology are due to Hinrichs et al. [41]. Causality or coupling directions in multimode laser dynamics is another field where the conditional mutual information was applied [57]. Having reviewed the relevant literature, we can state that the information-theoretic approach to the Granger causality plays an important, if not a dominant role in analyses of causal relationships in nonlinear systems.

8.2 Information Theory as a Tool for Causality Detection

8.2.1 Definitions of Basic Information Theoretic Functionals

We begin with the definition of differential entropy for a continuous random variable as it was introduced in 1948 by Shannon [83]. Let X be a random vector taking values in R^d with probability density function (pdf) $p(x)$, then its *differential entropy* is defined by

$$H(x) = -\int p(x) \log p(x) dx, \tag{8.5}$$

where log is natural logarithm. We assume that $H(x)$ is well-defined and finite. Let S be a discrete random variable having possible values $s_1, \ldots, s_m$, each with corresponding probability $p_i = p(s_i), i = 1, \ldots, m$. The average amount of information gained from a measurement that specifies one particular value s_i is given by *entropy* $H(S)$:

$$H(S) = -\sum_{i=1}^{m} p_i \log p_i. \tag{8.6}$$

More general term of entropy for which is Shannon differential entropy a special case, is *Rényi entropy*, defined for a continuous case as [75]

$$H_\alpha(x) = \frac{1}{1-\alpha} \int \log^\alpha p(x) dx \tag{8.7}$$

and for the discrete case

$$H_\alpha(S) = \frac{1}{1-\alpha} \log \sum_{i=1}^{n} p_i^\alpha, \tag{8.8}$$

where $\alpha > 0$, $\alpha \neq 1$. As $\alpha \to 1$, $H_\alpha(x)$ converges to $H(x)$ (or $H_\alpha(S)$ converges to $H(S)$), which is Shannon entropy. The *joint entropy* $H(X,Y)$ of two discrete random variables X and Y is

$$H(X,Y) = -\sum_{i=1}^{m_X} \sum_{j=1}^{m_Y} p(x_i, y_j) \log p(x_i, y_j), \tag{8.9}$$

where $p(x_i, y_j)$ denotes the joint probability that X is in state x_i and Y in state y_j. In general, the joint entropy may be expressed in terms of *conditional entropy* $H(X|Y)$ as follows

$$H(X,Y) = H(X|Y) + H(Y), \qquad \text{where} \tag{8.10}$$

$$H(X|Y) = -\sum_{i=1}^{m_X} \sum_{j=1}^{m_Y} p(x_i, y_j) \log p(x_i|y_j) \tag{8.11}$$

and $p(x_i|y_j)$ denotes the conditional probability. The *mutual information* $I(X,Y)$ between two random variables X and Y is then defined as [83]

$$I(X;Y) = H(X) + H(Y) - H(X,Y). \tag{8.12}$$

It reflects the mutual reduction in uncertainty of one by knowing the other variable. This measure is nonnegative since $H(X,Y) \leq H(X) + H(Y)$. The equality holds if and only if X and Y are statistically independent. The *conditional mutual information* [83] between random variables X and Y given Z is defined as

$$I(X,Y|Z) = H(X|Z) + H(Y|Z) - H(X,Y|Z). \tag{8.13}$$

For Z independent of X and Y we have $I(X,Y|Z) = I(X,Y)$. The *Kullback–Leibler divergence* (KLD, also called relative entropy or cross-entropy), introduced by Kullback and Leibler [51], is an alternative approach to mutual information. $K(p,p^0)$ between two probability distributions $\{p\}$ and p^0 is

$$K(p,p^0) = \sum_{i=1}^{m} p_i \log\left(\frac{p_i}{p_i^0}\right). \tag{8.14}$$

It can be interpreted as the information gain when an initial probability distribution p^0 is replaced by a final distribution p. This entropy is however not symmetric and therefore not a distance in the mathematical sense. The KLD is always nonnegative and is zero iff the distributions p and p^0 are identical. Mutual information is the Kullback–Leibler divergence of the product $P(X)P(Y)$ of two marginal probability distributions from the joint probability distribution $P(X,Y)$, see, i.e., [30]. So we can look at the results about Kullback–Leibler entropy as if they were applied to mutual information.

8.2.2 Coarse-Grained Entropy and Information Rates

A considerable amount of approaches to inferring causality from experimental time series have their roots in studies of synchronization of chaotic systems. A.N. Kolmogorov, who introduced the theoretical concept of classification of dynamical system by information rates [47], was inspired by information theory and together with Y.G. Sinai generalized the notion of entropy of an information source [47, 85]. Paluš [60] concentrated on attributes of dynamical systems studied in the ergodic theory, such as mixing and generating partitions, and demonstrated how they were reflected in the behaviour of information-theoretic functionals estimated from chaotic data. In order to obtain an asymptotic entropy estimate of an m-dimensional dynamical system, large amounts of data are necessary [60]. To avoid this, Paluš [60] proposed to compute "coarse-grained entropy rates" (CER's) as relative measures of "information creation" and of regularity and predictability of studied processes.

Let $\{x(t)\}$ be a time series considered as a realization of a stationary and ergodic stochastic process $\{X(t)\}$, $t = 1,2,3,\ldots$. We denote $x(t)$ as x and $x(t+\tau)$ as x_τ for simplicity. To define the simplest form of CER, we compute the mutual information

$I(x;x_\tau)$ for all analyzed datasets and find such τ_{max} that for $\tau' \geq \tau_{max}$: $I(x;x_{\tau'}) \approx 0$ for all the data sets. Then we define a *norm of the mutual information*

$$||I(x;x_\tau)|| = \frac{\Delta\tau}{\tau_{max} - \tau_{min} + \Delta\tau} \sum_{\tau=\tau_{min}}^{\tau_{max}} I(x;x_\tau) \tag{8.15}$$

with $\tau_{min} = \Delta\tau = 1$ sample as a usual choice. The CER h^1 is then defined as $h^1 = I(x,x_{\tau_0}) - ||I(x;x_\tau)||$. Since usually $\tau_0 = 0$ and $I(x;x) = H(X)$ which is given by the marginal probability distribution $p(x)$, the sole quantitative descriptor of the underlying dynamics is the mutual information norm (8.15). Paluš et al. [58] called this descriptor the *coarse-grained information rate* (CIR) of the process $\{X(t)\}$ and denoted by $i(X)$.

Now, consider two time series $\{x(t)\}$ and $\{y(t)\}$ regarded as realizations of two processes $\{X(t)\}$ and $\{Y(t)\}$ which represent two possibly linked (sub) systems. These two systems can be characterized by their respective CIR's $i(X)$ and $i(Y)$. In order to characterize an interaction of the two systems, in analogy with the above CIR, Paluš et al. [58] defined their *mutual coarse-grained information rate* (MCIR) by

$$i(X,Y) = \frac{1}{2\tau_{max}} \sum_{\tau=-\tau_{max}}^{\tau_{max};\tau\neq 0} I(x;y_\tau). \tag{8.16}$$

Due to the symmetry properties of $I(x;y_\tau)$ is the mutual CIR $i(X,Y)$ symmetric, i.e., $i(X,Y) = i(Y,X)$. Assessing the direction of coupling between the two systems, i.e., causality in their evolution, we ask how is the dynamics of one of the processes, say $\{X\}$, influenced by the other process $\{Y\}$. For the quantitative answer to this question, Paluš et al. [58] proposed to evaluate the *conditional coarse-grained information rate* CCIR $i_0(X|Y)$ of $\{X\}$ given $\{Y\}$:

$$i_0(X|Y) = \frac{1}{\tau_{max}} \sum_{\tau=1}^{\tau_{max}} I(x;x_\tau|y), \tag{8.17}$$

considering the usual choice $\tau_{min} = \Delta\tau = 1$ sample. For independent variables we have $i_0(X|Y) = i(X)$ for $\{X\}$ independent of $\{Y\}$, i.e., when the two systems are uncoupled. In order to have a measure which vanishes for an uncoupled system (although then it can acquire both positive and negative values), Paluš et al. [58] define

$$i(X|Y) = i_0(X|Y) - i(X). \tag{8.18}$$

For another approach to a directional information rate, let us consider the mutual information $I(y;x_\tau)$ measuring the average amount of information contained in the process $\{Y\}$ about the process $\{X\}$ in its future τ time units ahead (τ-future thereafter). This measure, however, could also contain an information about the τ-future of the process $\{X\}$ contained in this process itself if the processes $\{X\}$ and $\{Y\}$ are not independent, i.e., if $I(x;y) > 0$. In order to obtain the "net" information about the τ-future of the process $\{X\}$ contained in the process $\{Y\}$, we need the conditional mutual information $I(y;x_\tau|x)$. Next, we sum $I(y;x_\tau|x)$ over τ as above

$$i_1(X,Y|X) = \frac{1}{\tau_{max}} \sum_{\tau=1}^{\tau_{max}} I(y;x_\tau|x); \tag{8.19}$$

In order to obtain the "net asymmetric" information measure, we subtract the symmetric MCIR (8.16):

$$i_2(X,Y|X) = i_1(X,Y|X) - i(X,Y). \tag{8.20}$$

Using a simple manipulation, we find that $i_2(X,Y|X)$ is equal to $i(X|Y)$ defined in (8.18). By using two different ways for definition of a directional information rate, Paluš et al. [58] arrived to the same measure which they denoted by $i(X|Y)$ and called the *coarse-grained transinformation rate* (CTIR) of $\{X\}$ given $\{Y\}$. It is the average rate of the net amount of information "transferred" from the process $\{Y\}$ to the process $\{X\}$ or, in other words, the average rate of the net information flow by which the process $\{Y\}$ influences the process $\{X\}$.

Using several numerical examples of coupled chaotic systems, Paluš et al. [58] demonstrated that the CTIR is able to identify the coupling directionality from time series measured in coupled, but not yet fully synchronized systems. As a practical application, CTIR was used in analyses of electroencephalograms of patients suffering from epilepsy. Causal relations between EEG signals measured in different parts of the brain were identified. Paluš et al. demonstrated suitability of the conditional mutual information approach for analyzing causality in cardio-respiratory interaction [58].

8.2.3 Conditional Mutual Information and Transfer Entropy

The principal measure, used by Paluš et al. [58] for inferring causality relations, i.e., the directionality of coupling between the processes $\{X(t)\}$ and $\{Y(t)\}$, is the conditional mutual information $I(y;x_\tau|x)$ and $I(x;y_\tau|y)$. If the processes $\{X(t)\}$ and $\{Y(t)\}$ are substituted by dynamical systems evolving in measurable spaces of dimensions m and n, respectively, the variables x and y in $I(y;x_\tau|x)$ and $I(x;y_\tau|y)$ should be considered as n- and m-dimensional vectors. In experimental practice, however, usually only one observable is recorded for each system. Then, instead of the original components of the vectors $\mathbf{X}(t)$ and $\mathbf{Y}(t)$, the time delay embedding vectors according to Takens [87] are used. Then, back in time-series representation, we have

$$\begin{aligned} &I\big(\mathbf{Y}(t);\mathbf{X}(t+\tau)|\mathbf{X}(t)\big) \\ =&I\Big(\big(y(t),y(t-\rho),\ldots,y(t-(m-1)\rho)\big);x(t+\tau)\big|\big(x(t),x(t-\eta),\ldots,x(t-(n-1)\eta)\big)\Big), \end{aligned} \tag{8.21}$$

where η and ρ are time lags used for the embedding of systems $\mathbf{X}(t)$ and $\mathbf{Y}(t)$, respectively. For simplicity, only the information about one component $x(t+\tau)$ in the τ-future of the system $\mathbf{X}(t)$ is used. The opposite CMI $I\big(\mathbf{X}(t);\mathbf{Y}(t+\tau)|\mathbf{Y}(t)\big)$ is defined in the full analogy. Exactly the same formulation can be used for Markov processes of finite orders m and n. Using the idea of finite-order Markov processes, Schreiber [81] introduced a measure quantifying causal information transfer between systems evolving in time, based on appropriately conditioned transition probabilities. Assuming that the system under study can be approximated by a stationary Markov process of order k, the transition probabilities describing the evolution of the system are $p(i_{n+1}|i_n,...,i_{n-k+1})$. If two processes I and J are independent, then the generalized Markov property

$$p(i_{n+1}|i_n,...,i_{n-k+1}) = p\left(i_{n+1} \mid i_n^{(k)}, j_n^{(l)}\right), \tag{8.22}$$

holds, where $i_n^{(k)} = (i_n,...,i_{n-k+1})$ and $j_n^{(l)} = (j_n,...,j_{n-l+1})$ and l is the number of conditioning state from process J. Schreiber proposed using the KLD (8.14) to measure the deviation of the transition probabilities from the generalized Markov property (8.22) and got the definition

$$T_{J\to I} = \sum p(i_{n+1}, i_n^{(k)}, j_n^{(l)}) \log \frac{p(i_{n+1}|i_n^{(k)}, j_n^{(l)})}{p(i_{n+1}|i_n^{(k)})}, \tag{8.23}$$

denoted as *transfer entropy*. It can be understood as the excess amount of bits that must be used to encode the information of the state of the process by erroneously assuming that the actual transition probability distribution function is $p(i_{n+1}|i_n^{(k)})$, instead of $p(i_{n+1}|i^{(k)}, j_n^{(l)})$. It was shown (for example in [43]) that the transfer entropy is in fact an equivalent expression for the conditional mutual information.

8.2.4 Comparison of Coarse Grained Measures and Two Deterministic Measures for Causality Detection in Bivariate Time Series

A good causality detector in time series should have a low rate of false detections. Paluš and Vejmelka [64] experimentally analyzed causality detection for bivariate time series by coarse grained measures (CTIR, defined by (8.20)) and compared it to two common deterministic approaches. Numerous examples demonstrated what problems can appear in inference of causal relationship. This to the date unique comparative work on information-theoretic causality detectors to the deterministic ones definitely deserves our attention. Three approaches were compared. In all cases, the driving, autonomous system is denoted by X, and the driven, response system by Y. As Paluš et al. in [58] explain, the direction of coupling can be inferred from experimental data only when the underlying systems are coupled, but not yet synchronized. The CTIR defined by formula (8.20) was compared to the method from Le Van Quyen [73] and the method from Arnhold et al. [8] and Quian Quiroga [74] (belonging to the methods discussed in Sect. 8.3.4). The second method is based on cross-prediction using the idea of mutual neighbors. A neigborhood size δ is given. Considering a map from X to Y, a prediction is made for the value of y_{n+1} one step ahead using the formula

$$\widehat{y}_{n+1} = \frac{1}{|V_\delta(X_n)|} \sum_{j:X_j \in V_\delta(X_n)} y_{j+1}. \tag{8.24}$$

The volume $V_\delta(X_n) = \{X_{n'} : |X_{n'} - X_n| < \delta\}$ is δ neighborhood of X_n and $|V_\delta(X_n)|$ denotes the number of points in the neighborhood. Using data rescaled to the zero mean and the unit variance, the authors define a crosspredictability index by subtracting the root-mean-square prediction error from one

$$P(X \to Y) = 1 - \sqrt{\frac{1}{N}\sum_{n=1}^{N}(\widehat{y}_{n+1} - y_{n+1})^2}, \tag{8.25}$$

measuring how system X influences the future of system Y.

The third method from Arnhold et al. [8] and Quian Quiroga [74] uses mean square distances instead of the cross-predictions in order to quantify the closeness of points in both spaces. The time-delay embedding is first constructed in order to obtain state space vectors X and Y for both time series $\{x_i\}$ and $\{y_i\}$, respectively and then the mean squared distance to k nearest neighbors is defined for each X as

$$R_n^{(k)}(X) = \frac{1}{k}\sum_{j=1}^{k} |X_n - X_{r_{n,j}}|^2, \tag{8.26}$$

where $r_{n,j}$ the index of the j-th nearest neighbor of X_n. The Y-conditioned squared mean distance is defined by replacing the nearest neighbors of X_n by the equal time partners of the nearest neighbors of Y_n as

$$R_n^{(k)}(X|Y) = \frac{1}{k}\sum_{j=1}^{k} |X_n - X_{s_{n,j}}|^2, \tag{8.27}$$

where $s_{n,j}$ denotes the index of the j-th nearest neighbor of Y_n. Then the asymmetric measure

$$S^{(k)}(X|Y) = \frac{1}{N}\sum_{j=1}^{N} \frac{R_n^{(k)}(X)}{R_n^{(k)}(X|Y)} \tag{8.28}$$

should reflect the interdependence in the sense that closeness of the points in Y implies closeness of their equal time partners in X and the values of $S^{(k)}(X|Y)$ approach to one, while, in the case of X independent of Y, $S^{(k)}(X|Y) \ll 1$. The quantity $S^{(k)}(Y|X)$ measuring the influence of X on Y is defined in full analogy.

These three measures were tested on the examples of the Rössler system driving the Lorenz system and for the unidirectionally coupled Henon system and then on unidirectionally coupled Rössler systems. Neither the cross-predictability, nor the mutual nearest neighbours statistics gave consistent results when using three different examples of unidirectionally coupled systems. Only the coarse-grained transinformation rate correctly identified the direction of the causal influence in the above three examples as well as in many other systems of different origins (tested in other works from the authors). In the above mentioned examples of unidirectionally coupled systems, the used measures were generally non-zero in both directions even before the systems became synchronized and comparison of the values of such measures did not always reflect the true causality given by the unidirectional coupling of the systems. The intuitively understandable implication that the lower prediction error (better predictability) implies the stronger dependence cannot be in general applied to nonlinear systems. When the coupling of the systems is weaker than what is necessary for the emergence of synchronization, as used in the above examples, any smooth deterministic function between the states of the systems does not have to exist yet. However, there is already some statistical relation valid on the coarse-grained description level. Although the deterministic quantities are based on the existence of a smooth functional relation, when estimated with finite precision they usually give nonzero values influenced not only by the existing statistical dependence but also by the properties of the systems other then the coupling. Therefore it is necessary to use quantities proposed for measuring statistical dependence such as information-theoretic measures which have solid mathematical background. Conditional mutual information vanishes in the uncoupled direction in the case of unidirectional coupling so that the causal direction can be identified by its statistically significant digression from zero, while in the uncoupled direction it does not cross the borders of a statistical zero. From this respect has the CTIR based causality detector a special position in the comparison to the deterministic ones. Factors and influences, which can lead to either decreased test sensitivity or to false causality detections, were identified and concrete remedies proposed in [64] in order to perform tests with high sensitivity and low rate of false positive results.

8.2.5 Classification and Criteria for Methods for Entropy Estimation

The key problem for causality detection by means of conditional mutual information is to have a "good" estimator of mutual information. Most entropy estimators in the literature, which are designed for multi-dimensional spaces, can be applied to mutual information estimation. In the following, we adopt mathematical criteria for evaluation of the entropy estimators from Beirlant et al. [11].

8.2.6 Conditions and Criteria

If for the identically independent distributed (i.i.d.) sample $X_1, \ldots, X_n$, H_n is an estimate of $H(f)$, then the following types of consistencies can be considered:

Weak consistency: $\lim_{n\to\infty} H_n = H(f)$ in probability; *Mean square consistency*: $\lim_{n\to\infty} E(H_n - H(f))^2 = 0$; *Strong (universal) consistency*: $\lim_{n\to\infty} H_n = H(f)$ a.s. (almost sure); *Slow-rate convergence*: $\limsup_{n\to\infty} \frac{E|H_n - H|}{a_n} = \infty$ for any sequence of positive numbers $\{a_n\}$ converging to zero; *Root-n consistency* results are either of form of *asymptotic normality*, i.e., $\lim_{n\to\infty} n^{1/2}(H_n - H(f)) = N(0, \sigma^2)$ convergence in distribution, of L_2 *rate of convergence*: $\lim_{n\to\infty} nE(H_n - H(f))^2 = \sigma^2$ or for the *consistency in* L_2, $\lim_{n\to\infty} E(H_n - H(f))^2 = 0$.

The conditions on the underlying density f are: *Smoothness conditions*:$(S1)$ f is continuous. $(S2)$ f is k times differentiable. *Tail conditions*: $(T1)$ $H([X]) < \infty$, where $[X]$ is the integer part of X. $(T2)$ $\inf_{f(x)>0} f(x) > 0$. *Peak conditions*: $(P1)$ $\int f(\log f)^2 < \infty$. (This is also a mild tail condition.) $(P2)$ f is bounded.

Many probability distributions in statistics can be characterized as having maximum entropy and can be generally characterized by Kagan–Linnik–Rao theorem [45]. When dealing with the convergence properties of estimates, one needs the following definitions. *Asymptotically consistent* estimator means that the series of the approximants converge in infinity to the function to be approximated (see, i.e., [11]). *Asymptotically unbiased* estimator is that one which is unbiased in the limit.

In [43] we reviewed current methods for entropy estimation. Most of the methods were originally motivated by other questions than detection of causality: by learning theory questions, or by nonlinear dynamics applications. Many of them, although accurate in one or two dimension, become inapplicable in higher dimensional spaces (because of their computational complexity). Here we discuss these methods from their consistency point of view.

8.3 Non-Parametric Entropy Estimators

8.3.1 Plug-in Estimates

Plug-in estimates are based on a consistent density estimate f_n of f such that f_n depends on $X_1, \ldots, X_n$. Their name "plug-in" was introduced by Silverman [84]. A consistent probability density function estimator is substituted into the place of the pdf of a functional. The most used plug-in estimators are integral estimators, resubstitution estimates, splitting date estimates and cross-validation estimates. Strong consistency of integral estimators was proven by Dmitriev and Tarasenko [25] and by Prasaka Rao [71]. The resubstitution estimates have the mean square consistency which was proven by Ahmad and Lin [2]. Splitting data estimate have under some mild tail and smoothness condition on f strong consistency for general dimension d [38]. Ivanov and Rozhkova showed strong consistency of cross-validation estimates [44]. Convergence properties of discrete plug-in estimators were studied by Antos and Kontoyiannis [5] in a more general scope. They proved that for additive functionals, including the cases of the mean, entropy, Rényi entropy and mutual information, satisfying some mild conditions, the plug-in estimates are universally consistent and consistent in L_2 and he L_2-error of the plug-in estimate is of order $O(\frac{1}{n})$. For discrete estimators, the convergence results obtained by Antos and Kontoyiannis [5] are in agreement with the convergence results of the all above mentioned plug-in methods. On the other hand, for a wide class of other functionals, including entropy, it was shown that the universal convergence rates cannot be obtained for any sequence of estimators. Therefore, for positive rate-of-convergence results, additional conditions need to be placed on the class of considered distributions.

8.3.2 Entropy Estimates Based on the Observation Space Partitioning

8.3.2.1 Fixed Partitioning

These estimators divide the observation space into a set of partitions. The partition is generated either directly or recursively (iteratively). The algorithms employ a fixed scheme independent of the data distribution or an adaptive scheme which takes the actual distribution of the data into account. The most widely used methods with fixed partitioning are classical histogram methods, where the approximation of the probability distributions $p(x_i, y_j)$, $p(x_i)$ and $p(y_j)$ is by a histogram estimation [18]. These methods work well only up to three scalars. An insufficient amount of data, occurring especially in higher dimensions, leads to a limited occupancy of many histogram bins giving incorrect estimations of the probability distributions and consequently leads to heavily biased, usually overestimated values of mutual information. Consistency of histogram methods was analysed by Lugosi and Nobel in [53], who presented general sufficient conditions for the almost sure L_1-consistency of multivariate histogram density estimates based on data-dependent partitions. Analogous conditions guarantee the almost-sure risk consistency of histogram classification schemes based on data-dependent partitions.

8.3.2.2 Adaptive Partitioning

Marginal equiquantization

Any method for computation of mutual information based on partitioning of data space is always connected with the problem of quantization, i.e., a definition of finite-size boxes covering the state (data) space. The probability distribution is then estimated as relative frequencies of the occurrence of data samples in particular boxes (the histogram approach). A naive approach to estimate the mutual information of continuous variables would be to use the finest possible quantization, e.g., given by a computer memory or measurement precision. One must however keep in mind that a finite number N of data samples is available. Hence, using a quantization that is too fine, the estimation of entropies and mutual information can be heavily biased – we say that the data are overquantized.

As a simple data adaptive partitioning method, Paluš [59, 61] used a simple box-counting method with marginal equiquantization. The marginal boxes are not defined equidistantly but so that there is approximately the same number of data points in each marginal bin. The choice of the number of bins is, however, crucial. In [59] Paluš proposed that computing the mutual information I^n of n variables, the number of marginal bins should not exceed the $n+1$-st root of the number of the data samples, i.e., $q \leq \sqrt[n+1]{N}$. The equiquantization method effectively transforms each variable (in one dimension) into a uniform distribution, i.e., the individual (marginal) entropies are maximized and the MI is fully determined by the value of the joint entropy of the studied variable. This type of MI estimate, even in its coarse-grained version, is invariant against any monotonous (and nonlinear) transformation of the data [62]. Due to this property, MI, estimated using the marginal equiquantization method, is useful for quantifying dependence structures in data as well as for statistical tests for nonlinearity which are robust against static nonlinear transformations of the data [59].

Darbellay and Vajda [21] demonstrated that MI can be approximated arbitrarily closely in probability and proved the weak consistency. Their method was experimentally compared to maximum-likelihood estimators (Sect. 8.3.6). The partitioning scheme used by Darbellay and Vajda [21] was originally proposed by Fraser and Swinney [29] and in physics literature is referred to as the Fraser–Swinney algorithm, while in the information-theoretic literature as the Darbellay–Vajda algorithm.

8.3.3 Ranking

Pompe [70] proposed an estimator of dependencies of a time series based on second order Rényi entropy. Pompe noticed that if the time series is uniformly distributed, some of the desirable properties of Shannon entropy can be preserved for the second order Rényi entropy. Moreover, the second order Rényi entropy can be effectively estimated using the Grassberger–Procaccia–Takens Algorithm (GPTA) [36]. The idea of Pompe's entropy is in finding a transformation of an arbitrarily distributed time series to a uniform distribution and is accomplished by sorting the samples using some common fast sorting algorithm. There are no consistency results (even in their weakest form) known.

8.3.4 Estimates of Entropy and Mutual Information Based on Nearest Neighbor Search

Estimators of Shannon entropy based on k-nearest neighbor search in one-dimensional spaces were studied in statistics already almost 50 years ago by Dobrushin [26] but they cannot be directly generalized to higher dimensional spaces. For general multivariate densities, the nearest neighbor entropy estimate is defined as the sample average of the algorithms of the normalized nearest neighbor distances plus the Euler constant. Under the condition (P1) introduced in Sect. 8.2.6, Kozachenko and Leonenko [48] proved the mean square consistency for general $d \geq 1$. Tsybakov and van der Meulen [88] showed root-n rate of convergence for a truncated version of H_n in one dimension for a class of densities with unbounded support and exponential decreasing tails, such as the Gaussian density.

Leonenko et al. [52] studied a class of k-nearest-neighbor-based Rényi estimators for multidimensional densities. It was shown that Rényi entropy of any order can be estimated consistently with minimal assumptions on the probability density. For Shannon entropy, and for any $k > 0$ integer, the expected value of the k-nearest neighbor estimator (including the KSG algorithm described below) converges with the increasing size of data set N to infinity to the entropy of f if f is a bounded function (asymptotical unbiasedness). For any $k > 0$ integer, the k-nearest neighbor estimator converges for the Euclidean metric (L_2 rate of convergence), with the increasing size of data set N to infinity, to the entropy of f if f is a bounded function (consistency).

An improvement of KL algorithm for using in higher dimensions was proposed by Kraskov, Stögbauer and Grassberger (KSG) [49]. The estimator differs from the KL that it uses different distance scales in the joint and marginal spaces. Any consistency results are not known. A nearest neighbor approach to estimate Kullback–Leibler divergence was studied in [52] and by Wang et al. in [90] and its asymptotical consistency was proven.

8.3.5 Estimates Based on Learning Theory Methods

8.3.5.1 Motivated by Signal Processing Problems

Entropy and mutual information are often used as a criterion in learning theory. Entropy as a measure of dispersion is applied in many other areas, in control, search, or in the area of neural networks and supervised learning, i.e., [28, 68, 80]. Many of the developed methods belong as well to non-parametric plug-in estimators. Learning theory is interested in computationally simpler entropy estimators which are continuous and differentiable in terms of the samples, since the main objective is not to estimate the entropy itself but to use this estimate in optimizing the parameters of an adaptive (learning) system. The consistency properties of an estimator are not questioned strictly in this field since for relatively small data sets it is not critical to have a consistent or an inconsistent estimate of the entropy as long as the global optimum lies at the desired solution. These methods work in general also in higher dimensional spaces and therefore can be applicable to mutual information. From the variety of learning theory applications, we mention here the nonparametric estimator of Rényi entropy from Erdogmus [28], based on Parzen (window) estimate [66] and some neural network-based approaches. The former estimator is consistent if the Parzen windowing and the sample mean are consistent for the actual pdf of the iid samples.

8.3.5.2 Estimates by Neural Network Approaches

In the probabilistic networks, the nodes and connections are interpreted as the defining parameters of a stochastic process. The net input to a node determines its probability of being active rather than its level of activation. The distribution of states in a stochastic network of these nodes can be calculated with models from statistical mechanics by treating the net inputs as energy levels. A well-known example of this type of network is Boltzmann Machine (i.e., [42]). The entropy of the stochastic process can then be calculated from its parameters, and hence optimized. The non-parametric technique by Parzen or kernel density estimation leads to an entropy optimization algorithm in which the network adapts in response to the distance between pairs of data samples. Such entropy estimate is differentiable and can therefore be optimized in a neural network, allowing to avoid the limitations encountered with parametric methods and probabilistic networks. The consistency of such method depends on the optimization algorithm.

8.3.6 Entropy Estimates Based on Maximum Likelihood

When maximizing the likelihood, we may equivalently maximize the log of the likelihood and the number of calculations may be reduced. The log-likelihood is closely related to entropy and Fisher information. Popular methods for maximum likelihood are the Expectation-Maximization (EM) (i.e., Demster et al. [22]) and Improved Iterative Scaling (IIS) algorithms (Berger [13]). These methods are often used in classification tasks, especially in speech recognition. Paninski [65] used an exact local expansion of the entropy function and proved almost sure consistency (strong consistency) for three of the most commonly used discretized information estimators, namely the maximum likelihood (MLE) estimator the MLE with the so-called Miller–Madow bias correction [54], and for the jackknifed version of MLE from Efron and Stein [27].

8.3.7 Correction Methods and Bias Analysis in Undersampled Regime

These entropy estimates are mostly analytical and their bias can be computed. Most of them use Bayesian analysis and are asymptotically consistent (Miller [54],

Paninski [65], Nemenman et al. [56]) but there is also another approach from Grassberger [37] applying Poisson distribution.

8.3.8 Kernel Methods

Kernel density estimation methods (KDE)

Mutual information was first estimated by this approach by Moon et al. [55]. The KDE methods have more advantages than the classical histogram methods: they have a better mean square error rate of convergence of the estimate to the underlying density, are insensitive to the choice of origin and the window shapes are not limited to the rectangular window. Kernel density estimator introduced by Silverman [84] in one-dimensional space is defined

$$f(x) = \frac{1}{Nh}\sum_{i=1}^{N} K\left(\frac{x - x_i}{h}\right), \tag{8.29}$$

where h is the kernel width parameter and $K(x)$ the kernel function. It was shown by Kulkarni et al. in [50] that these estimators are consistent for any dimension.

Prichard and Theiler [72] introduced a method to compute information theoretic functionals based on mutual information using correlation integrals. Correlation integrals were introduced by Grassberger and Procaccia in [36]. Consistency of this method was proven by Borovkova et al. in [16]. Schreiber [81] proposed to compute the transfer entropy also using the correlation integrals [36].

8.4 Parametric Estimators

Some assumption about either the functional form of the density or about its smoothness can be appropriate in some cases. The most common is to assume that the density has a parametric form. This approach is preferred when there is confidence that the pdf underlying the samples belongs to a known parametric family of pdf's. It is effective when the assumed parametric family is accurate but it is not appropriate in adaptation scenarios where the constantly changing pdf of the data under consideration may not lie in a simple parametric family. Parametric entropy estimation is a two step process. First, the most probable density function is selected from the space of possible density functions. This often requires a search through parameter space (for example maximum likelihood methods). Second, the entropy of the most likely density is evaluated.

Verdugo Lazo and Rathie [89] computed a table of explicit Shannon entropy expressions for many commonly used univariate continuous pdfs. Ahmed and Gokhale [3] extended this table and results to the entropy of several families of multivariate distributions, including multivariate normal, normal, log-normal, logistic and Pareto distributions. Consistent estimators for the parametric entropy of all the above listed multivariate distributions can be formed by replacing the parameters with their consistent estimators (computed by Arnold [7]).

8.4.1 Entropy Estimators by Higher-Order Asymptotic Expansions

This class includes Fourier Expansion, Edgeworth Expansion and Gram–Charlier Expansion and other expansions [39]. These methods are recommended especially for distributions which are "close to the Gaussian one" [43]. The Edgeworth expansion, similarly as the Charlier–Gram expansion approximates a probability distribution in terms of its cumulants. All the three expansion types are consistent, i.e., in infinity converge to the function which they expand, Cramer [20].

8.5 Generalized Granger Causality

The classical approach of Granger causality as mentioned in Sect. 1.2 is intuitively based on the temporal properties, i.e., the past and present may cause the future but the future cannot cause the past [32]. Accordingly, the causality is expressed in terms of predictability: if the time series Y causally influences the time series X, then the knowledge of the past values of X and Y would improve a prediction of the present value of X compared to the knowledge of the past values of X alone. The causal influence in the opposite direction can likewise be checked by reversing the role of the two time series. Although this principle was originally formulated for wide classes of systems, both linear and nonlinear systems, the autoregressive modeling framework [Eq. (8.1)] proposed by Granger was basically a linear model, and such a choice was made primarily due to practical reasons [33]. Therefore, its direct application to nonlinear systems may or may not be appropriate.

8.5.1 Nonlinear Granger Causality

Ancona et al. [4] extended Granger's causality definition to nonlinear bivariate time series. To define linear Granger causality [32], the vector autoregressive model was modeled by radial basis neural networks [17]. A directionality index was introduced measuring the unidirectional, bidirectional influence or uncorrelation which was computed again by means of conditional mutual information applying generalized correlation integral [36].

8.5.2 Nonparametric Granger Causality

Despite the computational benefit of model-based (linear and/or nonlinear) Granger causality approaches, it should be noted that the selected model must be appropriately matched to the underlying dynamics, otherwise model mis-specification would arise, leading to spurious causality values. A suitable alternative would be to adopt nonparametric approaches which are free from model mismatch problems. We discuss here those nonparametric approaches which can be expressed in the information theoretic terms. Let us first reformulate the Granger causality in information theoretic terms [23, 24]: For a pair of stationary, weakly dependent, bivariate time series $\{X_t, Y_t\}$, Y is a Granger cause of X if the distribution of X_t given past observations of X and Y differs from the distribution of X_t given past observations of X only. Thus $\{Y_t\}$ is a Granger cause of $\{X_t\}$ if

$$F_{X_{t+1}}(x|F_X(t), F_Y(t)) \neq F_{X_{t+1}}(x|F_X(t)), \tag{8.30}$$

where $F_{X_{t+1}}$ represents the cumulative distribution function of X_{t+1} given F, and $F_X(t)$ and $F_Y(t)$ represents the information contained in past observations of X and Y up to and including time t. The idea of the Granger causality is to quantify the additional amount of information on X_{t+1} contained in $\mathbf{Y}_t$, given $\mathbf{X}_t$. Now, the average amount of information which a random variable X contains about another random variable Y can be expressed in terms of generalized correlation integrals [see the equivalent Eq. (8.9)] as $I_q(X,Y) = \log C_q(X,Y) - \log C_q(X) - \log C_q(Y)$ where the generalized correlation integral [36], C_q can be estimated by

$$C_q(\mathbf{X}, \varepsilon) = \frac{1}{N(N-1)^{q-1}} \sum_{j=1}^{N} \left[\sum_{i \neq j} \Theta(||X_j - X_i|| - \varepsilon) \right]^{q-1}; \tag{8.31}$$

Θ is the Heaviside function, $||.||$ a norm and the last term is related to kernel density estimation. The extra amount of information that $\mathbf{Y}_t$ contains about X_{t+1} in addition to the information already contained in $\mathbf{X}_t$ will be measured by the information theoretic measure of Granger causality: $I^{GC}_{Y \to X} = I(\mathbf{X}_t, \mathbf{Y}_t; X_{t+1}) - I(\mathbf{X}_t; X_{t+1}) = \log C(\mathbf{X}_t, \mathbf{Y}_t, X_{t+1}) - \log C(\mathbf{X}_t, X_{t+1}) - \log C(\mathbf{X}_t, \mathbf{Y}_t) + \log C(\mathbf{X}_t)$.

In order to obtain statistical significance, bootstrapping procedure is recommended to check if the statistic is significantly larger than zero [23].

Here the causality measure is based on conditional entropy, and unlike mutual or time-lagged information measures, can distinguish actually transported information from that produced as a response to a common driver or past history [81]. Interestingly, these entropies can be expressed in terms of generalized correlation integrals whose nonparametric estimation is well known. Correlation integral based nonparametric Granger causality test was originally proposed by Baek and Brock [9] and then later modified by Hiemstra and Jones [40] in the field of econometrics. More details to this method can be found in [43].

8.6 Conclusion

The main objective of this paper was to show that information theory and information theoretical measures, in particular conditional mutual information, can detect and measure causal link and information flow between observed variables. However, it opens a more difficult question: How to reliably estimate these measures from a finite data set? Research literature abounds with various estimators with a diverse range of assumptions and statistical properties. Theoretically, for a good entropy estimator, the condition of consistency seems to be important. However, it should be noted that the conditions for desired consistency might be too restrictive for an experimental environment. Accordingly, we also critically reviewed those methods which have surprisingly good overall performance (i.e., small systematic and statistical error for a wide class of pdfs) though their consistency properties are not yet known. Last but not least, let us mention some informal comments on the detection of causality which are relevant to any causality measure applied. One needs to be extra careful before claiming a causal relationship between observed variables. From the viewpoint of establishing new models, inferences and control strategies, establishing a causal relationship is always tempting. However, one has to first carefully scrutinize the statistical properties of the observed data sequences and the completeness of the model or the assumptions necessary for the estimation of the information theoretic measures. Otherwise, spurious results could often be obtained (i.e., as discussed in Sect. 8.2.4). Despite these precautionary remarks, we would like to stress again that there are enough good reasons, contrary to B. Russel's arguments [76], to investigate causality, offering numerous applications in natural and physical sciences.

Acknowledgements The author thanks her colleagues M. Paluš, M. Vejmelka and J. Bhattacharya for their valuable discussions and for the support of the Czech Grant Agency by project MSMT CR 2C06001 (Bayes).

References

1. Abarbanel, H.D.I.: Introduction to Nonlinear Dynamics for Physicists. In: Lecture Notes in Physics. World Scientific, Singapore (1993)
2. Ahmad, I.A., Lin, P.E.: A nonparametric estimation of the entropy for absolutely continuous distributions. IEEE Trans Inform Theory **22**, 372–375 (1976)
3. Ahmed, N.A., Gokhale, D.V.: Entropy expressions and their estimators for multivariate distributions. IEEE Trans Inform Theory **35**, 688–692 (1989)
4. Ancona, N., Marinazzo D., Stramaglia, S.: Radial basis function approach to nonlinear Granger causality of time series. Phys Rev E **70**, 056221 (2004)
5. Antos A., Kontoyiannis, I.: Convergence properties of functional estimates for discrete distributions. Random Struct Algor, Special issue: Average-Case Analysis of Algorithms **19**, 163–193 (2002)
6. Aparicio, F.M, Escribano, A.: Information-theoretic analysis of serial dependence and cointegration. Studies in Nonlinear Dynamics and Econometrics **3**, 119–140 (1998)

7. Arnold, B.C.: Pareto Distributions. International Co-Operative, Burtonsvile, MD, 1985
8. Arnhold, J., Grassberger, P., Lehnertz, K., Elger, C.E.: A robust method for detecting interdependences: Application to intracranially recorded EEG. Physica D **134**, 419–430 (1999)
9. Baek, E.G., Brock, W.A.: A general test for nonlinear Granger causality: Bivariate model, Working paper, Iowa State University and University of Wisconsin, Madison (1992)
10. Baghli, M.: A model-free characterization of causality. Econ Lett **91**, 380–388 (2006)
11. Beirlant, J., Dudewitz, E.J., Györfi, L., van der Meulen, E.C.: Nonparametric entropy estimation: An overview. Int J Math Stat Sci **6**, 17–39 (1997)
12. Bell, D., Kay, J., Malley, J.: A non-parametric approach to non-linear causality testing. Econ Lett **51** 7–18 (1996)
13. Berger, A.: The improved iterative scaling algorithm: A gentle introduction (http://www.cs.cmu.edu/afs/cs/user/aberger/www/ps/scaling.ps (1997)
14. Blinowska, K.J., Kuś, R., Kamiński, M.: Granger causality and information flow in multivariate processes. Phys Rev E **70**, 050902(R) (2004)
15. Boccaletti, S., Kurths, J., Osipov, G., Valladares, D.L., Zhou, C.S.: The synchronization of chaotic systems. Phys Rep **366**, 1–101 (2002)
16. Borovkova, S., Burton, R., Dehling, H.: Consistency of the Takens estimator for the correlation Dimension. Ann Appl Probab **9** 2, 376–390 (1999)
17. Broomhead, D.S., Lowe, D.: Multivariate functional interpolation and adaptive networks. Complex Syst **2**, 321–355 (1988)
18. Butte, A.J., Kohane, I.S.: Mutual information relevance networks: Functional genomic clustering using pairwise entropy measurements. Pac Symp Biocomput 418–429 (2000)
19. Chen, Y., Rangarajan, G., Feng, J., Ding, M.: Analyzing mulitiple nonlinear time series with extended Granger causality. Phys Lett A **324**, 26–35 (2004)
20. Cramer, H.: On the composition of elementary errors. Skand Aktuarietidskr **11**, 13–14 and 141–180 (1928)
21. Darbellay, G., Vajda, I.: Estimation of the information by an adaptive partitioning of the observation space. IEEE Trans Inform Theory **45**, 1315–1321 (1999)
22. Dempster, A., Laird, N., Rubin, D.: Maximum likelihood from incomplete data via the EM algorithm. J R Stat Soc B **39**, 1–38 (1977)
23. Diks, C., DeGoede, J.: A general nonparametric bootstrap test for Granger causality. In: Broer, Krauskopf and Vegter (eds.), Global Analysis of Dynamical Systems, Chapter 16. IoP, London (2001), 391–403
24. Diks, C., Panchenko, V.: A note on the Hiemstra–Jones test for Granger non-causality. Stud Nonlinear Dynamics Econometrics **9**(4), 1–7 (2005)
25. Dmitriev, Y.G., Tarasenko, F.P.: On the estimation functions of the probability density and its derivatives. Theory Probab Appl **18**, 628–633 (1973)
26. Dobrushin, R.L.: A simplified method of experimentally evaluating the entropy of a stationary sequence. Teoriya Veroyatnostei i ee Primeneniya **3**, 462–464 (1958)
27. Efron, B., Stein, C.: The jackknife estimate of variance. Ann Stat **9**, 586–596, (1981)
28. Erdogmus, D.: Information theoretic learning: Renyi's Entropy and its Application to Adaptive System Training, PhD thesis, University of Florida (2002)
29. Fraser, A., Swinney, H.: Independent coordinates for strange attractors from mutual information. Phys Rev A **33**, 1134–1140 (1986)
30. German, A., Carlin, J.B., Stern, H.S., Rubin, D.B.: Bayesian Data Analysis. Texts in Statistical Science Series. Chapman and Hall, London (2004)
31. Geweke, J.: Inference and causality in economic time series models. In: Griliches, Z., Intriligator, M.D. (eds.), Handbook of Econometrics, vol. 2, 1101–1144. North-Holland, Amsterdam (1984)
32. Granger, C.W.J.: Investigating causal relations by econometric and cross-spectral methods. Econometrica **37**, 424–438 (1969)
33. Granger, C.W.J., Newbold, P.: Forecasting Economic Time Series. Academic, New York (1977)

34. Granger, C.W.J.: Testing for causality: A personal viewpoint. J Econ Dyn Control **2**, 329–352 (1980)
35. Granger, C.W.J.: Time series analysis, cointegration, and applications. Nobel Lecture, December 8, 2003. In: Frängsmyr, T. (ed.), Les Prix Nobel. The Nobel Prizes 2003. Nobel Foundation, Stockholm (2004), pp. 360–366. http://nobelprize.org/nobel_prizes/economics/laureates/2003/granger-lecture.pdf
36. Grassberger, P., Procaccia, I.: Measuring of strangeness of strange attractors. Physica D **9**, 189–208 (1983)
37. Grassberger, P.: Finite sample corrections to entropy and dimension estimates. Phys Lett A **128**, 369–373 (1988)
38. Györfi, L., Van der Meulen, E.C.: On nonparametric estimation of entropy functionals. In: G. Roussas (ed.), Nonparametric Functional Estimation and Related Topics. Kluwer, Amsterdam (1990), pp. 81–95
39. Haykin, S.: Neural Networks: A Comprehensive Foundation, Second Edition. Prentice Hall, Englewood Cliffs (1998)
40. Hiemstra, C., Jones, J.D.: Testing for linear and nonlinear Granger causality in the stock price–volume relation. J Finance **49**, 1639–1664 (1994)
41. Hinrichs, H., Heinze, H.J., Schoenfeld, M.A.: Causal visual interactions as revealed by an information theoretic measure and fMRI. NeuroImage **31**, 1051–1060 (2006)
42. Hinton, G., Sejnowski, T.: Learning and relearning in Boltzmann machines. In: Rumelhart, D., J. McClelland J. (eds.), Parallel Distributed procesing, Vol. 1. MIT, Cambridge (1986), Chap. 7, pp. 282–317
43. Hlaváčková-Schindler, K., Paluš, M., Vejmelka, M., Bhattacharya, J.: Causality detection based on information-theoretic approaches in time series analysis. Phys Rep **441**(1), 1–46 (2007), doi:10.1016/j.physrep.2006.12.004
44. Ivanov, A.V., Rozhkova, A.: Properties of the statistical estimate of the entropy of a random vector with a probability density. Prob Inform Transmission **10**, 171–178 (1981)
45. Kagan, A.M., Linnik, Y.V., Rao, C.R.: Characterization Problems in Mathematical Statistics. Wiley, New York (1973)
46. Katura, T., Tanaka, N., Obata, A., Sato, H., Maki, A.: Quantitative evaluation of interrelations between spontaneous low-frequency oscillations in cerebral hemodynamics and systemic cardiovascular dynamics. NeuroImage **31**, 1592–1600 (2006)
47. Kolmogorov, A.N.: Entropy per unit time as a metric invariant of automorphism. Dokl Akad Nauk SSSR **124**, 754–755 (1959)
48. Kozachenko, L.F., Leonenko, N.N.: Sample estimate of the entropy of a random vector. Prob Inform Transmission **23**, 95–100 (1987)
49. Kraskov, A., Stögbauer H., Grassberger, P.: Estimation mutual information. Phys Rev E **69**, 066138 (2004)
50. Kulkarni, S.R., Posner, S.E., Sandilya, S.: Data-dependent $k-NN$ and kernel estimators consistent for arbitrary processes. IEEE Trans Inform Theory **48**(10), (2002)
51. Kullback, S., Leibler, R.A.: On information and sufficiency. Ann Math Stat **22**, 79–86 (1951)
52. Leonenko, N., Pronzato, L., Savani, V.: A class of Rényi information estimators for multidimensional densities. Laboratoire I3S, CNRS–Universit de Nice-Sophia Antipolis, Technical report I3S/RR-2005-14-FR (2005)
53. Lugosi, G., Nobel, A.: Consistency of data-driven histogram methods for density estimation and classification. Ann Stat **24**(2), 687–706 (1996)
54. Miller, G.: Note on the bias of information estimates. In: Quastler, H. (ed.), Information theory in psychology II-B. Free Press, Glencoe (1955), pp. 95–100
55. Moon, Y., Rajagopalan, B., Lall, U.: Estimation of mutual information using kernel density estimators. Phys Rev E **52**, 2318–2321 (1995)
56. Nemenman, I., Bialek, W., de Ruyter van Stevenick, R.: Entropy and information in neural spike trains: progress on sampling problem. Phys Rev E **69**, 056111 (2004)
57. Otsuka, K., Miyasaka, Y., Kubota, T.: Formation of an information network in a self-pulsating multimode laser. Phys Rev E **69**, 046201 (2004)

58. Paluš, M., Komárek, V., Hrnčíř, Z., Štěrbová, K.: Synchronization as adjustment of information rates: Detection from bivariate time series. Phys Rev E **63**, 046211 (2001)
59. Paluš, M.: Testing for nonlinearity using redudancies: Quantitative and qualitative aspects. Physica D **80**, 186–205 (1995)
60. Paluš, M.: Coarse-grained entropy rates for characterization of complex time series. Physica D **93**, 64–77 (1996)
61. Paluš, M.: Identifying and quantifying chaos by using information-theoretic functionals. In: Weigend, A.S., Gershenfeld, N.A. (eds.), Time series prediction: Forecasting the future and understanding the past. Santa Fe Institute Studies in the Sciences of Complexity, Proc. Vol. XV. Addison-Wesley, Reading, (1993), pp. 387–413
62. Paluš, M.: Detecting nonlinearity in multivariate time series. Phys Lett A **213**, 138–147 (1996)
63. Paluš, M., Hoyer, D.: Detecting nonlinearity and phase synchronization with surrogate data. IEEE Eng Med Biol **17**, 40–45 (1998)
64. Paluš, M., Vejmelka, M.: Directionality of coupling from bivariate time series: How to avoid false causalities and missed connections. Phys Rev E **75** 056211 (2007), doi:10.1103/PhysRevE.75.056211
65. Paninski, L.: Estimation of entropy and mutual information. Neural Comput **15**, 1191–1253 (2003)
66. Parzen, E.: On estimation of a probability density function and mode. In: Time Series Analysis Papers. Holden-Day, San Diego (1967)
67. Pearl, J.: Causality: Models, Reasoning and Inference. Cambridge University Press, New York (2000)
68. Peters, M.A., Iglesias, P.A.: Minimum entropy control for discrete-time varying systems. Automatica **33**, 591–605 (1997)
69. Pikovsky, A., Rosenblum, M., Kurths, J.: Synchronization. A Universal Concept in Nonlinear Sciences. Cambridge University Press, Cambridge (2001)
70. Pompe, B.: Measuring statistical dependencies in a time series. J Stat Phys **73**, 587–610 (1993)
71. Prasaka Rao, B.L.S.: Nonparametric Functional Estimation. Academic, New York (1983)
72. Prichard, D., Theiler, J.: Generalized redundancies for time series analysis. Physica D **84**, 476–493 (1995)
73. Le Van Quyen, M., Martinerie, J., Adam, C., Varela, F.J.: Nonlinear analyses of interictal EEG map the brain interdependences in human focal epilepsy. Physica D **127**, 250–266 (1999)
74. Quian Quiroga, R., J. Arnhold, J., Grassberger, P.: Learning driver-response relationships from synchronization patterns. Phys Rev E **61**(5), 5142–5148 (2000)
75. Rényi, A.: On measures of entropy and information, In: Proc. Fourth Berkeley Symp. Math. Stat. and Probability, Vol. 1. University of California Press, Berkeley (1961), pp. 547–561
76. Russel, B.: On the notion of cause. In: Proceedings of the Aristotelian Society, New Series **13**, 1–26 (1913)
77. Schäfer, C., Rosenblum, M.G., Kurths, J., Abel, H.H.: Heartbeat synchronized with ventilation. Nature **392**, 239–240 (1998)
78. Schiff, S.J., So, P., Chang, T., Burke, R.E., Sauer, T.: Detecting dynamical interdependence and generalized synchrony through mutual prediction in a neural ensemble. Phys Rev E **54**, 6708–6724 (1996)
79. Schmitz, A.: Measuring statistical dependence and coupling of subsystems. Phys Rev E **62**, 7508–7511 (2000)
80. Schraudolph, N.: Gradient-based manipulation of non-parametric entropy estimates. IEEE Trans Neural Netw **14**, 828–837 (2004)
81. Schreiber, T.: Measuring information transfer, Phys Rev Lett **85**, 461–464 (2000)
82. Selltiz, C., Wrightsman, L.S., Cook, S.W.: Research Methods in Social Relations. Holt, Rinehart and Winston, New York (1959)
83. Shannon, C.E.: A mathematical theory of communication. Bell System Tech J **27**, 379–423 (1948)
84. Silverman, B.W.: Density Estimation. Chapman and Hall, London (1986)

85. Sinai, Y.G.: On the concept of entropy for a dynamic system. Dokl Akad Nauk SSSR **124**, 768–771 (1959)
86. Suppes, P.: A Probabilistic Theory of Causality, North-Holland, Amsterdam (1970)
87. Takens, F.: In: Rand, D.A., Young, D.S. (eds.), Dynamical Systems and Turbulence, Warwick 1980. Lecture Notes in Mathematics, Vol. 898. Springer, Berlin, (1981), p. 365
88. Tsybakov A.B., van Meulen, E.C.: Root-n consistent estimators of entropy for densities with unbounded support. Scand J Stat **23**, 75–83 (1994)
89. Verdugo Lazo, A.C.G., Rathie, P.N.: On the entropy of continuous probability distributions. IEEE Trans Inform Theory **24**, 120–122 (1978)
90. Wang, Q., Kulkarni, S.R., Verdú, S.: A nearest-neighbor approach to estimating divergence between continuous random vectors. ISIT 2006, Seattle, USA, July 9–14 (2006)
91. Wiener, N.: The theory of prediction. In: Beckenbach, E.F. (ed.), Modern Mathematics for Engineers. McGraw-Hill, New York (1956)
92. http://en.wikipedia.org/wiki/Causality

Chapter 9
Information Theoretic Learning and Kernel Methods

Robert Jenssen

Abstract In this chapter, we discuss important connections between two different approaches to machine learning, namely Renyi entropy-based information theoretic learning and the Mercer kernel methods. We show that Parzen windowing for estimation of probability density functions reveals the connections, enabling the information theoretic criteria to be expressed in terms of mean vectors in a Mercer kernel feature space, or equivalently, in terms of kernel matrices. From this we learn not only that two until now separate paradigms in machine learning are related, it also enables us to interpret and understand methods developed in one paradigm in terms of the other, and to develop new sophisticated machine learning algorithms based on both approaches.

9.1 Introduction

In machine learning, the goal is to learn the parameters of a machine, or system, such that an optimal output is produced based on some input data samples. Optimality is measured by a *learning criterion*, which basically determines the type of problems which may be solved. For instance, traditional mean squared error (MSE) criteria may only capture the second order statistics of the data, thus rendering themselves unsuitable for solving problems where higher order statistical properties are needed. Examples of such problems are abundant, including non-linear clustering and classification, blind source separation, dimensionality reduction, etc. Figure 9.1 illustrates the machine learning process.

Recently, a new approach to machine learning has been developed, coined *information theoretic learning* (ITL) [25]. As the name suggests, in ITL, the learning criteria are defined in terms of certain information theoretic properties of the prob-

R. Jenssen
Department of Physics and Technology, University of Tromsø, 9037 Tromso, Norway
e-mail: robert.jenssen@phys.uit.no, URL: http://www.phys.uit.no/ robertj

F. Emmert-Streib, M. Dehmer (eds.), *Information Theory and Statistical Learning*, 209
DOI: 10.1007/978-0-387-84816-7_9,

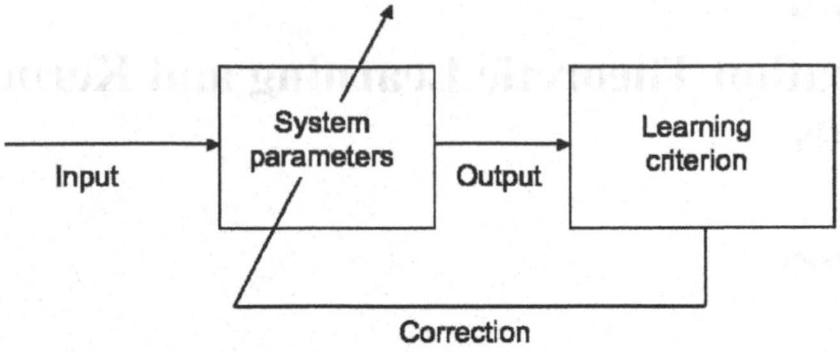

Fig. 9.1 The machine, or system, receives some input data samples and produces an output. The output depends on the current setting of the parameters of the system. The goal is to adjust, or learn, the parameters such that the problem at hand is solved, thus producing the optimal output. The adjustment of the parameters is typically performed in an iterative process. A learning criterion is invoked, which measures the goodness of the parameters at each iteration step, in terms of solving the problem. Based on the learning criterion, a correction term is fed back to the system, to guide the parameter adjustment

ability density function (pdf) of either the input data or the output data. Such properties are defined as functions over pdfs, and capture therefore *all* the statistical properties of the data.

The most fundamental information theoretic quantity is the entropy of a pdf. The entropy $H(p)$ of a pdf $p(\mathbf{x})$ is related to the shape of the pdf. A broad pdf, like the Gaussian, typically corresponds to large entropy, and vice versa. A versatile family of entropy functions, characterized by the parameter α, is given by Renyi's entropy [26]

$$H_\alpha(p) = \frac{1}{1-\alpha} \log \int p^\alpha(\mathbf{x}) d\mathbf{x}. \tag{9.1}$$

This definition of entropy contains Shannon's entropy [29] as a special case for $\alpha \to 1$. ITL is however based on Renyi's entropy of order $\alpha = 2$. The reason for this choice is that the resulting entropy function $H_2(p)$, called Renyi's quadratic entropy, may be elegantly estimated based on *samples* from the set $\mathscr{D} = \{\mathbf{x}_1, \ldots, \mathbf{x}_N\}$, assumed generated from the pdf $p(\mathbf{x})$. This is achieved by replacing the pdf by an estimate $\widehat{p}(\mathbf{x})$, obtained using the so-called Parzen window density estimator, as explained in Sect. 9.2. Thus, an estimate $H_2(\widehat{p})$ of the entropy is obtained.

For example, Renyis's quadratic entropy has been used a cost function in supervised neural networks training [7]. In this case, the training input data comes in pairs $\{\mathbf{x}_t, d_t\}$, $t = 1, \ldots, N$, where d_t is the label of $\mathbf{x}_t$. The network produces the output y_t for each $\mathbf{x}_t$, and an error term $e_t = d_t - y_t$. Instead of minimizing the mean of the squared errors, the Renyi quadratic entropy of the error pdf $p(e)$ is minimized with respect to the system weights. The minimization is implemented using gradient descent by estimating $p(e)$ by a Parzen window estimator. On real time series data, such a minimum error entropy neural network was able to predict future test data more accurately than networks trained with MSE.

The other fundamental information theoretic quantity is the divergence between two or more pdfs. A divergence measure between the pdfs $p_1(\mathbf{x}),\ldots,p_C(\mathbf{x})$ may be denoted $D(p_1,\ldots,p_C)$. It measures the "distance" between the pdfs, in the sense that it is zero only if $p_1(\mathbf{x}) = \cdots = p_C(\mathbf{x})$, and takes larger values the further "apart" the pdfs are. The Kullback–Leibler divergence [19] is the most well-known such measure. It is based on the Shannon entropy. In ITL, divergence measures based on Renyi's quadratic entropy have been proposed, in order to be able to derive sample-based estimators using the Parzen window method. One such measure is based on the Cauchy–Schwarz (CS) inequality and another is based on the integrated squared error (ISE). These will be reviewed in Sect. 9.2.

For example, the CS divergence has been used as a cost function for non-linear clustering [15]. Clustering refers to the process of partitioning a data set $\mathscr{D}$ into "natural" groups, or clusters $\mathscr{D}_1,\ldots,\mathscr{D}_C$ (assuming C clusters). The CS approach to clustering consists of partitioning the input data set such that the CS divergence between the resulting cluster pdfs $p_1(\mathbf{x}),\ldots,p_C(\mathbf{x})$ is maximized, where the pdf $p_i(\mathbf{x})$ is associated with $\mathscr{D}_i$. Hence, the partition represents the system weights, the cluster pdfs are the output and the divergence is the learning criterion, as illustrated in Fig. 9.2. Note that the output pdfs are estimated by the Parzen window method $\widehat{p}_1(\mathbf{x}),\ldots,\widehat{p}_C(\mathbf{x})$, such that the CS divergence measure can be evaluated as $D_{CS}(\widehat{p}_1,\ldots,\widehat{p}_C)$. In this approach therefore, we maximize the dissimilarity between the clusters by maximizing the dissimilarity between the cluster pdfs. In mathematical terms, this is obtained by $\arg\max_{\mathscr{D}_1,\ldots,\mathscr{D}_C} D_{CS}(\widehat{p}_1,\ldots,\widehat{p}_C)$.

Many other pattern recognition problems may be formulated as the optimization of an information theoretic quantity. ITL has therefore been pursued rigorously, and its importance and usefulness have been demonstrated with great success on problems ranging from clustering [15] and classification [18] to blind source separation [11], minimum error entropy neural networks prediction [7], feature extraction [12] and matched filtering [6]. See also the recent review paper [8].

Independently of ITL, another seemingly radically different approach to machine learning has been developed, namely the celebrated Mercer kernel methods. The kernel methods [27, 30] are based on the idea of non-linearly mapping the input data points $\mathbf{x}_t \in \mathscr{D}$ into a potentially infinite dimensional data space, obtaining the

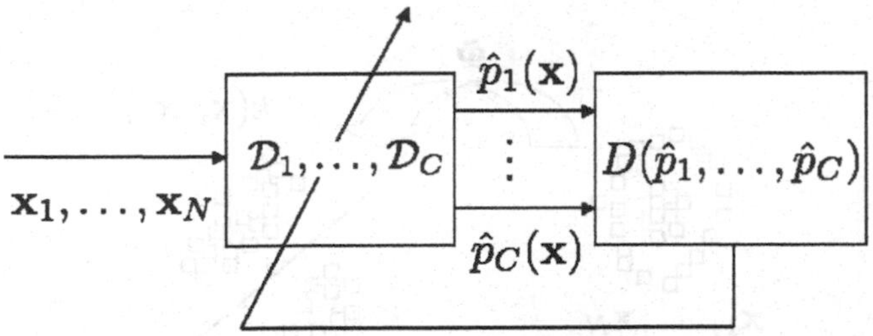

Fig. 9.2 An information theoretic approach to clustering consists of partitioning the data such that the divergence between the resulting Parzen window estimated cluster pdfs is maximized. The partition $\mathscr{D}_1,\ldots,\mathscr{D}_C$ is adjustable (system weights) and the divergence is the learning criterion

new data points $\Phi(\mathbf{x}_t)$, $t = 1,\ldots,N$. The actual learning algorithm is now executed on this new data set. One reason often stated for justifying this mapping in the context of classification, is that by Cover's theorem [4], the data set is more likely to be linearly separable in the new representation.

A special feature of the learning algorithms used by these methods, is that they are linear in nature in terms of the Φ-transformed data and expressed solely in terms of inner-products. This is very convenient, since inner-products in the high-dimensional space may be easily computed using a Mercer kernel function $k(\cdot,\cdot)$, by [20]

$$k(\mathbf{x}_t,\mathbf{x}_{t'}) = \langle \Phi(\mathbf{x}_t), \Phi(\mathbf{x}_{t'}) \rangle \,. \tag{9.2}$$

This is the so-called "kernel-trick," which is at the core of the kernel methods. The Mercer kernel is by definition positive semi-definite. The most commonly used Mercer kernel is the radial basis function

$$k(\mathbf{x}_t,\mathbf{x}_{t'}) = \exp\left\{-\frac{1}{2\sigma^2}\left\|\mathbf{x}_t - \mathbf{x}_{t'}\right\|^2\right\}, \tag{9.3}$$

where σ is a scale parameter. Many other choices also exist [30]. One reason why the kernel-trick is so important, is that the learning algorithm doesn't have to operate directly on the Φ-transformed data, it only needs to be able to compute inner-products, which is enabled by the kernel function. Hence, the learning algorithm operates *implicitly* in the kernel feature space, in general not even knowing the actual mapping Φ. Since the kernel feature space is non-linearly related to the input space, the kernel methods are non-linear in terms of the input space. Figure 9.3 illustrates the mapping to kernel feature space.

The perhaps most well-known kernel method is the support vector machine [3]. Subsequently, methods like kernel C-means [10], kernel principal component analysis [28], kernel independent component analysis [2], kernel canonical correlation analysis [9] and kernel Fisher discriminant analysis [22] have been proposed. See also the review paper [24].

Perhaps surprisingly, it has recently been shown that there are important connections between these two different approaches to machine learning [16, 17]. In this chapter, we review these relationships in detail. We show that Parzen windowing for

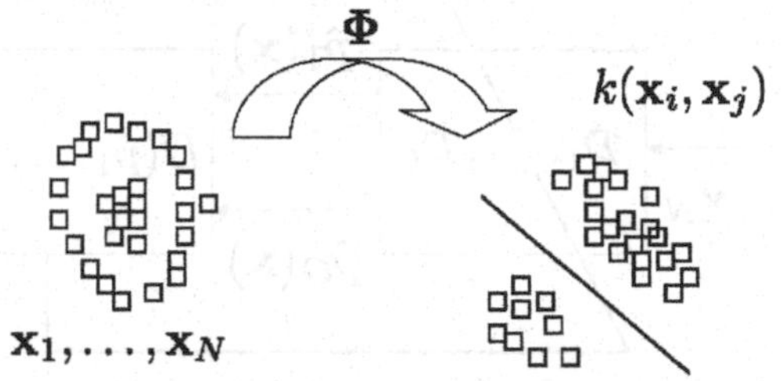

Fig. 9.3 Mapping the input data to a kernel feature space by the mapping Φ increases the likelihood of linear separability by Cover's theorem. Kernel feature space inner-products are computed using a Mercer kernel $k(\cdot,\cdot)$

estimation of probability density functions reveals the connections, enabling the information theoretic criteria to be expressed in terms of mean vectors in a Mercer kernel feature space, or equivalently, in terms of kernel matrices. From this we learn not only that two until now separate paradigms in machine learning are related, it also enables us to interpret and understand methods developed in one paradigm in terms of the other, and to develop new sophisticated machine learning algorithms based on both approaches. Specifically, we provide an interpretation of CS divergence-based clustering in terms of kernel methods and of kernel C-means clustering in terms of ITL. Finally, we discuss a new data transformation and dimensionality reduction technique called kernel entropy component analysis, which has flavors from both machine learning approaches.

This chapter is organized as follows. In Sect. 9.2, we review the basic components of information theoretic learning. Thereafter, in Sect. 9.3, the connections between information theoretic learning and the kernel methods are discussed in detail. Section 9.4 is devoted to new interpretations of methods in information theoretic learning in terms of kernel methods, and vice versa. Section 9.5 concludes the chapter.

9.2 Information Theoretic Learning Using Parzen Windowing

In this section, the basic components of information theoretic learning is briefly reviewed. The review follows to a large degree [25], where this theory was introduced.

9.2.1 Renyi Entropy

The key quantity in information theoretic learning is Renyi's entropy of order $\alpha = 2$. This is called the Renyi quadratic entropy

$$H_2(p) = -\log \int p^2(\mathbf{x})d\mathbf{x}. \tag{9.4}$$

Note that since the logarithm is a monotonic function, we may concentrate on the quantity $V(p) = \int p^2(\mathbf{x})d\mathbf{x}$. For convenience, in the following we refer to the Renyi quadratic entropy simply as Renyi entropy. In order for the Renyi entropy, equivalently $V(p)$, to be useful as a machine learning criterion, we need to be able to estimate it based on the samples alone.

Assume that the data samples $\mathscr{D} = \{\mathbf{x}_1, \ldots, \mathbf{x}_N\}$ generated from $p(\mathbf{x})$ is available. One strategy for estimating $V(p)$ is to replace $p(\mathbf{x})$ by a sample-based density estimator $\widehat{p}(\mathbf{x})$. This estimator should preferably be smooth, in order to be able to differentiate $V(p)$ with respect to system weights. One readily available density estimator is provided by the Parzen window method [23], also known as kernel density estimation. The Parzen window estimator is given by

$$\widehat{p}(\mathbf{x}) = \frac{1}{N} \sum_{\mathbf{x}_t \in \mathscr{D}} W_\sigma(\mathbf{x}, \mathbf{x}_t), \tag{9.5}$$

where $W_\sigma(\cdot,\cdot)$ is the Parzen window, whose width is governed by the parameter σ. In order for the estimator to be smooth, the window function should be smooth. Also, in order for $\widehat{p}(\mathbf{x})$ to be non-negative and integrate to one, $W_\sigma(\cdot,\cdot)$ should be non-negative and integrate to one.

There are many valid Parzen windows. The Gaussian function is perhaps the most well-known and widely used window function. In that case,

$$W_\sigma(\mathbf{x}, \mathbf{x}_t) = \frac{1}{(2\pi\sigma^2)^{\frac{d}{2}}} \exp\left\{-\frac{1}{2\sigma^2} \|\mathbf{x} - \mathbf{x}_t\|^2\right\}, \tag{9.6}$$

where d is the dimension of the data. For a fixed kernel size, it is well known that [23]

$$E\{\widehat{p}(\mathbf{x})\} = \lim_{N\to\infty} \widehat{p}(\mathbf{x}) = p(\mathbf{x}) \star W_\sigma(\mathbf{x}), \tag{9.7}$$

where $\star$ denotes convolution. The Parzen estimator is however asymptotically unbiased and consistent given a suitable annealing rate for the kernel size as $N \to \infty$. In practice though, the number of samples are limited, and the kernel size must be chosen in a trade-off between estimation bias and variance. A small kernel size generally favors low bias, at the expense of increased variance. There are many proposed criteria for kernel size selection available in the statistics literature [31], especially suitable for low-dimensional data. Figure 9.4 illustrates the process of Parzen window density estimation in the one-dimensional case.

If we assume a Gaussian Parzen window, the sample-based estimator $V(\widehat{p})$ is given by

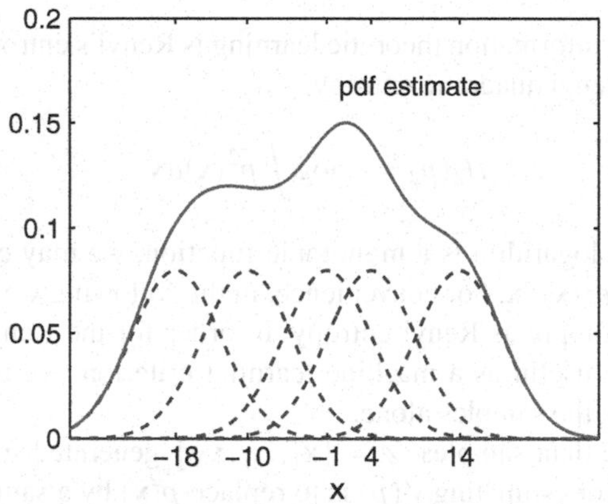

Fig. 9.4 Illustration of Parzen windowing in the one-dimensional case: The data samples $\{-18, -10, -1, 4, 14\}$ are available. A Parzen window is centered at every sample. The pdf estimate at any x is given by the sum of the Parzen windows at that location x (divided by the number of samples)

$$
\begin{aligned}
V(\widehat{p}) &= \int \frac{1}{N} \sum_{\mathbf{x}_t \in \mathscr{D}} W_\sigma(\mathbf{x}, \mathbf{x}_t) \frac{1}{N} \sum_{\mathbf{x}_{t'} \in \mathscr{D}} W_\sigma(\mathbf{x}, \mathbf{x}_{t'}) d\mathbf{x} \\
&= \frac{1}{N^2} \sum_{\mathbf{x}_t \in \mathscr{D}} \sum_{\mathbf{x}_{t'} \in \mathscr{D}} \int W_\sigma(\mathbf{x}, \mathbf{x}_t) W_\sigma(\mathbf{x}, \mathbf{x}_{t'}) d\mathbf{x} \\
&= \frac{1}{N^2} \sum_{\mathbf{x}_t \in \mathscr{D}} \sum_{\mathbf{x}_{t'} \in \mathscr{D}} W_{\sqrt{2}\sigma}(\mathbf{x}_t, \mathbf{x}_{t'}),
\end{aligned}
\tag{9.8}
$$

where in the last line we have used the property that the convolution of two Gaussian functions is itself a Gaussian function, but with increased width $\tilde{\sigma} = \sqrt{2}\sigma$. Of course, an estimator for the Renyi entropy is obtained by taking the negative logarithm of this expression. This shows that we may formulate a sample-based estimator for the Renyi entropy involving no approximations or assumptions besides the density estimation itself. This is not possible using the Shannon entropy, and was therefore one of the key arguments for using the Renyi entropy as a machine learning cost function.

A Gaussian Parzen window is however a special case. Other Parzen windows may be used by approximating the expectation operator by the sample mean. Note that $V(p) = E_p\{p(\mathbf{x})\}$, i.e., the expectation of $p(\mathbf{x})$ with respect to $p(\mathbf{x})$. Using $E_p\{p(\mathbf{x})\} \approx \frac{1}{N}\sum_{t=1}^{N} p(\mathbf{x}_t)$, and replacing $p(\mathbf{x}_t)$ by $\widehat{p}(\mathbf{x}_t)$, we obtain

$$
V(\widehat{p}) = \frac{1}{N^2} \sum_{\mathbf{x}_t \in \mathscr{D}} \sum_{\mathbf{x}_{t'} \in \mathscr{D}} W_\sigma(\mathbf{x}_t, \mathbf{x}_{t'}). \tag{9.9}
$$

In the following, we leave the choice of Parzen window open.

9.2.2 Renyi Entropy-Based Divergence Measures

Assume that the data set $\mathscr{D}$ consists of subgroups $\mathscr{D}_1, \ldots, \mathscr{D}_C$. Furthermore, assume that the N_i data points which comprise $\mathscr{D}_i$ are generated from a pdf $p_i(\mathbf{x})$. In the following, we review two Renyi entropy-based divergence measures between pdfs. These are the Cauchy–Schwarz divergence and the integrated squared error divergence.

9.2.2.1 Cauchy–Schwarz Divergence

One Renyi entropy-based divergence measure is derived from the Cauchy–Schwarz inequality, accordingly called the Cauchy–Schwarz divergence. It is given by

$$
D_{CS}(p_i, p_j) = -\log \frac{\int p_i(\mathbf{x}) p_j(\mathbf{x}) d\mathbf{x}}{\sqrt{\int p_i^2(\mathbf{x}) d\mathbf{x} \int p_j^2(\mathbf{x}) d\mathbf{x}}}. \tag{9.10}
$$

This measure if zero only if $p_i(\mathbf{x}) = p_j(\mathbf{x})$, and increases towards infinity as the two pdfs are further and further "apart," that is $D_{CS}(p_i, p_j) \in [0, \infty)$. It is also symmetric. The CS divergence is based on the Renyi entropy, since

$$D_{CS}(p_i, p_j) = -\log \int p_i(\mathbf{x}) p_j(\mathbf{x}) d\mathbf{x} - \frac{1}{2} H_2(p_i) - \frac{1}{2} H_2(p_j), \tag{9.11}$$

where $-\log \int p_i(\mathbf{x}) p_j(\mathbf{x}) d\mathbf{x}$ can be considered a Renyi "cross entropy."

Since the logarithm is a monotonic function, we may focus on the argument of the log in (9.10), which we may denote $V_{CS}(p_i, p_j)$. Now, Parzen window estimators $\widehat{p}_i(\mathbf{x})$ and $\widehat{p}_j(\mathbf{x})$ may be inserted into the expression for $V_{CS}(p_i, p_j)$, where

$$\widehat{p}_i(\mathbf{x}) = \frac{1}{N_i} \sum_{\mathbf{x}_n \in \mathscr{D}_i} W_\sigma(\mathbf{x}, \mathbf{x}_n), \qquad \widehat{p}_j(\mathbf{x}) = \frac{1}{N_j} \sum_{\mathbf{x}_m \in \mathscr{D}_j} W_\sigma(\mathbf{x}, \mathbf{x}_m), \tag{9.12}$$

obtaining the estimator

$$V_{CS}(\widehat{p}_i, \widehat{p}_j) = \frac{\frac{1}{N_i N_j} \sum_{\mathbf{x}_n \in \mathscr{D}_i} \sum_{\mathbf{x}_m \in \mathscr{D}_j} W_\sigma(\mathbf{x}_n, \mathbf{x}_m)}{\sqrt{\frac{1}{N_i^2} \sum_{\mathbf{x}_n \in \mathscr{D}_i} \sum_{\mathbf{x}_{n'} \in \mathscr{D}_i} W_\sigma(\mathbf{x}_n, \mathbf{x}_{n'}) \frac{1}{N_j^2} \sum_{\mathbf{x}_m \in \mathscr{D}_j} \sum_{\mathbf{x}_{m'} \in \mathscr{D}_j} W_\sigma(\mathbf{x}_m, \mathbf{x}_{m'})}}. \tag{9.13}$$

The estimator of the CS divergence is correspondingly $D_{CS}(\widehat{p}_i, \widehat{p}_j) = -\log V(\widehat{p}_i, \widehat{p}_j)$.

The CS divergence may be extended, so that instead of measuring the divergence between pairs of pdfs, it measures the divergence between C pdfs simultaneously, as

$$D_{CS}(p_1, \ldots, p_C) = -\log \frac{1}{\kappa} \sum_{i=1}^{C-1} \sum_{j>i} \frac{\int p_i(\mathbf{x}) p_j(\mathbf{x}) d\mathbf{x}}{\sqrt{\int p_i^2(\mathbf{x}) d\mathbf{x} \int p_j^2(\mathbf{x}) d\mathbf{x}}}, \tag{9.14}$$

where $\kappa = \sum_{c=1}^{C-1} c$. A sample-based estimator of this quantity is readily obtained, by replacing the actual pdfs by their Parzen window estimators.

9.2.2.2 Integrated Squared Error Divergence

Another Renyi entropy-based divergence measure is obtained as the integrated squared error between the pdfs. The ISE divergence is given by

$$D_{ISE}(p_i, p_j) = \int [p_i(\mathbf{x}) - p_j(\mathbf{x})]^2 d\mathbf{x}. \tag{9.15}$$

This measure also vanishes if $p_i(\mathbf{x}) = p_j(\mathbf{x})$, $D_{ISE}(p_1, p_2) \in [0, \infty)$, and it is symmetric. It is based on $V(p)$, since

$$D_{ISE}(p_i, p_j) = V(p_i) - 2 \int p_i(\mathbf{x}) p_j(\mathbf{x}) d\mathbf{x} + V(p_j). \tag{9.16}$$

By replacing $p_i(\mathbf{x})$ and $p_j(\mathbf{x})$ by $\widehat{p}_i(\mathbf{x})$ and $\widehat{p}_j(\mathbf{x})$, the ISE divergence estimator is given by

$$D_{ISE}(\widehat{p}_i, \widehat{p}_j) = \frac{1}{N_i^2} \sum_{\mathbf{x}_n \in \mathscr{D}_i} \sum_{\mathbf{x}_{n'} \in \mathscr{D}_i} W_\sigma(\mathbf{x}_n, \mathbf{x}_{n'}) - \frac{2}{N_i N_j} \sum_{\mathbf{x}_n \in \mathscr{D}_i} \sum_{\mathbf{x}_m \in \mathscr{D}_j} W_\sigma(\mathbf{x}_n, \mathbf{x}_m)$$
$$+ \frac{1}{N_j^2} \sum_{\mathbf{x}_m \in \mathscr{D}_j} \sum_{\mathbf{x}_{m'} \in \mathscr{D}_j} W_\sigma(\mathbf{x}_m, \mathbf{x}_{m'}). \tag{9.17}$$

The ISE divergence may also be extended, as follows

$$D_{ISE}(p_1, \ldots, p_C) = \sum_{i=1}^{C-1} \sum_{j>i} \int [p_i(\mathbf{x}) - p_j(\mathbf{x})]^2 d\mathbf{x}. \tag{9.18}$$

9.3 Information Theoretic Learning and the Connection to Kernel Methods

In this section, we outline the fundamental connections between information theoretic learning and the kernel methods. We first relate the Parzen window to the kernel trick, for then to show that the information theoretic learning criteria may be expressed in terms of mean vectors in a Mercer kernel feature space.

9.3.1 Parzen Window as a Kernel Trick

All the Renyi entropy-based quantities used in information theoretic learning are expressed in terms of sums of Parzen windows, $W_\sigma(\cdot,\cdot)$, as a result of the Parzen window estimation. In the following, we assume that the Parzen window, $W_\sigma(\cdot,\cdot)$, is a positive semi-definite function. This is true for instance for the Gaussian Parzen window. The implication of this assumption, is that the Parzen window obeys Mercer's conditions, and hence computes an inner product in some kernel induced feature space

$$W_\sigma(\cdot,\cdot) = k(\cdot,\cdot) = \langle \Phi(\cdot), \Phi(\cdot) \rangle . \tag{9.19}$$

It is now possible to interpret the Renyi-entropy based information theoretic quantities in terms of Mercer kernel feature space quantities.

9.3.2 Renyi Entropy as Length of Mean Vector

The Parzen window-based estimator of the quantity $V(p)$, and hence the Renyi entropy, is given by (9.9). By replacing $W_\sigma(\cdot,\cdot)$ by the inner-product $\langle \Phi(\cdot), \Phi(\cdot) \rangle$, the following alternative expression is obtained

$$
\begin{aligned}
V(\widehat{p}) &= \frac{1}{N^2} \sum_{\mathbf{x}_t \in \mathscr{D}} \sum_{\mathbf{x}_{t'} \in \mathscr{D}} \langle \Phi(\mathbf{x}_t), \Phi(\mathbf{x}_{t'}) \rangle \\
&= \left\langle \frac{1}{N} \sum_{\mathbf{x}_t \in \mathscr{D}} \Phi(\mathbf{x}_t), \frac{1}{N} \sum_{\mathbf{x}_{t'} \in \mathscr{D}} \Phi(\mathbf{x}_{t'}) \right\rangle \\
&= \mathbf{m}^T \mathbf{m} = ||\mathbf{m}||^2,
\end{aligned}
\tag{9.20}
$$

where $\mathbf{m}$ is the mean vector of the Φ-transformed data

$$
\mathbf{m} = \frac{1}{N} \sum_{\mathbf{x}_t \in \mathscr{D}} \Phi(\mathbf{x}_t). \tag{9.21}
$$

It thus turns out that the quantity $V(\widehat{p})$ may be related to the squared Euclidean length of the Mercer kernel feature space mean vector of the transformed data points $\Phi(\mathbf{x}_t)$, $t = 1, \ldots, N$. The Renyi entropy estimator is thus $H_2(\widehat{p}) = -\log V(\widehat{p}) = -\log ||\mathbf{m}||^2$.

9.3.3 CS Divergence as Angle Between Mean Vectors

Based on (9.13), the Parzen window based estimator of the CS divergence may be re-written as

$$
\begin{aligned}
V_{CS}(\widehat{p}_i, \widehat{p}_j) &= \frac{\frac{1}{N_i N_j} \sum_{\mathbf{x}_n \in \mathscr{D}_i} \sum_{\mathbf{x}_m \in \mathscr{D}_j} \langle \Phi(\mathbf{x}_n), \Phi(\mathbf{x}_m) \rangle}{\sqrt{\frac{1}{N_i^2} \sum_{\mathbf{x}_n \in \mathscr{D}_i} \sum_{\mathbf{x}_{n'} \in \mathscr{D}_i} \langle \Phi(\mathbf{x}_n), \Phi(\mathbf{x}_{n'}) \rangle \frac{1}{N_j^2} \sum_{\mathbf{x}_m \in \mathscr{D}_j} \sum_{\mathbf{x}_{m'} \in \mathscr{D}_j} \langle \Phi(\mathbf{x}_m), \Phi(\mathbf{x}_{m'}) \rangle}} \\
&= \frac{\left\langle \frac{1}{N_i} \sum_{\mathbf{x}_n \in \mathscr{D}_i} \Phi(\mathbf{x}_n), \frac{1}{N_j} \sum_{\mathbf{x}_m \in \mathscr{D}_j} \Phi(\mathbf{x}_m) \right\rangle}{\sqrt{\left\langle \frac{1}{N_i} \sum_{\mathbf{x}_n \in \mathscr{D}_i} \Phi(\mathbf{x}_n), \frac{1}{N_i} \sum_{\mathbf{x}_{n'} \in \mathscr{D}_i} \Phi(\mathbf{x}_{n'}) \right\rangle \left\langle \frac{1}{N_j} \sum_{\mathbf{x}_m \in \mathscr{D}_j} \Phi(\mathbf{x}_m), \frac{1}{N_j} \sum_{\mathbf{x}_{m'} \in \mathscr{D}_j} \Phi(\mathbf{x}_{m'}) \right\rangle}} \\
&= \frac{\langle \mathbf{m}_i, \mathbf{m}_j \rangle}{\sqrt{\langle \mathbf{m}_i, \mathbf{m}_i \rangle \langle \mathbf{m}_j, \mathbf{m}_j \rangle}} = \cos \angle(\mathbf{m}_i, \mathbf{m}_j),
\end{aligned}
\tag{9.22}
$$

where $\mathbf{m}_i = \frac{1}{N_i} \sum_{\mathbf{x}_n \in \mathscr{D}_i} \Phi(\mathbf{x}_n)$ and $\mathbf{m}_j = \frac{1}{N_j} \sum_{\mathbf{x}_m \in \mathscr{D}_j} \Phi(\mathbf{x}_m)$ can be considered mean vectors of feature space data clusters corresponding to the data points associated with $p_i(\mathbf{x})$ and $p_j(\mathbf{x})$, respectively. The CS divergence estimator is expressed as $D_{CS}(\widehat{p}_i, \widehat{p}_j) = -\log V_{CS}(\widehat{p}_i, \widehat{p}_2) = -\log \cos \angle(\mathbf{m}_i, \mathbf{m}_j)$.

The above relationship shows that the CS divergence estimator is directly associated with the *angle* between the Mercer kernel feature space mean vectors $\mathbf{m}_i$ and $\mathbf{m}_j$.

When measuring the divergence between C pdfs simultaneously, the corresponding CS estimator may be expressed in terms of sums of cosines between pairs of kernel feature space mean vectors, as

$$V_{CS}(\widehat{p}_1,\ldots,\widehat{p}_C) = \frac{1}{\kappa}\sum_{i=1}^{C-1}\sum_{j>i}\cos\angle(\mathbf{m}_i,\mathbf{m}_j). \tag{9.23}$$

9.3.4 ISE Divergence as Length Between Mean Vectors

Following the derivation outlined in the previous subsection, then, based on (9.17), the ISE divergence estimator may be re-written as

$$D_{ISE}(\widehat{p}_i,\widehat{p}_j) = ||\mathbf{m}_i - \mathbf{m}_j||^2. \tag{9.24}$$

Hence, the ISE divergence estimator corresponds to measuring the squared Euclidean length *between* the mean vectors $\mathbf{m}_i$ and $\mathbf{m}_j$.

When measuring the divergence between C pdfs simultaneously, the corresponding ISE estimator may be expressed as

$$D_{ISE}(\widehat{p}_1,\ldots,\widehat{p}_C) = \sum_{i=1}^{C-1}\sum_{j>i}||\mathbf{m}_i - \mathbf{m}_j||^2. \tag{9.25}$$

The geometrical interpretations explained above are illustrated in Fig. 9.5.

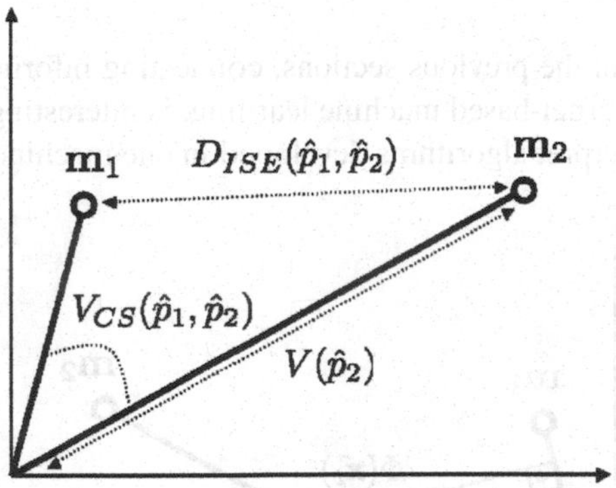

Fig. 9.5 Illustration of the connection between Parzen window-based estimators of Renyi entropy, CS divergence, ISE divergence and mean vectors in Mercer kernel feature space. Only two mean vectors $\mathbf{m}_1$ and $\mathbf{m}_2$ are considered. The Renyi entropy corresponds to the squared Euclidean length of a mean vector, the CS divergence corresponds to the angle between the mean vectors and the ISE divergence corresponds to the squared Euclidean length between the mean vectors

9.3.5 Projections onto Mean Vectors in Kernel Feature Space

We now study projections in the kernel feature space onto the mean vector $\mathbf{m}_i = \frac{1}{N_i}\sum_{\mathbf{x}_n \in \mathscr{D}_i} \Phi(\mathbf{x}_n)$ based on $\mathbf{x}_n \in \mathscr{D}_i$ generated from $p_i(\mathbf{x})$, $i = 1, \ldots, C$. When a data point $\Phi(\mathbf{x}_t)$, $\mathbf{x}_t \in \mathscr{D}$, is projected onto such a mean vector, the following relationship is obtained.

$$\begin{aligned}
\langle \Phi(\mathbf{x}_t), \mathbf{m}_i \rangle &= \left\langle \Phi(\mathbf{x}_t), \frac{1}{N_i} \sum_{\mathbf{x}_n \in \mathscr{D}_i} \Phi(\mathbf{x}_n) \right\rangle \\
&= \frac{1}{N_i} \sum_{\mathbf{x}_n \in \mathscr{D}_i} \langle \Phi(\mathbf{x}_t), \Phi(\mathbf{x}_n) \rangle \\
&= \frac{1}{N_i} \sum_{\mathbf{x}_n \in \mathscr{D}_i} W_\sigma(\mathbf{x}_t, \mathbf{x}_n) \\
&= \widehat{p}_i(\mathbf{x}_t).
\end{aligned} \tag{9.26}$$

This shows that the length of the projection of $\Phi(\mathbf{x}_t)$ onto $\mathbf{m}_i$ is the Parzen window estimate $\widehat{p}_i(\mathbf{x})$ evaluated at $\mathbf{x}_t$, i.e., $\widehat{p}_i(\mathbf{x}_t)$.

Hence, any kernel method which is expressed in terms of inner-products between individual samples and mean vectors in kernel feature space, may be interpreted in terms of the Parzen window estimate of input space pdfs at that individual sample, and vice versa. This is illustrated in Fig. 9.6.

9.4 New Insights and Algorithms

The theory outlined in the previous sections, connecting information theoretic machine learning with kernel-based machine learning, is interesting in its own right. It also enables us to interpret algorithms developed in one machine learning paradigm

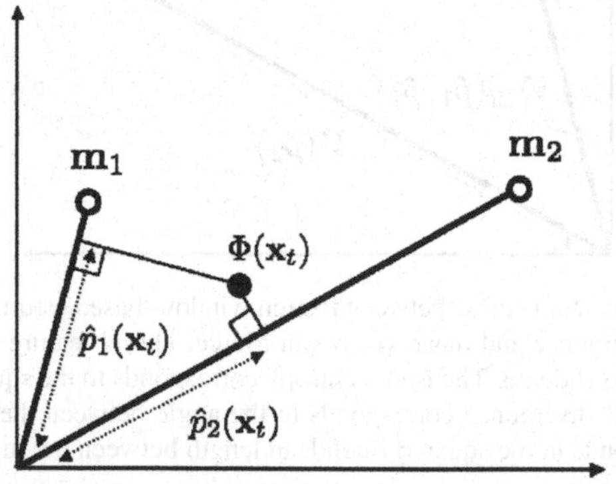

Fig. 9.6 Illustration of the projection of a data point $\Phi(\mathbf{x}_t)$ onto the mean vectors $\mathbf{m}_1$ and $\mathbf{m}_2$. The length of the projection corresponds to the Parzen window estimate at $\mathbf{x}_t$ of the pdf associated with each of the mean vectors

in terms of the other, thus in a sense enhancing our understanding of the algorithm in question. In the following, we provide two such examples.

We also review a recent data transformation and dimensionality reduction method which has flavors from both paradigms. This method is called kernel entropy component analysis, and is closely related to kernel principal component analysis.

9.4.1 Clustering by Maximizing Divergence

In the Introduction, we mentioned as an example the CS divergence used as a cost function for clustering. In [15], this cost function, estimated using Parzen windowing, was implemented, and positive results were obtained also on non-linear clustering problems.

The optimization in [15] is carried out with respect to the partition of the data set, i.e., the clusters, and the goal is to maximize the CS divergence

$$\max_{\mathscr{D}_1,\ldots,\mathscr{D}_C} D_{CS}(\widehat{p}_1,\ldots,\widehat{p}_C). \tag{9.27}$$

Using the technique of Lagrange multipliers and gradient descent, the optimal partition may be found.

The connections between ITL and kernel methods provided here, enables us to look at this clustering algorithm from a totally different perspective. Since the Parzen window estimated CS divergence corresponds to a measure of angles in kernel feature space, the CS divergence criterion function may alternatively be expressed as

$$\max_{\mathscr{D}_1,\ldots,\mathscr{D}_C} -\log \frac{1}{\kappa} \sum_{i=1}^{C-1} \sum_{j>i} \cos\angle(\mathbf{m}_i,\mathbf{m}_j). \tag{9.28}$$

That is, the clustering is performed in such a way that the angles between the resulting kernel feature space cluster mean vectors is maximized. This makes perfect sense.

9.4.2 Kernel C-Means

Recently, it has been proposed to execute the popular C-means algorithm [5] in the kernel feature space instead of in the input space [10]. The resulting algorithm is referred to as kernel C-means. Positive results have been obtained, also on non-linear clustering problems. The relationships between information theoretic learning and kernel methods enable us to provide a better understanding of the kernel C-means algorithm. For a more detailed exposition, see [13].

The algorithm is given by:

1. Initialize the kernel feature space cluster mean vectors, $\mathbf{m}_i$, $i = 1,\ldots,C$
2. For all $\mathbf{x}_t \in \mathscr{D}$:

$$\begin{aligned}\mathbf{x}_t \to \mathscr{D}_i \ : \quad & \|\Phi(\mathbf{x}_t) - \mathbf{m}_i\|^2 \leq \left\|\Phi(\mathbf{x}_t) - \mathbf{m}_j\right\|^2 \\ \Leftrightarrow \quad & \left\langle \mathbf{w}_{ij}, \Phi(\mathbf{x}_t)\right\rangle + b_{ij} \geq 0,\end{aligned} \tag{9.29}$$

where $\mathbf{w}_{ij} = \mathbf{m}_i - \mathbf{m}_j$ and $b_{ij} = \frac{1}{2}\left[\left\|\mathbf{m}_j\right\|^2 - \|\mathbf{m}_i\|^2\right], \forall j \neq i$
3. Update $\mathbf{m}_i$, $i = 1,\ldots,C$
4. Repeat steps 2 and 3 until convergence

This is a hyperplane assignment rule based on squared Euclidean distance.

The kernel C-means algorithm minimizes a sum-of-squared-error criterion function in terms of the kernel feature space. This criterion equivalently corresponds to the maximization of the following quantity [13]

$$J = \sum_{i=1}^{C} N_i \|\mathbf{m}_i - \mathbf{m}\|^2, \tag{9.30}$$

where as before $\mathbf{m}_i = \frac{1}{N_i}\sum_{\mathbf{x}_n \in \mathscr{D}_i} \Phi(\mathbf{x}_n)$ and $\mathbf{m} = \frac{1}{N}\sum_{\mathbf{x}_t \in \mathscr{D}} \Phi(\mathbf{x}_t)$.

We now analyze both the kernel C-means assignment rule and the cost function J in terms of ITL.

Firstly, note that the hyperplane assignment rule involves the projection of $\Phi(\mathbf{x}_t)$ onto $\mathbf{w}_{ij}$. Since $\mathbf{w}_{ij} = \mathbf{m}_i - \mathbf{m}_j$, we have

$$\left\langle \mathbf{w}_{ij}, \Phi(\mathbf{x}_t)\right\rangle = \langle \mathbf{m}_i, \Phi(\mathbf{x}_t)\rangle - \langle \mathbf{m}_j, \Phi(\mathbf{x}_t)\rangle = \widehat{p}_i(\mathbf{x}_t) - \widehat{p}_j(\mathbf{x}_t), \tag{9.31}$$

based on the discussion in Sect. 9.3.5. Hence, the projection of $\Phi(\mathbf{x}_t)$ onto $\mathbf{w}_{ij}$ corresponds to calculating the difference between the Parzen window estimated densities evaluated at the point of interest, i.e., $\mathbf{x}_t$. Furthermore, the threshold which $\widehat{p}_i(\mathbf{x}_t) - \widehat{p}_j(\mathbf{x}_t)$ is compared against in order to assign $\mathbf{x}_t$ to a cluster, is based on the Renyi entropies of the clusters, since

$$b_{ij} = \frac{1}{2}\left[\left\|\mathbf{m}_j\right\|^2 - \|\mathbf{m}_i\|^2\right] = \frac{1}{2}\left[V\left(\widehat{p}_2\right) - V\left(\widehat{p}_1\right)\right], \tag{9.32}$$

based on the discussion in Sect. 9.3.2. Hence, the kernel C-means assignment rule may be expressed in terms of *input space* quantities as

$$\mathbf{x}_t \to \mathscr{D}_i: \quad \widehat{p}_i(\mathbf{x}_t) - \widehat{p}_j(\mathbf{x}_t) + b_{ij} \geq 0, \qquad \forall j \neq i. \tag{9.33}$$

This discussion also reveals that kernel C-means tends to favor the large entropy cluster. To see this, note that if $V\left(\widehat{p}_i\right) < V\left(\widehat{p}_j\right)$ such that in terms of Renyi entropies $H_{R_2}\left(\widehat{p}_i\right) > H_{R_2}\left(\widehat{p}_j\right)$, then $\mathbf{x}_t \to \mathscr{D}_i$ potentially even if $\widehat{p}_i(\mathbf{x}_t) < \widehat{p}_j(\mathbf{x}_t)$. This favors the large entropy cluster $\mathscr{D}_i$.

Secondly, note that the cost function J is expressed in terms of the squared Euclidean distance between kernel feature space mean vectors. In fact, it is easily shown that the integrated squared error between $p_i(\mathbf{x})$ and $p(\mathbf{x})$, using Parzen windowing to estimate $p_i(\mathbf{x})$ based on $\mathbf{x}_n \in \mathscr{D}_i$ and $p(\mathbf{x})$ based on $\mathbf{x}_t \in \mathscr{D}$, is expressed as

$$D_{ISE}(\widehat{p}_i, \widehat{p}) = \|\mathbf{m}_i - \mathbf{m}\|^2. \tag{9.34}$$

Hence, we have

$$J = \sum_{i=1}^{C} N_i D_{ISE}(\widehat{p}_i, \widehat{p}). \tag{9.35}$$

This shows that the kernel C-means criterion function has an alternative representation in terms of ISE divergence between the cluster pdfs and the overall pdf of the data.

9.4.3 Kernel Entropy Component Analysis

The Parzen window Renyi entropy-based information theoretic quantities may also be expressed in terms of kernel matrices. In this section, we focus on the Renyi entropy. With the Renyi entropy as a starting point, we review a recent data transformation and dimensionality reduction technique [14], referred here to as kernel entropy component analysis (kernel ECA).

The kernel matrix, $\mathbf{K}$, is constructed such that element (t,t') of $\mathbf{K}$ equals $W_\sigma(\mathbf{x}_t, \mathbf{x}_{t'}) = k(\mathbf{x}_t, \mathbf{x}_{t'})$, $t,t' = 1,\ldots,N$. For example, we have $V(\widehat{p}) = \|\mathbf{m}\|^2$ by (9.9), but also

$$V(\widehat{p}) = \frac{1}{N^2}\mathbf{1}^T\mathbf{K}\mathbf{1}, \tag{9.36}$$

where $\mathbf{1}$ is a $(N \times 1)$ vector where each element equals one. Note that element (t,t') of $\mathbf{K}$ corresponds to an inner-product in the kernel feature space, that is $\langle \Phi(\mathbf{x}_t), \Phi(\mathbf{x}_{t'}) \rangle$. The matrix $\mathbf{K}$ is therefore an inner-product matrix in terms of the kernel feature space.

It is possible to obtain a data set $\Phi_\mathbf{x}$ in terms of the kernel feature space, representing these inner-products, in the sense that $\Phi_\mathbf{x}^T \Phi_\mathbf{x} = \mathbf{K}$. Let $\mathbf{K}$ be eigendecomposed as $\mathbf{K} = \mathbf{E}\mathbf{D}\mathbf{E}^T$, where we assume that the eigenvalues are stored in the diagonal matrix $\mathbf{D} = diag(\lambda_1, \ldots, \lambda_N)$ such that $\lambda_1 \geq \cdots \geq \lambda_N$. The corresponding eigenvectors are stored in the $(N \times N)$ matrix $\mathbf{E} = [\mathbf{e}_1, \ldots, \mathbf{e}_N]$. Observe now that the data set

$$\Phi_\mathbf{x} = \mathbf{D}^{\frac{1}{2}}\mathbf{E}^T \tag{9.37}$$

is such that $\Phi_\mathbf{x}^T \Phi_\mathbf{x} = (\mathbf{D}^{\frac{1}{2}}\mathbf{E}^T)^T \mathbf{D}^{\frac{1}{2}}\mathbf{E}^T = \mathbf{E}\mathbf{D}\mathbf{E}^T = \mathbf{K}$.

In terms of Renyi entropy, the kernel feature space data set $\Phi_\mathbf{x}$ may be considered an *equivalent* representation of the input space data set $\mathscr{D} = \{\mathbf{x}_1, \ldots, \mathbf{x}_N\}$, since

$$V(\widehat{p}) = \frac{1}{N^2}\mathbf{1}^T \Phi_\mathbf{x}^T \Phi_\mathbf{x} \mathbf{1} = \frac{1}{N^2}\mathbf{1}^T\mathbf{K}\mathbf{1}. \tag{9.38}$$

In fact, the t-th column of $\Phi_{\mathbf{x}} = \mathbf{D}^{\frac{1}{2}}\mathbf{E}^T$ corresponds to the projection of the kernel feature space data point $\Phi(\mathbf{x}_t)$, $t = 1,\ldots,N$, onto the space spanned by all the *principal axes* in that space. This result follows from the kernel principal component analysis (kernel PCA) developed in [28]. The k-th principal axis $\mathbf{v}_k$ in kernel feature space is given by

$$\lambda_k \mathbf{v}_k = \mathbf{C}\mathbf{v}_k, \tag{9.39}$$

where the correlation matrix is defined as

$$\mathbf{C} = \frac{1}{N}\sum_{t=1}^{N} \Phi(\mathbf{x}_t)\Phi(\mathbf{x}_t)^T. \tag{9.40}$$

The projection of *all* the data points $\Phi(\mathbf{x}_t)$, $t = 1,\ldots,N$, onto $\mathbf{v}_k$ is now given by the $(1 \times N)$ vector $\sqrt{\lambda_k}\mathbf{e}_k^T$, where λ_k and $\mathbf{e}_k$ are the k-th eigenvalue and eigenvector of $\mathbf{K}$. The projection onto the space spanned by all the kernel feature space principal axes is therefore given by $\Phi_{\mathbf{x}}$.

Let Φ_{pca} be a matrix where each column corresponds to the PCA projection of the data points $\Phi(\mathbf{x}_t)$, $t = 1,\ldots,N$, onto the *subspace* spanned by the kernel space principal axes corresponding to the l *largest* eigenvalues. Then,

$$\Phi_{pca} = \mathbf{D}_l^{\frac{1}{2}}\mathbf{E}_l^T, \tag{9.41}$$

where the $(l \times l)$ matrix $\mathbf{D}_l$ stores the l largest eigenvalues, and the $(N \times l)$ matrix $\mathbf{E}_l$ stores the corresponding eigenvectors. This is the kernel PCA transformed data set. Since ordinary PCA preserves variance in terms of the input data set, kernel PCA thus preserves variance in terms of the kernel induced feature space. In kernel PCA, one often assume that the data is centered in the kernel feature space. This can be achieved by a centering operation on the kernel matrix [28].

If the set $\mathscr{D}$ consists of subgroups $\mathscr{D}_1,\ldots,\mathscr{D}_C$, one often choose $l = C$. The reason is that for the "ideal" kernel there will be only C non-zero eigenvalues of the kernel matrix $\mathbf{K}$ [14]. The "ideal" kernel is such that $k(\mathbf{x}_n,\mathbf{x}_{n'}) = 1$ (or some other constant) for $\mathbf{x}_n,\mathbf{x}_{n'}$ in the same subgroup, and $k(\mathbf{x}_n,\mathbf{x}_m) = 0$ for $\mathbf{x}_n,\mathbf{x}_m$ in different subgroups. Moreover, in this "ideal" case, it has been shown that the kernel PCA transformed data set will consist of C clusters which are mutually orthogonal in the sense that the cluster mean vectors are orthogonal [14]. In practice however, there is no such ideal kernel.

Let us consider a different mapping of the data points $\Phi(\mathbf{x}_t)$, $t = 1,\ldots,N$, onto a subspace spanned by k kernel space principal axes *not necessarily corresponding to the k largest eigenvalues*. The resulting data set may be denoted

$$\Phi_{eca} = \mathbf{D}_k^{\frac{1}{2}}\mathbf{E}_k^T. \tag{9.42}$$

The inner-product matrix corresponding to this data set is thus $\mathbf{K}_{eca} = \Phi_{eca}^T\Phi_{eca} = \mathbf{E}_k\mathbf{D}_k\mathbf{E}_k^T$.

Note that $\mathbf{K}_{eca}$ corresponds to an input space kernel matrix, such that $\frac{1}{N^2}\mathbf{1}^T\mathbf{K}_{eca}\mathbf{1}$ corresponds to the Renyi entropy of some input space data set which we may denote $\mathscr{D}' = \{\mathbf{x}'_1, \ldots, \mathbf{x}'_N\}$. Let the goal be to obtain Φ_{eca} such that the entropy related to $\mathscr{D}'$ is as similar as possible to the entropy related to $\mathscr{D}$ according to

$$\min_{\lambda_1,\mathbf{e}_1,\ldots,\lambda_N,\mathbf{e}_N} \left[\frac{1}{N^2}\mathbf{1}^T\mathbf{K}\mathbf{1} - \frac{1}{N^2}\mathbf{1}^T\mathbf{K}_{eca}\mathbf{1}\right] = \min_{\lambda_1,\mathbf{e}_1,\ldots,\lambda_N,\mathbf{e}_N} \frac{1}{N^2}\mathbf{1}^T\left(\mathbf{K} - \mathbf{K}_{eca}\right)\mathbf{1}. \quad (9.43)$$

Which k eigenvalues and eigenvectors of $\mathbf{K}$ should be used in $\Phi_{eca} = \mathbf{D}_k^{\frac{1}{2}}\mathbf{E}_k^T$ in order to achieve this?

Note that

$$V(\widehat{p}) = \frac{1}{N^2}\sum_{t=1}^{N}\lambda_t(\mathbf{1}^T\mathbf{e}_t)^2 = \frac{1}{N^2}\sum_{t=1}^{N}\lambda_t\gamma_t^2, \quad (9.44)$$

where $\mathbf{1}^T\mathbf{e}_i = \gamma_i$. We assume in (9.44) that the products $\lambda_i\gamma_i^2$ have been sorted in decreasing order, such that $\lambda_1\gamma_1^2 \geq \cdots \geq \lambda_N\gamma_N^2$. Now, we see that

$$\min_{\lambda_1,\mathbf{e}_1,\ldots,\lambda_N,\mathbf{e}_N} \frac{1}{N^2}\mathbf{1}^T\left(\mathbf{K} - \mathbf{K}_{eca}\right)\mathbf{1} = \frac{1}{N^2}\sum_{t=k+1}^{N}\lambda_t\gamma_t^2, \quad (9.45)$$

if we select $\Phi_{eca} = \mathbf{D}_k^{\frac{1}{2}}\mathbf{E}_k^T$ using the k eigenvalues and eigenvectors of $\mathbf{K}$ corresponding to the k largest products $\lambda_i\gamma_i^2$. This is called kernel entropy component analysis, or kernel ECA for short. Note that this is in general not the same as kernel PCA, although the two methods are closely related. Also note that centering of the kernel matrix does not make sense in kernel ECA, since it would correspond to zero mean kernel feature space data, i.e., $\mathbf{m} = \mathbf{0}$. This again corresponds to infinite Renyi entropy, since $H_2(\widehat{p}) = -\log\|\mathbf{m}\|^2$. Therefore, the kernel matrix used in kernel ECA is not centered.

Figure 9.7 illustrates the difference between kernel PCA and kernel ECA. To the left, an illustration of a kernel PCA mapping is shown. Kernel PCA projects in this

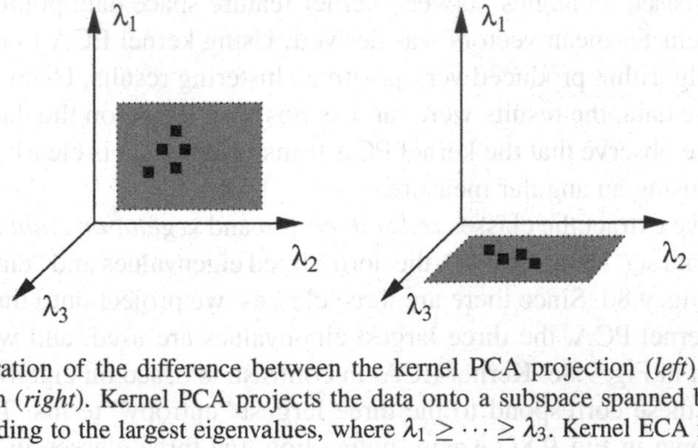

Fig. 9.7 Illustration of the difference between the kernel PCA projection (*left*) and the kernel ECA projection (*right*). Kernel PCA projects the data onto a subspace spanned by the principal axes corresponding to the largest eigenvalues, where $\lambda_1 \geq \cdots \geq \lambda_3$. Kernel ECA may project the kernel feature space data onto a different subspace, not necessarily corresponding to the largest eigenvalues

case the data onto a subspace spanned by the principal axes corresponding to the two largest eigenvalues, where in this case $\lambda_1 \geq \cdots \geq \lambda_3$. As shown to the right, kernel ECA may project the kernel feature space data onto a subspace spanned by principal axes, *not* necessarily corresponding to the largest eigenvalues. Rather, it is the value of the components $\lambda_i \gamma_i^2$ which determines the mapping. In this example, it means that $\lambda_1 \gamma_1^2$ is less than $\lambda_2 \gamma_2^2$ and $\lambda_3 \gamma_3^2$.

As an example, we consider the Landsat multi-spectral satellite image studied in [21], obtained from the UCI repository [1]. Each pixel is represented by 36 spectral values. The labels of the pixels are available, since the image scene has been manually labeled into classes such as *red soil, cotton, vegetation stubble*, etc. We use a sample of this data set consisting of 2,000 data points.

Firstly, we extract the classes *cotton* and *vegetation stubble* in order to generate a two-class data set. We create the kernel matrix using a Gaussian kernel with $\sigma = 35$. Thereafter, the kernel matrix is eigendecomposed. The vertical lines in Fig. 9.8a show the ten largest (normalized) eigenvalues of the kernel matrix. The bars show the corresponding "entropy" values $\lambda_1 \gamma_1^2, \ldots, \lambda_{10} \gamma_{10}^2$ (also normalized). Projecting the two-class data set onto the principal axes in the kernel feature space corresponding to the two largest eigenvalues produces the kernel PCA data set shown in Fig. 9.8b (using a centered kernel matrix). The classes are marked by different symbols for clarity. In kernel ECA, we also project onto a subspace defined by two principal axes. However, instead of using the two largest eigenvalues as in kernel PCA, kernel ECA projects the kernel feature space data onto the principal axes corresponding to eigenvalue λ_1 and λ_3, since the terms $\lambda_1 \gamma_1^2$ and $\lambda_3 \gamma_3^2$ are the two largest "entropy" terms as shown by the bars in Fig. 9.8a. The resulting kernel ECA transformed data set is shown in Fig. 9.8c. Notice that the structure of the kernel ECA data is radically different than the structure of the kernel PCA data. In kernel ECA, the two classes are distributed along two different angular directions. Moreover, these directions are orthogonal. In fact, the angle between the class mean vectors is basically 90 degrees. Kernel ECA thus seems to create a data set which resembles the "ideal" situation discussed above. The results shown here are not unique to the kernel size $\sigma = 35$. Similar results are obtained over a wide range of kernel sizes.

As an example of the usefulness of kernel ECA, we mention that in [14] a clustering algorithm based on angles between kernel feature space data points and kernel feature space cluster mean vectors was derived. Using kernel ECA to represent the data set, this algorithm produced very positive clustering results. Using kernel PCA to represent the data, the results were far less positive. Based on the data set shown in Fig. 9.8b, we observe that the kernel PCA transformed data is clearly not suitable for clustering using an angular measure.

Secondly, we extract the classes *red soil, cotton* and *vegetation stubble*, obtaining a three-class data set. Using $\sigma = 40$, the normalized eigenvalues and "entropy" terms are shown in Fig. 9.8d. Since there are three classes, we project onto three principal axes. Using kernel PCA, the three largest eigenvalues are used, and we obtain the data set shown in Fig. 9.8e. Kernel ECA, in contrast, is based on eigenvalues λ_1, λ_2 and λ_5, since these correspond to the three largest "entropy" terms. The resulting data set is shown in Fig. 9.8f. Again, notice how the three classes in kernel ECA

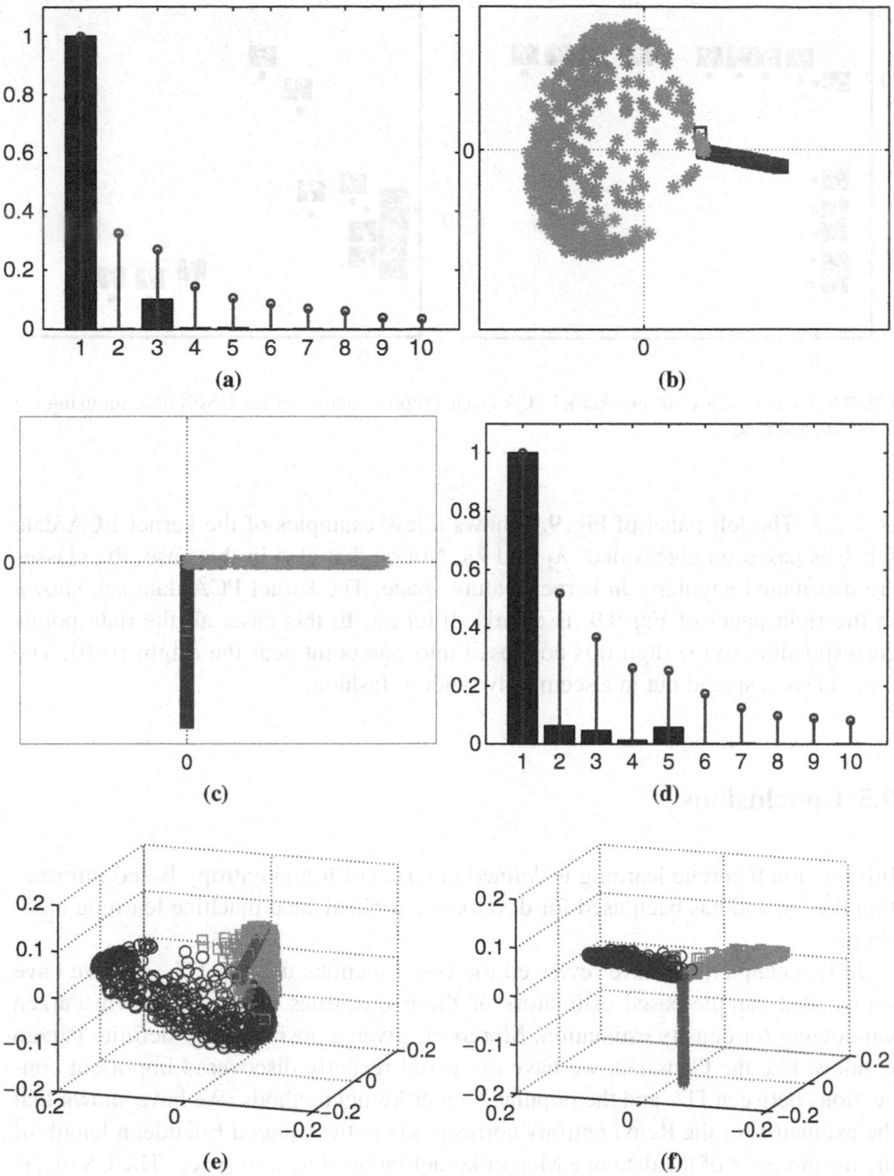

Fig. 9.8 Kernel ECA and kernel PCA representation of Landsat image. Two classes (**a**)–(**c**). Three classes (**d**)–(**f**)

are distributed along different mutually orthogonal angular directions, in contrast to kernel PCA.

Finally, we demonstrate kernel ECA on the 256-dimensional USPS handwritten digits data set [1], using the digits six and nine. The kernel size used here is

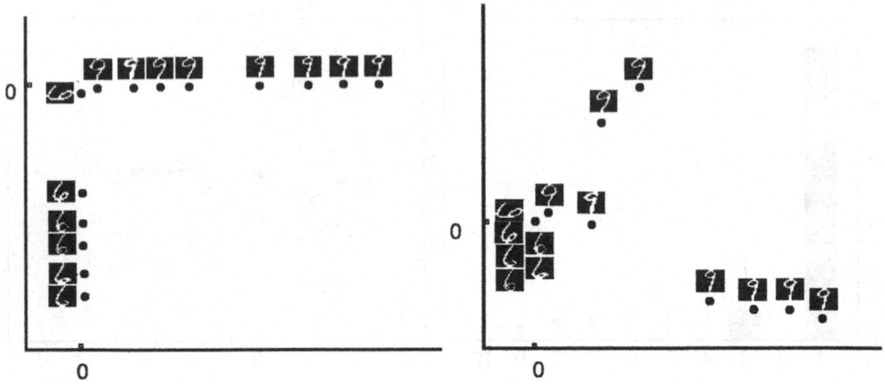

Fig. 9.9 Kernel ECA (*left*) and kernel PCA (*right*) representation of the USPS data set using the digits six and nine

$\sigma = 2.5$. The left panel of Fig. 9.9 shows a few examples of the kernel ECA data set. It is based on eigenvalues λ_1 and λ_9. Notice that also in this case, the classes are distributed angularly in kernel feature space. The kernel PCA data set, shown in the right panel of Fig. 9.9, is clearly different. In this case, all the data points corresponding to the digit 6 is collapsed into one point near the origin $(0,0)$. The other class is spread out in a seemingly random fashion.

9.5 Conclusions

Information theoretic learning is defined in terms of Renyi entropy-based information theory, and has been used for developing sophisticated machine learning algorithms.

In this chapter, we have reviewed the cost functions used in ITL, and we have shown that sample-based estimators of these quantities are obtained via Parzen windowing for density estimation. Moreover, given a positive semi-definite Parzen window, like the Gaussian, we have discussed recently discovered important connections between ITL and the popular Mercer kernel methods. We have shown that the estimator for the Renyi entropy corresponds to the squared Euclidean length of the mean vector of the data in a Mercer kernel induced feature space. The CS divergence estimator has an equivalent expression as a measure of the cosine of the angle between kernel feature space cluster mean vectors, while the ISE divergence estimator measures the squared Euclidean difference between the mean vectors. The projection of a kernel feature space data point onto a kernel feature space cluster mean vector has also been shown to correspond to the Parzen window estimated cluster density, evaluated at the data point in question.

From these new insights, we not only learn that these two seemingly radically different machine learning paradigms are related, they also enable us to understand

and interpret ITL in terms of kernel methods and kernel methods in terms of ITL. We have for example interpreted a recent clustering method which maximizes the CS divergence between the cluster pdfs as a procedure which labels the data in such a way that the angle between the resulting kernel feature space cluster mean vectors is maximized. Kernel *C*-means have been interpreted in terms of Parzen window estimated cluster densities, Renyi entropy and the ISE divergence. These analysis clearly enhances our understanding of the methods in question.

Finally, we have discussed a recent data transformation and dimensionality reduction method, namely kernel ECA. This method brings together elements from both information theoretic learning and kernel methods into a new and sophisticated machine learning algorithm. Kernel ECA is closely related to kernel PCA, but may produce strikingly different data sets. Kernel ECA produces a data set with an angular structure, which is especially suitable for further processing using angular cost functions, for example for clustering.

Acknowledgements The author would like to thank research assistant Ola K. Storås for help generating Fig. 9.9, and Torbjørn Eltoft, Deniz Erdogmus and Jose C. Principe for many discussions on the connections between information theoretic learning and kernel methods. This work was supported by the Research Council of Norway grant NFR 171125/V30 "New Information Theoretic Methods for Intelligent Data Analysis."

References

1. Asuncion, A., Newman, D.J.: UCI Machine Learning Repository. http://www.ics.uci.edu/~mlearn/MLRepository.html. University of California, School of Information and Computer Science, Irvine (2007)
2. Bach, F., Jordan, M.I.: Kernel independent component analysis. Journal of Machine Learning Research **2**, 121–167 (2003)
3. Cortes, C, Vapnik, V.N.: Support vector networks. Machine Learning **20**, 273–297 (1995)
4. Cover, T.M., Thomas, J.A.: Elements of Information Theory. Wiley, New York (1991)
5. Duda, R.O., Hart, P.E, Stork, D.G.: Pattern Classification. Wiley, New York (2001)
6. Erdogmus, D., Agrawal, R., Principe, J.C.: A mutual information extension to the matched filter. Signal Processing **85**(7), 927–935 (2005)
7. Erdogmus, D., Principe, J.C.: An error-entropy minimization algorithm for supervised training of nonlinear adaptive systems. IEEE Transactions on Signal Processing **50**(7), 1780–1786 (2002)
8. Erdogmus, D., Principe, J.C.: From linear adaptive filtering to nonlinear information processing. IEEE Signal Processing Magazine **23**(6), 14–33 (2006)
9. Fyfe, C., Lai, P.L.: Kernel and nonlinear canonical correlation analysis. International Journal of Neural Systems **10**, 365–374 (2001)
10. Girolami, M.: Mercer kernel-based clustering in feature space. IEEE Transactions on Neural Networks **13**(3), 780–784 (2002)
11. Hild, K.E., Erdogmus, D., Principe, J.C.: Blind source separation using Renyi's mutual information. IEEE Signal Processing Letters **8**(6), 174–176 (2001)
12. Hild, K.E., Erdogmus, D., Torkkola, K., Principe, J.C.: Feature extraction using information-theoretic learning. IEEE Transactions on Pattern Analysis and Machine Intelligence **28**(9), 1385–1392 (2006)

13. Jenssen, R., Eltoft, T.: A new information theoretic analysis of sum-of-squared-error kernel clustering. Neurocomputing, to appear (2008)
14. Jenssen, R., Eltoft, T., Girolami, M., Erdogmus, D.: Kernel maximum entropy data transformation and an enhanced spectral clustering algorithm. In: Schölkopf, B., Platt, J., Hoffmann, T. (eds.) Advances in Neural Information Processing Systems 19, pp. 633–640. MIT, Cambridge (2007)
15. Jenssen, R., Erdogmus, D., Hild, K.E., Principe, J.C., Eltoft, T.: Information cut for clustering using a gradient descent approach. Pattern Recognition **40**, 796–80 (2007)
16. Jenssen, R., Erdogmus, D., Principe, J.C., Eltoft, T.: The Laplacian pdf distance: A cost function for clustering in a kernel feature space. In: Saul, L.K., Weiss, Y., Bottou, L. (eds.) Advances in Neural Information Processing Systems 17, pp. 625–633. MIT, Cambridge (2005)
17. Jenssen, R., Erdogmus, D., Principe, J.C., Eltoft, T.: Some equivalences between kernel methods and information theoretic methods. Journal of VLSI Signal Processing **45**, 49–65 (2006)
18. Jenssen, R., Erdogmus, D., Principe, J.C., Eltoft, T.: The Laplacian classifier. IEEE Transactions on Signal Processing **55**(7), 3262–3271 (2007)
19. Kullback, S., Leibler, R.A.: On information and sufficiency. The Annals of Mathematical Statistics **22**(1), 79–86 (1951)
20. Mercer, J.: Functions of positive and negative type and their connection with the theory of integral equations. Philosophical Transactions of the Royal Society London **A**, 415–446 (1909)
21. Michie, D., Spiegelhalter, D.J., Taylor, C.C.: Machine Learning, Neural and Statistical Classification. Ellis Horwood, New York (1994)
22. Mika, S., Rätsch, G., Weston, J., Schölkopf, B., Müller, K.-R.: Fisher discriminant analysis with kernels. In: Hu, Y.-H., Larsen, J., Wilson, E., Douglas, S (eds.) Neural Networks for Signal Processing 9, pp. 41–48. IEEE (1999)
23. Parzen, E.: On the estimation of a probability density function and its mode. The Annals of Mathematical Statistics **32**, 1065–1076 (1962)
24. Perez-Cruz, F., Bousquet, O.: Kernel methods and their potential use in signal processing, IEEE Signal Processing Magazine **21**(3), 57–65 (2004)
25. Principe, J.C., Xu, D., Fisher, J.: Information theoretic learning. In: Haykin, S. (ed.) Unsupervised Adaptive Filtering, vol. I, pp. 265–319. Wiley, New York (2000)
26. Renyi, A.: On measures of entropy and information. Selected Papers of Alfred Renyi, Akademiai Kiado, Budapest, **2**, 565–580 (1976)
27. Schölkopf, B., Smola, A.: Learning with Kernels. MIT, Cambridge (2002)
28. Schölkopf, B., Smola, A., Müller, K.-R.: Nonlinear component analysis as a kernel eigenvalue problem. Neural Computation **10**, 1299–1319 (1998)
29. Shannon, C.E.: A mathematical theory of communication. Bell System Technical Journal **27**, 379–423, 623–653 (1948)
30. Shawe-Taylor, J., Cristianini, N.: Kernel Methods for Pattern Analysis. Cambridge University Press, Cambridge (2004)
31. Silverman, B.W.: Density Estimation for Statistics and Data Analysis. Chapman and Hall, London (1986)

Chapter 10
Information-Theoretic Causal Power

Kevin B. Korb, Lucas R. Hope, and Erik P. Nyberg

Abstract The *causal power* of C over E is (roughly) the degree to which changes in C cause changes in E. A formal measure of causal power would be very useful, as an aid to understanding and modeling complex stochastic systems. Previous attempts to measure causal power, such as those of Good [16], Cheng [3], and Glymour [15], while useful, suffer from one fundamental flaw: they only give sensible results when applied to very restricted types of causal system, all of which exhibit causal transitivity. Causal Bayesian networks, however, are not in general transitive. We develop an information-theoretic alternative, *causal information,* which applies to any kind of causal Bayesian network. Causal information is based upon three ideas. First, we assume that the system can be represented causally as a Bayesian network. Second, we use hypothetical interventions to select the causal from the non-causal paths connecting C to E. Third, we use a variation on the information-theoretic measure *mutual information* to summarize the total causal influence of C on E. Our measure gives sensible results for a much wider variety of complex stochastic systems than previous attempts and promises to simplify the interpretation and application of Bayesian networks.

10.1 Introduction

In some systems, the relationship between a cause C and an effect E is like True Love: deterministic and simple, and not affected by other mediating and modifying variables. In such happy relationships the only question is precisely what causal law

K.B. Korb (✉)
Clayton School of IT, Monash University, Clayton 3600, Australia
e-mail: kevin.korb@infotech.monash.edu.au

L.R. Hope
Bayesian Intelligence Pty. Ltd.
e-mail: lhope@bayesian-intelligence.com

E.P. Nyberg
School of Philosophy, University of Melbourne, Parkville 3052, Australia
e-mail: e.nyberg@pgrad.unimelb.edu.au

F. Emmert-Streib, M. Dehmer (eds.), *Information Theory and Statistical Learning,* 231
DOI: 10.1007/978-0-387-84816-7_10,

binds C to E, for better or for worse. But in many systems that are worthy of scientific study, the relationship between C and E may be like Real Life: stochastic and complex, encompassing multiple indirect paths and various peripheral influences. Such systems can be found in diverse scientific disciplines, including Medicine, Psychology, Meteorology, Ecology, and Economics.

In such cases, conventional Statistics offers ways to summarize the relationship numerically. If C and E are continuous scalar variables, then the correlation coefficient provides some indication of how much increasing C results in increasing E. With multiple variables, an Analysis of Variance (ANOVA) does a similar job. But such crude summaries are deficient in several ways: they fail to indicate any causal direction; they fail to map any complicated causal structure linking many variables; they fail to capture any complicated non-linear relationships; and they are not applicable to some kinds of variable (these two measures are inapplicable to non-scalar variables).

There is a new paradigm for modeling complex stochastic systems that is rapidly increasing in popularity, both in practical scientific applications and in the theoretical developments of AI and the Philosophy of Science. *Causal Bayesian networks* offer a simple and general way to capture all the nuances of variable relationships, without any of the defects listed above. They consist of directed graphs, in which variables (such as C and E) are represented by nodes, and any direct probabilistic dependency (such as the dependency of E on C) is represented by an arc (such as $C \rightarrow E$). Thus, they provide a nice visual summary of the causal paths in the system, particularly if the graph is sparse. The arcs themselves do not indicate the precise nature of each connection, but this is encoded numerically in conditional probability tables, which can be inspected as required. We present a more detailed account in Sect. 10.3.

Bayesian networks were not originally intended to be interpreted *causally*: they were simply maps of probabilistic dependence, in which the arcs might be oriented in an anti-causal direction (e.g., $C \leftarrow E$). But in a causal Bayesian network the arcs are also supposed to reflect the direction of causation, and this interpretation has become increasingly important. In AI, many causal discovery algorithms have been developed to learn causal Bayesian networks from data, and they have been quite successful [30, 44, 47]. In addition, networks have been "knowledge engineered" from a combination of expert opinion and data, whose performance is better than that of the experts alone (e.g., [22]).

In Philosophy, there is a 2,000 year history of debate over the epistemology, metaphysics and semantics of causation, most of which assumed that causation involves physical necessitation. In the twentieth century this assumption eventually yielded to a serious discussion of *probabilistic causation,* in which a cause need not completely determine an effect, but may instead only influence its probability (e.g., [40]). There has also been controversy over the nature of good causal attributions and scientific explanations. In the last few years there has been a flurry of activity, both to apply useful philosophical ideas to Bayesian networks, and conversely, to use Bayesian networks to clarify the philosophy (e.g., [17, 18, 46]). We outline some relevant philosophical issues in Sect. 10.2.

Paradoxically, the strength of causal Bayesian networks – accurately modeling complex stochastic relationships – poses a problem for human users. It can be important to understand the causal influence of C on E, particularly when considering a possible intervention on C to try to influence E. But complexity makes this understanding difficult: both in the subtleties of the individual links, and in the presence of multiple paths between C and E (some of which may not transmit causal influence all the way from C to E). Thus, we need a good summary statistic – like a correlation coefficient, but both causally directed and more widely applicable – to measure the "causal power" of C over E: roughly, the degree to which changes in C cause changes in E. In scientific applications, a good measure of causal power would provide a guide to understanding and intervening upon real systems. In AI, it would provide a guide to developing and explaining Bayesian networks. In Philosophy, it would provide a guide to good causal attributions and scientific explanations.

There have been several previous attempts to provide a measure of causal power, such as the work of Wright [49], Good [16], Cheng [3], and Glymour [15]. Unfortunately, all these attempts suffer from one fundamental flaw: they only give sensible results when applied to very restricted classes of Bayesian network. They either cannot be applied to other classes of network, or else applying them does not generate sensible results. For example, all the attempts just mentioned work only for networks in which causation is *transitive*: if C influences D, and D influences E, then C must influence E. But causation is not always transitive, as Hitchcock [18] and others have shown.

We present a new measure of causal power, which applies to all causal Bayesian networks and generates appetizing results. The first ingredient is the causal Bayesian networks themselves, since they encode the true causal structure connecting C and E. The second ingredient is accurately representing possible interventions on C, since this is a convenient way to sift the causal from the non-causal paths to E. The third ingredient is the information-theoretic measure "mutual information", since this provides a useful summary of the net effect of C on E.

10.2 Probabilistic Causality

Philosophers like to contrast "types" (classes of thing that share some general property) with "tokens" (things that are members of the class, and hence are particular instances of the general property). Thus, "smoking" is a type of event, whereas "Susan smoking" is a token event of this type. Similarly, "smoking causes cancer" is an example of "type causation", whereas "Susan's smoking caused her cancer" is an example of"token causation".[1]

[1] Here "Susan smoking" refers to one event (e.g., one cigarette), while "Susan's smoking" refers to a fusion of such events. But in either case, what Susan does is a token instantiation of a general type of behavior – whether it is simple or complex.

There is more than one way to interpret Bayesian networks in terms of types and tokens. We shall assume the following "type" interpretation. Suppose, for example, that we have data from a longitudinal medical study concerning the causes of lung problems. One variable, say C, represents smoking behavior. For each individual in the study, it takes one of three mutually exclusive values: c_0 if they do not smoke, c_1 if they smoke a little, or c_2 if they smoke heavily. Another variable, say E, represents their lung health outcomes in ten years time. The data consists of observations of many individuals, with the unfortunate Susan being just one participant. Once the study is complete, we can use such data to construct an appropriate causal Bayesian network to model the system. For example, we can calculate the probability that an individual will contract lung cancer given only that they smoke heavily, $P(e_1|c_2)$, and compare it to the probability that an individual will contract lung cancer given only that they do not smoke, $P(e_1|c_0)$. This is a typical way to apply Bayesian networks to real data analysis. Note that in this model, each variable value represents an event *type* (e.g., heavy smoking). The tokens of this type are the observed behaviors of individuals (e.g., Susan's heavy smoking). The variables represent sets of mutually exclusive event types (e.g., various types of smoking behavior). Our approach will measure the causal power of both variables and variable values. But in models like this, both these things concern *type* causation.

Our "type" interpretation should not be confused with the alternative "token" semantics used in [17]. They are primarily concerned with token causation, so in their scheme, each variable value represents an event token: c_2 for "Susan smokes heavily", and c_0 for "Susan does not smoke".[2] Each variable value is instantiated either once (for actual values) or not at all (for counterfactual values). Therefore, there is no data set of multiple observations for inferring probabilistic connections, and any such connections in the network must be postulated based on theoretical knowledge (such as our longitudinal results!). Each variable in this scheme represents a class of event tokens.[3] Both the "type" and "token" interpretations are viable, and therefore it is a mistake to equate the type-token distinction with the variable-value distinction. Variables need not be types (and vice versa), and values need not be tokens (and vice versa).

Despite our choice of the "type" interpretation, there is an obvious connection between studies of type causation and their tokens. Suppose that our medical study concludes that heavy smoking greatly increases the probability of contracting lung cancer, and furthermore, all we know about Susan is that she is a heavy smoker (who participated in the study). In that case, our best estimate of the probability that

[2] A similar result would be obtained under our interpretation if the event types were specified so restrictively that there was only one token of each type.

[3] Perhaps such variables also represent a corresponding event type. But "Susan's smoking behavior" is a rather unnatural event type, since it includes such diverse token events as heavy smoking and not smoking at all. Moreover, the variable values are not supposed to be *only* tokens of this type (instances of Susan's smoking behavior); they are supposed to have distinct, causally important properties. So even in the "token" interpretation, the variable-value relationship is not just a type-token relationship.

Susan contracted lung cancer is simply $P(e_1|c_2)$. Indeed, we expect our measure of causal power to be applicable to analyses of token causation, but we will not argue that point here.[4]

"Smoking causes cancer" implies a probabilistic connection, in which smoking raises the probability of contracting cancer, rather than leading inevitably to it. Many philosophical accounts of causation (especially token causation) have been deterministic (e.g., [26]). But logically, such deterministic causal connections can be viewed as extreme cases of probabilistic causal connections (in which the probabilities are 0 or 1). The possibility of less extreme probabilistic connections has been recognized and discussed by philosophers since Reichenbach [35]. Representing such connections in causal Bayesian networks can be seen as a technical advance in the discussions: it builds on their prior work, while also rendering some of the non-technical philosophical debate obsolete.

The primary evidence for probabilistic causation is an observed change in probability, e.g., the increase in the probability of lung cancer when someone is a smoker.[5] Such changes are often summarized by measures of "statistical relevance". These various competing measures are all plausible candidates for measuring causal power, so it is worth explaining the issues involved in choosing an appropriate measure, and why we prefer mutual information instead.

$$SR(c_0, e_1) = P(e_1|c_0) - P(e_1), \tag{10.1a}$$

$$\Delta P = P(e_1|c_0) - P(e_1|\neg c_0), \tag{10.1b}$$

$$BR(c_0, e_1) = \frac{P(e_1|c_0)}{P(e_1)}, \tag{10.1c}$$

$$IR(c_0, e_1) = -\left[\log P(e_1|c_0) - \log P(e_1)\right]. \tag{10.1d}$$

Equation (10.1a) is the standard formulation for Statistical Relevance (SR) in Philosophy (e.g., [38]), whereas (10.1b) is more common in Psychology (e.g., [3]). Equations (10.1c) and (10.1d) are not standard measures of statistical relevance, but they are plausible alternatives. Equation (10.1c) uses the same probabilities as (10.1a), but measures the *proportional* change in probability rather than the *absolute* difference. It can be used in Bayesian fashion as a multiplicative factor to update the probability $P(e_1)$ upon learning that c_0, so we dub it Bayesian Relevance (BR). Equation (10.1d) is simply the negative log of (10.1c), which makes

[4] Strangely, the study of token causation has often been regarded as quite separate from the study of type causation (e.g., [8]). This has been encouraged by the fact that probabilistic type judgments are often made prospectively about propensities (e.g., the chance that a patient will contract cancer). In contrast, token judgments are usually made retrospectively about counterfactual possibilities (e.g., while Susan is suing the cigarette company, she maintains that she would not have contracted cancer without smoking). Thus, philosophical discussions of token causation have been dominated by deterministic analyses of counterfactual problem cases, and these raise issues we cannot properly address here. However, there have been recent encouraging moves to incorporate type relationships into the attribution of causal blame (e.g., [27]).

[5] Technically, we observe differences in sample frequencies and use these to estimate population probabilities.

the expression more similar to information-theoretic measures like mutual information (see Sect. 10.5), so we dub it Information-theoretic Relevance (IR).

All these measures share some basic features that are intuitively attractive. There is a natural point of division (0, or 1 for (10.1c)) between causes that promote an effect (and have positive relevance, or factors greater than 1), and causes that prevent it (and have negative relevance, or factors less than 1). They also agree that c_0 promotes e_1 more strongly when $P(e_1|c_0)$ is greater (provided the other terms remain constant).[6] Moreover, they are all easy to calculate for one pair of variables. But each measure differs in the exact number it assigns, and these shared virtues do not provide any reason to prefer one number to another.

One critical issue is the choice of *comparative probability*. All the measures use $P(e_1|c_0)$, but there are two different probabilities for e_1 that can plausibly be used to measure an increase or decrease. Equations (10.1a), (10.1c), and (10.1d) use $P(e_1)$, i.e., some overall probability for e_1 – which implies that there is also some overall probability for both c_0 and $\neg c_0$. In some situations this will be appropriate: e.g., if we observe that someone is a non-smoker (or intervene to make them a non-smoker!), then we might wish to know how much difference this makes to the probability that they will contract lung cancer. On this measure, the statistical relevance of not smoking increases if more people are smokers. In contrast, (10.1b) compares $P(e_1|c_0)$ to the alternative condition $P(e_1|\neg c_0)$, without assuming any probability distribution over these two conditions upon C (although there must still be a probability distribution over c_1 and c_2). This measure is appropriate for answering a slightly different question: what is the difference in lung cancer probability between smokers and non-smokers? Thus, the right choice of comparison probability depends somewhat on the situation: the information we have available and the question we wish to ask.

Another critical issue is the choice of *weighting function*. All these measures merely relate a single variable value of C to a single variable value of E. But to provide a more general measure of causal power, we want also to relate single values to variables, and variables to variables. For example, it is reasonable to ask how much heavy smoking (c_2) affects all lung cancer outcomes (E), or how much smoking behavior (C) affects all lung cancer outcomes (E). Again, the right choice of weighting function depends somewhat on the situation. But none of the measures listed give any indication of the appropriate weighting function. Mutual information offers a coherent way to compare and weight changes in probability, which can be adjusted appropriately for the type of situation, and exhibits attractive mathematical properties in more complicated examples.

One remaining problem is that all these measures, including mutual information, are symmetric: if C is relevant to E, then E is relevant to C. Yet we know that causal direction is crucial to measuring causal power. So how can we tell, for example, if smoking causes lung cancer or if lung cancer causes smoking?

---

[6] Hence, they support a kind of egalitarianism for causes: it doesn't matter what the variable represents, or how indirect its path of influence, since only the resulting increase in probability matters.

The most obvious reply in this case is to appeal to temporal order. It is generally accepted that causes precede their effects.[7] The data should show that smoking *earlier* in life increases the risk of lung cancer *later* in life (whereas the converse does not hold). So this time difference rules out lung cancer as a cause of smoking. However, such temporal information is not always available.

In Science and statistical analysis, it has long been recognized that the gold standard for the discovery of causal power (and hence causal direction) is *intervention*. If an experimenter can manipulate one variable and produce a change in another variable, then the first variable causally influences the second. Thus, intervention is central to the concept of causation and its utility, as Woodward [48] has recently argued. We use *hypothetical* interventions as a way of measuring the causal power implied by a model, as detailed in Sects. 10.3 and 10.5. Of course, *actual* intervention data is not always obtainable.

A more indirect way to determine causal direction or causal structure is to appeal to background knowledge. Statistical relevance shows that there is *some* causal process linking smoking and cancer, however indirect, and whatever the causal direction.[8] But is there a plausible causal process leading from smoking to cancer? Providing biochemical evidence of the causal mechanism was crucial to killing off the tobacco lobby's defense. Causal process accounts (e.g., [8, 40]) emphasize this aspect of causation.

Automated methods for discovering causal structure from data rely less upon background knowledge and more upon a presumption of Simplicity, or her kissing cousin, Probability. This raises philosophical issues that are too great to address here, but we refer the reader to Korb and Nyberg [25] for a better indication of how this union works. The success of causal discovery algorithms in using these criteria illustrates that causal structure can be discovered by using more than just a local, bottom-up approach in which individual connections are assembled into a network.[9] Rather, there are global aspects that can be used for top-down causal discovery, in which the best total network for describing a system can dictate our beliefs about local causal structure.[10]

We have listed some interesting features of causation and its discovery – precedence, intervention, process, and simplicity – but all these features apply primarily to *developing* a causal Bayesian network. Our measure of causal power assumes that the network is *given* and simply summarizes its causal implications.

---

[7] Some philosophers (e.g., [34, 35]) have argued that precedence should not be a necessary criterion for causation. Their motivation is either to avoid using time as a fundamental property, or else to allow for some logical exceptions (e.g., physical or fictional time travel). But nobody disputes that precedence is a good practical guide!

[8] To suppose that there is persistent correlation without underlying causation seems, at first glance, to be tantamount to believing in magic (or an incredible run of luck). Bayesian networks are usually constructed to satisfy the opposite assumption, satisfying the "Markov property". Nonetheless, there are some troubling putative counterexamples, such as the rising water levels in Venice and the rising price of fish in China. But we shall plumb those depths another time.

[9] Although this is the explicit approach of conditional dependence learners, e.g., IC [47].

[10] An approach that is emphasized in our own metric learner, CaMML [24, chap. 8].

When a network is given, it makes some previously troubling philosophical questions easy to answer. Sometimes C may be statistically relevant to E, but neither is a cause of the other: e.g., because some other variable B influences them both. Philosophers have grappled with the problem of how to identify such "spurious correlations" between C and E, and "fix the background" so that it disappears (e.g., [2, 13, 39, 45]). But given a network, it is easy to identify "common causes" like B, and spurious relevance can be removed by controlling for the value of B (either by observing the value or by manipulating it). Conversely, it is impossible to identify causally relevant variables without also having some idea of the correct causal network. Therefore, the problem of spurious correlations has now been subsumed by the study of causal discovery algorithms and knowledge engineering, in order to correctly model the domain of interest [46].

Similarly, philosophers have worried that to correctly measure statistical relevance, we need to use the right reference class. Smoking may promote lung cancer more in females than in males, so to assess Susan's risk most accurately we need to condition upon her sex. But we should not condition upon every variable, since some do not really affect the connection between smoking and cancer! Given a network, it is easy to identify all the other variables that impact upon lung cancer, and condition upon these to create a suitable "homogenous reference class".

Some philosophers have argued that genuine causes, like smoking, should be context independent: they should promote their effects across almost all homogenous reference classes.[11] Others have argued that fickle, context-sensitive causes are logically possible.[12] The structure of causal Bayesian networks certainly allows for highly context-sensitive causal influences. We will not engage in the debate over their plausibility or admissibility here, but simply point out that, whichever kind of causal network is given to it, our measure of causal power will happily report its causal implications.

10.3 Causal Bayesian Networks

We now present a more detailed description of Bayesian networks, their causal interpretation, and how to model interventions upon them.

10.3.1 Bayesian Networks

Bayesian networks, popularized by Pearl [32], Neapolitan [28] and Jensen [21], are graphical representations of the probabilistic relationships between random variables. A Bayesian network is a directed acyclic graph, e.g., $A \rightarrow B \rightarrow C$. Each node,

[11] This is the Contextual Unanimity Thesis (CUT), supported by Cartwright [2], Skyrms [42], Eells and Sober [13] and Humphreys [20].

[12] This is the Objective Homogeneity Thesis (OHT), supported by Salmon [37], Eells [12] and Twardy and Korb [46].

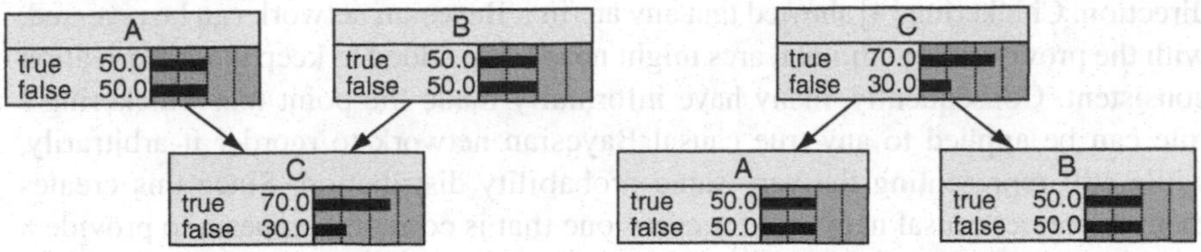

a Comman effect **b** Comman cause

Fig. 10.1 Two Bayesian networks representing different factorizations of $P(A,B,C)$. *Left*: Common effect. *Right*: Common case

say B, represents a domain variable. Each arc between nodes, say $A \rightarrow B$, encodes an elementary conditional dependence relation between the parent variable and its child, e.g., the probability distribution over the values of B depends upon the value of A. So there is an elementary conditional probability function $P(B|\pi_B)$ associated with each node, which specifies a probability for its variable, B, that depends only upon its parents, π_B (here only A), and not upon other variables (such as C).

The network entails a joint probability distribution $P(A,B,C)$ which is simply the product of these elementary probability functions: $P(A,B,C) = P(A)P(B|A)P(C|B)$. The Markov assumption (which we make throughout) is that this scheme is sufficient to capture all the true probabilistic dependencies between the variables. Note that the joint distribution does imply various specific probability distributions, such as $P(B|A,C)$, in which the probability distribution over a variable, here B, can depend upon its children, here C. Figure 10.1 shows two other networks factoring $P(A,B,C)$. Figure 10.1a represents the factorization $P(A)P(B)P(C|A,B)$, while Fig. 10.1b represents $P(C)P(A|C)P(B|C)$.

While in the worst case Bayesian networks are intractable for probabilistic reasoning [5], with a sparse network the computational savings over dealing with a full joint probability table can be considerable. For example: assuming binary variables, the full joint distribution for the networks in Fig. 10.1 would take ($2^3 = 8$) eight parameters to specify. The V-structure of Fig. 10.1a saves two parameters ($1+1+4=6$), and Fig. 10.1b saves three ($1+2+2=5$). The computational advantages of Bayesian networks are one significant reason for their current popularity; another is their perspicuous display of dependence relations for human eyes; and another is their potential causal interpretation.

10.3.2 Causal Interpretation

Bayesian networks are causal when each arc also represents some causal process, through which the parent variable makes the corresponding probabilistic difference to its child. However, there are always some Bayesian networks that can mirror the true joint probability distribution without orienting all their arcs in the true causal direction. Chickering [4] showed that any arc in a Bayesian network can be reversed, with the proviso that additional arcs might need to be added to keep the factorization consistent. Consequently, many have informally made the point that Chickering's rule can be applied to any true causal Bayesian network to reorder it arbitrarily, while still representing the very same probability distribution. Since this creates many incorrect causal networks for every one that is correct, it appears to provide a reason against making causal interpretations.

In fact, there are several good replies to this coffee house argument. Ontologically, probabilistic dependencies must arise from underlying causal connections (a view expressed in the Common Cause Principle of [35]), even if the epistemic challenge is to infer causality from probability. Therefore, there is always a correct causal model to be discovered, and causal interpretations are certainly not illegitimate.

Furthermore, there are good reasons for thinking that such discovery is feasible. If a causal network's arcs are necessary and sufficient for representing its dependencies, then applying Chickering's rule can only make the network more complex. Indeed, repeated application of this rule will generally lead to a fully connected network! So Ockham's razor suggests that we should reject many of the spurious networks that can be created through Chickering transformations. Mathematically, each of these more complex networks generally requires some of its parameters to exactly match, which makes them individually much less probable than the simple truth. Finally, practice trumps theory: causal discovery algorithms, such as CaMML [31], are based on a preference for minimal networks – and they have been demonstrably successful in recovering the true causal model from data sets. For all of these reasons the common, and sensible, practice of knowledge engineers is to ask domain experts for the *causal* relations between variables, rather than just the probabilistic dependencies.

10.3.3 Observation vs. Intervention

While the causal interpretation of Bayesian networks is becoming more widely accepted, the difference between modeling observation and intervention is still often confused. This is particularly true in areas where regression models, rather than Bayesian networks, are the norm – since ordinary regression models simply cannot model interventions.

We illustrate the difference between intervention and observation with a simple example. Figure 10.2 presents a three-variable causal model of heart attack risk. A patient's exercise routine at 40 affects her or his heart attack risk in two ways: directly and indirectly. More exercise tends to reduce blood pressure at 50 (which signifies that the heart does not need to work as hard), which in turn reduces the risk of heart attack by 60. Thus, exercise indirectly reduces heart attack risk. But exercise also reduces heart attack risk in other ways (such as by improving the blood supply to the heart muscle), and this additional effect is indicated by its direct connection to the heart attack node.

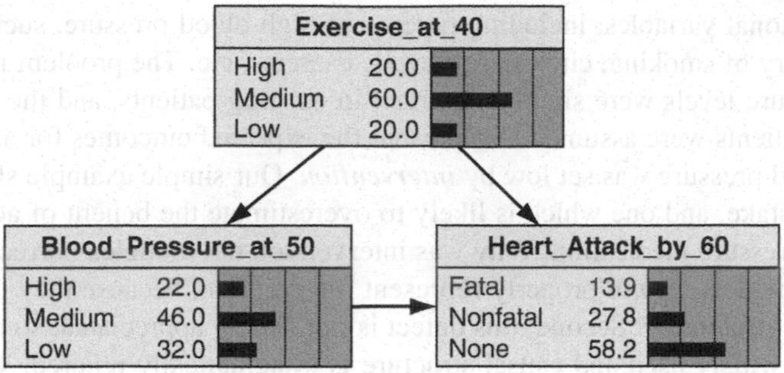

Fig. 10.2 A simple causal Bayesian network linking exercise at age 40 and blood pressure at 50 with the risk of heart attack by 60

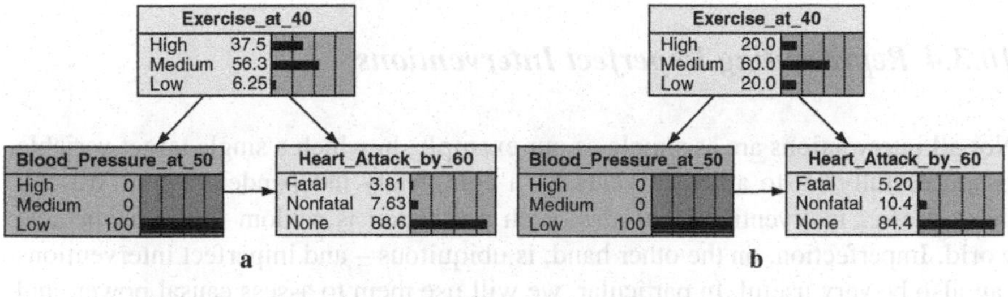

Fig. 10.3 The blood pressure causal network where blood pressure is **a** observed to be low **b** set low by intervention

Suppose we observe that a group of patients have low blood pressure at 50. Then the new probabilities for *Exercise* and *Heart Attack* are shown in Fig. 10.3a. Their probability of avoiding a heart attack has increased, as you'd expect. But it is also more likely that they were doing high levels of exercise at 40, since that would explain the low blood pressure we are now observing.

In contrast, suppose that we intervene at 50 by giving some random patients medication to lower their blood pressures. The new probabilities are shown in Fig. 10.3b. This time the probabilities for *Exercise* have not changed, since we cannot retrospectively affect their exercise routines ten years ago! Their probability of avoiding a heart attack has still increased, but this time by a lesser amount. Why? By intervening on blood pressure at 50, we have only changed the direct effect of this variable. Since it is *not* more likely that they were doing high levels of exercise at 40, we can no longer count on this factor to also directly reduce heart attack risk. In short, it is better to have achieved low blood pressure *naturally* than to suddenly achieve it by artificial means! This is revealed in the causal model by the difference between observation and intervention.

A real example of people confusing observation with intervention stems from the widespread use of regression models in public health. Regression models of the famous Framingham data on heart disease have been used to assess, for example, the expected value of intervening on blood pressure [1, 7]. The models incorporate many additional variables, including parents of high blood pressure, such as a patient's history of smoking, cholesterol levels, exercise, etc. The problem is that low blood pressure levels were simply *observed* in existing patients, and the outcomes for these patients were assumed to represent the expected outcomes for any patient whose blood pressure was set low by *intervention*. Our simple example shows why this is a mistake, and one which is likely to overestimate the benefit of administering blood pressure medication. Why was intervention not modeled correctly? First, regression models *cannot* properly represent intervention, because they do not encode causal structure.[13] Second, this defect is not widely appreciated, so regression models are widely used and causal structure is systematically ignored. Tragically, despite important high-quality data, poor data analysis may well be causing bad public policy decisions and bad medical advice.

10.3.4 Representing Imperfect Interventions

Not all interventions are as simple as our example, in which a single target variable is successfully set to an exact value by a completely independent agent. We call these *perfect* interventions, but alas, such perfection is seldom found in the real world. Imperfection, on the other hand, is ubiquitous – and imperfect interventions can also be very useful. In particular, we will use them to assess causal power, and therefore we will explain how they can be modeled.[14]

Given a causal model M, we can represent a wide variety of interventions within the following scheme:

1. We construct a larger augmented graph, M^*, which includes:
   a. M as a subgraph
   b. A new intervention arc, that impacts upon a variable in M, say $\rightarrow C$
   c. A new intervention node, from which this arc originates, say I_C
2. The intervention node I_C includes:
   a. At least one state, say "No", which represents the fact that no intervention was attempted
   b. At least one state, say "Yes", which represents the fact that an intervention was attempted

[13] In order to be *capable* of representing interventions we require a graphical model in which the parental effects upon an intervened-upon variable can be altered. Minimally, this requires moving from ordinary regression models to path models or structural equation models, if not to Bayesian networks.

[14] Many authors, including Spirtes et al. [43], Pearl [33], Eberhardt et al. [9] and Eberhardt et al. [10], concentrate mainly upon perfect or near-perfect interventions. But there has also been some work on less perfect interventions, including Spirtes et al. [43] (under "rigid indistinguishability"), Pearl [33] (under "instrumental variables" and "imperfect experiments"), Eberhardt and Scheines [11], Fell [14], and Korb and Nyberg [25]. The many varieties of imperfect intervention were the main topic of discussion in Korb et al. [23].

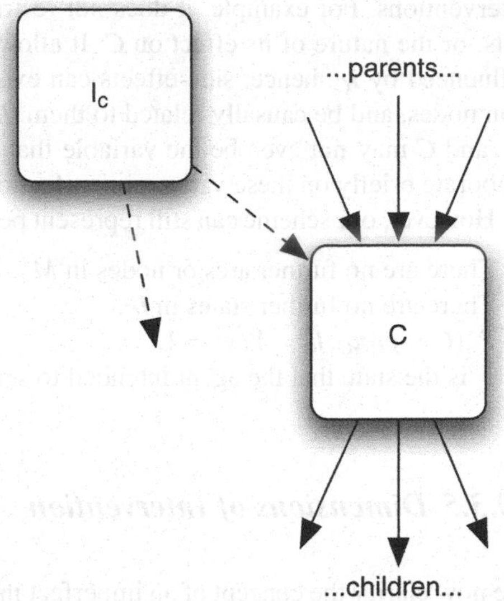

Fig. 10.4 Fragment of a generic augmented causal network

3. We define a new probability distribution over M^*, namely $P^*(\cdot)$, which includes the features:

   a. $P^*(\cdot|I_C = No) = P(\cdot)$ over the variables in M
   b. $P^*(C|\pi_C, I_C = Yes) \neq P(C|\pi_C)$

Figure 10.4 depicts the relevant fragment of M^*.

We motivate this scheme as follows. The augmented graph is intended to represent the original system M prior to intervention. Therefore, it includes M as a subgraph, with its original probability distribution intact when intervention is not attempted. The augmented graph is also intended to represent M after intervention. In ordinary usage, an intervention represents an intentional influence on a causal system which is extraneous to that system. For example, if Susan's everyday lifestyle is viewed as a system, then a medical intervention (such as advice to curtail cigarette smoking) is regarded as coming from outside that system. To represent changes in external circumstances, we can add any number of new nodes. But minimally, we can assume that there is an intervention node I_C to represent the decision to intervene. To represent the effects of intervention, we can add any number of new arcs between nodes. But minimally, we can assume that the decision to intervene has some impact upon at least one variable in M. The fact that I_C is directly connected to C does not really constrain the nature of the intervention, since this single arc can represent any tortuously complex causal process. Our concern is not to accurately represent the wider system for its own sake, only to include enough of it to represent M both before and after intervention. The result of the intervention is a new probability distribution over C's states.

Our scheme therefore specifies some very basic, minimal requirements for any intervention. It leaves open all sorts of possibilities to represent complex, imperfect interventions. For example, it does not restrict I_C's interaction with C's other parents, or the nature of its effect on C. It allows variables other than C to be directly influenced by I_C; hence, side-effects can exist. I_C may coexist with other intervention nodes, and be causally related to them. I_C may even be affected by variables in M; and C may not even be the variable that the agent intended to effect in M! We elaborate briefly on these various imperfections below.

However, our scheme can still represent perfect interventions. We simply specify:

1. There are no further arcs or nodes in M^*.
2. There are no further states in I_C.
3. $P^*(C = c_i|\pi_C, I_C = Yes) = 1$.
4. c_i is the state that the agent intended to set.

10.3.5 Dimensions of Intervention

We now clarify the concept of an imperfect intervention, by listing five distinct ways to be imperfect. In each case we define a clear dichotomy between perfection and imperfection. But since each imperfection can come in degrees, and is independent of the others, they form orthogonal axes in a five-dimensional space.

Dimension 1. Overwhelming vs. underwhelming:

1. An overwhelming intervention results in a distribution over C that makes it completely independent of its original parents. Thus:

$$P^*(C|\pi_C, I_C) = P^*(C|I_C). \tag{10.2}$$

2. An underwhelming intervention results in a distribution over C that still leaves it dependent upon its original parents.

An overwhelming intervention on C "cuts it off" from its parents, and the effect of this intervention is sometimes represented by removing the other parental arcs altogether. Underwhelming interventions can interact with the other parents to produce any kind of elementary function: linear, noisy-OR, or a complex, non-linear interaction. These are precisely the kinds of dependency that Bayesian networks model already, so it is no extension to the semantics of Bayesian networks to incorporate them. A real-world intervention will often interact with the existing complex of causal processes in somewhat unpredictable ways. Medical interventions, for example, often fail – Susan refused to stop smoking because her peers also smoked; yet with different friends, she might have heeded her doctor's warning.

Dimension 2. Deterministic vs. stochastic:

1. A deterministic intervention leaves the target variable in one particular state.
2. A stochastic intervention leaves the target variable with a positive probability over two or more states.

The perfectly deterministic intervention has only one possible result (after taking into account any interaction). Ideally, Susan's doctor would have liked to *make certain* that she stopped smoking. But there are some very useful interventions that are deliberately stochastic: in scientific experiments, subjects are often divided randomly into treatment and control groups, in the hope that this assignment will be independent of all other factors, and reveal the causal influence of the randomized variable.

Dimension 3. Targeted vs. indiscriminate:

1. A targeted intervention affects only one node.
2. An indiscriminate intervention affects more than one node.

Ideally, many medical interventions would be targeted with surgical precision (e.g., to affect only one organ). Unfortunately, side-effects are very common, and may or may not be anticipated. Experimental protocols are often designed with painstaking care, so that randomization only affects the target variable.

Dimension 4. Independent vs. interactive:

1. An independent intervention node I_C is not affected by any other variables.
2. An interactive intervention node I_C is affected by at least one other variable (either new or original).

Idealistically, we tend to think of our own interventions as completely independent decisions. But sometimes our actions may be coordinated with the decisions of others; and sometimes actually influenced by the state of M. For example, surgeons would not intervene to remove Susan's lung unless they were influenced by the fact that it was already cancerous.[15]

Dimension 5. Intended vs. unintended:

1. An intended intervention has precisely the effect upon M that the agent intended (in any of the preceding respects).
2. An unintended intervention has some effect upon M that the agent did not intend.

Of course, we are not offering a definition of either intention or agency, just applying this concept to interventions. We note that other external variables can frequently play a similar role to I_C in perturbing M, and hence facilitating causal discovery, as Spirtes et al. [44] make clear. Following ordinary language, we do not count such cases as interventions, since they lack intention.[16]

---

[15] In fact, most practical interventions incorporate some prior feedback about the state of the target variable, or more precisely, its immediate antecedent – but this is a subtlety we can usually afford to ignore.

[16] We should point out that perfect intervention nodes can be used to model all the varieties of imperfection discussed here, by encoding those imperfections in new nodes mediating between the intervention node and the original network. But this hardly makes the intervention perfect!

10.4 Alternative Measures of Causal Power

Intuitively, causal power is the strength of the connection from a cause to an effect. Over the last century, several influential analyses of causal power have been proposed, in which this basic idea is translated into a formal measure.

In fact, a measure of causal power was already implicit in the earliest graphical models of causal networks: the linear path models of Wright [49]. Wright's models relate scalar variables in a linear way through path coefficients, which are closely related to correlation coefficients. Thus, the size of a path coefficient can be used as a measure of causal power. Moreover, the coefficients for sub-paths can easily be combined to calculate the coefficient for a longer directed path.

The earliest explicit proposal for a causal power measure was that of Good [16]. Good did not restrict its application to any particular class of models. He offered an additional statistic, besides whatever calculations were required to implement the model itself, and also a method for combining the statistics of sub-paths to calculate the causal power of a longer directed path. His measure seems to be inspired by calculations of conductivity and impedance in electrical circuits (although it also has some similarity to information-theoretic measures). While interesting, Good's measure can lead to some very strange results when applied to longer causal paths. As pointed out in [39], two different paths can result in two different end-to-end dependencies, and yet applying Good's combinatorial method will assess their causal power as the same.

We shall confine ourselves to briefly describing the measure more recently proposed by Patricia Cheng [3]. Cheng offered her measure as an improvement over Rescorla and Wagner [36], and it was further developed by Glymour [15]. It serves to illustrate the problems associated with all these rival measures.

10.4.1 Cheng's Measure

Cheng begins with the ΔP measure of statistical relevance, mentioned in Sect. 10.2. As she puts it, a "positive probabilistic contrast"

$$\Delta P_c = P(e|c) - P(e|\neg c) > 0$$

indicates "candidate generative causation". In this respect, Cheng echoes the work of Suppes [45], who called this difference *prima facie causation*. c is only a prima facie cause because the probabilistic contrast may actually be caused by a common ancestor that raises the probability of c and e occurring together. Suppes' theory goes on to lay down conditions ruling such cases out. The continuing research program on probabilistic causality is largely concerned with the further refinement of such

Probabilistically, the net effect of the intervention on the system remains the same; while graphically, the net result of intervention is to add *all* these nodes, not just the perfect intervention node. So this maneuver retains only a fig-leaf of perfection, at the high cost of distracting complexity.

conditions, which in the end have been subsumed by technical developments using Bayesian networks to represent conditional independence (cf. [46]).

Like Suppes, Cheng needs to exercise such troublesome ancestors in order to make her theory work. But she does so by laying down some very stringent requirements for the causal relationships permitted in her models. First, c itself must be independent of these other parents. This implies either a limited causal structure in which there are no causal paths between c and these parents, or that the effect of these paths can be removed by fixing some background variables (which is not possible in some graphs). Second, the co-variation between c and e must be independent of the co-variation between e and any other parent. In other words, the effect of c on e cannot be altered by the state of any other parent; there must be no causal interaction between any parents of E.[17]

Given these restrictions, it is clear that the probabilistic contrast must be caused by c. In other words, the occurrences of c must be "generating" the additional occurrences of e. Cheng now defines the causal power of c over e as *the probability that any given occurrence of c will generate e*. This causal power of c is labelled p_c, leaving e implicit. Her basic insight is that ΔP is not a fair measure of p_c. There is a specific background rate at which e occurs even without c, namely $P(e|\neg c)$. This means that we can only detect the effect of c on the *remaining* instances of E: by how many background occurrences of $\neg e$ are converted to e. Now, $\neg e$ occurs with a background frequency of $1 - P(e|\neg c)$; it is converted with a frequency of ΔP; and therefore, the success rate of c must be the ratio of these two quantities. Formally, she derives:

$$p_c = \frac{\Delta P_c}{1 - P(e|\neg c)} \,. \tag{10.3}$$

In contrast, a negative ΔP indicates "candidate preventative causation", in which c appears to prevent e from occurring. To analyse this, Cheng places the same stringent restrictions on the parental relationships. She then defines the causal power of c to prevent e in an analogous way, as *the probability that any given occurrence of c will prevent e*. To distinguish prevention from generation, we write preventive powers as $\overline{p_c}$. By similar reasoning, we can only detect the success rate of c against the background rate of e, namely $P(e|\neg c)$. Thus:

$$\overline{p_c} = \frac{-\Delta P_c}{P(e|\neg c)} \,. \tag{10.4}$$

Cheng claims that these formulae are a significant improvement on previous theories, such as that of Rescorla and Wagner [36], because (among other reasons) the formula for p_c provides the correct answer when e always occurs. If e always occurs, then the value for p_c is undefined, rather than a power of zero, as Rescorla and Wagner had suggested. Cheng deems leaving p_c unspecified to be correct because we should be unable to statistically assess the candidate causes of a universal event.

17 She also insists that the causal power of c cannot be affected by the frequency of c, which for Bayesian networks is an unnecessary stipulation, and that whenever c occurs it must be preceded by some promoting cause, which is an unnecessary concession to determinism.

Similarly, the value of $\overline{p_c}$ is undefined when e never occurs. However, we do not see this feature of her theory as a significant advantage or disadvantage. Rescorla and Wagner might reply that no candidate cause could demonstrate any statistical power over a universal event, and therefore in such cases zero is a reasonable statistical assessment of causal power. In the absence of strong intuitions on this issue, there are far more serious problems with both their measures!

10.4.2 Problems with Cheng's Measure

The main problem with Cheng's measure is that it has an extremely limited range of application. It is only applicable to questions about causal relations between values, as opposed to the variables themselves. Structurally, the restrictions upon parental connections are very strong, and will not be met by many Bayesian networks. But perhaps the most severe restriction is to the probability distributions: her blanket ban on any causal interactions between parent variables. These restrictions are necessary to make her derivations of (10.3) and (10.4) work, but as Glymour [15] has shown, it limits Cheng's theory to linear and noisy-OR Bayesian networks.

One notable consequence is that Cheng's causal power necessarily exhibits a form of transitivity: if c causes d causes e, then c must cause e. This follows from her model restrictions and power measure. But the same is true of all the other rival measures. For Wright, it follows from the use of linear models and path coefficients. For Good, it follows not from any model restrictions, but simply from the additivity of electrical – or causal – resistance. Yet any account of causal power that entails transitivity is misleading, since causation in general is not transitive – a fact which is reflected in other types of Bayesian network. Take, for example, Richard Neapolitan's case of finesteride [29]. Finesteride reduces testosterone levels; lowered testosterone levels can lead to erectile dysfunction. However, finesteride fails to reduce testosterone levels *sufficiently* to cause erectile dysfunction.[18] Such threshold effects do not occur in linear or noisy-OR networks, but they are common elsewhere.

10.4.3 Desiderata for Causal Power

Examining these rival measures suggests a number of principles that an ideal causal power measure would uphold:

1. The measure should be applicable to all kinds of Bayesian network.

[18] This result was reported in at least one scientific study. Whether or not it is true generally, the point is that it *could* be true, and provides a neat illustration. Incidentally, this case does *not* appear in Neapolitan [30] because the publisher thought the example too challenging for its delicate readership!

2. The theory should generalize over linear path models. Thus, they should report causal powers that are directly comparable to the correlations in Wright's models generated by causal paths.
3. The measure should not entail transitivity – simply because causation is not, in general, transitive. Of course, the measure needs to reflect transitivity when it appears.
4. The measure should be compatible with intervention. It should support the fundamental idea that interventions test causal power, and it should be able to assess the power of setting a variable to a particular value.
5. The measure should have an information-theoretic interpretation. Causality gives rise to probabilistic relationships, which should lead to a reasonable interpretation under Shannon's theory of information.

Prior measures, such as Cheng's, fulfill some of these requirements, but none of them successfully fulfill them all.

10.5 Causal Information

Now we present our solution to the problem of measuring causal power, which satisfies our requirements above.[19] We assume that we are given a causal Bayesian network; the problem is to state the causal power of one variable, C, over another variable, E, which is implied by that network.

10.5.1 Background Conditions and Active Paths

10.5.1.1 Background Conditions: φ_h

We may wish to ask causal questions in the context of specific background conditions. We suppose that such conditions can be specified by identifying a set of network variables, Φ, whose values are given, $\Phi = \varphi_h$. Thus, all the probabilities discussed in the following sections will implicitly be conditional probabilities of the form $P(\cdot|\varphi_h)$, but for brevity we will omit the condition φ_h in our formulae for causal power.

If we simply wanted information about E, then the usual procedure would be to condition upon all our available knowledge, say $\Psi = \psi_g$, and the network would provide the most informative probability distribution over E. We could then ask what additional information would be provided in these circumstances by also conditioning upon C. However, when asking causal questions, it is not always appropriate to condition upon ψ_g. There are two general reasons for this. First, we may be interested in understanding causal processes by examining possible scenarios other

[19] Note that causal information was first introduced by Hope and Korb [19].

than our current situation. These scenarios may be counterfactual, such as the causal process that *would have existed* between C and E if some known antecedent condition had been different. They may be past scenarios, such as the causal process that *did exist* between C and E before some known condition resulted from this process (such as $E = e_k$). Or they may be future scenarios, such as the causal process that *will exist* if some condition comes to pass. The second general reason is more subtle: we are only asking about the *causal* influence of C on E, not the *information* that C provides about E. Hence, we may need to select our background conditions carefully, so that only causal paths are "active", as we shall now explain.

10.5.1.2 Causal Paths

Consider the graph shown in Fig. 10.5. This shows the same relationships we discussed earlier between exercise at 40, blood pressure at 50, and heart attack by 60. But another variable has been added: whether or not a patient is referred by their doctor for an echocardiogram (a scan of their heart) by the time they are 60 years old. We suppose that two of the original variables causally influence the probability of such a referral. A patient is quite likely to be referred if they exhibit high blood pressure at 50, certain to be referred after they have a heart attack and survive, and will never be referred without either of those factors (in our simplified example).

Now, any causal path from C to E is mono-directional, with all the arcs along the path pointing from C to E. Here there are two causal paths from *Exercise* to *Heart Attack*: the direct path *Exercise* → *Heart Attack*, and the indirect path *Exercise* → *Blood Pressure* → *Heart Attack*. Causal paths are inactive if we condition upon any of their variables. For example, how much do variations in exercise routine affect

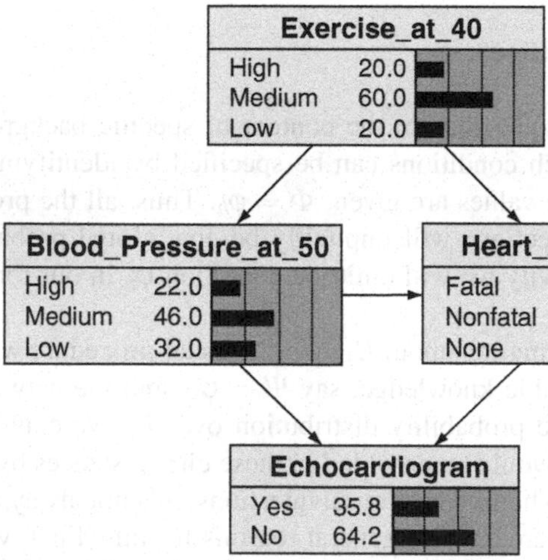

Fig. 10.5 A four variable Bayesian network linking exercise, blood pressure, heart attacks, and echocardiograms

heart attack outcomes? In posing this question, it would obviously be a mistake to stipulate that only patients with the same exercise routine should be considered (conditioning upon one particular value for this variable). If no variation in exercise routine is permitted, then heart attacks obviously cannot be affected by any such variation! Similarly, we cannot condition upon a particular heart attack outcome. A less obvious mistake would be to stipulate that only patients with the same blood pressure should be considered (perhaps with the idea of removing the influence of a confusing background variable). The problem is that some of the causal influence of *Exercise* is exerted precisely by changing *Blood Pressure* (a possibility which is made clear by presenting the system as a causal Bayesian network). So fixing *Blood Pressure* makes it impossible to detect such indirect causal influence. In general, if we want to include the causal power of C on E via any particular path, then we had better ensure that this path is active, by not conditioning upon C, E, or any intermediate variable on this path.

10.5.1.3 Common Effect Paths

Now we turn to non-causal but informative paths from C to E: they include at least one arc that does not point from C to E. Consider in particular those paths that begin with an arc "forwards" out of C. Since they reverse direction at some point, such paths must include at least one "common effect" structure, such as *Blood Pressure* $\rightarrow$ *Echocardiogram* $\leftarrow$ *Heart Attack*. Common effect paths have this peculiarity: they are active only if we know something about the common effect, here *Echocardiogram*. For example, suppose we learn that a patient has received an echocardiogram. This increases the probability that they had high blood pressure at 50, since high blood pressure would explain why they were referred for this scan. But it also increases the probability that they had a non-fatal heart attack, since this would also explain the referral. If we now learn that the patient had low blood pressure at 50, then we can infer that they *certainly* had a non-fatal heart attack, as illustrated in Fig. 10.6. This indirect connection between blood pressure and heart attack is not causal (since their low blood pressure didn't cause their heart attack), but it is nonetheless very informative. In contrast, if we do not know whether the patient has received an echocardiogram, then this indirect inference is not possible.

Suppose we now ask a causal question: "How much did this patient's blood pressure causally influence their chances of a heart attack?" There is a direct causal link between *Blood Pressure* and *Heart Attack*, and in particular, low blood pressure tends to *prevent* heart attacks. The graph will show this reduced probability – provided that we do not mistakenly include the fact that the patient received an echocardiogram! If we include this fact, then we activate the common cause path, and low blood pressure increases the probability of a heart attack (which does not accurately reflect the causal story). In general, if we want to know only about the causal influence of C on E, then we had better make sure that all non-causal paths are inactive. So to measure causal power, we make this stipulation: for each non-causal path between C and E that begins with an arc forwards out of C, the background conditions,

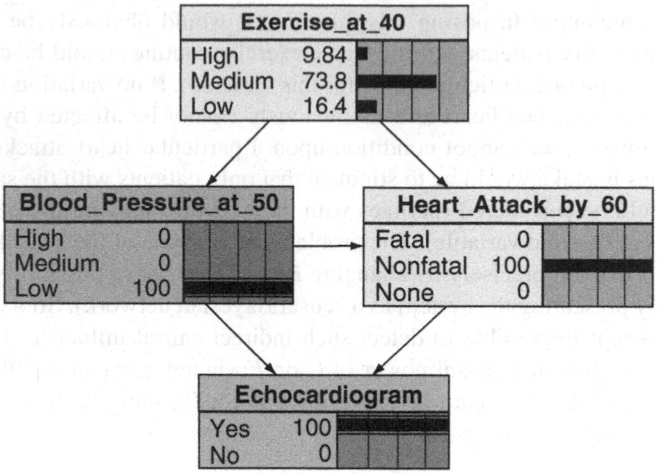

Fig. 10.6 Given an echocardiogram, discovering low blood pressure proves that the patient had a non-fatal heart attack

φ_h, must deactivate this path. If necessary, this should be achieved by excluding a common effect variable on that path, and any descendants of this variable. Any such alteration to φ_h will ensure that the path is inactive (without deactivating any causal path).

This means that under our measure, causal power is not defined under background conditions that include such common effects or their descendants. Since our knowledge of the system, ψ_g, may include such variables, this appears to be a significant restriction. However, such variables are always causal consequences of the variables in the causal process. Therefore, there are good independent reasons for excluding such variables from φ_h. If we include them, then we are asking about restricted aspects of the causal process (given the restriction that it led to these specific results, even though these results played no role in the causal process). If we exclude them, then we are asking about the entire causal process (given only causally significant antecedents, and not pre-supposing any specific results). For example, the preceding question was a retrospective one, concerning the causal process that previously existed between the patient's blood pressure and their chances of a heart attack. Therefore, it was a mistake to include information about their echocardiogram, simply because this was a later result of the causal process.

10.5.1.4 Common Cause Paths

Now consider those non-causal paths that begin with an arc "backwards" out of C. To reverse direction at some point, such paths must include at least one "common cause" structure, such as *Blood Pressure* $\leftarrow$ *Exercise* $\rightarrow$ *Heart Attack*. Common cause paths are active only if we do not know the value of the common cause variable, here *Exercise*. For example, suppose that a patient has low blood pressure.

This lowers the probability that they will have a heart attack, for two distinct reasons, as explained in Sect. 10.3.3. First, low blood pressure has a direct causal influence on heart attack. Second, we can infer that if a patient has low blood pressure, then they are more likely to do a lot of exercise. Since exercise also directly prevents heart attack, there will be an additional correlation between low blood pressure and low risk of heart attack. This indirect connection is not causal (since their low blood pressure didn't cause them to exercise), but it is nonetheless informative. In contrast, if we *already know* the patient's exercise regime, then this indirect inference is not possible. The only remaining source of probabilistic dependence between blood pressure and heart attack will be the direct causal connection.

One way to ensure that such common cause paths are inactive is to make another stipulation about φ_h: for each path between C and E that begins with an arc backwards out of C, the background conditions must deactivate this path. If necessary, this should be achieved by including the first common effect variable on that path, say B. (Alternatively, we could include the first parent of C on the path.) Any such alteration to φ_h will ensure that the path is inactive (without deactivating any causal path).

But there are two serious problems with this suggestion. First, any alteration to φ_h adds a variable B whose value is not known. So rather than one conditioning value $B = b_i$ and one corresponding dependency between C and E, there is a set of possible values and a set of corresponding dependencies. For example, conditioning upon each type of exercise routine may produce a different dependency between *Blood Pressure* and *Heart Attack*. The results of conditioning upon high exercise are depicted in Fig. 10.7. It is not obvious how such a set should be combined to form some reasonable overall measure of how much C affects E, c affects e, etc.

Second, each of these conditions $B = b_i$ only captures the extent to which C can affect E when B is fixed. Thus, a straightforward average of these individual

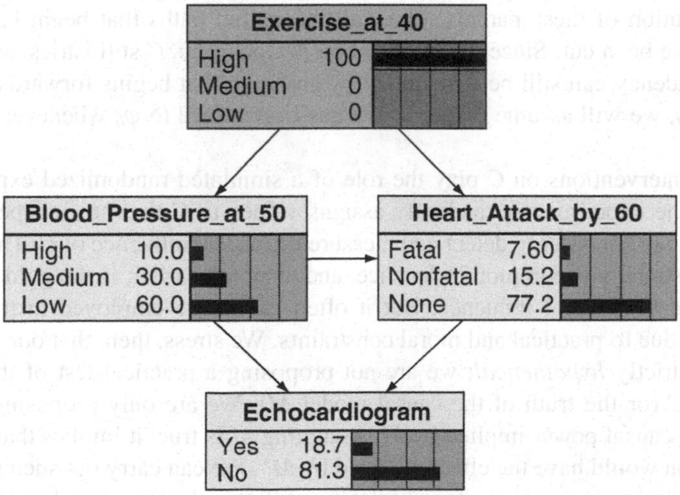

Fig. 10.7 Trying to isolate the effect of blood pressure on heart attack by conditioning upon exercise

dependencies (weighted according to $P(b_i)$) is clearly incorrect as an overall measure of how much C affects E. It fails to capture the dependency between C and E that results from variation in B. For example, conditioning upon each type of exercise routine may greatly reduce the variation in *Blood Pressure*: high levels of exercise are strongly associated with low blood pressure, etc. Therefore, within each exercise category, the apparent effect of *Blood Pressure* on *Heart Attack* may be low. Yet overall, *Exercise* may vary greatly and strongly affect *Blood Pressure*, which varies greatly and strongly affects *Heart Attack*. So just measuring the effect of blood pressure on heart attacks within each exercise category (and taking a weighted average) can grossly underestimate the effect of blood pressure in general.

Given these difficulties, rather than inactivating common cause paths by adding background conditions, we have chosen an alternative solution: hypothetical interventions on C.

10.5.2 Hypothetical Interventions: $P^*(C)$

10.5.2.1 Why Use $P^*(C)$?

Intervention upon C provides a straightforward way to distinguish between paths to E that begin backwards (which must be non-causal) and paths to E that begin forwards (which may be causal). For simplicity, we shall suppose that the intervention is intended, independent, targeted, overwhelming and stochastic (four out of five perfections!). So we augment the model M to M^*, with just one new intervention node and arc $I_C \rightarrow C$, and just one new elementary conditional probability function $P^*(C|\pi_C, I_C)$ over C, instead of the original function $P(C|\pi_C)$. Since the intervention is overwhelming, when $I_C = Yes$ then C is no longer directly dependent upon its old parents, and so discovering that $C = c_i$ no longer allows any direct inference about the distribution of these parents. Thus, all inferential paths that begin backwards from C have been cut. Since the intervention is stochastic, C still varies, and therefore dependency can still be transmitted by any path that begins forwards from C. For brevity, we will assume that $I_C = Yes$ has been added to φ_h whenever we refer to $P^*(\cdot)$.

These interventions on C play the role of a simulated randomized experiment, in which the experimenter randomly assigns values to C that are independent of other variables, in order to detect and measure the causal influence of C. This causal discovery strategy is common in Science, and as noted earlier, is the gold standard for determining causal influence. But it often cannot be employed in this idealized form, due to practical and moral constraints. We stress, then, that our interventions are strictly *hypothetical*; we are not proposing a practical test of the causal power of C (or the truth of the causal model M). We are only proposing a measure of the causal power implied by M. Assuming M is true, it implies that such an intervention would have the effect modeled by M^*. We can carry out such idealized thought experiments or computer simulations unfettered by the usual practical and moral constraints. For example, to measure the causal power of *Blood Pressure* over

Heart Attack, imagine that an experimenter assigns some blood pressure at random to each patient, and then administers drugs to achieve this particular blood pressure. Also, they exclude any information about echocardiograms. Any resulting dependency between *Blood Pressure* and *Heart Attack* must be due to their direct causal connection.

Given a hypothetical intervention upon C, it is natural to ask how much this intervention has affected E. To answer this question, we could compare the old distribution $P(E)$ to the new distribution $P^*(E)$. We could also apply information-theoretic measures to provide a summary of the effect of intervening, such as computing the Kullback–Leibler distance between the two distributions. However, this is not the question we are addressing! The aim of experimentally randomizing *Blood Pressure* is not simply to see how much the marginal distribution of *Heart Attack* has changed (e.g., to see how many additional heart attacks have been created by the experimenter!). The aim is to see how much the contrasting values of *Blood Pressure* affect *Heart Attack* in this experimental situation. In short, the aim is not to measure the power of the experimental intervention, but to measure the power of C given the intervention. Nevertheless, the results of such exploratory interventions can subsequently be used to guide utilitarian ones. For example, the randomization of *Blood Pressure* will include some patients whose blood pressure is low. Their distribution of heart attacks shows what setting low blood pressure can do; so we may later seek the same distribution in a perfect and utilitarian fashion, by setting a nice low blood pressure for every patient.

But what intervention distribution should be imposed upon C? There are three alternative choices that strike us as reasonable, each serving a slightly different purpose.

10.5.2.2 Original $P^*(C)$

Experimental interventions are open to a generic objection: if the experimental system, modeled by M^*, differs from the real system, modeled by M, then we are not measuring reality – we are measuring an artificial, counterfactual construct. So why should we be interested in the causal power of C over E in M^*?

The correct reply is that in some important respects the differences between M and M^* are kept to a minimum. M^* is not merely a counterfactual model that differs extensively and arbitrarily from reality. The experimenter tries to preserve in M^* the key features of M that are under investigation. In our simulated experiment, this is particularly easy. M^* is identical to M in most ways, regardless of the intervention distribution chosen. In particular, all the causal paths between C and E are preserved, and the elementary conditional probability functions for every variable except C are preserved. Thus, in important respects, the quantitative power of C over E is preserved. The similarity between M and M^* means that the causal power of C over E in M^* genuinely reflects the original situation in M.

In accordance with this principle, one plausible choice for the intervention distribution on C is to copy the original distribution over C. For example, if high blood pressure is rare in the general population, then it can be equally rare in the intervention distribution. To be precise, we can copy the original conditional probability function $P(C|\pi_C)$. If the background conditions, φ_h, include some of the parents of C, then these parental conditions entail the distribution $P(C|\pi_C \cap \varphi_h)$. Now, for each value c_i, the same probability can be assigned in the intervention probability function $P^*(C)$. Once we include the rest of the background conditions, the distribution $P^*(C|\varphi_h)$ will also be identical to the original distribution $P(C|\varphi_h)$. Note, however, that $P^*(C)$ does not incorporate the same intricate dependence upon π_C, so any further information about the value of C will not tell us anything further about the values of its parents.

The main attraction of using the original distribution over C is that it seems to minimize the difference between M and M^*. Thus, it promises a measure of causal power that is closest to the effect of C on E in the original model. For example, we can ask, "Given the variation in blood pressure among the general population, how much is this variable affecting heart attack outcomes?" We could answer this question by using the original distribution over *Blood Pressure*, and measuring the causal power of this variable over the *Heart Attack* variable, according to the formula given below.

There are two notable features of this answer. First, we are able to give a single variable-to-variable answer, rather than a large table of all possible blood pressures and the corresponding probabilities for all possible heart attack outcomes. Second, we are able to measure the overall power the variable is actually exerting within the given population. For example, suppose we are considering some subpopulation of Sweden, where most people do nothing but eat pickled herring and go cross-country skiing, and consequently most people have admirably low blood pressure (we shall refer to them as "Swedish" for short). Nevertheless, heart attack outcomes still vary, for genetic and other reasons. There is a valid sense in which the blood pressure variable will not be exerting much influence over heart attack outcomes within this population, simply because there is not much variation in the former. Thus, discovering a Swedish person's blood pressure seldom yields any useful information about their heart attack risk. We can only measure the lack of impact of Swedish blood pressure if we model the Swedish population distribution.

We should note, however, that even if we impose the original distribution upon C, the resulting distribution upon E may still be considerably different in M^* than it was in M. This is because (as intended) C is no longer dependent on its parents. Also, there are other valid senses in which Swedish blood pressure may still be exerting a strong influence. This is evident if we switch from variables to their values: if a Swedish person has high blood pressure, then this will still make a big difference to their chances of having a heart attack. Alternatively, we could contemplate a counterfactual distribution on Swedish blood pressure.

10.5.2.3 Uniform $P^*(C)$

We may not always wish to measure causal power relative to the original distribution over C. The connection between Swedish blood pressure and heart attack outcomes is concealed by their universally healthy lifestyle. So one way to bring out this latent feature of M is to consider a different intervention distribution over C, even though it is not the naturally occurring distribution in M. For example, an experimental investigation into the power of Swedish blood pressure would not randomize its subjects so that they all fell into the low blood pressure group.

One plausible choice is a uniform distribution on C, so that there are equal numbers of subjects in every blood pressure group. In comparing the effects of different blood pressures, this provides a "level playing field" in which the results are not biased by different frequencies for these blood pressures. Similarly, in comparing the influence of variables, it provides a standard distribution for comparison.

To be precise, we would impose some intervention distribution $P^*(C)$ such that after we take into account the background conditions, the resulting distribution $P^*(C|\varphi_h)$ is uniform. That is, $P^*(c_i|\varphi_h) = \frac{1}{|C|}$ for each c_i. We note that to achieve this, $P^*(C)$ itself will not always be uniform.

10.5.2.4 Maximizing $P^*(C)$

It is also reasonable to ask about the maximum impact that C could potentially have on E, according to M. To be precise, we can search the space of possible intervention distributions $P^*(C)$, to find those that maximize our causal power measure given the background conditions. This will not always be the uniform distribution on C. For example, suppose that there are only three blood pressure categories, and while both low and medium blood pressures result in a similar risk of heart attack, high blood pressure results in a much higher risk. Then the maximum probabilistic dependence between *Blood Pressure* and *Heart Attack* will result from a distribution in which nearly 50% of subjects have high blood pressure, rather than 33%.

These three possible intervention distributions seem to be complementary, in that they attempt to measure three slightly different forms of causal power of C over E: the original causal power, the standardized causal power, and the maximum causal power. In the formulae that follow we leave open the choice of intervention distribution, which is simply denoted $P^*(C)$. But to illustrate their application, we can imagine that a uniform distribution has been imposed upon *Blood Pressure* to measure its causal power over *Heart Attack*, as illustrated in Fig. 10.8.

10.5.3 Formulae: CI

10.5.3.1 Two Values: c and e

We begin with the simplest formula, and work our way towards the most complicated. What is the causal power of one value, c, to affect another value, e?

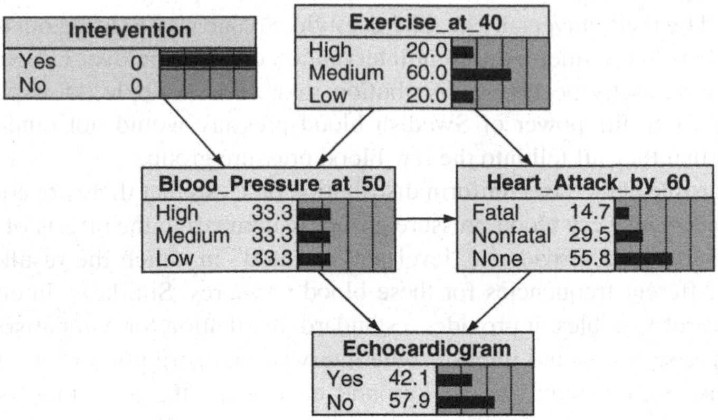

Fig. 10.8 A uniform intervention on blood pressure, to measure its causal power

$$CI(c,e) = P^*(e|c)\log\frac{P^*(e|c)}{P^*(e)}\,. \tag{10.5}$$

In information theory, this formula gives the information about e that is provided by the discovery that $C = c$ (compared to knowing the distribution $P^*(C)$). Given that only causal paths are active, we suggest that this formula can also serve as a good measure of the causal power of $C = c$ to affect the probability of $E = e$ (compared to the distribution $P^*(C)$). For example, suppose we observe that a patient has high blood pressure, c. This increases their probability of having a fatal heart attack, e, from the average probability $P^*(e) = 0.147$ to $P^*(e|c) = 0.230$. So $P^*(e|c)/P^*(e) = 1.565$. This is converted to a natural logarithm, which takes the positive value 0.448. It is multiplied by the probability of having a heart attack given high blood pressure, 0.230. So the causal power of high blood pressure to promote heart attack is 0.103.

Value-to-value questions such as "If I have high blood pressure, then how much does this affect my risk of having a heart attack?" are frequently asked, and may appear straightforward. But this is deceptive, since (as we noted in Sect. 10.2) there are many plausible ways to measure the increased probability of e that results from conditioning upon c. It is worth comparing our causal information measure to the measures of statistical relevance canvassed there. As a comparison probability, causal information uses the marginal probability $P^*(e)$ rather than the complementary probability $P^*(e|\neg c)$. This is similar to the standard formula for statistical relevance used in Philosophy (SR), rather than the standard formula used in Psychology (ΔP). Causal information initially compares these two probabilities as a ratio rather than a difference, thus measuring the proportional change (like the Bayesian updating factor, BR) rather than the absolute change (like SR). This proportion is converted to a logarithm, which may seem an odd way of re-scaling the change, but it is the usual format in information measures, since it facilitates appropriate averaging. This logarithm is then weighted by the absolute probability $P^*(e|c)$, and therefore the final number is actually more similar to the absolute change in probability ($SR = 0.083$) than to the proportional change in probability ($BR = 1.565$). The CI(c,e) measure is positive for promoting causes and negative for preventative causes, just like SR and ΔP.

Note that in this formula the probability of high blood pressure, $P^*(c)$, does not feature as a weighting factor. $C = c$ is treated as a given, as in the example question "If I have high blood pressure, ...", so we set $P^*(c) = 1$.

10.5.3.2 Various Causes: C and e

What is the causal power of one variable, C, to affect a particular value, e?

$$CI(C,e) = \sum_{c \in C} P^*(c)P^*(e|c)\log\frac{P^*(e|c)}{P^*(e)}. \tag{10.6}$$

In information theory, this formula gives the expected information about e that will be provided by discovering the value of C, whatever that turns out to be (compared to knowing the distribution $P^*(C)$). The difference between this and (10.5) is that the value of C is no longer treated as a given. Instead, we take the information (or power) from each individual value c_i, and weight this by the probability $P^*(c_i)$ to calculate the expected value. We suggest that this formula can also serve as a good measure of the causal power of C to affect the probability of $E = e$. For example, "How much does variation in blood pressure affect the risk of having a heart attack?" appears to be a variable-to-value question.

Note that some of the individual figures for causal power will be positive, and other figures will be negative. If we took a weighted average of the absolute magnitudes, then this would be the expected magnitude of the causal power exerted when C takes a specific value. However, the information-theoretic formula given above does not use absolute magnitudes, and the negative individual powers will partially offset the positive ones. Therefore, the magnitude of CI(C,e) should not be directly compared to the magnitude of CI(c,e). The CI(C,e) measure will always be positive, provided that C has some effect, i.e., $\exists i,j: \; P^*(e|c_i) \neq P^*(e|c_j)$, and otherwise it will be zero.

10.5.3.3 Various Effects: c and E

What is the causal power of one particular value, c, to affect a variable, E?

$$CI(c,E) = \sum_{e \in E} P^*(e|c)\log\frac{P^*(e|c)}{P^*(e)}. \tag{10.7}$$

In information theory, this formula gives the total information about E that is provided by the discovery that $C = c$ (compared to knowing the distribution $P^*(C)$). The difference between this and (10.5) is that we are interested in all the values of

E, not just one e. So we take the information from c for each individual value e_i, and add them to calculate the total value. We suggest that this formula can also serve as a good measure of the total causal power of c to affect the probability of E. For example, "How much does having high blood pressure affect heart attack outcomes?" is a value-to-variable question. Again, note that our information-theoretic formula does not use absolute magnitudes, and the negative individual powers will partially offset the positive ones. The CI(c,E) measure is equivalent to the Kullback–Leibler divergence between $P^*(E|c)$ and $P^*(E)$.

10.5.3.4 Two Variables: C and E

What is the causal power of one variable, C, to affect another variable, E?

$$CI(C,E) = \sum_{c \in C, e \in E} P^*(c)P^*(e|c)\log\frac{P^*(e|c)}{P^*(e)}\,. \tag{10.8}$$

In information theory, this formula gives the expected information about E that will be provided by discovering the value of C, whatever that turns out to be. It uses both the weighted average over the values of C and the sum over the values of E. We suggest that this formula can also serve as a good measure of the total causal power of C to affect the probability of E. For example, "How much does variation in blood pressure affect heart attack outcomes?" is a variable-to-variable question. Again, the negative individual powers will partially offset the positive ones, but the CI(C,E) measure will always be positive, provided that C has some effect on E.

The number of alternative formulae illustrate that there are several related questions about the power of C over E. So it is important to disambiguate informal queries such as "How much does blood pressure affect heart attacks?"

10.5.4 Mutual Information

This last equation can be transformed as follows:

$$CI(C,E) = \sum_{c \in C, e \in E} P^*(c)P^*(e|c)\log\frac{P^*(e|c)}{P^*(e)} \tag{10.9a}$$

$$= \sum_{c \in C, e \in E} P^*(c,e)\log\frac{P^*(e|c)}{P^*(e)} \tag{10.9b}$$

$$= \sum_{c \in C, e \in E} P^*(c,e)\log\frac{P^*(c,e)}{P^*(c)P^*(e)} \tag{10.9c}$$

$$= MI(C,E). \tag{10.9d}$$

This shows that causal information is identical to the information-theoretic quantity *mutual information* (MI), when applied to the two variables C and E, and given the intervention upon C. The mutual information formula looks a little different. It compares the probability that c and e will occur together, $P^*(c,e)$, to the probability that they *would* occur together *if the two variables were independent*, $P^*(c)P^*(e)$. Thus, it measures the amount of dependency that exists between each pair of variable values. The accumulated dependency for the two variables is obtained by weighting these ratios according to the probability that this pair of values will actually arise, $P^*(c,e)$. In fact, the causal information formula does the same job, but it has been expressed in an asymmetrical fashion to suit the asymmetry between cause and effect.

By definition, mutual information is the expected amount of information that one variable provides about another (or the loss of information that arises by falsely assuming that they are independent).[20] But, as above, it can also be interpreted as the amount of dependency between them. Therefore, it would be a good measure of causal power – except that some of this dependency can arise from non-causal links. Causal information corrects this defect.

10.5.5 Entropy

Mutual information is also closely related to the entropy measure of randomness. The information entropy on the variable X is defined as follows [6]:[21]

$$H(E) = -\sum_{e \in E} P(e) \log P(e). \tag{10.10}$$

Entropy is zero when $P(x_i) = 1$ for some value x_i, when there is no uncertainty about the value of X. It is maximized when $P(X)$ is uniform across all the possible values of X, when uncertainty is highest.

Similarly, conditional entropy measures the randomness of one variable given knowledge of another:

$$H(E|C) = -\sum_{c \in C} \sum_{e \in E} P(c,e) \log P(c|e). \tag{10.11}$$

Thus:

$$MI(C,E) = H(E) - H(E|C). \tag{10.12}$$

[20] From Shannon [41], the negative log of the probability of an event is the optimal code length to describe that event. Hence, mutual information can also be interpreted as the expected excess code length involved in recording the values of X and Y while wrongly assuming that they are independent.

[21] Entropy is defined subject to the common assumption that $0 \log 0 = 0$, which is justified by continuity arguments.

This supports the interpretation of mutual information as the reduction in the uncertainty of E due to the knowledge of C.

10.5.6 Relative Advantages

Causal information has some clear advantages over rival measures of causal power.

- Causal information is well-defined for all causal Bayesian networks. This includes all the restricted classes of network for which other measures were designed: linear models, Cheng models and their extensions, and whatever models Good had in mind. But it also includes classes of network for which these rival measures are not well-defined: e.g., ones with interactive causes, intransitivity, or multinomial (discrete) variables.
- Causal information is well-defined for a wider variety of questions. It relates any causal variable or value (either observed or observable) to any effect variable or value. It does so with a uniform approach, unlike Cheng's measure (for example), which uses a different formula for promoting and preventative causes.
- Causal information yields appropriate results in all the restricted classes of network, where it mirrors the local properties. For example, in any network that exhibits causal transitivity, $C \rightarrow D \rightarrow E$ implies that E is dependent upon C. But it follows immediately that $CI(C,D) \neq 0$, $CI(D,E) \neq 0$, and $CI(C,E) \neq 0$. So causal information itself exhibits causal transitivity, simply by accurately summarizing the true amount of dependency. Similarly, in linear path models, causal information is an increasing function of the magnitude of correlation. Therefore, the fact that other measures are *necessarily* transitive (or have other local properties built-in) offers no advantage, even when they are applied to their own preferred class of network.
- Causal information yields appropriate results in the other classes of network, where it does not impose inappropriate properties. For example, in any network that exhibits causal intransitivity or interaction, causal information itself exhibits intransitivity or interaction, for the reason specified above. It follows that causal information can be applied uniformly, without making assumptions about the local properties of the network. In contrast, even if the rival measures are well-defined for these networks, they will exhibit inappropriate properties that do not match the system. For example, Good's measure always exhibits a certain form of transitivity, but he does not restrict its application to models where this property is present.

10.6 Conclusions

Causal information, our new measure of causal power, is theoretically well-founded. Causal Bayesian networks provide a very general and powerful way to represent complex stochastic systems. Hypothetical interventions, when properly modeled in causal Bayesian networks, provide a clear separation of causal from non-causal paths. In mutual information, information theory provides an appropriate summary measure for cumulative causal influence, which applies to all sorts of networks and interventions, and can be tailored to specific purposes. The combination of the two, interventions and mutual information, yields causal information.

The result is a measure of causal power that has much wider application than previous accounts. Causal information can be applied to a wider variety of systems, including those with non-linear probabilistic influences and intricate structural relationships between variables. In such cases it still yields sensible results, unlike the alternative measures put forward by Cheng [3], Glymour [15], and Good [16]. These alternative measures were designed for simpler cases, such as noisy-OR networks that exhibit causal transitivity. But in these cases, too, our measure still yields appropriate results. And causal information is the only measure that is well-defined for relating any combination of values and variables.

We look forward to applying causal information to theoretical problems in Philosophy and AI. Causal information is also a promising measure for summarizing explanatory information encoded in a Bayesian network and so offers new means for simplifying the interpretation of complex Bayesian networks.

References

1. Anderson, K. M., P. M. Odell, P. W. Wilson, and W. B. Kannel (1991). Cardiovascular disease risk profiles. *American Heart Journal 121*, 293–298.
2. Cartwright, N. (1983). Causal laws and effective strategies. In *How the Laws of Physics Lie*, pp. 21–43. New York: Oxford University Press.
3. Cheng, P. W. (1997). From covariation to causation: A causal power theory. *Psychological Review 104*(2), 367–405.
4. Chickering, D. (1995). A transformational characterization of equivalent Bayesian network structures. In D. P. P. Besnard and S. Hanks (Eds.), *Proc of the 11th Conference on Uncertainty in AI*, San Fransisco, CA, pp. 87–98. Morgan Kaufmann.
5. Cooper, G. F. (1990). The computational complexity of probabilistic inference using Bayesian belief networks. *Artificial Intelligence 42*, 393–405.
6. Cover, T. M. and J. A. Thomas (1991). *Elements of Information Theory*. New York: Wiley.
7. D'Agostino, R., M. Russell, and D. Huse (2000). Primary and subsequent coronary risk appraisal: New results from the Framingham study. *American Heart Journal 139*, 272–81.
8. Dowe, P. (2000). *Physical Causation*. Cambridge studies in probability, induction, and decision theory. New York: Cambridge.
9. Eberhardt, F., C. Glymour, and R. Scheines (2005). On the number of experiments sufficient and in the worst case necessary to identify all causal relations among N variables. In F. Bacchus and T. Jaakkola (Eds.), *21st Conference on Uncertainty and Artificial Intelligence*, pp. 178–184.
10. Eberhardt, F., C. Glymour, and R. Scheines (2006). N-1 experiments suffice to determine the causal relations among N variables. In D. Holmes and L. Jain (Eds.), *Innovations in Machine Learning*, Volume 194. Berlin: Springer.
11. Eberhardt, F. and R. Scheines (2006). Interventions and causal inference. In *Philosophy of Science Association*.

12. Eells, E. (1988). Probabilistic causal levels. In B. Skyrms and W. Harper (Eds.), *Causation, Chance and Credence*, pp. 109–133. Dordrecht: Kluwer.
13. Eells, E. and E. Sober (1983). Probabilistic causality and the question of transitivity. *Philosophy of Science 50*, 35–57.
14. Fell, C. (2006). Causal discovery: The incorporation of latent variables in causal discovery using experimental data. Honours Thesis, Clayton School of IT, Monash University.
15. Glymour, C. (2001). *The Mind's Arrows: Bayes Nets and Graphical Causal Models in Psychology*. Cambridge: MIT.
16. Good, I. J. (1961). A causal calculus. *British Journal for the Philosophy of Science 11*, 305–318.
17. Halpern, J. Y. and J. Pearl (2001). Causes and explanations: A structural-model approach – Part I: Causes. In J. Breese and D. Koller (Eds.), *Uncertainty in Artificial Intelligence*, pp. 194–202.
18. Hitchcock, C. R. (2001). The intransitivity of causation revealed in equations and graphs. *Journal of Philosophy XCVIII*(6), 273–299.
19. Hope, L. R. and K. B. Korb (2005). An information-theoretic causal power theory. In *Lecture Notes in Artificial Intelligence*, pp. 805–811. Berlin: Springer.
20. Humphreys, P. (1989). *The Chances of Explanation*. Princeton: Princeton University Press.
21. Jensen, F. (1996). *An Introduction to Bayesian Networks*. University College, London: UCL Press.
22. Kennett, R. J., K. B. Korb, and A. E. Nicholson (2001). Seabreeze prediction using Bayesian networks. In *PAKDD'01 – Proceedings of the Fourth Pacific-Asia Conference on Knowledge Discovery and Data Mining*, Hong Kong, pp. 148–153.
23. Korb, K. B., L. R. Hope, A. E. Nicholson, and K. Axnick (2004). Varieties of causal intervention. In *PRICAI'04 – Proceedings of the 8th Pacific Rim International Conference on Artificial Intelligence*, Auckland, New Zealand, pp. 322–331.
24. Korb, K. B. and A. E. Nicholson (2004). *Bayesian Artificial Intelligence*. Boca Raton: Chapman and Hall / CRC.
25. Korb, K. B. and E. Nyberg (2006). The power of intervention. *Minds and Machines 16*, 289–302.
26. Lewis, D. (1973). Causation. *Journal of Philosophy 70*, 556–567.
27. Menzies, P. (2004). Difference making in context. In J. Collins, N. Hall, and L. Paul (Eds.), *Counterfactuals and Causation*, pp. 139–180. Cambridge: MIT.
28. Neapolitan, R. E. (1990). *Probabilistic Reasoning in Expert Systems*. New York: Wiley.
29. Neapolitan, R. E. (2003). Stochastic causality. In *International Conference on Cognitive Science*, Sydney, Australia.
30. Neapolitan, R. E. (2004). *Learning Bayesian Networks*. New York: Prentice-Hall.
31. Neil, J. R., C. S. Wallace, and K. B. Korb (1999). Learning Bayesian networks with restricted causal interactions. In K. B. Laskey and H. Prade (Eds.), *Proceedings of the Fifteenth Conference on Uncertainty in Artificial Intelligence (UAI-99)*, Stockholm, Sweden, pp. 486–493. UAI: Morgan Kaufmann, San Francisco, CA, USA.
32. Pearl, J. (1988). *Probabilistic Reasoning in Intelligent Systems*. San Mateo: Morgan Kaufmann.
33. Pearl, J. (2000). *Causality: Models, Reasoning and Inference*. Cambridge: Cambridge University Press.
34. Price, H. (1996). *Time's Arrow and Archimedes' Point: New Directions for the Physics of Time*. New York: Oxford University Press.
35. Reichenbach, H. (1956). *The Direction of Time*. Berkeley: University of California Press.
36. Rescorla, R. A. and A. R. Wagner (1972). A theory of Pavlovian conditioning. In A. H. Black and W. Prokasy (Eds.), *Classical Conditioning II: Current Theory and Research*, pp. 64–99. New York: Appleton-Century-Crofts.
37. Salmon, W. C. (1970). Statistical explanation. In R. G. Colodny (Ed.), *The Nature and Function of Scientific Theories*, pp. 173–231. Pittsburgh: University of Pittsburgh Press.

38. Salmon, W. C. (1974). *Statistical Explanation and Statistical Relevance*. Pittsburgh: University of Pittsburgh Press.
39. Salmon, W. C. (1980). Probabilistic causality. *Pacific Philosophical Quarterly 61*, 50–74.
40. Salmon, W. C. (1984). *Scientific Explanation and the Causal Structure of the World*. Princeton: Princeton University Press.
41. Shannon, C. (1948). A mathematical theory of communication. *The Bell System Technical Journal 27*, 379–423, 623–656.
42. Skyrms, B. (1980). *Causal Necessity*. New Haven: Yale University Press.
43. Spirtes, P., C. Glymour, and R. Scheines (1993). *Causation, Prediction and Search*. Number 81 in Lecture Notes in Statistics. Heidelberg: Springer.
44. Spirtes, P., C. Glymour, and R. Scheines (2000). *Causation, Prediction and Search: Second Edition*. Cambridge: MIT.
45. Suppes, P. (1970). *A Probabilistic Theory of Causality*. Amsterdam: North-Holland.
46. Twardy, C. R. and K. B. Korb (2004). A criterion of probabilistic causality. *Philosophy of Science 71*, 241–262.
47. Verma, T. and J. Pearl (1990). Equivalence and synthesis of causal models. In *Proceedings of the sixth conference on uncertainty in artificial intelligence*, San Francisco, pp. 462–470. UAI: Morgan Kaufmann.
48. Woodward, J. (2003). *Making Things Happen*. Oxford: Oxford University Press.
49. Wright, S. (1934). The method of path coefficients. *Annals of Mathematical Statistics 5*, 161–215.

Chapter 11
Information Flows in Complex Networks

João Barros

Abstract We give an overview on some of the main results in network information theory, that is the branch of Shannon theory that deals with the fundamental limits of information flow in complex networks. Particular emphasis is given to the fact that classical information-theoretic arguments, which yield the capacity of point-to-point channels, and standard network flow techniques, which are suitable for transport networks, do not necessarily apply when it comes to describing the behavior of information flows over complex networks that feature phenomena such as interference, cooperation or feedback. Notwithstanding this observation, we provide examples of information flow problems where max-flow min-cut type of arguments do prove useful for establishing performance bounds for complex networks and illustrate how mixing different flows through network coding may hold the key towards achieving those bounds.

11.1 Introduction

Since Shannon's *A Mathematical Theory of Communication* [43] a lot has been accomplished in terms of characterizing and achieving the maximum achievable rate (i.e., the capacity) at which two partners can communicate over a noisy channel with arbitrarily small probability of error. Not only do we know how to compute the capacity for many channel models that are relevant in practice, such as the additive white Gaussian channel (solved already in [43]) and wireless fading channel models [6], but feasible code constructions with performance close to the channel capacity are readily available, for example Turbo Codes [5] and Low-Density Parity Check (LDPC) codes [20, 32].

J. Barros
Instituto de Telecomunicações, Universidade do Porto, Porto, Portugal
e-mail: barros@dcc.fc.up.pt

F. Emmert-Streib, M. Dehmer (eds.), *Information Theory and Statistical Learning*,
DOI: 10.1007/978-0-387-84816-7_11,

Given the success of information theory in mastering point-to-point communications, one would be tempted to believe that a complete treatment of information flows over networks with multiple communicating partners should not take more than a small step. As it turns out, establishing the fundamental limits of communication over complex networks remains a formidable task, requiring a giant leap in terms of conceptual tools and mathematical sophistication.

The goal of this chapter is to illuminate this state of affairs by providing an introductory overview of some of the main tools, results and challenges that characterize the general area of network information theory. Focusing on discrete memoryless sources and channels, Sect. 11.2 establishes the notation and revisits Shannon's theorems for the point-to-point to problem. Section 11.3 then describes some of the most well-known problems in network information theory, which include distributed source coding, multiple access communications, broadcast channels and relay transmissions. The relationship between these problems and classical max-flow min-cut analysis is addressed in Sect. 11.4, both when the information sources are statistically independent and when they are correlated. Section 11.5 offers a simplified treatment of network coding as a technique that achieves the max-flow min-cut bound in multicast networks. The impact of topology on network information flow is highlighted in Sect. 11.6, which considers the max-flow min-cut capacity of various classes of random graphs. The chapter concludes in Sect. 11.6 with some final remarks.

Because this chapter is intended for the general scientific public, it is self-contained and aims mostly at providing intuition. The mathematically inclined reader will find all the proofs and technical details in the many references provided along the way.

11.2 Information-Theoretic Concepts

We start with a brief overview of some of the fundamental concepts of information theory. This will allow us to establish some notation and set the stage for the main results presented in subsequent sections. For a comprehensive introduction to the fundamental concepts and methods of information theory we refer to the treatises by Gallager [21], Cover and Thomas [13], and Yeung [48]. In [14] Csiszár and Körner offer a panoply of mathematical tools for discrete memoryless sources and channels.

11.2.1 The Point-to-Point Communications Problem

The foundations of information theory were laid by Claude E. Shannon in his 1948 paper entitled "A Mathematical Theory of Communication" [43]. In his own words: *the fundamental problem of communication is that of reproducing at one point either exactly or approximately a message selected at another point.* If the message – for

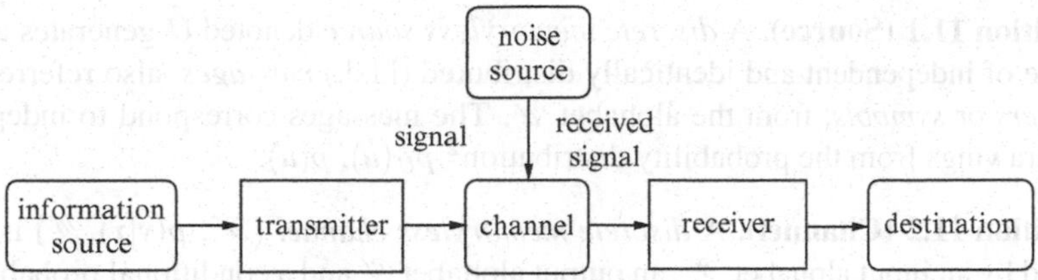

Fig. 11.1 Shannon's communications model (from [43])

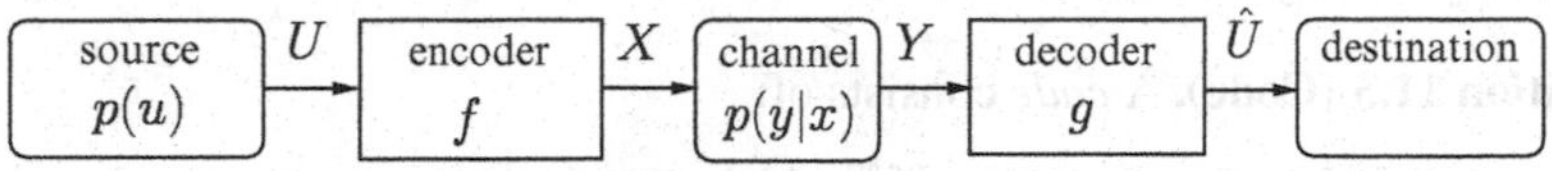

Fig. 11.2 Mathematical model of a communications system

example a letter from the alphabet, the gray level of a pixel or some physical quantity measured by a sensor – is to be reproduced at a remote location with a certain fidelity, some amount of information must be transmitted over a physical channel. This observation is the crux of Shannon's general model for *point-to-point* communication reproduced in Fig. 11.1. It consists of the following parts:

- The *information source* generates messages at a given rate according to some random process.
- The *transmitter* observes this messages and forms a signal to be sent over the channel.
- The *channel* is governed by a *noise source* which corrupts the original input signal. This models the physical constraints of a communications system, e.g., thermal noise in electronic circuits or multipath fading in a wireless medium.
- The *receiver* takes the received signal, forms a reconstructed version of the original message, and delivers the result to the *destination*.

Given the statistical properties of the information source and the noisy channel, the goal of the communications engineer is to design the transmitter and the receiver in a way that allows the sent information to reach its destination in a *reliable* way. Information theory can help us achieve this goal by characterizing the fundamental mechanisms behind communications systems and providing us with precise mathematical conditions under which reliable communication is possible.

To give a more precise formulation of the point-to-point communications problem, we require rigorous definitions for each of its constituent parts.[1] We assume that the *source* and the *channel* are described by discrete-time random processes, and we determine that the receiver and the transmitter agree on a common *code*, specified by an *encoder* and *decoder* pair. The basic relationship between these entities is illustrated in Fig. 11.2 and described rigorously in the following lines.

[1] We point out that although in this chapter we are mostly concerned with discrete memoryless sources and channels, many of the results presented here can be extended to account for continuous-valued alphabets, as well as sources and channels with memory.

Definition 11.1 (Source). A *discrete memoryless source* denoted U generates a sequence of independent and identically distributed (i.i.d.) *messages*, also referred to as *letters* or *symbols*, from the alphabet $\mathscr{U}$. The messages correspond to independent drawings from the probability distribution[2] $p_U(u)$. $p(u)$.

Definition 11.2 (Channel). A *discrete memoryless channel* $(\mathscr{X}, p(y|x), \mathscr{Y})$ is described by an input alphabet $\mathscr{X}$, an output alphabet $\mathscr{Y}$ and a conditional probability distribution $p(y|x)$, such that X and Y denote the *channel input* and the *channel output*, respectively.

Definition 11.3 (Code). A *code* consists of:

1. An encoding function $f : \mathscr{U} \to \mathscr{X}^N$ which maps a message u to a *codeword* x^N with N symbols.
2. A decoding function $g : \mathscr{Y}^N \to \widehat{\mathscr{U}}$, which maps a block of N channel outputs y^N to a message $\hat{u}$ from y^N the reconstruction alphabet $\widehat{\mathscr{U}}$. For simplicity, we assume that $\widehat{\mathscr{U}} = \mathscr{U}$, i.e., source and reconstruction alphabets are identical.

The *rate* of the code is given by $R = (1/N)\log_2 |\mathscr{U}|$ in *bits per channel use*, where $|\mathscr{U}|$ denotes the size of the alphabet $\mathscr{U}$.

To give a precise statement of the problem, we require one more definition:

Definition 11.4 (Reliable Communication). Given the rate R, reliable communication of the source $U \sim p(u)$ over the channel $(\mathscr{X}, p(y|x), \mathscr{Y})$ is possible if there exists a code $x^N(u)$ with rate R and with decoding function $g(y^N)$ such that, as $N \to \infty$,

$$P_N = p\{g(Y^N) \neq U\} \to 0,$$

i.e., the source messages are reconstructed with *arbitrarily small probability of error*. If reliable communication is possible at rate R then R is an *achievable rate.*

The main goal of the problem is to give precise conditions for reliable communication based on single-letter information-theoretic quantities that depend only on the given probability distributions and not on the block length N.

Remark 11.1. Notice that the classical information-theoretic formulation of the point-to-point communications problem does not put any constraints neither on the computational complexity nor on the delay of the encoding and decoding procedures. In other words, the goal is to describe the fundamental limits of communications systems irrespective of their technological limitations.

[2] In the sequel we follow the convention that subscripts of a probability distribution are dropped if the subscript is the capitalized version of the argument, i.e., we simply write $p(u)$ for the probability distribution $p_U(u)$.

11.2.2 Information-Theoretic Proof Techniques

The typical proofs in information theory are concerned with the existence of codes with certain asymptotic properties. A theorem that confirms the existence of codes for a class of achievable rates is often referred to as a *direct result* and the arguments that lead to this result constitute the *achievability proof*. On the other hand, when a theorem asserts that codes with certain properties do not exist, we speak of a *converse result* and a *converse proof*. To prove a complete *coding theorem* [14] we are required to provide both the achievability part and the converse proof.

One of the most important mathematical tools in information theory is the asymptotic equipartition property (AEP), which essentiality states that if we build sufficiently large sequences of random symbols X drawn independently and identically distributed according to some probability distribution $p(x)$, then with high probably the resulting sequences will belong to the so called *typical set*. Under the assumption of arbitrarily large sequence length N, these *typical sequences* share several fundamental properties: (a) the total probability of the typical set is arbitrarily close to 1, (b) the probability of a typical sequence is about $2^{-NH(X)}$ and (c) the number of strongly typical sequences is approximately $2^{NH(X)}$. Here, $H(X)$ denotes the *Shannon entropy* of the random variable X given by

$$H(X) = -\sum_{x} p(x) \log p(x),$$

where the summation is taken over the support of $p(x)$ and the logarithm is taken to base two.[3] The generalization of the AEP for two random variables X and Y can be obtained in a straightforward manner using the joint entropy

$$H(XY) = -\sum_{x}\sum_{y} p(xy) \log p(xy),$$

where the summation is once again carried out over the support of $p(xy)$. For large N, it follows that there exist around $2^{NH(XY)}$ jointly typical sequences x^N and y^N.

Conceptually, the entropy $H(X)$ can be viewed as a measure of the average amount of information contained in X or, equivalently, the amount of uncertainty that subsists until the outcome of X is revealed. Other useful information-theoretic measures include the conditional entropy of X given Y defined as

$$H(X|Y) = H(XY) - H(Y),$$

describing the amount of uncertainty that remains about X when Y is revealed, and the mutual information

$$I(X;Y) = H(X) - H(X|Y),$$

[3] Unless otherwise specified, all logarithms in this thesis are taken to base two.

which can be interpreted as the reduction in uncertainty about X when Y is given. The relationship between the aforementioned information-theoretic quantities is well explained in [48, Sect. 2.2].

Consider once again the formal statement of the point-to-point communications problem in Sect. 11.2.1. In Shannon's mathematical model a block of messages is mapped to a sequence of channel input symbols, also called *codeword*. The set of codewords builds the core of the code used by the transmitter and the receiver to communicate reliably over the channel.

Since information theory is primarily concerned with the fundamental limits of reliable communication, it is often useful to prove the existence of codes with certain properties without having to search for explicit code constructions. A simple way to accomplish this task is to perform a *random selection* of codewords. Random selection is often used in mathematics to prove the existence of mathematical objects without actually constructing them. For example, if we want to prove that a real-valued function $h(n)$ takes a value less than c for some n in a given set $\mathscr{S}$, then it suffices to introduce a uniform probability distribution on $\mathscr{S}$ and show that the mean value of $h(n)$ is less than c. When this technique is applied to prove the existence of codes with certain properties, we speak of *random coding*. Based on this simple idea, we can construct a random code for the system model shown in Fig. 11.2 by drawing codewords X^N at random X^N according to the probability distribution $\prod_{i=1}^{N} p(x_i)$. Then, if we want to prove that there exists a code such that the error probability goes to zero for N sufficiently large, it suffices to show that the average of the probability of error taken over all possible random codebooks goes to zero for N sufficiently large – in that case there exists at least one code whose probability of error is below the average.

A different coding technique, which is particularly useful in information-theoretic problems with multiple sources, consists of *throwing* sequences $u^N \in \mathscr{U}^N$ into a finite set of bins, such that the sequences that land in the same bin share a common bin index. If each sequence is assigned a bin at random according to a uniform distribution, then we refer to this procedure as *random binning*. By partitioning the set of sequences into equiprobable bins, we can rest assure that, as long as the number of bins is much larger than the number of typical sequences, the probability that there is more than one typical sequence in the same bin is very, very small [13, pp. 410–411]. This in turn means that each typical sequence is uniquely determined by its corresponding bin index. If *side information* is available and we can distinguish between different typical sequences in the same bin – e.g., when we are given a sequence w^N that is jointly typical with u^N – then we can decrease the number of bins, or equivalently the number of bin indices, and increase the efficiency of our coding scheme (see, e.g., [9] and [47]).

The large majority of converse proofs in network information theory uses Fano's inequality, which can be explained in very simple terms. Suppose that X is a random variable and that $\widehat{X}$ is an estimate of X taking values in the same alphabet $\mathscr{X}$. Fano's lemma gives a precise description of the relationship between the conditional entropy $H(X|\widehat{X})$ and the probability of error $P_e = p\{X \neq \widehat{X}\}$.

Lemma 11.1 (Fano's Inequality). *Let X and $\widehat{X}$ be two random variables with the same alphabet $\mathscr{X}$. Then*

$$H(X|\widehat{X}) \leq H_b(P_e) + P_e \log(|\mathscr{X}| - 1),$$

where $H_b(P_e)$ is the binary entropy function computed according to

$$H_b(P_e) = -P_e \log P_e - (1 - P_e) \log(1 - P_e)$$

Unfortunately, this lemma, which is key towards proving Shannon's channel coding theorem and computing the capacity region of the multiple access channel, has proven to be insufficient for obtaining tight converse results for other seemingly simple networks, such as the broadcast channel and the relay channel. The failure to find powerful alternatives to Fano's inequality is at the heart of why these problems, discussed in the next section, have been open for more than two decades.

11.2.3 Shannon's Coding Theorems

Having discussed some of the basic proof techniques in information theory, we will turn to Shannon's fundamental coding theorems. These results form the basis of classical information theory and are of great use in several information flow problems.

The *channel coding theorem* gives a complete solution (achievability and converse) for the point-to-point communications problem stated in Sect. 11.2.1. According to the problem statement, we use a code of rate R to transmit the messages produced by source U over a discrete memoryless channel $(\mathscr{X}, p(y|x), \mathscr{Y})$. If reliable communication is possible at rate R, i.e., the average error probability P_N goes to zero as the block length N goes to infinity, then we say the rate R is achievable. As it turns out, one simple condition is sufficient to fully characterize the set of achievable rates:

Theorem 11.1 (Channel Coding Theorem [48, Sect. 8.2]). *A rate R is achievable for a discrete memoryless channel $p(y|x)$ if and only if $R \leq C$, where C is the channel capacity given by*

$$C = \max_{p(x)} I(X;Y).$$

This remarkable result guarantees the existence of a code with arbitrarily small probability of error for all rates below the capacity of the channel. The latter equals the maximum mutual information between channel input X and channel output Y, where the maximization is carried out over all possible input probability distributions $p(x)$.

There are several ways to prove the channel coding theorem and details can be found, e.g., [13, Chap. 8] and [48, Chap. 8].

When the channel is noiseless, i.e., $Y = X$, it may still be useful to encode the messages produced by the source. In this case, the purpose of the code is not to compensate for the impairments caused by the channel, but to achieve a more efficient representation of the source information in terms of bits per message or equivalently bits per source symbol – this procedure is called *source coding* or *data compression*. The main idea is to consider only a subset $\mathscr{B}$ of all possible source sequences $\mathscr{U}^N$, and assign a different index $i \in \{1,2,\ldots,|\mathscr{B}|\}$ to each of the sequences u^N in $\mathscr{B}$. If the sequence produces a source sequence $u^N \in \mathscr{B}$, then the encoder outputs the corresponding index i, otherwise i is set to some predefined constant. The decoder receives the index i and outputs the corresponding sequence in $\mathscr{B}$. The rate of the resulting *source code* can be computed according to $R = (1/N)\log|\mathscr{B}|$. The following result gives the minimum rate R at which we can encode the data and still guarantee that the messages can be perfectly reconstructed.

Theorem 11.2 (Source Coding Theorem [48, Sect. 4.2]). *Let U be an information source drawn i.i.d. $\sim p(u)$. For N sufficiently large, there exists a code with arbitrarily small probability of error, whose coding rate R is arbitrarily close to the entropy $H(U)$. Conversely, if $R < H(U)$ the error probability goes to one, as N goes to infinity.*

A full proof can be found, e.g., in [48, Sect. 4.2]. The main idea behind the source coding theorem can be stated in very simple terms: since for large N it can be shown that any sequence produced by the source U belongs with high probability to the typical set $\mathscr{A}_\varepsilon^N(U)$, we only need to index the approximately $2^{NH(U)}$ typical sequences to achieve arbitrarily small probability of error. Thus, setting $\mathscr{B} = \mathscr{A}_\varepsilon^N(U)$, we get $R \approx H(U)$.

Alternatively, the theorem can be proved using a simple random binning argument: if we randomly assign each source sequence to one of a finite number of bins, then as long as the number of bins is larger than $2^{NH(U)}$ we know that the probability of finding more than one typical sequence in the same bin is very small [13, pp. 410–411]. Since each typical sequence is mapped to a different bin index, arbitrarily small probability of error can be easily achieved by letting the decoder output the typical sequence that corresponds to the received index.

11.3 Network Information Theory

The previous results help us characterize the fundamental limits of communication between two users, i.e., one sender and one receiver. However, in many communications scenarios – for example, satellite broadcasting, cellular telephony, the internet or wireless sensor networks – the information is sent by one or more transmitting nodes to one or more receiving nodes over more or less intricate communication networks. The interactions between the users of said networks introduce a whole new range of fundamental communications aspects that are not present in the classical point-to-point problem, such as *interference*, *user cooperation* and *feedback*.

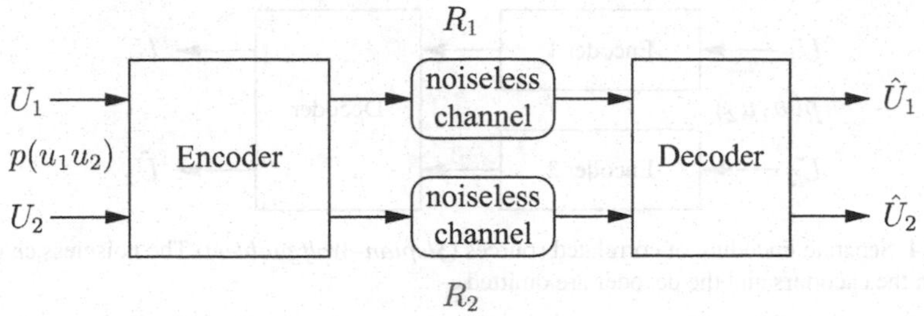

Fig. 11.3 Joint encoding of correlated sources

The central goal of *network information theory* is to provide a thorough understanding of these basic mechanisms, by characterizing the fundamental limits of complex networks with multiple users.

11.3.1 Distributed Source Coding

Assume that two sources U_1 and U_2 drawn i.i.d. $\sim p(u_1u_2)$ are to be processed by a joint encoder and transmitted to a common destination over two noiseless channels, as shown in Fig. 11.3. In general, $p(u_1u_2) \neq p(u_1)p(u_2)$, such that the messages produced by U_1 and U_2 at any given point in time are statistically dependent – we refer to U_1 and U_2 as *correlated sources*. Since the channels do not introduce any errors, we may ask the following question: at what rates R_1 and R_2 can we transmit information generated by U_1 and U_2 with arbitrarily small probability of error? Not surprisingly, since we have a common encoder and a common decoder, this problem reduces to the classical point-to-point problem and the solution follows naturally from Shannon's source coding theorem: the messages can be perfectly reconstructed at the receiver if and only if

$$R_1 + R_2 > H(U_1U_2),$$

i.e., the sum rate must be greater than the joint entropy of U_1 and U_2.

The problem becomes considerably more challenging if instead of a joint encoder we have two *separate* encoders, as shown in Fig. 11.4. Here, each encoder observes only the realizations of the one source it is assigned to and does not know the output symbols of the other source. In this case, it is not immediately clear which encoding rates guarantee perfect reconstruction at the receiver. If we encode U_1 at rate $R_1 > H(U_1)$ and U_2 at rate $R_2 > H(U_2)$, then the source coding theorem guarantees once again that arbitrarily small probability of error is possible. But, in this case, the sum rate amounts to $R_1 + R_2 > H(U_1) + H(U_2)$, which in general is greater than the joint entropy $H(U_1U_2)$.

In their landmark paper [44], Slepian and Wolf come to a surprising conclusion: the sum rate required by two separate encoders is the same as that required by a

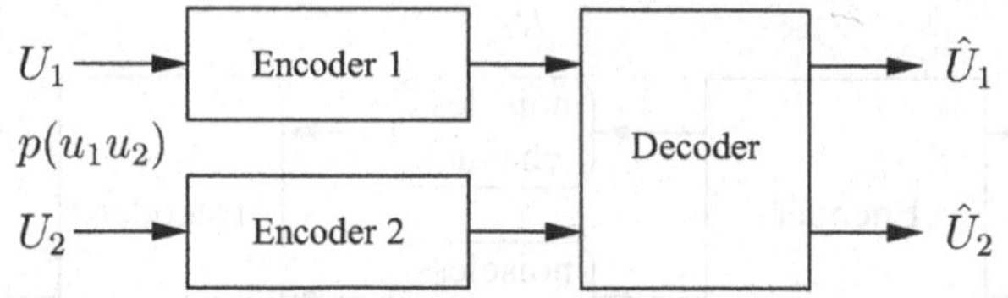

Fig. 11.4 Separate encoding of correlated sources (*Slepian–Wolf problem*) The noiseless channels between the encoders and the decoder are omitted

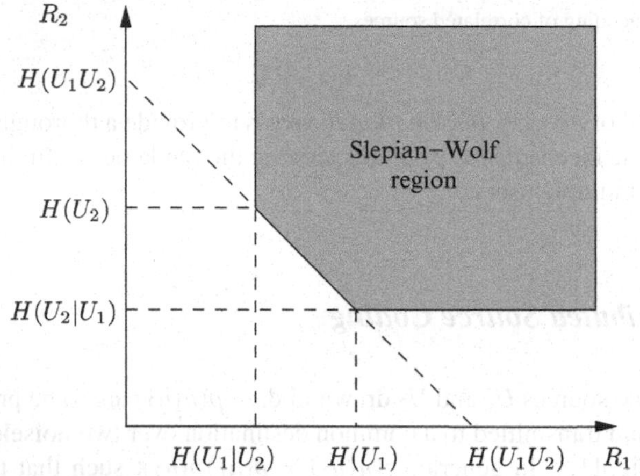

Fig. 11.5 The Slepian–Wolf region of achievable rates for separate encoding of correlated sources

joint encoder, i.e., $R_1 + R_2 > H(U_1U_2)$ is sufficient for perfect reconstruction to be possible. In other words, there is no loss in overall compression efficiency due to the fact that the encoders can only observe the realizations of the one source they have been assigned to. However, it is important to point out that the decoder does require a minimum amount of rate from each encoder, specifically the average remaining uncertainty about the messages of one source given the messages of the other source, i.e., $H(U_1|U_2)$ and $H(U_2|U_1)$. The region of achievable compression rate pairs (R_1, R_2), shown in Fig. 11.5, is thus fully characterized by the following theorem.

Theorem 11.3 (Slepian–Wolf Theorem, [44]). *Let (U_1U_2) be two correlated sources drawn i.i.d. $\sim p(u_1u_2)$. The compression rates (R_1, R_2) are achievable if and only if*

$$R_1 \geq H(U_1|U_2),$$
$$R_2 \geq H(U_2|U_1),$$
$$R_1 + R_2 \geq H(U_1U_2).$$

The Slepian–Wolf theorem can be easily generalized to more than two sources yielding the following result.

Theorem 11.4 (Slepian–Wolf with many sources [13, p. 409]). *Let $U_1U_2\ldots U_M$ denote a set of correlated sources drawn i.i.d. $\sim p(u_1u_2\ldots u_M)$. M $p(u_1u_2\ldots u_M)$ The set of achievable rates is given by*

$$R(S) > H(U(S)|U(S^c))$$

for all $S \subseteq \{1,2,\ldots,M\}$, where $R(S) = \sum_{i \in S} R_i$, S^c denotes the complement of S, and $U(S) = \{U_j : j \in S\}$.

Proof. The proof goes along the lines of the case with two sources. Details can be found in [13, Sect. 14.4].

11.3.2 Multiple Access Communications

In the previous problem, we assumed that the information generated by multiple sources is transmitted over noiseless channels. If this data is to be communicated over a common noisy channel to a single destination, we call this type of channel a *multiple access channel*. The resulting information-theoretic problem, illustrated in Fig. 11.6, takes into account not only the noise at the receiver, but also the interference caused by different users communicating over a common channel – the mathematical subtlety lies in allowing the channel output Y to depend on the channel inputs X_1 and X_2 according to the conditional probability distribution $p(y|x_1x_2)$.

The set of achievable rates at which the different encoders can transmit their data reliably is called the *capacity region* of the multiple access channel.

Assuming independent messages, i.e., $p(u_1u_2) = p(u_1)p(u_2)$, and independent encoders, Ahlswede [2] and Liao [30] were independently able to prove the following result which fully characterizes the set of achievable rates.

Theorem 11.5 (Multiple Access Channel [2, 30]). *The capacity region of the discrete multiple access channel is given by the convex hull of the set of points (R_1, R_2) satisfying*

$$R_1 \leq I(X_1;Y|X_2), \tag{11.1}$$
$$R_2 \leq I(X_2;Y|X_1), \tag{11.2}$$
$$R_1 + R_2 \leq I(X_1X_2;Y), \tag{11.3}$$

for some joint distribution $p(x_1)p(x_2)$.

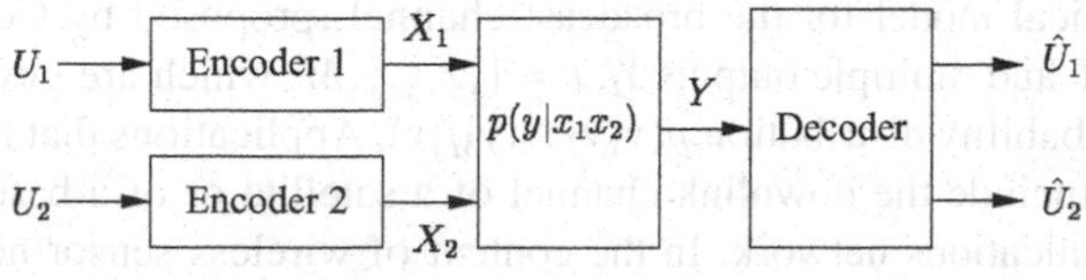

Fig. 11.6 The multiple access channel

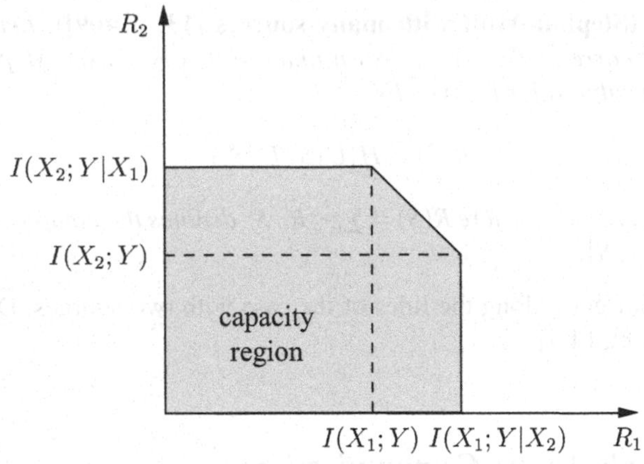

Fig. 11.7 The capacity region of the multiple access channel

The boundaries of the capacity region, shown in Fig. 11.7, can be explained in a very intuitive way. When encoder 1 views the signals sent by encoder 2 as noise, its maximum achievable rate is given by $R_1 \approx I(X_1;Y)$ – a direct consequence of the channel coding theorem. Then, the decoder can estimate the sent codeword x_1^N and subtract it from the channel output sequence y_1^N, thus allowing encoder 2 to achieve a maximum rate of $R_2 \approx I(X_2;Y|X_1)$. This procedure, sometimes referred to as *successive cancellation* [34], leads to the upper corner point of the capacity region. The lower corner point corresponds to the symmetric case and a time-sharing argument yields the remaining points in the segment between them.

It is also worth noting that conditions (11.1)–(11.3) can be easily generalized for more than two sources. In this case, the capacity region is given by

$$R(S) \leq I(X(S);Y|X(S^c))$$

for all $S \subseteq \{1,2,\ldots,M\}$, where $R(S) = \sum_{i \in S} R_i$, S^c denotes the complement of S, and $X(S) = \{X_j : j \in S\}$ [13, Chap. 14.3].

11.3.3 Broadcast Channels

While in the multiple access scenario we have multiple sources and one destination, in the *broadcast case* the information of one source is transmitted to multiple users. Thus, the classical model for the broadcast channel (proposed by Cover in [11]), has one input X and multiple outputs Y_i, $i = 1,2,\ldots,M$, which are governed by the conditional probability distribution $p(y_1 y_2 \ldots y_M|x)$. Applications that fall under this system model include the downlink channel of a satellite or of a base station in a mobile communications network. In the context of wireless sensor networks, it is conceivable that a remote control center broadcasts messages to the sensor nodes on the field in order to coordinate their transmissions or change their configurations.

As in many other fundamental problems of network information theory, determining the capacity of the broadcast channel turns out to be a very difficult task. Consequently, a complete characterization of the achievable rates is only known for a few special cases, e.g., the physically *degraded* broadcast channel in which $p(y_1y_2|x)$ factors to $p(y_1|x)p(y_2|y_1)$ [13, Sect. 14.6] or, most recently, the multiple-input multiple-output Gaussian broadcast channel [46]. For a survey on other interesting results, we refer the reader to Cover's survey [11].

11.3.4 Relay Channels

In wireless communications, fading of the signals transmitted due to multipath propagation is one of the major impairments that a communications system has to deal with. A natural way to deal with these impairments is by the use of *diversity*: redundant signals are transmitted over essentially independent channels and can then be combined at the receiver to average out distortion/noise effects induced by the independent channels [38]. If two or more transmitters are allowed to exchange information and coordinate their transmissions, they can exploit the resulting *spatial diversity* to improve the reliability and the efficiency of their communications. An information-theoretic abstraction of this user cooperation problem is the so called *relay channel*. At time i a relay node observes a noisy version $Y_R(i)$ of the symbol $X(i)$ transmitted by the sender and forms a symbol $X_R(i)$, which depends on all previously observed channel outputs $Y_R(1),\ldots,Y_R(i-1)$. The receiver observes the channel output $Y(i)$, whose relationship with $X(i)$, $X_R(i)$ and $Y_R(i)$ is characterized by the conditional $p(yy_R|xx_R)$. Once again, the capacity region is only known in special cases (see, e.g., [13, Sect. 14.7] and [26]). Recently, several contributions appeared, which connect the insights gained from the classical relay problem with practical wireless communications, most notably the papers of Narula et al. [36], Laneman et al. [27], Sendonaris et al. [41], and Dawy et al. [15].

11.3.5 The Two-Way Channel

The previous problems are instances of the general *two-way channel* proposed by Shannon in [42]. In its original formulation, two users both with transmitting and receiving capability, send information to each other over a common channel, as shown in Fig. 11.8. The channel outputs Y_1 and Y_2, depend on the channel input symbols X_1 and X_2 according to $p(y_1y_2|x_1x_2)$ (see Fig. 11.8). Since encoder 1 can decide on the next symbol X_1 to send based on the received channel symbol Y_2, the two-way channel introduces a new important aspect in the study of communications networks: *transmission feedback*. Unfortunately, the capacity of the two-way channel is only known in the Gaussian case, which decomposes into two independent channels [13, p. 383].

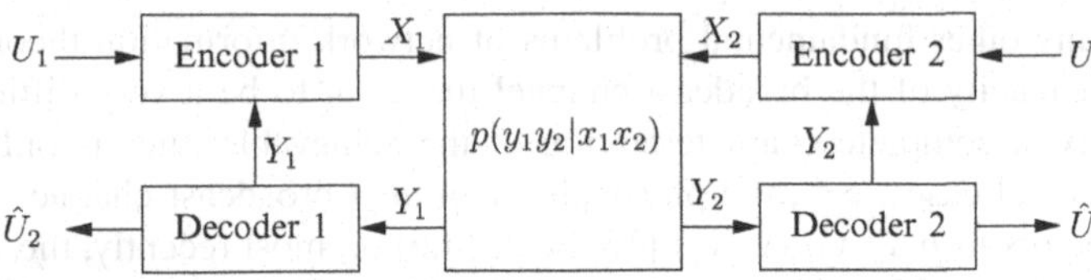

Fig. 11.8 The two-way channel

11.4 Max-Flow Min-Cut Bounds for Information Flows

Having described the basic building blocks of network information theory, it is only natural to ask how we can characterize the information flows in a general network with M users communicating over a channel, in which the dependencies between channel outputs $y_1,\ldots,y_M$ and channel inputs $x_1,\ldots,x_M$ is governed by the conditional probability distribution $p(y_1,\ldots,y_M|x_1,\ldots,x_M)$. With that goal in mind, it seems reasonable to exploit the rich body of network flow theory that successfully describes the behavior of fluids in networks of pipes.

In the case where the network has one or more independent sources of information but only one sink, it is known that routing offers an optimal solution for transporting messages [28] – in this case the transmitted information does behave like *water in pipes* and the capacity can be obtained by classical network flow methods. Specifically, the capacity of this network will then follow from the well-known Ford–Fulkerson *max-flow min-cut* theorem [18], which asserts that the maximal amount of a flow (provided by the network) is equal to the capacity of a minimal cut, i.e., a nontrivial partition of the graph node set V into two parts such that the sum of the capacities of the edges connecting the two parts (the cut capacity) is minimum.

Another problem in which network flow techniques have been found useful is that of finding the maximum stable throughput in certain networks. In this problem, posed by Gupta and Kumar in [22], the goal is to determine the maximum rate at which nodes can inject bits into a network, while keeping the system stable. The problem was reformulated by Peraki and Servetto as a multicommodity flow problem, for which tight bounds were obtained using elementary counting techniques [37].

For the general network information flow problem with multiple independent sources, it is shown in [13] that max-flow min-cut arguments yield a set of necessary conditions for reliable communication over a general network. More specifically, the rates R_{ij} at which two arbitrary nodes i and j communicate across the network must satisfy

$$\sum_{i\in S, j\in S^c} R_{ij} \leq I(X_S;Y_S|X_{S^c}), \tag{11.4}$$

for all subsets $S \subseteq \{1,2,\ldots,M\}$, where $S \neq \emptyset$, $X_S = \{x : x \in S\}$, $Y_S = \{y : y \in S\}$ and S^c denotes the complement of S. From an intuitive point of view, (11.4) basically states that, for every possible cut in the network, the total rate at which information flows across the cut cannot exceed the mutual information between the channel inputs on one side of the cut and the channel outputs on the other side. As intuitive and satisfying this interpretation may seem, this upper bound is not tight except for a few special cases of multiple-access, broadcast and relay channels. We are thus forced to conclude that proving general coding theorems for information flow in communication networks with interference, cooperation and feedback requires more powerful mathematical tools.

Suppose now that the sources of information are not independent, in other words their messages $U_1,\ldots,U_M$ are drawn i.i.d. from a probability distribution $p(u_1,\ldots,u_M)$ that does not factorize into the product of its marginals. This assumption is perfectly reasonable in certain scenarios for example when we consider the correlated measurements collected by a large number of nodes sensing a physical process within a confined area. Since this correlation may be exploited for efficient encoding and decoding, determining necessary and sufficient conditions for reliable communication with correlated sources requires a substantially different treatment. For instance, the max-flow min-cut argument of (11.4) is only valid for independent sources and cannot be applied here.

Motivated by a sensor networking application, in [4] Barros and Servetto formulated and solved a general network information flow problem with correlated sources. The network is modeled as the complete graph on $M+1$ nodes. For each $(v_i,v_j) \in E$ $(0 \le i,j \le M)$, there is a discrete memoryless channel $(\mathscr{X}_{ij}, p_{ij}(y|x), \mathscr{Y}_{ij})$, with capacity $C_{ij} = \max_{p_{ij}(x)} I(X_{ij};Y_{ij})$.[4] At each node $v_i \in V$, a random variable U_i is observed ($i = 0\ldots M$), drawn iid from a known joint distribution $p(U_0U_1\ldots U_M)$. Node v_0 is the *decoder* – the question in this problem is to find conditions under which $U_1\ldots U_M$ can be reproduced reliably at v_0. The answer is provided by the following theorem.

Theorem 11.6. *Let S denote a non-empty subset of node indices that does not contain node 0: $S \subseteq \{0\ldots M\}$, $S \neq \emptyset$, $0 \in S^c$. Then, it is possible to communicate $U_1\ldots U_M$ reliably to v_0 if and only if, for all S as above,*

$$H(U_S|U_{S^c}) < \sum_{i\in S, j\in S^c} C_{ij}. \tag{11.5}$$

Notwithstanding the inadequacy of network flow methods for general networks with interference, this result shows that in networks of independent channels the properties of Shannon information are exactly identical to those of water in pipes – here information is a flow. It follows also that, as in the point-to-point problem, separating source coding from channel coding (and, in this case, routing) is an optimal coding strategy, and thus there is nothing to lose from a layered architecture, provided that there is only one sink and interference is mitigated by partitioning the channel into independent sub-channels. The assumption of independence among sub-channels is crucial, because well-known counterexamples hold without it [12].

[4] Note that C_{ij} could potentially be zero, thus assuming a complete graph does not mean necessarily that any node can send messages to any other node in one hop.

11.5 Mixing Information Flows with Network Coding

The previous results show that when the network problem under consideration features only one sink, routing the messages along adequate paths is sufficient to achieve the max-flow min-cut bound for information flows, irrespective of whether the sources are independent or correlated. In general multicast networks, in which a single source broadcasts a number of messages to a set of sinks, this is no longer true. In [3] it is shown that mixing different information flows through coding operations carried out by intermediate nodes (i.e., *network coding*) is key towards achieving the max-flow/min-cut bound of the network. It turns out that if k messages are to be sent then the minimum cut between the source and each sink must be of size at least k.

The intuition behind this result is well illustrated by the butterfly network shown in Fig. 11.9, where each edge is assumed to have unitary capacity. If node 1 wishes to send a multicast flow to sinks 6 and 7 at the max-flow min-cut bound, which in this case is 2, the only way to overcome the bottleneck between nodes 4 and 5 is for node 4 to combine the incoming symbols through an XOR operation. Sinks 6 and 7 can then use the symbols they receive directly from nodes 2 and 3, respectively, to revert this XOR operation and thus reconstruct the desired multicast flow.

A converse proof for this problem, known as the *network information flow problem,* was provided by [7], whereas linear network codes were proposed and discussed in [29] and [25]. Since then, a rich body of work has emerged on the

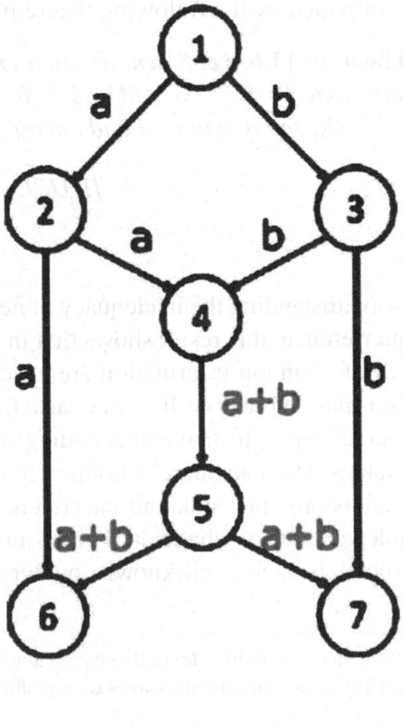

Fig. 11.9 Canonical network coding example

benefits of network coding for various scenarios. In particular, Random Linear Network Coding provides a fully distributed methodology for performing network coding [24], whereby each node in the network selects independently and randomly a set of coefficients and uses them to form linear combinations of the data symbols (or packets) it receives. These linear combinations are then sent over the outgoing links until the receivers are able to decode the original data using Gaussian elimination. It was shown that if the coefficients are chosen at random from a large enough field, it is very likely that the transfer matrix for the network can be inverted, thus allowing the receivers to revert the linear operations performed throughout the network and recover the original information. In [31], RLNC is studied from the point of view of asynchronous packet networks and it is shown that RLNC is capacity-achieving even on lossy packet networks. The benefits of RLNC in wireless environments with rare and limited connectivity, either due to mobility or battery scarcity are highlighted in [19].

11.6 Capacity of Complex Networks

When it comes to capacity not all networks are equal. Intuitively, different topologies have different cut sets and it is therefore reasonable to assume that the behavior of information flows in complex networks is deeply influenced by the actual network configuration. A natural approach to gain some understanding on the impact of topology on information flow is to focus on random graph models that are relevant in practice.

Among the most common instances used in communications research, we find Erdös–Rényi graphs, in which edges are drawn randomly with probability p for a fixed set of vertices, and random geometric graphs, in which the nodes in the network are positioned randomly in a prescribed area and any two nodes are connected if and only if they are within a certain distance of each other. Max-flow min-cut capacity bounds for these two classes of graphs can be found in [39].

The combination of strong local connectivity and long-range shortcut links renders *small-world* [45] topologies increasingly popular in various contexts, most notably in sociology, biology, statistical physics and man-made networks. Well-known examples include such disparate instances as Milgram's "six degrees of separation" between common people [35], the U.S. electric power grid, the nervous system of a nematode worm [1] and the World Wide Web [8].

The term small-world graph itself was coined by Watts and Strogatz, who in their seminal paper [45] defined a class of models which interpolate between regular lattices and random Erdös–Rényi graphs by adding shortcuts or rewiring edges with a given probability p (see Figs. 11.10 and 11.11). The most striking feature of these models is that for increasing values of p the average shortest-path length diminishes sharply, whereas the clustering coefficient, defined as the expected value of the number of links between the neighbors of a node divided by the total number of links that could exist between them, remains practically constant during this transition.

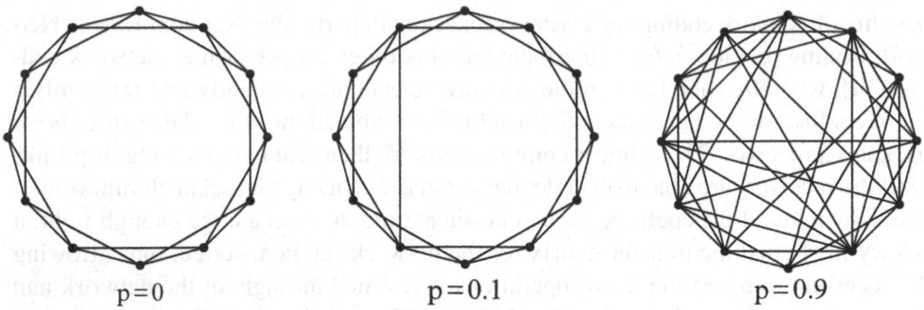

Fig. 11.10 Small-world model with shortcuts for different values of the adding probability p

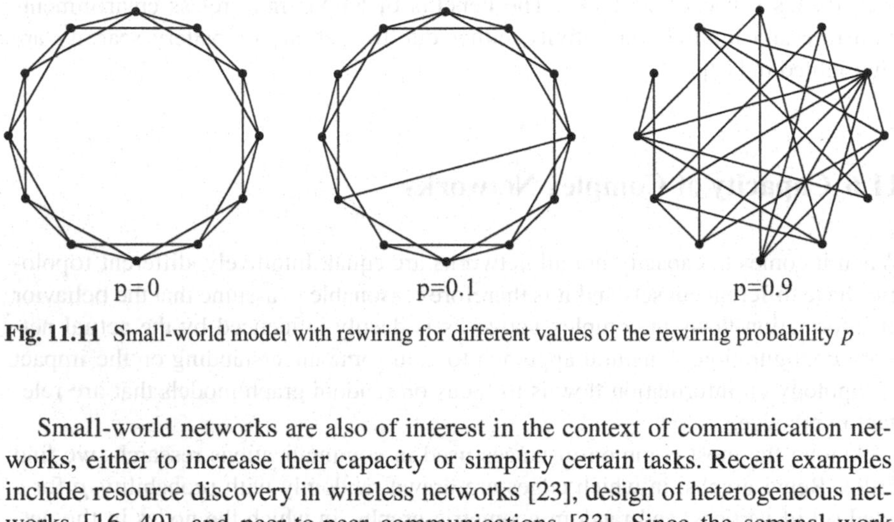

Fig. 11.11 Small-world model with rewiring for different values of the rewiring probability p

Small-world networks are also of interest in the context of communication networks, either to increase their capacity or simplify certain tasks. Recent examples include resource discovery in wireless networks [23], design of heterogeneous networks [16, 40], and peer-to-peer communications [33]. Since the seminal work of [45], key properties of small-world networks, such as clustering coefficient, characteristic path length, and node degree distribution, have been studied by several authors (see, e.g., [17] and references therein), however their max-flow min-cut capacity has only recently received some attention. Costa and Barros [10] combine classical network flow arguments and concentration results from random sampling in graphs in order to provide a set of upper and lower bounds for the max-flow min-cut capacity of several classes of small-world networks. The following theorem gives a flavor of the type of bounds that can be computed using these techniques.

Theorem 11.7. *The capacity of a Small-World Network with Shortcuts, denoted c_s, satisfies the following inequalities:*

$$c_s > (1-\varepsilon)[k+(n-1-k)p] \quad \textit{with probability } 1-O\left(\frac{1}{n^{2d}}\right),$$

$$c_s < (1+\varepsilon)[k+(n-1-k)p] \quad \textit{with probability } 1-O\left(\frac{1}{n^{4nd}}\right),$$

for $\varepsilon = \sqrt{\frac{2(n-2)d\ln(n-2)}{p^2}}$ *and* $d > 0$*, with* n*,* k *and* p *denoting the total number of nodes, the number of neighbors for each node in the base lattice, and the probability of adding a shortcut edge, respectively.*

Other results include navigable topologies, for which highly efficient distributed routing algorithms are known to exist and distributed network coding strategies are likely to be found.

11.7 Conclusions

We reviewed some of the most relevant contributions towards a thorough understanding of information flows in complex networks. The previous examples show that network information theory offers a myriad of very challenging problems, some of which have been open for more than two decades. Nevertheless, in the past few years we have witnessed considerable progress in this field, partly motivated by the remarkable advancements of mobile communications systems and, more recently, wireless sensor networks.

Although still at an infant stage and certainly not an easy task, the development of a comprehensive theory of information networks is likely to have a very strong impact on the design of contemporary communications systems and may find application in other areas where complex networks abound, such as statistical learning, computational biology, neuroscience, sociology and statistical physics.

Acknowledgements The author gratefully acknowledges many insightful discussions with his collaborators Sergio D. Servetto (deceased in July 2007), Rui Costa, Sergio Crisóstomo, Luísa Lima and Muriel Médard. This chapter is dedicated to Sergio D. Servetto in the deepest respect and admiration.

References

1. Achacoso, T., Yamamoto, W.: AY's Neuroanatomy of *C. elegans* for Computation. CRC, Boca Raton (1992)
2. Ahlswede, R.: Multi-way communication channels. In: Proc. 2nd International Symposium on Information Theory, pp. 23–52. Prague (1971)
3. Ahlswede, R., Cai, N., Li, S.Y.R., Yeung, R.W.: Network Information Flow. IEEE Trans Inform Theory **46**(4), 1204–1216 (2000)
4. Barros, J., Servetto, S.D.: Network information flow with correlated sources. IEEE Trans Inform Theory **52**(1), 155–170 (2006)
5. Berrou, C., Glavieux, A., Thitimajshima, P.: Near Shannon limit error-correcting coding and decoding: Turbo-codes. In: Proceedings of the IEEE International Conference on Communications. Geneva (1993)
6. Biglieri, E., Proakis, J., Shamai, S., di Elettronica, D.: Fading channels: information-theoretic and communications aspects. IEEE Trans Inform Theory **44**(6), 2619–2692 (1998)

7. Borade, S.: Network information flow: Limits and achievability. In: Proc. IEEE Int. Symp. Inform. Theory (ISIT). Lausanne, Switzerland (2002)
8. Broder, A.: Graph structure in the web. Comput Netw **33**, 309–320 (2000)
9. Costa, M.H.M.: Writing on Dirty Paper. IEEE Trans Inform Theory **IT-29**(3), 439–441 (1983)
10. Costa, R.A., Barros, J.: Network information flow in *navigable* small-world networks. In: Proceedings of the IEEE Workshop in Network Coding, Theory and Applications. Boston, MA, USA (2006)
11. Cover, T.M.: Comments on broadcast channels. IEEE Trans Inform Theory **44**, 2524–2530 (1998)
12. Cover, T.M., El Gamal, A.A., Salehi, M.: Multiple access channels with arbitrarily correlated sources. IEEE Trans Inform Theory **IT-26**(6), 648–657 (1980)
13. Cover, T.M., Thomas, J.: Elements of Information Theory. Wiley, New York (1991)
14. Csiszár, I., Körner, J.: Information Theory. Academic, New York (1981)
15. Dawy, Z., Kamoun, H.: The general Gaussian relay channel: Analysis and insights. In: 5th International ITG Conference on Source and Channel Coding (SCC'04). Erlangen, Germany (2004)
16. Dixit, S., Yanmaz, E., Tonguz, O.K.: On the design of self-organized cellular wireless networks. IEEE Commun Mag **43**(7), 86–93 (2005)
17. Dorogovtsev, S., Mendes, J.: Evolution of Networks: From Biological Nets to the Internet and WWW. Oxford University Press, Oxford (2003)
18. Ford, L., Fulkerson, D.: Flows in Networks. Princeton University Press, Princeton (1962)
19. Fragouli, C., Boudec, J.Y.L., Widmer, J.: Network coding: an instant primer. SIGCOMM Comput Commun Rev **36**(1), 63–68 (2006)
20. Gallager, R.: Low-density parity-check codes. IEEE Trans Inform Theory **8**(1), 21–28 (1962)
21. Gallager, R.G.: Information Theory and Reliable Communication. Wiley, New York (1968)
22. Gupta, P., Kumar, P.R.: The Capacity of Wireless Networks. IEEE Trans Inform Theory **46**(2), 388–404 (2000)
23. Helmy, A.: Small worlds in wireless networks. IEEE Commun Lett **7**(10), 490–492 (2003)
24. Ho, T., Medard, M., Shi, J., Effros, M., Karger, D.: On randomized network coding. In: Proceedings of 41st Annual Allerton Conference on Communication, Control, and Computing (2003)
25. Koetter, R., Médard, M.: An algebraic approach to network coding. IEEE/ACM Trans Netw **11**(5), 782–795 (2003)
26. Kramer, G., Gastpar, M., Gupta, P.: Capacity theorems for wireless relay channels. In: 41st Annual Allerton Conf. on Commun., Control and Comp. Allerton, IL, USA (2003)
27. Laneman, J., Tse, D., Wornell, G.: Cooperative diversity in wireless networks: Efficient protocols and outage behavior. IEEE Trans Inform Theory **50**(12), 3062–3080 (2004)
28. Lehman, A.R., Lehman, E.: Complexity classification of network information flow problems. In: Proceedings of the 15th annual ACM-SIAM symposium on Discrete algorithms, pp. 142–150. Society for Industrial and Applied Mathematics, New Orleans, LA, USA (2004)
29. Li, S.Y.R., Yeung, R.W., Cai, N.: Linear network coding. IEEE Trans Inform Theory **49**(2), 371–381 (2003)
30. Liao, H.: Multiple access channels. Ph.D. thesis, Department of Electrical Engineering, University of Hawaii, Honolulu, Hawaii (1972)
31. Lun, D., Medard, M., Koetter, R., Effros, M.: Further results on coding for reliable communication over packet networks. In: Proceedings of the IEEE International Symposium on Information Theory, pp. 1848–1852 (2005)
32. MacKay, D., Neal, R.: Near Shannon limit performance of low density parity check codes. Electron Lett **33**(6), 457–458 (1997)
33. Manku, G.S., Naor, M., Wieder, U.: Know thy neighbor's neighbor: The power of lookahead in randomized p2p networks. In: Proceedings of the 36th ACM Symposium on Theory of Computing. Chicago, IL, USA (2004)
34. Mecking, M.: Multiple-access with stripping receivers. In: Proc. 4th European Mobile Communication Conference (EPMCC). Vienna, Austria (2001)

35. Milgram, S.: Psychology today. Phys Rev Lett **2**, 60–67 (1967)
36. Narula, A., Trott, M.D., Wornell, G.W.: Performance limits of coded diversity methods for transmitter antenna arrays. IEEE Trans Inform Theory **45**(7), 2418–2433 (1999)
37. Peraki, C., Servetto, S.: On the maximum stable throughput problem in random networks with directional antennas. In: Proceedings of the 4th ACM International Symposium on Mobile Ad Hoc Networking and Computing, pp. 76–87. ACM, New York (2003)
38. Proakis, J.G.: Digital Communications, 4th edn. McGraw-Hill, New York (2001)
39. Ramamoorthy, A., Shi, J., Wesel, R.D.: On the capacity of network coding for random networks. IEEE Trans Inform Theory **51**, 2878–2885 (2005)
40. Reznik, A., Kulkarni, S.R., Verdú, S.: A small world approach to heterogeneous networks. Commun Inform Syst **3**(4), 325–348 (2004)
41. Sendonaris, A., Erkip, E., Aazhang, B.: User cooperation diversity – part 2: Implementation aspects and performance analysis. IEEE Trans Commun **51**(11) (2003)
42. Shannon, C.: Two-way communication channels. In: Proc. 4th Berkeley Symp. Math. Stat. Prob., vol. 1, pp. 611–644. University California Press, Berkeley (1961)
43. Shannon, C.E.: A mathematical theory of communication. Bell Syst Tech J **27**, 379–423, 623–656 (1948)
44. Slepian, D., Wolf, J.K.: A coding theorem for multiple access channels with correlated sources. Bell Syst Tech J **52**(7), 1037–1076 (1973)
45. Watts, D.J., Strogatz, S.H.: Collective dynamics of 'small-world' networks. Nature **393**(6684) (1998)
46. Weingarten, H., Steinberg, Y., Shamai, S.: The capacity region of the Gaussian MIMO broadcast channel. In: Proceedings of the 2004 IEEE International Symposium on Information Theory (ISIT 2004). Chicago, IL, USA (2004)
47. Wyner, A.D., Ziv, J.: The rate-distortion function for source coding with side information at the decoder. IEEE Trans Inform Theory **IT-22**(1), 1–10 (1976)
48. Yeung, R.W.: A First Course in Information Theory. Kluwer, Dordrecht (2002)

Chapter 12
Models of Information Processing in the Sensorimotor Loop

Daniel Polani and Marco Möller

Abstract We present a framework to study agent-environment systems from an information-theoretical perspective. For this, we use the formalism of Causal Bayesian Networks to model the probabilistic and causal dependencies of various system variables. This allows one to formulate a consistent informational view of how an agent extracts information from the environment, including the role of its actions as a natural part of the model. The model is motivated by increasing evidence of the importance of Shannon information for the behaviour of living organisms. We relate the model to existing views on information maximization and parsimony principles and apply it to a simple scenario demonstrating the discovery of implicit structured environment models by an agent with only a strongly limited and purely local sensorimotor embodiment. Further variations of the model are briefly introduced and discussed. The chapter concludes with an indication of relevant contributions for further research.

12.1 Introduction

The success of biological organisms in managing the complexity of the tasks of survival poses a significant challenge to science. One is faced with a rich set of abilities, of adaptivity and of self-organization that organisms make use of to survive in a difficult environment. What makes this a question of particular scientific interest is that the richness of the occurring phenomena makes it difficult to formulate a systematic characterization of the phenomena at hand: is there anything in common in the phenomena observed in living beings or does biology always have to reinvent the wheel?

D. Polani and M. Möller
Adaptive Systems Research Group, School of Computer Science, University of Hertfordshire, Hatfield, UK
e-mail: d.polani@herts.ac.uk

F. Emmert-Streib, M. Dehmer (eds.), *Information Theory and Statistical Learning,* 289
DOI: 10.1007/978-0-387-84816-7_12,

First, obviously, all biological organisms have to adhere to the laws of physics and consequently chemistry, so that principles of energy and mass conservation, and, more narrowly, the numerical relations pertaining to chemical reactions need to be respected; these principles can be modeled, e.g., with artificial chemistries [16].

Still, the complexity emerging even with these limitations is enormous, and there is the question how evolutionary, developmental and finally adaptive processes manage to cope with the daunting task of identifying the "right" or, at least, most promising paths of success. One important argument is that the evolutionary process has an enormous potential for parallel evaluation. This, while undoubtedly correct, does not account for the even more enormous space of potentially unsuccessful solutions. It has therefore been argued that the process of biological evolution makes significant use of principles of self-organization that allow the organisms to stay in the region of viability with a high probability.

Self-organization processes are well known in the physical world, for instance in the form of Turing patterns or Bnard cells [5, 20, 23] and it has been argued that they play an important role in the formation of patterns for living organisms or for guiding morphogenesis [35, 50].

There is no doubt that such principles are helpful to massively restrict the space of possible realizations and may indeed form building blocks from which the higher levels of complexity of living beings are constructed. How, however, are these building blocks combined? Is there the possibility for local gradients that guide particularly favourable constructs?

On first sight, the proposal that there are any concepts that could act as guiding principles is difficult to reconciliate with the biological principle that evolution is essentially historical, that, in particular, the genome is historical, too, and that extensive knowledge is encoded not only in the genome but in the whole machinery that creates and "runs" an organism: in other words, the machinery is today highly specialized and optimized.

Nevertheless, even if one ignores the difficult question of the first steps from anorganic matter to living beings [30], one cannot fail to wonder at how successful living beings are at evolving to formerly unexperienced situations. An instructive example is the evolution of sensors from one type of use to another [29], the acquisition of novel sensoric modalities [19], and their seamless integration in existing neural substrates [36]. It has been argued that there are indications for significant evolutionary developments having been triggered by the discovery of novel sensoric modalities [37]. Whether or not this turns out to be indeed the case, the emergence of new sensoric capabilities can create a coevolutionary pressure on the actuators (and morphology) of the organism, the most obvious examples being the actuators that allow bats to emit ultrasound or electric fish to create electrical fields around them [22].

In fact, the richness of the sensory equipment of living beings, of which we have mentioned a few examples above, indicates that sensors seem to provide a strong selector for successful organisms. While, as we have seen, there is a great diversity of sensoric modalities and channels to which a sensor may respond to, the central commodity that sensors deal with is *information.*

No matter what modality a sensor taps into, and how it is physically and biologically realized, the sensor provides its carrier with information and to make this a more precise statement, it would be useful to treat the concept of information quantitatively. A well-known quantification of information is provided by Shannon's notion [44], defined in Sect. 12.3. But before this can be done, several conceptual issues need to be resolved.

12.2 Shannon Information in Biology

In the following, when we talk about *information*, we will always imply Shannon's notion. Information considers the transmission of symbols from a sender to a receiver. Using the emission probability of a symbol-emitting source, it quantifies the most compact way these symbols can be transported (or, the other way around, it provides limits on the transport rate of such symbols). Information is universal in the sense that it applies to any system that can be formulated using probability theory. This is a very general assumption and any system, artificial or biological, for which this holds, obeys these limitations.

One problem in using this concept is that Shannon's notion is inherently devoid of semantics. There is no provision in Shannon's theory to endow the stream of symbols with any meaning whatsoever. Information theory will just quantify the bandwidth expense to transmit these symbols, whether useful to the receiver or not. Since the inception of information theory, this has been perceived as a conceptual challenge that for a long time hampered the application of information theory to biological organisms[1] (which obviously need to distinguish between important and unimportant signals). It even prompted the suspicion that information theory may not be suited at all to model biological scenarios, since (to paraphrase Gibson) "the environment is not a sender that aims to send a message to a receiver organism" [18].

The second problem is that Shannon's theory deals with limits. It quantifies how much a sender *can* transmit to a receiver, not how much it actually *does*. If a channel is given, its actual use can not exceed Shannon's limits, but it may well be entirely underused. Interestingly, it turns out that exactly this seems not to be the case in biology. In fact, biological systems seem to operate close to this limit. There are many examples that biological systems utilize available information channels to the limit: the human ear can operate close to the thermal noise level [15]. After adaptation, the human eye has been shown to react to small numbers of photons [21]; toad receptors can identify individual photons [7]. Information processing often operates at a tight trade-off with the available metabolic energy [31]. There are many indications that this may be a universal property of biological information processing [10].

Why should this be the case? One interpretation that prompted the pioneering work introducing Linsker's Infomax principle is that the early layers of the sensorics

[1] Except for the relatively simple typical application as a correlation measure.

do not have enough knowledge of which external information to process. Therefore these layers' most unbiased "guess" would be to maximize the total information they transmit for further processing [33]. Related to that, but with a slightly different slant, there is another view: information processing is expensive in terms of life energy. The eye consumes roughly 10% of a fly's energy [32] and the human brain can consume 20% of the resting energy [24]. In view of this, it makes sense to propose a principle of parsimony [41]: over evolution or adaptation, a biological information channel that is not used optimally will tend to degenerate to a level matching its actual use. There is also a converse implication: if a channel operates close to the optimum, then, e.g., mutations or other fluctuations might be able to probe whether a suitable increase in channel capacity may confer additional fitness advantages. This process then could provide a driver towards the "discovery" of novel modes of operation, or, in the case of sensors, novel sensoric modalities.

Thus, assuming its validity can be empirically sustained, the principle of parsimony implies that, in the case of a well-balanced adaptation (i.e., in the case when the organism is well adapted to the current contingencies of the environment), biologically relevant information channels will be operating close to the maximal information capacity allowed by the Shannon limit. This gives us a handle for a quantitative treatment of biological information. More than that, as will be seen in Sect. 12.5, the parsimony principle in conjunction with the information processing requirements of an organism also provides a path towards endowing Shannon's "indifferent" measure of data volume with semantic flavour.

12.3 Notation

We will repeatedly refer to a number of quantities and symbols as follows: Consider random variables, denoted by capital letters $X,Y,Z,\dots$ which take on values, denoted by lowercase letters $x,y,z,\dots$ in corresponding sets $\mathscr{X},\mathscr{Y},\mathscr{Z},\dots$. For the probability that a random variable, say X, assumes a value $x \in \mathscr{X}$, we will write $\Pr(X = x)$, or equivalently, if there is no ambiguity, simply $p(x)$. For the probability distribution of the joint variable (X,Y), we will write simply $\Pr(X = x, Y = y) \equiv p(x,y)$, and similarly for more than two variables. For the probability of Y given X, we write $\Pr(Y = y|X = x) \equiv p(y|x)$. The *entropy* of a random variable is defined as $H(X) = -\sum_{x\in\mathscr{X}} p(x)\log p(x)$ where we will sometimes omit the set $\mathscr{X}$ summed over if clear from the context and we will always implicitly assume the logarithm to the basis of 2, measuring entropy and related quantities in *bits*. The conditional entropy of two random variables is given by $H(Y|X) = \sum_x p(x)H(Y|X = x) = \sum_x p(x)\sum_y p(y|x)\log p(y|x)$. Shannon's mutual information is defined as

$$I(X;Y) = H(Y) - H(Y|X). \tag{12.1}$$

It is symmetric with respect to exchange of X and Y. The conditional mutual information is defined as $I(X;Y|Z) = H(Y|Z) - H(Y|X,Z)$.

12.4 Historical Remarks

Notwithstanding the aforementioned difficulties in applying information theory to characterize the information processing of an organism or agent, already early post-Shannon work identified its potential. Among them, we find the Law of Requisite Variety [1] which quantifies the amount of information a system has to take in if it is to reduce a certain amount of entropy in the environment. In addition, it has been suggested that Shannon information and related optimization principles could be useful to characterize information transmission in biological systems [2, 6].

Ashby's classical result has been rediscovered in slightly different form in recent work [48, 49], highlighting the renewed interest in the field. A shift in the view of the field is has also been initiated by the novel *information bottleneck* concept [47] that has spawned a significant number of ramifications [13, 17, 43]. For the matter at hand, the important aspect of the information bottleneck concept is its ability to capture the concept of relevance in information.

The information bottleneck principle operates as follows: in Fig. 12.1, let X be some accessible random state variable which is jointly distributed with the *relevance indicator variable* Y. While the latter is the actual quantity of interest, it cannot be accessed directly, but only via X. With the joint distribution between X and Y, we can quantify the amount of information that X provides about Y by the mutual information $I(X;Y)$. Note that Y defines all that is relevant in the current scenario, thus identifying the relevant information that X captures. In the information bottleneck formalism one now proceeds to extract from X into a variable $\tilde{X}$ as much information about Y as possible, "squeezing out" irrelevant information in X.

The information bottleneck formalism shows how relevance can be seamlessly integrated into the information-theoretic framework. Additionally, it does not treat information as a bulk quantity, but as a notion from which different components can be extracted, not unlike the perspective provided by Independent Component Analysis (ICA) [14] which, however, does not address the issue of relevance.

12.5 Structure and Information

Not only does the information bottleneck principle demonstrate that it is possible to imbue information with relevance, but it also indicates that information may have an intrinsic structure which then one can attempt to unravel. The information

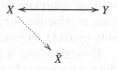

Fig. 12.1 The information bottleneck principle. See text for details

bottleneck methods and its factorized variants [17] demonstrate, in addition to ICA, how do that. However, a no less interesting question is where such structure arises from in the first place. Related to this question is how information can give rise to structure.

In Linsker's work, the Infomax principle was applied to a feed-forward neural network whose inputs was a two-dimensional set of random images. As a result of the Infomax process, different unit types emerged. Among them were center surround cells (cells which react to dark ring around a white center or vice versa) and other unit types which can be found in early stages of biological visual processing.

Note that these structures do not emerge from the Infomax principle alone, but require additional structure which was imposed in the architecture of the system: First of all, the neural network was structured into layers. The neurons in a given layer (except for the input layer) received their inputs from neurons in the previous layer. In addition, their *receptive fields*, i.e., the area of neurons in the previous layer from which they received their input (i.e., the *receptive field*) is *localized*. Localization requires the assumption that neurons in a given layer are spatially organized, i.e., that there is a concept of which neurons are close to each other and which ones are not. In other words, Infomax requires the additional structure of layeredness as well as localized receptive fields to evoke the aforementioned structured characteristics. This constitutes a certain amount of structure on the information processing architecture and we will see that it is possible to get away with even less. Furthermore, we will see that, in a structured scenario Infomax-type principles can indeed promote structuring also of the information itself. What exactly is meant by that will be the topic of the coming sections.

12.6 Utility-Induced Information Structure

We begin by giving a brief section indicating how structure can be induced by the presence of a utility function that governing the action selection of an agent. The information bottleneck method mentioned in Sect. 12.5 teaches us how to "tag" information according to the given relevance indicator variable. It provides a way of distinguishing relevant from irrelevant information. In the original bottleneck model, the relevance indicator variable is external, corresponding to a supervised labeling.

However, if we consider an agent, then there is a more natural way of constructing the relevance indicator variable. Consider for simplicity that the agent observes the world through a sensor variable S (a random variable) and assume for simplicity of exposition that S indeed encompasses the complete world state. What is then the relevant information in this sensoric variable? If we have no concept of relevance then, in principle, the agent would have to suitably process whole extent of sensoric entropy $H(S)$. In general, this will be far too much.

However, there is a natural selector for the information relevant to an agent: as the name *agent* indicates (Lat. agere: acting), an agent is an entity that *acts*. Any information that an agent attains from its environment only plays a role with respect to what actions it can decide upon, given this information. This suggests that the action A should be used as a relevance indicator variable. For this, however, it is necessary to specify how the actuation should be distributed with the state of the sensor S. While the distribution of the sensor S is given a priori,[2] there is no canonical distribution of actions; rather, the actions should be obtained from the *policy* adopted by the agent, i.e., via an action-selection mechanism $p(a|s)$. If such a mechanism is not present a priori, then the natural way to specify it is via the existence of a utility function $U : \mathscr{S} \times \mathscr{A} \to \mathbb{R}$ that assigns a real value to any concrete (world or sensor) state s and action a, and which the agent aims to maximize. Once such a policy $p(a|s)$ has been formulated, the bottleneck formalism suggests that the relevant information is simply given by $I(S;A)$, where the joint distribution of S and A is given by $p(s)p(a|s)$, where $p(s)$ is the assumed a priori distribution of the world (sensor) states and $p(a|s)$ is the policy.

This approach was suggested in [39] and later refined in [40]. The latter model considers the utility arising in a Markovian Decision Process through rewards additively cumulated over time on selecting certain actions in given states. It is well known from the theory of Dynamical Programming and Reinforcement Learning how to identify an optimal policy for a given reward [45]. The optimal policies are not necessarily unique. In that case, it makes sense to select an optimal policy $p(a|s)$ for which $I(S;A)$ becomes minimal. The reason is that $I(S;A)$ can be considered a measure for the processing effort required to implement a given policy. In line with the principle of parsimony discussed in Sect. 12.2, if we restrict ourselves to consider only optimal policies, among all of these, according to the principle of parsimony, a biologically plausible model will adapt to the informationally cheapest one. This can be elegantly formulated and solved as a rate-distortion problem [11]. In fact even more can be done: instead of asking for an informationally optimal solution, if one is ready to sacrifice some utility, one can get away with even less relevant information to be processed.[3] In other words, one can reduce the amount of (relevant) information that has to be processed by the agent if one is ready to forgo some of the utility. In some scenarios, it is possible to reduce the (relevant) sensoric information to 0 while sustaining only moderate utility losses [40], by use of a judiciously randomized action strategy.

The bottom line of these observations is: once a utility (or a reward structure) is specified, it is possible to imbue an agent with a natural characterization of relevance. This takes place by instantiating the actions as relevance indicator variables. The joint distribution between state S and the relevance indicator variable A is governed by the policies induced on A through the utilities. Thus, we see how a single step in the perception–action loop, together with a reward structure, can endow the information processed by an agent with relevance structure.

A problem remains: utilities have to be somehow specified and it is not clear that there is a simple natural choice for them. This is particularly conspicuous in the

---

[2] This assumption is not entirely correct, as the actuation will affect the distribution of S, but for the transparency of the argument, we will make this simplifying assumption.

[3] How to achieve this is an interesting, but subtle and slightly technical point. See [40] for details.

biological case where one does not necessarily have access to a reward model, and where a consistent quantitative concept of fitness is still difficult to formulate. Part of the difficulty is due to rewards typically being significantly delayed in biological settings. Some recent work manages to do this and relates fitness to informational considerations [8, 46].

Rewards, however, are not the only driver of information structuring in an agent. Another powerful driver is embodiment [34]. This becomes most apparent if one considers the information flow through the perception–action loop of an agent through time. In the following, we will do that, and consider the processing of information in an unrolled perception–action loop, as opposed to the single timestep models employed in the present section or in Ashby's or Touchette and Lloyds model. To achieve a consistent treatment of information processing in such a setting, however, requires a suitable formalism which is provided by the concept of *Causal Bayesian Networks* which we will briefly introduce next.

12.7 Causal Bayesian Networks

We first introduce some notions. Because of limited space, exposition will be kept to a minimum. A *directed graph* $(\mathscr{V},\mathscr{E})$ consists of a finite set $\mathscr{V}$ of *vertices* or *nodes* and of a set $\mathscr{E} \subseteq \mathscr{V} \times \mathscr{V}$ of edges. We assume only proper edges, i.e., $v \neq w$ for $(v,w) \in \mathscr{E}$. A *path* is an ordered sequence $v_1, v_2, \ldots, v_n$ of vertices such that $(v_k, v_{k+1}) \in \mathscr{E}$ for $k = 1 \ldots n-1$. A *loop* is a closed path, i.e., a path where $v_n = v_1$. In the following, we consider only *directed acyclic graphs* (DAG), i.e., graphs without loops. For a vertex $v \in \mathscr{V}$, its *parent set* is the set $\mathsf{Pa}(v) = \{u \mid (u,v) \in \mathscr{E}\}$ of predecessors in the graph.

A Bayesian Network is a graph $(\mathscr{V},\mathscr{E})$ associated with an additional structure that associates with each vertex v a random variable X_v with values in $\mathscr{X}_v$ whose probability distribution is determined as follows: for each random variable $X_v, v \in \mathscr{V}$, a *kernel* p_v is defined which is interpreted as specifying the probability of observing a particular value of X_v if certain values have been observed on the set of its parents: $p_v(x_v \mid x_{\mathsf{Pa}(v)})$. Per convention, we identify $p_v(x_v) \equiv p_v(x_v \mid x_{\mathsf{Pa}(v)})$ if the parent set $\mathsf{Pa}(v)$ for v is empty. The kernels then induce a joint probability on the whole set of random variables via

$$p(x_1, \ldots, x_{|\mathscr{V}|}) = \prod_{v \in \mathscr{V}} p_v\big(x_v \mid x_{\mathsf{Pa}(v)}\big). \tag{12.2}$$

While a Bayesian Network provides a compact notation for possibly composite probability distributions, it does not, in general, provide a unique representation for modelling a given probability distribution on a set of random variables X_v. Although the notation of a node as successor of its parent nodes is seductive, this is a purely observational description and does in no way imply any causal relation. For this, additional structure, so-called *intervention* is required [38].

A *Causal Bayesian Network* (CBN) $(\mathscr{V}, \mathscr{E})$ with kernels p_v is a Bayesian Network imbued with an additional structure, namely a semantics of intervention.[4] An *intervention* at a node v defines a family of new Bayesian Networks as follows:

1. In the resulting networks the graph is modified to $(\mathscr{V}, \mathscr{E}_v)$ where in $\mathscr{E}_v$ all the edges from the parents of v to v have been removed.
2. For an arbitrarily defined kernel $\widehat{p}_v(x_v)$ on v, a new probability is computed according to (12.2), using the modified parents and kernel of v.

The formalism generalizes naturally to interventions at sets of nodes. Using intervention, one can *impose* a particular value x_v on the node v and observe how it affects another node x_w. In Pearl's interventional notation the probability of observing x_w if x_v is imposed is written as $p(x_w \,|\, \widehat{x}_v)$.

The interventional concept is significantly different from the non-causal interpretation of Bayesian Networks: the latter are purely observational. In CBNs, however, one has a precise semantics of intervention which allows one to describe how active modification of the state of a node (as opposed to observational selection of particular states) will affect the rest of the system. The edges in this network are not merely indicators of correlations, but they are to be interpreted as causal mechanisms. The concept of CBNs turns out to be a powerful tool for the modelling of the informational dynamics of agents once one turns to cases that are more involved than the ones discussed in Sects. 12.5 and 12.6.

12.8 Modelling the Information Dynamics in Agents

Using CBNs, a generic model of an agent acting in an environment can be formulated as in Fig. 12.2, with the arrows representing the directed edges of the graph [25, 26]. The figure shows the perception–action loop of an agent unrolled in time. Note that, consistent with the semantics of a CBN, the arrows denote actual causal mechanisms. To understand the diagram, consider first the arrows $S_t \to A_t$. These

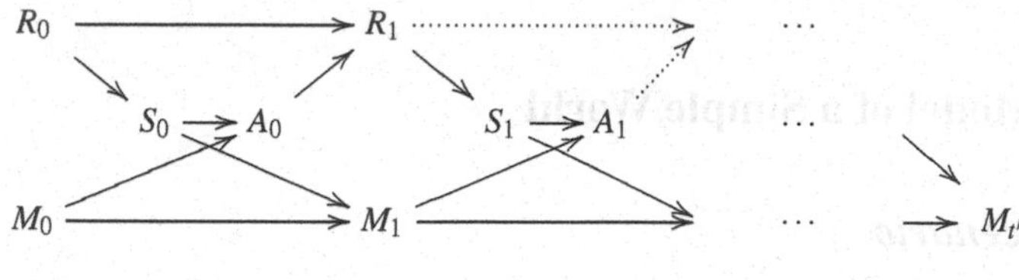

Fig. 12.2 The perception–action loop of an agent modeled using a CBN. The S_t denote the random variables representing the sensors, the A_t the actuators, the M_t the agent "memory" state and R_t the state of the "rest of the world" at different time steps t. The *arrows* indicate the causal edges (mechanisms) of the Bayesian Network

[4] We give a slightly different, but conceptually equivalent presentation from the definitions used in [3, 38].

denote the mechanisms of the agent policy, namely the probability $p(a_t|s_t)$ of selecting an action a_t if a sensor state s_t is observed at a time t. The arrows $R_t \to S_t$ indicate what the sensors capture about the state of the "rest of the world". Similarly, the arrows $A_t \to R_{t+1}$ denote the influence of the agent's actions on the world. Finally, the agent is equipped with a brain which is realized as a Mealy machine with memory M_t. It modulates how an action is selected depending on the sensor input and, at the same time defines the subsequent memory state depending on the current sensor input and memory state.[5]

The result is a highly generic model for a single agent. To see this, one should note that, while S_t denotes a single sensor, it actually refers to a "sensor complex", so that any composition of subsensors is encompassed in this description. It is easy to modify the CBN in Fig. 12.2 to denote two separate sensors $S_t^{(1)} \otimes S_t^{(2)}$ feeding into the actuators and the memory. In fact, the case of separate sensors can be construed as a special case of a single sensor, as long as the size of the state space for the single sensor is large enough to contain the total size of the space for the composite sensor, i.e., as long as $|\mathscr{S}_t| \geq |\mathscr{S}_t^{(1)}||\mathscr{S}_t^{(2)}|$. The composite sensor assumes more structure and is therefore more specific than the single sensor model. Analogous statements hold for the other variables.

In particular this implies that very little is assumed about the control architecture (i.e., M_t and the associated arrows) of the agent. In particular, as long as the architecture is fully deterministic, Fig. 12.2 shows the most general case. This observation should be kept in mind for the discussions on the formation of structured controllers in Sect. 12.10.

Finally, it should be noted that the formalism also allows to model the interaction of several agents through the environment [27] or hierarchical memory structures (not further discussed here). Similar models have been used to characterize autonomy [9] or to provide quantitative models for universal utilities [28]. In the following, we will restrict ourselves to demonstrate how the formalism, together with Infomax principles (equivalent to parsimony-type arguments), gives rise to structured processing architectures, only by virtue of the agent's embodiment; in the CBN model this is the dynamics by which the agent is linked into its environment, i.e., the arrows $R_t \to S_t$ and $A_t \to R_{t+1}$.

12.9 A Model of a Simple World

12.9.1 Scenario

We will study how the earlier mentioned principles combine with an agent embodied in a simple grid world. The world under consideration consists of an infinite grid. At certain positions, there are "chemical" sources that generate a "chemical" gradient

[5] The map used here is not the most general possible, but will be sufficient for the subsequent discussion.

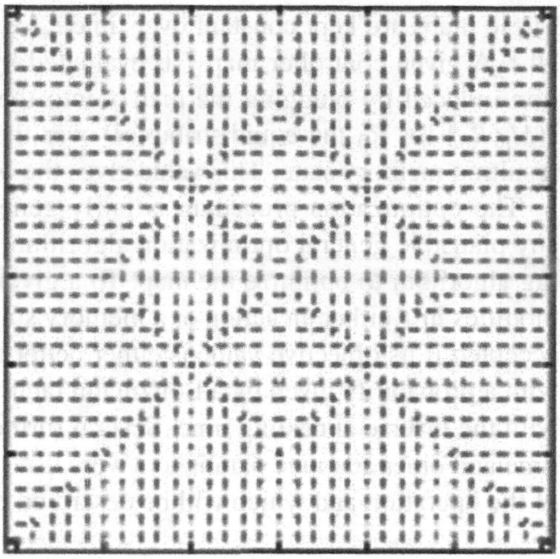

Fig. 12.3 Returned sensor direction at different positions in the world grid

over the grid world that can be sensed by the agent's sensors. The agent, located at a given grid position in the world, has a sensor which detects the strength of the signal at the position adjacent to the north, east, south and west (n/e/s/w) of the agent. The sensor variable S_t will return the direction n/e/s/w with the strongest signal experienced at a certain time t. Thus, the maximal possible information conveyed to the agent by its sensor at a given time t is 2 bit.

The start position of the agent is equally distributed at positions $\{-10,\ldots,10\} \times \{-10,\ldots,10\}$ in the world, and the sources are placed at positions $\{-5,5\} \times \{-5,5\}$. Figure 12.3 shows the gradient direction returned by the sensor at different positions in the world. The center of the image is the position $(0,0)$. Note that the gridworld continues infinitely outside of the image.

12.9.2 Information Maximization in the Perception–Action Loop

We now plug the sensor and actuator model from Sect. 12.9.1 into the generic agent model from in Fig. 12.2 and apply an Infomax principle. Among the selection of possible principles one could opt for with some justification, here we use the objective function from [25], namely the maximization of the information $I(M_t;R_0)$ that the agent memory M_t has managed to collect about the initial state R_0 of the world at a given time t (note that the state R of the world essentially consists of the position of the agent – in other words, we ask the final memory state to reconstruct the original starting position of the agent as well as possible).

For this purpose, we consider the agent controller which would be in general represented by the probabilistic map $p(a_t, m_{t+1}|s_t, m_t)$. Since we are maximizing $I(M_t;R_0)$, and any probabilistic mapping would introduce new noise into the system, we limit ourselves to deterministic controller maps $f : \mathscr{S}_t \times \mathscr{M}_t \to \mathscr{A}_t \times \mathscr{M}_{t+1}$.

These maps are kept constant over time t, i.e., we use the same controller map for all t. A given controller map f is evaluated by calculating the probability distribution induced by plugging the map f into the CBN from Fig. 12.2 and evaluating $I(M_t;R_0)$ for R_0 the equidistribution over the initial square, M_0 fixed and M_t the resulting memory state. This value gives the objective that we wish to optimize for.

The optimization procedure is fairly generic and could be implemented using any plausible general optimization algorithm. We use a standard simulated annealing algorithm to optimize the map. To a given map f, a mutation is applied which modifies in f either an action or a memory state f is mapping to. The new map f' is accepted if its objective value is either better than the old one or, if it is lower, then with a probability given by a Boltzmann factor. A temperature parameter is reduced with the inverse annealing time. Several further heuristics to determine optimal run time and other adaptive parameters were used which are not essential to the success of the operation and which will be described elsewhere.

12.9.3 Results

We show results maximizing $I(M_t;R_0)$ for agent runs of length $t = 15$ using a memory size of $|\mathscr{M}_t| = 12$ (kept constant for all time steps). corresponding to approximately 3.6 bit The resulting memory representation captures approximately 3.31 bit about the original position information. This is significantly less than the more than 8.7 bit that the initial position contains, but it comes close to the theoretical maximum that the available memory size can achieve.

A direct investigation of the controllers reveals that their detailed operation is quite difficult to interpret, even for relatively small controllers and the simple scenario from Sect. 12.9.1. Therefore, to understand what the Infomax principle manages to achieve, better insights are obtained by investigating what original world states R_0 the memory state M_t in the final state of the agent run actually represents. More precisely, we are interested in the probability distribution $p(r_0|m_t)$.

The distributions are shown in Fig. 12.4. The figure is to be read as follows. Each of the squares in Fig. 12.4 represents one concrete memory state m_t at the final time step; more precisely, it represents the complete distribution $p(r_0|m_t)$ of initial positions r_0 of the agent if one is in the memory state m_t (the square covers the whole area which R_0 is uniformly distributed over at the beginning of each agent run). White-colored areas stand for zero probability of having started in the corresponding position, black stands for a high probability of having started in that position. Note that, because of massive variations in the different images, the squares are individually normalized, so the color black only indicates the maximum conditional probability $p(r_0|m_t)$ for the given memory state, which is in general not the same for different memory states.

As an example, consider m_t to be found to be the third state (third square from the left, in the upper row in Fig. 12.4). This indicates that the agent must have started in one of the triangular regions south of any of the four sources. Similarly, finding

Fig. 12.4 Memory representation of the world $p(r_0|m_t)$ for the agent optimized in the four-sources world. See text for details

the fifth memory state indicates that the agent must have started in the triangular areas west of the sources. The detailed insights from the results will be discussed in Sect. 12.10.

12.9.4 Factorized Representations

One tool that is occasionally useful for the analysis of the resulting representations is the *factorization of representations* found according to the method described in Sects. 12.9.2 and 12.9.3. The idea of the factorization is to impose an informationally "orthogonal" coordinate system on the memory representation. The factorization method is closely related to the principle of Independent Component Analysis [14] and multivariate information bottleneck [17]. Our situation of a CBN-modeled agent requires a modified treatment and we follow here the ideas from [25].

Basically, one is interested in decomposing the final memory M_t into two "independent" components $M^{(1)}$ and $M^{(2)}$, formulating this decomposition in the information-theoretic language which we have focused upon throughout the paper.[6] We wish the decomposition to respect a number of principles:

1. M-decomposition: the components should be directly causally derived from M. That is, the CBN describing the components together with the rest of the network should have the form

$$\begin{array}{ccc} & & M^{(1)} \\ & & \nearrow \\ R_0 \cdots\!\!> M_t & & \\ & & \searrow \\ & & M^{(2)}, \end{array} \qquad (12.3)$$

where the dotted arrow encapsulates the full agent-environment CBN lying between the initial R_0 and the final memory M_t; the solid arrows denote direct causal mechanisms.

[6] The present discussion generalizes immediately to decomposition into multiple components.

2. Completeness: ideally, the decomposition should capture all of the information in R_0. This can be generalized to imperfect decompositions (e.g., decompositions under additional constraints) by asking for the decomposition to maximize $I(R_0;M^{(1)},M^{(2)})$. Since the $M^{(i)}$ derive directly from M_t, this implies that, if the memory is smaller than the initial world state space, these components are likely to capture most, if not all of M_t.
3. Independence: one would like one component to capture independent qualities from the other. This can be quantified by asking for $I(M^{(1)};M^{(2)})$ to be as small as possible.
4. Disentangled representation of R_0: it would be desirable to ensure that both "coordinate" components $M^{(i)}$ would contribute as independently as possible to the reconstruction of R_0. A counterexample would be an XOR-like combination of the components where only the knowledge of both components would allow one to reconstruct anything about R_0. The earlier criteria, including independence, cannot prevent this from happening, so we require a third objective, namely that $I(M^{(1)};M^{(2)}|R_0)$ be as small as possible. Conversely, any XOR-like "entanglement" or "synergetic" coding schemes [42] of the two components would instead produce a large value of this conditional information.

Searching (again with a Simulated Annealing algorithm and deterministic mappings) for a factorization mapping that maximizes the criterium in (2) and minimizes those in (3) and (4) (all objectives weighted equally in a linear combination), we can construct a corresponding factorization for a given memory map $p(m_t|r_0)$. If we do this for the map given in Fig. 12.4, for components of size 4 and 5, respectively, we obtain a result as in Fig. 12.5. Again, the diagrams show the conditionals $p(r_0|m^{(i)})$ for $i = 1, 2$. The factorization is almost perfect with respect to the criteria above, in that it is informationally almost complete, independent, and disentangled.

Note that as the 12 states of M_t are factorized into the two variables, the four-sized component is fully exploited, however from the five-sized component, only three states are used. This is consistent with the numerical result that the components split the available states of M_t fully and perfectly.

In the following section, we will now discuss in more detail the implications of the experiments.

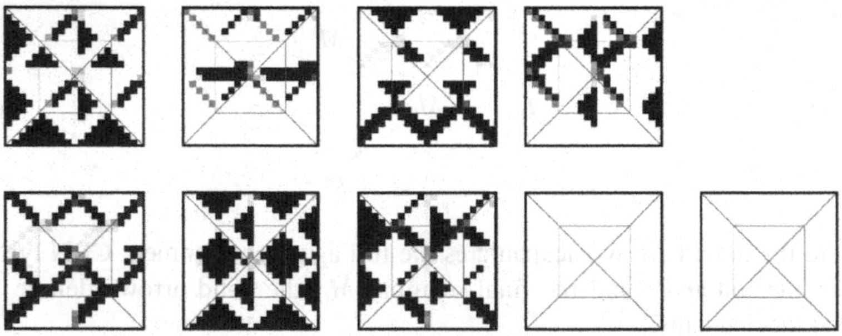

Fig. 12.5 Factorized memory structure for the agent optimized in the four-sources world

12.10 Discussion

The results of implementing the Infomax principle in the agent model introduced in Fig. 12.2 and specialized in Sect. 12.9.1 reveal some striking properties. While the agent only possesses a limited 2-bit sensor denoting the local gradient, the agent memory manages to extract around 3.3 bit of information into the memory of size of around 3.6 bit, thus exploiting the available memory to a relatively high level. This is consistent with the perspective that M_t is able to increasingly collect the information about R_0 through the sequence of sensoric inputs $(S_{t'})_{t'=0,1,2,\ldots,t} \equiv S_{[0,t]}$. This sequence determines all that the agent can learn about R_0, so any information the agent learns about R_0 is bounded above by $I(R_0;S_{[0,t]})$ which, in turn, is upper bounded by the entropy $H(R_0)$ of R_0. In some scenarios, $I(R_0;S_{[0,t]})$ indeed converges towards this upper limit for $t \longrightarrow \infty$.

In addition to the reconstruction of R_0 from the series of sensors, the present model has several important aspects. The limited size of the memory limits how much information processing (and storage) capacity is available at each time step. It is not possible to "squeeze" sequences of arbitrary complexity through the sensors to reconstruct the original state R_0 of the agent, rather each time step provides only limited processing bandwidth to "digest" past information for the future. The information-theoretic characterization of this bandwidth is attractive because it provides a universal currency which allows, e.g., the treatment of trade-offs of bandwidth capacities allocated between various tasks in an agent.

In addition, the agent does not just passively collect the sensoric information about R_0 as it "drips in" through $S_{t'}$, but is able to select actions that will extract a larger amount of information. For instance, if the actions are not optimized for, but instead a fixed "gradient following" actuator is used which just takes a step along the current gradient (sensors and memory mappings are optimized as usual), the resulting agent is able to extract only 2.2 bit of information. Thus the CBN-based agent model provides a natural approach to study active sensing.

The spatial representations in Fig. 12.4 capture coherent features about the state space. Typical structures involve the "triangular" regions of memory states 3,5,7 and 11 (read Fig. 12.4 left to right, top to bottom) capturing whether the agent started inside the areas south, west, north and east of the sources. Other features denote essentially diagonals which separate different triangular regions (states 2, 6, 8, 10, and 12), and a few other, less explicit features, such as in 1, 4, 9 which mix different aspects). Common to all, though, is that, even while the agent and its memory representation have no prior concept of space, the application of the Infomax principle in the context of this embodied agent is sufficient to identify and bring out the most salient aspects of the geometrical structure of the environment. Note, however, that the spatial structures are not an *explicit* concept for the optimized agent. The representational mapping $p(r_0|m_t)$ is an interpretation of an external observer as to the *implicit* "meaning" of a memory state.

If we compare the results to Linsker's Infomax studies, the CBN-based modeling approach requires significantly weaker architectural assumptions to achieve a structured representation of the environment. Linsker still had to assume a layered network architecture and the presence of localized receptive fields. The model from Fig. 12.2 does not make any assumptions about the processing architecture; from the standpoint of the formalism, the memory forms an unstructured "blob". Any structure is then introduced into the architecture by the interaction of the Infomax principle with the embodiment of the agent. It is purely the embodiment that imprints structure on the memory M_t of the agent. However, any agent operating in some kind of structured environment will have an embodiment, and thus this condition constitutes a much weaker set of assumptions than those used in Linsker's original Infomax principle.

There is another view of the results from Sect. 12.9.3. The mappings $p(r_0|m_t)$ can be interpreted as highly impoverished Bayesian models for the agent's initial position r_0: instead of full Bayesian models with their rich and expensive representations of continuous valued probability distribution, our agent possesses a model of very restricted representation and processing bandwidth. On the other hand, in the context of the original motivation for the present work, it would not be plausible that organisms would employ fully developed Bayesian models for their orientation in their environment, even if the resulting behaviours may end up to be qualitatively similar [51]. Rather, we can expect realistic models of biologically relevant information processing to be likely to involve informationally limited, and thus suitably impoverished models of the environment, closer to the spirit of the present work.

The present work shows how actions can be treated, equivalently to sensors, as information transmitting components (in our case, they "inject" information from the agent into the environment) and establish a picture of information processing that treats all components of the system, sensors, memory, actuators *and the environment* in a conceptually coherent way. In this model, there is no clear-cut separation into passive sensing and active acting; rather sensing and acting are intimately intertwined and, in particular, active sensing becomes the regular mode of operation.

While the information gathered in M_t is achieved purely by means internal to the agent, the Infomax principle still requires the optimization according to the external criterium $I(R_0;M_t)$. Is this plausible? There are two ways of looking at that: first, the external criterium could stem from some adaptation on a meta-level, say evolution, which would favour better informed individuals [40, 46]. The informedness about the objective function $I(R_0;M_t)$ could arise, for instance, from an actual search on the levels of populations rather than individuals. Second, it is indeed possible to reconstruct aspects of the structure of the external world by purely intrinsic information [12]. Note that, while we implemented Infomax as a strong principle, at this point no founded hypothesis can be made whether biological systems would indeed implement a strong or a weak Infomax principle, i.e., whether they actively strive to maximize certain information transmission rates, or whether Infomax principles just emerge as a side effect of other adaptation processes.

Finally, some remarks on the factorization: the factorized memory representations are interesting from a number of points. The components of the factorization capture different aspects of the system. These are not always easy to interpret, but sometimes they provide immediately interpretable higher-level concepts [25]. While this is not always the case (cf. Fig. 12.5), the principle of obtaining a factorization through other variables (such as obtaining $M^{(i)}, i = 1,2$ about R_0 through M_t) is a powerful generalization of ICA and multivariate bottleneck methods. There are indications that a systematic use of the factorization method through the perception–action loop in various combinations may provide a path towards principles for the discovery of generalizing concepts with respect to the agent's embodiment.

This concept-formation perspective is appealing from a (human) observer view; however, to properly fit into the framework developed above, it also should have a justification in informational terms. Indeed, it does: space does not permit us to go into details here, but one important aspect is the complexity cost of actually finding or constructing the mappings (i.e., arrows) in Fig. 12.2. For large memories, the complexity of the mappings (and thus the cost of finding or adapting them) can be quite considerable and, even if we limit ourselves just to deterministic mappings, they grow exponentially with the size of the involved states. A memory that is pre-structured, e.g., via factorization, would allow one to limit adaptation of mappings to significantly smaller spaces and thus would open the chance for a much more efficient map construction. The price to be paid is that any mapping respecting the components will be significantly more specific than one found over the whole memory structure. However, we have grounds for hope that an adequate pre-factorization will allow us to find compositions that are sufficiently preadapted to the scenario such as to reduce possible performance losses due to the limitation to component-preserving mappings.

12.11 Conclusions and Outlook

We have presented a formalism based on Causal Bayesian Networks to create a model of the informational interactions between agents and their environment. The formalism is general in the sense that it allows the selection of the specific level of detail; it is easy to specify, e.g., structural constraints or multiple agents in the model. The model allows the specification of generalized forms of Linsker's Infomax principles. In a simple scenario, we have demonstrated the emergence of implicitly structured state representations in an unstructured memory using Infomax. The structures emerged through the particular embodiment of the agent, i.e., the particular form of the coupling of the agent with its environment. It was shown how the emerged structures can be factorized into smaller, independent components, opening the way for selective processing of partial aspects of the acquired information.

The formalism discussed in the present paper is flexible and provides a useful framework for the informational modeling of information processing in agents aimed for. The long-term quality of a formalism, however, is determined from the directions it spawns which were not obvious at its inception. Several such lines emerged recently from this work: from the desire to track Shannon information through the perception–action loop and several studies on composite Markovian networks [4, 52] emerged the idea of developing a causal concept of information flow for general CBNs [3]. Another promising line of research pursues the formulation of a class of *universal utilities* based on the external channel capacity of a perception–action loop [28], a notion which requires the CBN formalism for its precise definition in the general case. Further fundamental questions on the information processing in agents which could be addressed using the formalism are already appearing at the horizon. These indicate that the framework provides a versatile tool for the study of the informational dynamics of the perception–action loop, contributing to the increasingly powerful and rich set of tools in the field.

References

1. Ashby, W. R., (1956). *An Introduction to Cybernetics*. London: Chapman and Hall.
2. Attneave, F., (1954). Informational aspects of visual perception. *Psychol. Rev.*, 61:183–193.
3. Ay, N., and Polani, D., (2007). Information flows in causal networks. *Adv. Complex Syst.*, 11(1):17–41.
4. Ay, N., and Wennekers, T., (2003). Dynamical properties of strongly interacting Markov chains. *Neural Netw.*, 16(10):1483–1497.
5. Bar-Yam, Y., (1997). *Dynamics of Complex Systems*. Studies in Nonlinearity. Boulder: Westview.
6. Barlow, H. B., (1959). Possible principles underlying the transformations of sensory messages. In: Rosenblith, W. A., editor, *Sensory Communication: Contributions to the Symposium on Principles of Sensory Communication*, 217–234. Cambridge: MIT.
7. Baylor, D., Lamb, T., and Yau, K., (1979). Response of retinal rods to single photons. *J. Physiol. London*, 288:613–634.
8. Bergstrom, C. T., and Lachmann, M., (2004). Shannon information and biological fitness. In: *Information Theory Workshop*, 50–54. IEEE.
9. Bertschinger, N., Olbrich, E., Ay, N., and Jost, J., (2006). Autonomy: an information theoretic perspective. In: *Proc. Workshop on Artificial Autonomy at Alife X*, Bloomington, Indiana, 7–12.
10. Bialek, W., de Ruyter van Steveninck, R. R., and Tishby, N., (2007). Efficient representation as a design principle for neural coding and computation. arXiv.org:0712.4381 [q-bio.NC].
11. Blahut, R., (1972). Computation of Channel Capacity and Rate Distortion Functions. *IEEE Transactions on Information Theory*, 18(4):460–473.
12. Capdepuy, P., Polani, D., and Nehaniv, C. L., (2007). Constructing the basic umwelt of artificial agents: an information-theoretic approach. In: Almeida e Costa, F., Rocha, L. M., Costa, E., Harvey, I., and Coutinho, A., editors, (2007). *Advances in Artificial Life (Proc. ECAL 2007, Lisbon)*, vol. 4648 of *LNCS*, Berlin: Springer.
13. Chechik, G., Globerson, A., Tishby, N., and Weiss, Y., (2005). Information Bottleneck for Gaussian Variables. *J Machine Learn. Res.*, 6:165–188.
14. Comon, P., (1991). Independent Component Analysis. In: *Proc. Intl. Signal Processing Workshop on Higher-order Statistics*, Chamrousse, France, 111–120.
15. Denk, W., and Webb, W. W., (1989). Thermal-noise-limited transduction observed in mechanosensory receptors of the inner ear. *Phys. Rev. Lett.*, 63(2):207–210.
16. di Fenizio, P. S., and Dittrich, P., (2007). Chemical organizations at different spatial scales. In: Almeida e Costa, F., Rocha, L. M., Costa, E., Harvey, I., and Coutinho, A., editors, (2007). *Advances in Artificial Life (Proc. ECAL 2007, Lisbon)*, vol. 4648 of *LNCS*, Berlin: Springer, 1–11.

17. Friedman, N., Mosenzon, O., Slonim, N., and Tishby, N., (2001). Multivariate information bottleneck. In: *Uncertainty in Artificial Intelligence: Proceedings of the Seventeenth Conference (UAI-2001)*, 152–161. San Francisco: Morgan Kaufmann.
18. Gibson, J. J., (1979). *The Ecological Approach to Visual Perception*. Boston: Houghton Mifflin.
19. Griffin, D. R., Webster, F. A., and Michael, C. R., (1960). The echolocation of flying insects by bats. *Anim. Behav.*, 8:141.
20. Haken, H., (1983). *Advanced Synergetics*. Berlin: Springer.
21. Hecht, S., Schlaer, S., and Pirenne, M., (1942). Energy, quanta and vision. *J. Opt. Soc. Am.*, 38:196–208.
22. Heiligenberg, W., (1991). *Neural Nets in Electric Fish*. Cambridge: MIT.
23. Hoyle, R., (2006). *Pattern Formation*. Cambridge: Cambridge University Press.
24. Kandel, E. R., Schwartz, J. H., and Jessell, T. M., (1991). *Principles of Neural Science*, Third edition. New York: McGraw-Hill.
25. Klyubin, A., Polani, D., and Nehaniv, C., (2007). Representations of space and time in the maximization of information flow in the perception-action loop. *Neural Comput.*, 19(9):2387–2432.
26. Klyubin, A. S., Polani, D., and Nehaniv, C. L., (2004). Organization of the information flow in the perception-action loop of evolved agents. In: *Proceedings of 2004 NASA/DoD Conference on Evolvable Hardware*, 177–180. IEEE Computer Society.
27. Klyubin, A. S., Polani, D., and Nehaniv, C. L., (2004). Tracking information flow through the environment: simple cases of stigmergy. In: Pollack, J., Bedau, M., Husbands, P., Ikegami, T., and Watson, R. A., editors, *Artificial Life IX: Proceedings of the Ninth International Conference on Artificial Life*, 563–568. Cambridge: MIT.
28. Klyubin, A. S., Polani, D., and Nehaniv, C. L., (2005). All else being equal be empowered. In: *Advances in Artificial Life, European Conference on Artificial Life (ECAL 2005)*, vol. 3630 of *LNAI*, 744–753. Berlin: Springer.
29. Lakes-Harlan, R., and Heller, K.-G., (1992). Ultrasound-sensitive ears in a parasitoid fly. *Naturwissenschaften*, 79:224–226.
30. Langton, C., (1991). Life at the edge of chaos. In: Langton, C. G., Taylor, C., Farmer, J. D., and Rasmussen, S., editors, *Artificial Life II*, Santa Fe Institute Studies in the Sciences of Complexity. Redwood City: Addison-Wesley.
31. Laughlin, S. B., (2001). Energy as a constraint on the coding and processing of sensory information. *Curr. Opin. Neurobiol.*, 11:475–480.
32. Laughlin, S. B., de Ruyter van Steveninck, R. R., and Anderson, J. C., (1998). The metabolic cost of neural information. *Nat. Neurosci.*, 1(1):36–41.
33. Linsker, R., (1988). Self-organization in a perceptual network. *Computer*, 21(3):105–117.
34. Lungarella, M., and Sporns, O., (2006). Mapping information flow in sensorimotor networks. *PLoS Comput. Biol.*, 2(10).
35. Meinhardt, H., (1972). A theory of biological pattern formation. *Kybernetik*, 12:30–39.
36. Newman, E. A., and Hartline, P. H., (1981). Integration of visual and infrared information in bimodal neurons in the rattlesnake optic tectum. *Science*, 213:789–791.
37. Parker, A., (2004). *In the Blink of an Eye: How Vision Kick-started the Big Bang of Evolution*. London: Free Press.
38. Pearl, J., (2000). *Causality: Models, Reasoning and Inference*. Cambridge: Cambridge University Press.
39. Polani, D., Martinetz, T., and Kim, J., (2001). An information-theoretic approach for the quantification of relevance. In: Kelemen, J., and Sosik, P., editors, *Advances in Artificial Life (Proc. 6th European Conference on Artificial Life)*, vol. 2159 of *LNAI*, 704–713. Berlin: Springer.
40. Polani, D., Nehaniv, C., Martinetz, T., and Kim, J. T., (2006). Relevant information in optimized persistence vs. progeny strategies. In: Rocha, L. M., Bedau, M., Floreano, D., Goldstone, R., Vespignani, A., and Yaeger, L., editors, *Proc. Artificial Life X*, 337–343.
41. Polani, D., Sporns, O., and Lungarella, M., (2007). How information and embodiment shape intelligent information processing. In: Lungarella, M., Iida, F., Bongard, J., and Pfeifer, R.,

editors, *Proc. 50th Anniversary Summit of Artificial Intelligence*. Berlin, Heidelberg, New York: Springer.
42. Schneidman, E., Bialek, W., and Berry II, M. J., (2003). Synergy, redundancy, and independence in population codes. *J. Neurosci.*, 23(37):11539–11553.
43. Shalizi, C. R., and Crutchfield, J. P., (2002). Information bottlenecks, causal states, and statistical relevance bases: How to represent relevant information in memoryless transduction. *Adv. Complex Syst.*, 5(1):91–95.
44. Shannon, C. E., (1949). The mathematical theory of communication. In: Shannon, C. E., and Weaver, W., editors, *The Mathematical Theory of Communication*. Urbana: The University of Illinois Press.
45. Sutton, R. S., and Barto, A. G., (1998). *Reinforcement Learning*. Cambridge: MIT.
46. Taylor, S. F., Tishby, N., and Bialek, W., (2007). Information and Fitness. arXiv.org:0712.4382 [q-bio.PE].
47. Tishby, N., Pereira, F. C., and Bialek, W., (1999). The information bottleneck method. In: *Proc. 37th Annual Allerton Conference on Communication, Control and Computing, Illinois*. Urbana-Champaign.
48. Touchette, H., and Lloyd, S., (2000). Information-theoretic limits of control. *Phys. Rev. Lett.*, 84:1156.
49. Touchette, H., and Lloyd, S., (2004). Information-theoretic approach to the study of control systems. *Physica A*, 331:140–172.
50. Turing, A. M., (1952). The chemical basis of morphogenesis. *Philos. Trans. R. Soc. London B*, 327:37–72.
51. Vergassola, M., Villermaux, E., and Shraiman, B. I., (2007). 'Infotaxis' as a strategy for searching without gradients. *Nature*, 445:406–409.
52. Wennekers, T., and Ay, N., (2005). Finite state automata resulting from temporal information maximization. *Neural Comput.*, 17(10):2258–2290.

Chapter 13
Information Divergence Geometry and the Application to Statistical Machine Learning

Shinto Eguchi

Abstract This chapter presents intuitive understandings for statistical learning from an information geometric point of view. We discuss a wide class of information divergence indices that express quantitatively a departure between any two probability density functions. In general, the information divergence leads to a statistical method by minimization which is based on the empirical data available. We discuss the association between the information divergence and a Riemannian metric and a pair of conjugate linear connections for a family of probability density functions. The most familiar example is the Kullback–Leibler divergence, which leads to the maximum likelihood method associated with the information metric and the pair of the exponential and mixture connections. For the class of statistical methods obtained by minimizing the divergence we discuss statistical properties focusing on its robustness. As applications to statistical learning we discuss the minimum divergence method for the principal component analysis, independent component analysis and for statistical pattern recognition.

13.1 Introduction

Statistical machine learning approach are very successful in providing powerful and efficient methods for inductive reasoning in information spaces with uncertainty [22, 43]. In the following we assume a geometric view for learning algorithms and the statistical discussion about statistical learning methods. In this context, a challenging problem was solved to answer the question to what extent geometry and the maximum likelihood method are associated with each other, see [1, 2]. It is known that the linear Gaussian regression associates with a Euclidean geometry in which the least squares method is characterized by projection of the observed data

S. Eguchi
Institute of Statistical Mathematics, 4-6-7 Minami-Azabu, Minato-ku, Tokyo 106-8569, Japan
e-mail: eguchi@ism.ac.jp

F. Emmert-Streib, M. Dehmer (eds.), *Information Theory and Statistical Learning*,
DOI: 10.1007/978-0-387-84816-7_13,

point onto the linear hull of explanatory data vectors. This is only a special example of the geometry associated with maximum likelihood. In general, the geometry is elucidated by a dualistic Riemannian geometry such that the information metric is introduced as Riemannian metric and two linear connections, called e-connection and m-connection. In this framework two connections are conjugate with respect to the information metric. The optimality of the maximum likelihood is characterized by m-projection onto the e-geodesic model. We will discuss this structure in addition to the extension to a class of minimum divergence methods.

We review a close relation between the maximum likelihood method and the Kullback–Leibler divergence. Let p and q be probability density functions on a data space $\mathscr{X}$. The Kullback–Leibler divergence is defined by

$$D_{\mathrm{KL}}(p,q) = \int_{\mathscr{X}} p(x)\{\log p(x) - \log q(x)\}\Lambda(dx),$$

where Λ is a carrier measure. Consider a statistical situation in which p is an underlying density function for data and q is a model density function. In this context we define a statistical model by a parametric family of probability density functions

$$M = \{q_\theta(x), \theta \in \Theta\}.$$

Then the log-likelihood function for the model based on a given dataset is approximated by $D_{\mathrm{KL}}(p,q_\theta)$ neglecting a constant in θ, where p is the underlying density function of the dataset. Hence we observe that the minimization of $D_{\mathrm{KL}}(p,q)$ in q of M is almost surely equivalent to the maximum likelihood method. Since the principle of maximum likelihood was proposed by Fisher [14], it has been applied to a vast of datasets with various forms from almost all scientific fields. In general the maximum likelihood estimator is supported by several points such as the invariance under one-to-one data transformations, the covariance by a parameter transformation, the asymptotically consistency and the asymptotical efficiency. Furthermore, several advantageous properties in theoretical aspects are proven on the assumption of an exponential family, see [3] for a detailed discussion.

13.2 Class of Information Divergence

The principle of maximum likelihood is established on the basis of a specific property for a pair of elementary functions. One is an exponential function that defines an exponential model; the other is a logarithmic function that defines a log-likelihood function. It is well known that they are connected by conjugate convexity,

$$\log(s) = \operatorname*{argmax}_{t\in\mathbf{R}}\{ts - \exp(t)\}. \tag{13.1}$$

We will see that this convexity leads to the Kullback–Leibler divergence and the Boltzmann–Shannon entropy, which lead us to deeper understanding about the relation between the exponential family and the log-likelihood function. In a subsequent discussion the pair of the exponential and logarithmic functions will be extended to a pair of a convex positive function $u(t)$ and the inverse function $\xi(t)$.

Let $\mathscr{F}$ be a space of all integrable functions with respect to a carrier measure Λ on a data space $\mathscr{Z}$ and let $\mathscr{M}$ be the space of nonnegative functions in $\mathscr{F}$. We also discuss the space of probability density functions,

$$\mathscr{P} = \{p \in \mathscr{M} : \int_{\mathscr{Z}} p(z)\Lambda(dz) = 1\}.$$

In subsequent sections the data variable z in $\mathscr{Z}$ will be expressed by x in $\mathscr{X}$ in unsupervised learning, and will be expressed by (x,y) in $\mathscr{X} \times \mathscr{Y}$ in supervised learning. Let us overview the geometry that expresses a natural structure in an information space. We call D an information divergence on $\mathscr{M}$ if D satisfies the first axiom of distance: $D(\mu,\nu) \geq 0$ for any μ and ν in $\mathscr{M}$ with equality if and only if $\mu = \nu$ a.e. Λ.

Let U be a convex function on a real line. Then the conjugate convex function Ξ is given by

$$\Xi(s) = \max_{t \in \mathbf{R}} \{st - U(t)\} \tag{13.2}$$

which is written by $\Xi(s) = s\xi(s) - U(\xi(s))$, where $\xi(s)$ is the inverse of the derivative of U. Thus $\Xi(s)$ is a primitive function of ξ. We construct an information divergence over a function space $\mathscr{M}$ employing the conjugate convexity associated with U. For an arbitrarily fixed f and g in $\mathscr{F}$ we define an index expressing the difference between f and g by

$$d_U(f,g) = \Big\langle U(f) - U(g) - U'(g)(f-g) \Big\rangle,$$

where $\langle\ \rangle$ denotes the integration over the data space. It follows from the convexity of U that $d_U(f,g)$ satisfies the first axiom of distance. Take a mapping $\varphi : \mathscr{M} \to \mathscr{F}$, and then we define $D_\varphi(\mu,\nu) = d_U(\varphi(\mu),\varphi(\nu))$ for μ and ν in $\mathscr{M}$. By definition, $D_\varphi(\mu,\nu)$ becomes an information divergence on $\mathscr{M}$. Assume that the convex function U has a positive derivative $u = U'$. Then the inverse function ξ of u can be viewed as a mapping from $\mathscr{M}$ to $\mathscr{F}$. Hence, taking as $\varphi = \xi$ we get a specific divergence

$$\begin{aligned} D_U(\mu,\nu) &= d_U(\xi(\nu),\xi(\mu)) \\ &= \Big\langle U(\xi(\nu)) - U(\xi(\mu)) \Big\rangle - \Big\langle \mu, \xi(\nu) - \xi(\mu) \Big\rangle, \end{aligned} \tag{13.3}$$

which is called U-divergence, where $\langle A,B \rangle = \langle AB \rangle$. By definition $u(\varphi(\mu)) = \mu$, which is essential to produce a variety of empirical loss functions.

We next introduce an idea of entropy associated with the information divergence. Define U-cross entropy by

$$C_U(\mu,\nu) = \langle U(\xi(\nu))\rangle - \langle \mu, \xi(\nu)\rangle \tag{13.4}$$

and U-entropy by

$$H_U(\mu) = C_U(\mu,\mu) = \langle U(\xi(\mu))\rangle - \langle \mu, \xi(\mu)\rangle. \tag{13.5}$$

Then we observe that $D_U(\mu,\nu) = C_U(\mu,\nu) - H_U(\mu)$, which leads to

$$C_U(\mu,\nu) \geq H_U(\mu) \tag{13.6}$$

since $H_U(\mu) = \langle -\Xi(\mu)\rangle$.

One of the most typical example is $U(t) = \exp(t)$ with the derivative $u(t) = \exp(t)$ and the inverse derivative $\xi(u) = \log(u)$. Thus the U-divergence corresponding to this choice is

$$D_U(\mu,\nu) = \langle \nu - \mu\rangle - \langle \mu, \log(\nu) - \log(\mu)\rangle, \tag{13.7}$$

which is nothing but the Kullback–Leibler divergence. We note that the first term in (13.7) vanishes if we restrict D_U as a functional on $\mathscr{P} \times \mathscr{P}$.

The second typical example is

$$U_\beta(t) = \frac{1}{\beta+1}(1+\beta t)^{\frac{\beta+1}{\beta}}$$

and thus the derivative and the inverse derivative are

$$u_\beta(t) = (1+\beta t)^{\frac{1}{\beta}},\ \xi_\beta(u) = \frac{u^\beta - 1}{\beta}.$$

Noting $U_\beta(\xi_\beta(u)) = u^{\beta+1}/(\beta+1)$ we get the resulting divergence

$$D_\beta(\mu,\nu) = \frac{\langle \nu^{\beta+1} - \mu^{\beta+1}\rangle}{\beta+1} - \frac{\langle \mu, \nu^\beta - \mu^\beta\rangle}{\beta},$$

which is called β-divergence [4, 33]. We notice that $\lim_{\beta\downarrow 0} U_\beta = \exp$, which implies the limit of β-divergence reduces to the Kullback–Leibler divergence as β goes to 0. Alternatively, when $\beta = 1$, $D_\beta(\mu,\nu) = \frac{1}{2}\|\mu - \nu\|^2$, where $\|\ \|$ denotes the L_2-norm in $\mathscr{M}$. See [40] for the statistical argument.

These properties have a good analogy with the Box-Cox data transformation such that a non-negative variable t is transformed by $(t^\beta - 1)/\beta$ with $\log t$ as the limit of β into 0, cf. [7]. Thus the transformation is the same as ξ_β, from which we can say that the β-divergence is derived by the Box-Cox transformation to density functions.

On the other hand, α-divergence is defined by

$$D_\alpha(\mu,\nu) = \frac{4}{\alpha^2-1}\langle \mu - \nu^{\frac{\alpha-1}{2}}\mu^{\frac{1+\alpha}{2}}\rangle + \frac{2}{\alpha+1}\langle \nu - \mu\rangle,$$

see [1] for the relation of α-connection. There is no exact relation of α-divergence with β-divergence except for the case of $(\alpha,\beta) = (1,0)$, which leads to the Kullback–Leibler divergence. Thus α-divergence cannot directly apply a statistical analysis because the empirical form is not feasible without density estimation, See [5] for estimation method by α-divergence with $\alpha = 0$, or equivalently the squared Hellinger distance. In this aspect we will see that any U-divergence is directly applicable for a statistical analysis because U-cross entropy $C_U(\mu,\nu)$ is a linear in μ as a functional.

In fact, α-divergence belongs to a class of f-divergence

$$D_f(\mu,\nu) = \langle \mu f(\nu/\mu) - f'(1)(\nu-\mu) \rangle, \tag{13.8}$$

where f is a convex function. See [8] for more extensive discussion. Note that the integrand in the right-hand side of (13.8) is nonnegative, which shows that $D_f(\mu,\nu)$ is an information divergence.

In a subsequent discussion we will introduce a statistical method using U-divergence for statistical learning including the principal component analysis, independent component analysis, statistical pattern recognition.

13.2.1 Information Geometry of U-Divergence

We discuss a geometry associated with minimum divergence. Let us consider a parametric model

$$N = \{\mu(z,\theta) : \theta \in \Theta\}$$

embedded in the space $\mathscr{M}$ of all nonnegative functions on the data space. We regard a model N as a differentiable manifold of dimension d with the coordinate θ in the coordinate space Θ, see [36] for the pioneering work to this approach. For guarantee of smoothness we assume some regularity conditions for $\mu(z,\theta)$ such as the differentiability under the integral sign.

Let D be an information divergence and assume that if the function D is restricted to $N \times N$, then D is differentiable on $N \times N$. Then we see that D associate with a Riemannian metric $g^{(D)}$, two linear connection $\nabla^{(D)}$ and ${}^*\nabla^{(D)}$ as follows [10, 11]. First,

$$g^{(D)}(X,Y)(\mu) = -D(X|Y)(\mu) \quad (\forall \mu \in N).$$

Here and hereafter we write $D(X|Y)$, or in general for vector fields $X,Y,\ldots,Z,W,\ldots$ on N, by

$$D(X,Y,\cdots|Z,W,\cdots)(\mu) = X_\mu Y_\mu \cdots Z_\nu W_\nu \cdots D(\mu,\nu)\Big|_{\nu=\mu}.$$

Secondly, we define vector field $\nabla_X^{(D)}Y$ and ${}^*\nabla_X^{(D)}Y$ to satisfy

$$g^{(D)}(\nabla_X^{(D)}Y, Z) = -D(XY|Z), \qquad g^{(D)}({}^*\nabla_X^{(D)}Y, Z) = -D(Z|XY)$$

for all vector fields Z. Note that this formulation leads $\nabla_X^{(D)}Y$ and ${}^*\nabla_X^{(D)}Y$ to linear connections because of the non-degeneracy of the metric $g^{(D)}$. In general $\nabla_X^{(D)}Y$ and ${}^*\nabla_X^{(D)}Y$ are both not metric in the sense of $g^{(D)}$, however we observe that

$$\bar{\nabla}^{(D)} = \frac{1}{2}(\nabla^{(D)} + {}^*\nabla^{(D)})$$

is the Riemannian connection with respect to $g^{(D)}$. Thus $\nabla^{(D)}$ and ${}^*\nabla^{(D)}$ are said to be conjugate.

The direct application of this formula to U-divergence defined in (13.3) yields $g^{(U)}, \nabla^{(U)}, {}^*\nabla^{(U)}$ as follows:

$$\begin{aligned} g^{(U)}(X,Y)(\mu) &= \langle X\mu, Y\xi(\mu)\rangle, \\ g^{(U)}(\nabla_X^{(U)}Y, Z)(\mu) &= \langle XY\mu, Z\xi(\mu)\rangle, \\ g^{(U)}({}^*\nabla_X^{(U)}Y, Z)(\mu) &= \langle Z\mu, XY\xi(\mu)\rangle, \end{aligned}$$

for $\mu \in N$. Thus geometric quantities associated with U-divergence is derived, which depend only on N and ξ, where ξ is the inverse derivative of U.

As a special case, the KL divergence is led to $U = \exp$, which implies, noting $\xi = \log$, the information metric g, m-connection ∇, e-connection ∇^* are written by

$$(g^{(\exp)}, \nabla^{(\exp)}, {}^*\nabla^{(\exp)}) = (g, \nabla, \nabla^*).$$

See [1, 2] for the original definition of g, ∇, ∇^*. As the most characteristic of the triple $(g^{(U)},, {}^*\nabla^{(U)})$ we state that

$$\nabla^{(U)} = \nabla$$

for any function U. This surprising property will suggest that if we employ U-divergence, then the minimum divergence method directly associates with a simple form of the empirical loss function.

13.2.2 Minimum U-Divergence Estimator

Consider a statistical model $M = \{p(z;\theta) : \theta \in \Theta\}$ in the space $\mathscr{P}$ of all the probability density functions. We define U-loss function by

$$\ell_U(\theta, p) = C_U(p, p(\cdot,\theta)) = b_U(\theta) - E_p\{\xi(p(z;\theta))\}, \tag{13.9}$$

where C_U is U-cross entropy defined in (13.4) and E_p denotes the expectation with respect to p. Then for a given data set $z_1,\ldots,z_n$ the empirical U-loss function is given by

$$\ell_U^{(\mathrm{emp})}(\theta) = b_U(\theta) - \frac{1}{n}\sum_{i=1}^{n}\xi(p(z_i;\theta)), \tag{13.10}$$

where $b_U(\theta) = \langle U(\xi(p(z;\theta)))\rangle$, and we call

$$\widehat{\theta}_U = \underset{\theta\in\Theta}{\operatorname{argmin}}\, \ell_U^{(\mathrm{emp})}(\theta)$$

the minimum U-divergence estimator. Thus, $\widehat{\theta}_U$ is asymptotically consistent since the inequality $C_U(p,q) \geq H_U(p)$ holds with equality only when $q = p$ (Λ-a.e.). Because θ_U minimizing $C_U(p, p(\cdot;\theta'))$ in θ' equals θ if $p = p(\cdot,\theta)$ The estimating function is

$$s_U(z;\theta) = \frac{\partial}{\partial\theta}\xi(p(z;\theta)) - \frac{\partial}{\partial\theta}b_U(\theta). \tag{13.11}$$

Hence we observe the unbiasedness of the estimating function, that is $E_p\{s_U(z;\theta)\}$ $= 0$ for $p = p(\cdot;\theta)$ since

$$E_p\{s_U(z;\theta)\} = \langle p - p(\cdot;\theta), \frac{\partial}{\partial\theta}\xi(p(\cdot;\theta))\rangle.$$

Thus all the minimum U-divergence estimators is the same asymptotics as the maximum likelihood estimator in the sense of consistency.

We next investigate the asymptotic normality. By definition the U-estimator $\widehat{\theta}_U$ satisfies the estimation equation $n^{-1}\sum_{i=1}^{n} s_U(z_i;\theta) = 0$. Hence the Taylor approximation gives

$$\sqrt{n}(\widehat{\theta}_U - \theta) = J_\theta^{-1}\frac{1}{\sqrt{n}}\sum_{i=1}^{n} s_U(z_i;\theta) + o_P(1).$$

By a direct application of the central limit theorem we obtain that the limiting distribution of $\sqrt{n}(\widehat{\theta}_U - \theta)$ becomes a normal distribution with the mean vector 0 and the variance matrix $J_\theta^{-1}V_\theta J_\theta^{-1}$, where

$$J_\theta = E_{p(\cdot;\theta)}\{(\partial/\partial\theta)s_U^{\mathrm{T}}(z,\theta)\} \quad V_\theta = \mathrm{var}(s_U(z;\theta)). \tag{13.12}$$

13.2.3 Γ-Minimax

In this subsection we introduce an idea on Γ-minimaxity in which the log-loss is extended to the U-loss function (13.9). See [19] for Bayesian discussion on

Γ-minimax solution, and also [20] for a more general discussion from a game theoretic point of view. We will see that U-entropy maximization is equivalent to the minimax game between Nature and a decision maker. For simplicity we consider a situation in which Nature supposes a probability distribution with a mean-value restriction, that is,

$$\Gamma_\tau = \{p \in \mathscr{P} : E_p\{t(z)\} = \tau\}.$$

Here let us assume that τ is an interior point of the convex hull of the range $t(\mathscr{Z})$ of the k-variate statistic $t(z)$. Then if Nature restricts p to be in Γ_τ, the maximization of U-entropy $H_U(p)$ defined in (13.5) is given as follows. When we take a variation to the Lagrangian

$$\mathscr{L}(p,\theta,\kappa) = H_U(p) - \langle \theta^{\mathrm{T}}\{t-\tau\}, p\rangle - \kappa\{\langle p\rangle - 1\}$$

the equilibrium state satisfies that $\xi(p^*) = \theta^{\mathrm{T}} t - \kappa$, that is,

$$p^*(z) = u\big(\theta^{\mathrm{T}} t(z) - \kappa\big), \tag{13.13}$$

where θ and κ are uniquely determined by constraints

$$\big\langle u\big(\theta^{\mathrm{T}} t - \kappa\big)\big\rangle = 1, \qquad \big\langle t, u\big(\theta^{\mathrm{T}} t - \kappa\big)\big\rangle = \tau.$$

In fact, for any $p \in \Gamma_\tau$ the difference between U-entropy of p^* and that of p is

$$H_U(p^*) - H_U(p) = D_U(p, p^*).$$

Consequently we conclude the maximum U-entropy distribution p^* because of the inequality (13.3). We note that the identity

$$C_U(p, p^*) = H_U(p^*) \quad (\forall p \in \Gamma_\tau)$$

plays an essential role on this discussion.

We next consider a minimax game. The inequality (13.6) leads us to

$$H_U(p^*) \le C_U(p^*, q) \le \max_{p \in \Gamma_\tau} C_U(p, q),$$

which implies

$$H_U(p^*) \le \min_{q \in \mathscr{P}} C_U(p^*, q) \le \min_{q \in \mathscr{P}} \max_{p \in \Gamma_\tau} C_U(p, q).$$

Similarly we get $\max_{p \in \Gamma_\tau} C_U(p, p^*) = H_U(p^*)$. Hence, we obtain

$$\max_{p \in \Gamma_\tau} \min_{q \in \mathscr{P}} C_U(p, q) = H_U(p^*) = \min_{q \in \mathscr{P}} \max_{p \in \Gamma_\tau} C_U(p, q).$$

This is the Γ-minimax theorem for the game with U-loss function. In this way we observe that any function U associates with Γ-minimaxity. In essence it is a natural consequence with the linearity of U-loss function in p. We get the fundamental property of Γ-minimax, which is closely related with the property that U-divergence satisfies $\nabla^{(U)} = \nabla$.

13.2.4 U-Model

It is well known that an exponential model satisfies the maximization of the Boltzmann–Shannon entropy. The maximum likelihood works elegant performance on the exponential model with minimal sufficiency, efficiency and unbias. In this we define a U-model including the exponential model, in which we investigate a basic property of the minimum U-divergence estimation.

Consider a foliation

$$\bigcup_{\tau \in \mathrm{Con}(t(\mathscr{Z}))} \Gamma_\tau, \tag{13.14}$$

where $\mathrm{Con}(A)$ denote the convex closure of A. Thus the parameter τ is a ray combining a leaf Γ_τ, and simultaneously models a maximum U-entropy distribution (13.13) as

$$M_U = \{p_U(z;\theta) = u(\theta^{\mathrm{T}} t(z) - \kappa_\theta) : \theta \in \Theta\}, \tag{13.15}$$

which we call a U-model, where κ_θ is normalizing constant to satisfy $\langle u(\theta^{\mathrm{T}} t - \kappa_\theta)\rangle = 1$

For example β-entropy

$$H_\beta(p) = \frac{\langle p^{\beta+1}\rangle - \beta - 1}{\beta(1+\beta)}$$

is maximized by

$$p_\beta(z;\theta) = [1 + \beta\{\theta^{\mathrm{T}} t(z) - \kappa_{\beta\,\theta}\}]^{\frac{1}{\beta}}.$$

We assume that the canonical parameter θ and the mean-value parameter $\tau = E_{p_U(\cdot;\theta)}\{t(z)\}$ are connected by a one-to-one transformation. This suffices to require that

$$\frac{\partial \tau}{\partial \theta} = \left\langle \left(t - \frac{\partial \kappa_\theta}{\partial \theta}\right)\left(t - \frac{\partial \kappa_\theta}{\partial \theta}\right)^{\mathrm{T}} u'(\theta^{\mathrm{T}} t - \kappa_\theta)\right\rangle$$

is non-singular. Then two parameters θ and τ become affine parameters with respect to ∇ and ${}^*\nabla^{(U)}$, respectively. For a given data set $z_1, \ldots, z_n$ the minimum U-divergence estimator for the parameter τ is $\widehat{\tau}_U = n^{-1}\sum_{i=1}^n t(z_i)$ since it has a estimating function defined by

$$s_U(z;\theta) = t - \frac{\partial \kappa_\theta}{\partial \theta} - E_{p_U(\cdot;\theta)}\Big(t - \frac{\partial \kappa_\theta}{\partial \theta}\Big) = t(z) - \tau.$$

This leads to the unbias of $\widehat{\tau}_U$ and sufficiency of statistic $t(z)$. When we view $\widehat{\tau}_U$ as a statistical functional defined on $\mathscr{P}$, then we can write $\widehat{\tau}_U(p) = E_p\{t(z)\}$, which is equivalent to the foliation (13.14) such that any leaf Γ_τ has a singleton τ as the image $\widehat{\tau}_U(\Gamma_\tau)$. However, the estimator $\widehat{\tau}_U$ is neither efficient nor asymptotically efficient, while the maximum likelihood is asymptotically efficient.

We discuss the projection onto a U-model in the sense of U-divergence. Let p be in $\mathscr{P}$ and let

$$q^* = \underset{q \in M_U}{\operatorname{argmin}}\, D_U(p,q).$$

Then we observe the Pythagorean theorem

$$D_U(p,q) - \{D_U(p,q^*) + D_U(q^*,q)\} = 0. \tag{13.16}$$

In fact, the left-hand side of (13.16) is $\langle p - q^*, \xi(q^*) - \xi(q)\rangle$, which implies that

$$(\theta^* - \theta)^{\mathrm{T}}\big[E_p\{t(z)\} - E_{q^*}\{t(z)\}\big]$$

becomes to vanish on account of $E_p\{t(z)\} = E_{q^*}\{t(z)\} = \tau$, where $\tau = E_p\{t(z)\}$ and θ^* is a parameter to express q^* according to the definition (13.15).

We focus on the minimum U-divergence estimator for the model N embedded in $\mathscr{M}$ as discussed in Sect. 13.2.1. In fact the U-loss function is defined by U-cross entropy $C(p,\nu)$ defined in (13.4) by substitution of ν into $\mu(\cdot;\theta)$ of the model N. Consider U-model with a shift in $\mathscr{M}$ as

$$M_U^{\text{shift}} = \{u(\theta^{\mathrm{T}}\{t(z) - \tau\}) : \theta \in \Theta\}, \tag{13.17}$$

where $\tau = E_p\{t(z)\}$. We note that the model is not a standard model in the sense that the model itself depends on the data distribution p. Then, noting that U-loss function (13.9) is written as $\langle p, \theta^{\mathrm{T}}\{t - \tau\}\rangle = 0$, we get

$$\ell_U(\theta, p) = \big\langle U\big(\theta^{\mathrm{T}}\{t(z) - \tau\}\big)\big\rangle. \tag{13.18}$$

Hence the U-estimator for the parameter

$$\tau^{\text{shift}} = \frac{\big\langle t, u\big(\theta^{\mathrm{T}}\{t(z) - \tau\}\big)\big\rangle}{\big\langle u\big(\theta^{\mathrm{T}}\{t(z) - \tau\}\big)\big\rangle}$$

is obtained by $n^{-1}\sum_{i=1}^n t(z_i)$. In a subsequent discussion AdaBoost algorithm sequentially learns U-model, in which it implements the sequential minimization of the exponential loss (13.17) with $U = \exp$. Thus we got a variety of estimation methods and models in which we discuss an effective method with high robustness in a subsequent section.

13.2.5 Robustness

We introduce a specific procedure with robustness in the class of minimum U-divergence methods. See [25] for general discussion. It is known that the maximum likelihood method is rather sensitive to even a small portion of outliners contaminated in a hull of observations, which will occur by a small degree of perturbations for the assumed model.

Let us consider a statistical model $M = \{p(z;\theta) : \theta \in \Theta\}$ with the parameter θ of interests in which the data distribution p is not in general in M. We write the score function on M by $s_\theta(z;\theta) = (\partial/\partial\theta)\log p(z;\theta)$. Then the estimating function (13.11) for the minimum U-divergence estimator is written by

$$s_U(z;\theta) = w(z;\theta)s(z;\theta) - E_{p(\cdot;\theta)}\{w(z;\theta)s(z;\theta)\}, \tag{13.19}$$

where $w(z;\theta)$ is a nonnegative function defined by $p(z;\theta)\xi'(p(z;\theta))$. This expression suggests that $s_U(z;\theta)$ is a weighted score function with the weight function $w(z;\theta)$. For example, β-divergence leads to the weight function $w_\beta(z;\theta) = p(z;\theta)^\beta$. Consider a situation in which the maximum likelihood estimator arises a large bias in the presence of one outlier $z_{\rm out}$. Then we observe that the weight function $w_\beta(z;\theta)$ becomes much smaller at $z = z_{\rm out}$ because the likelihood function $p(z;\theta)$ has much smaller contribution at $z = z_{\rm out}$. Thus the estimating function for the minimum β-divergence estimator automatically eliminates the effect of $z_{\rm out}$. Obviously, the maximum likelihood function is sensitive to $z_{\rm out}$ because the weight function is constant for all the observations. This is an intuitive explanation for robustness of the minimum β-divergence estimator. We need to choose the value of β since the weight function $w_\beta(z;\theta)$ becomes a constant function 1 as β goes to 0. See [17, 18] for further discussion and extension of β-divergence.

In robust statistics the influence function is established to quantitatively assesses robustness for a statistic, cf. [21]. In almost all the cases an estimator θ can be viewed as a functional of the empirical distribution function P_n, say $\widehat{\theta}(P_n)$ because P_n is sufficient in a nonparametric manner. We assume Fisher-consistency for $\widehat{\theta}(P_n)$, that is $\widehat{\theta}(P_\theta) = \theta$, where P_θ denotes the distribution induced by $p(z;\theta)$. Then the influence function for $\widehat{\theta}(P_n)$ is defined by

$$\mathrm{IF}(\widehat{\theta}, z) = \frac{\partial}{\partial\varepsilon}\widehat{\theta}((1-\varepsilon)P_\theta + \varepsilon\delta_z)\Big|_{\varepsilon=0},$$

where δ_z denotes a point-mass distribution at z. This measure expresses the behavior of $\widehat{\theta}(P)$ when the underlying distribution P is deviate from the model distribution P_θ and hence P is written by $(1-\varepsilon)P_\theta + \varepsilon\delta_z$ with a small ε that is a contamination proportion of the outlier z into the hull of data population. For example, consider the estimation of a mean parameter μ of a normal distribution. Then the influence function of the sample mean is $z - \mu$, while that of the sample median is $\mathrm{sgn}(z - \theta)$. Thus we see that the influence function of the sample mean is unbounded; that of the sample median is bounded.

For the minimum U-divergence estimator $\widehat{\theta}_U$ the influence function is

$$\text{IF}(\widehat{\theta}_U, z) = J_\theta^{-1} s_U(z;\theta), \tag{13.20}$$

where J_θ is a constant matrix defined in (13.12), $s_U(z;\theta)$ is the estimating function (13.19). Let us return the case of the minimum β-divergence estimator. For simplicity we discuss a situation such that the model distribution is an exponential model. Then the influence function of the minimum β-divergence estimator is bounded because the influence function is expressed essentially by $f(t) = \exp(-\beta|t|)|t|$ and f is a bounded function.

13.3 Applications to Unsupervised Learning

Principal component analysis (PCA) and independent component analysis (ICA) are both widely used for unsupervised learning in a number of research fields ranging from social sciences and natural sciences. The goal is feature extraction and data dimension reduction from an observation vector of high dimension. Both analyses play a role to be complemented each other.

Let $x = (x_1, \ldots, x_p)^{\mathrm{T}}$ be an observation vector of dimension p. Then PCA aims to search a matrix W of $k \times p$ that components of $y = Wx$ are uncorrelated while ICA aims to search a matrix W of $k \times p$ that those of $y = Wx$ are independent. It is known that independence implies uncorrelation, but the reverse does not hold. Hence there is an essential difference between aims of PCA and ICA. In fact, ICA is frequently applied after sphering, which avoids any information with PCA. However in many situations one does not know which procedure is effective for the feature extraction and dimension reduction or not, and thus both application results are important to strengthen the understanding through the comparison.

We will make a direct application of minimum U-divergence method to PCA and ICA with the learning algorithm. In particular we introduce a special U-divergence that leads to a robust method according to the general discussion of Sect. 13.2.5.

13.3.1 PCA

We consider an aspect of data dimension reduction in PCA for unified discussion on PCA and ICA. For this we review the conventional PCA procedure. PCA is characterized by minimization problem to find a projection matrix W from p-dimensional Euclidean space to k-dimensional Euclidean space to minimize

$$\min\left\{E\{\|x-\mu\|^2 - \|W(x-\mu)\|^2\}\right\} \tag{13.21}$$

in W such that $WW^{\mathrm{T}} = I_k$ (k-identity matrix). See [6] for the associations with neural network models. Actually we observe that

$$E\{\|x-\mu\|^2 - \|W(x-\mu)\|^2\} = \text{trace}\{V(x)\} - \sum_{i=1}^{k} w_k V(x) w_k^{\mathrm{T}}, \qquad (13.22)$$

where $V(x)$ denotes the variance matrix of x and $w_1,\ldots,w_k$ is the set of law vectors of W. Hence a unique solution W of the optimization problem (13.21) is given by the matrix formed by eigenvectors of $V(x)$ corresponding to k largest eigenvalues. Thus, the constraint of orthonormality follows from the property of eigenvectors of a symmetric matrix.

On the other hand we can formulate another variant of PCA using a minimum U-divergence method. This is proposed by minimization of

$$\ell_U(W,\mu) = -E\{\psi(\|x-\mu\|^2 - \|W(x-\mu)\|^2)\}$$

with constraints $WW^{\mathrm{T}} = I_k$, where $\psi(t) = \xi(\exp(t))$ with ξ being the inverse derivative of U. We note that if a general U-loss function (13.9) is applied this context, the second term becomes a constant of W. The choice of $\xi = \log$ reduces $\ell_U(W,\mu)$ to the classical case given by (13.22) In general this minimization problem is numerically solved by the learning algorithm from updating (μ, W) to (μ^*, W^*) as

$$\mu^* = \sum_{i=1}^{n} w(x_i;\mu,W)x_i, W^* = \text{Eigen}_k(S(\mu,W)),$$

where $\text{Eigen}_k(A)$ is the matrix of k dominant eigenvectors of a matrix A. Here $S(\mu,W)$ is a weighted matrix

$$S(\mu,W) = \sum_{i=1}^{n} p(x_i;\mu,W)(x_i-\mu)(x_i-\mu)^{\mathrm{T}}$$

with a weight function $p(x_i;\mu,W)$ being

$$\frac{\psi'(\|x_i-\mu\|^2 - \|W(x_i-\mu)\|^2)}{\sum_j^n \psi'(\|x_j-\mu\|^2 - \|W(x_j-\mu)\|^2)}.$$

This algorithm has an advantageous aspect such that the objective function is uniformly decreasing in any iteration as

$$\ell_U(W^*,\mu^*) < \ell_U(W,\mu),$$

which is of EM algorithm with a stable convergence. The influence function is given by a special case of the general formula (13.20), see [23, 27] for detailed form of the influence function and also [24] for a method of adaptive tuning parameter.

13.3.2 ICA

We next discuss ICA in the point of robustness focusing on an instant mixture model, which is one of the most basic models for blind source separation. Let x an input vector of dimension p and let A a nonsingular matrix of $p \times p$. See [26] for extensive discussion. We assume a simple model by

$$x = As, \tag{13.23}$$

where s is a random vector of p dimension with independent components. Thus the model (13.23) implies a linear combination of unobservable components of s by the matrix A. For model identifiability we add to an assumption of non-Gaussianity of the components.

A goal of ICA is to learn a matrix W of size $p \times p$ such that Wx has independent components for a given empirical examples $x_1, \ldots, x_n$. For example, if we know A, then it suffices to take the inverse of A, but it is not necessary. In fact it is sufficient that WA is a diagonal matrix.

By the assumption of independence the recovered source vector Wx has a distribution with the density function decomposed into as

$$q(y) = q_1(y_1) \cdots q_m(y_m),$$

which implies that the density function of x is

$$p(x; W, \mu) = |\det(W)| q_1(w_1 x - \mu_1) \cdots q_m(w_m x - \mu_m),$$

where $w_1, \ldots, w_m$ denote law vectors of W. A standard method appropriately chooses $q_1, \ldots, q_m$ and derives the quasi-likelihood function, based on the data set $x_1, \ldots, x_n$,

$$L(W, \mu) = \frac{1}{n} \sum_{i=1}^{n} \log p(x_i; W, \mu). \tag{13.24}$$

In this way we get the maximum quasi-likelihood estimator W, μ by maximization of (13.24).

On the other hand, we apply the minimum U-divergence method by the use of U-divergence (13.3). From a general formula the empirical U-loss function is given by

$$\ell_U^{\text{emp}}(W, \mu) = -\frac{1}{n} \sum_{i=1}^{n} \xi(p(x_i; W, \mu)) + \int U(\xi(p(z; W, \mu))) dz. \tag{13.25}$$

In general, the second term of (13.25) is a function of only $\det(W)$ as given $c_\beta \det(W)^\beta$ for the case of β-divergence with a constant c_β. See [33] for more detailed equations. The estimating function to find a maximizer of the function (13.24) is

$$s(x;W,\mu) = \begin{bmatrix} \{I_m - h(Wx-\mu)(Wx)^{\mathrm{T}}\}W^{-\mathrm{T}} \\ h(Wx-\mu) \end{bmatrix},$$

where $h(y) = \left((\partial/\partial y_i)\log q_i(y_i)\right)_{i=1}^{m}$. Alternatively, the minimum β-divergence method gives an estimating function

$$s_\beta(x;W,\mu) = p^\beta(x;W,\mu)s(x;W,\mu) + \begin{bmatrix} \beta c_\beta |\det(W)|^\beta W^{-\mathrm{T}} \\ 0 \end{bmatrix}.$$

Therefore, we observe that if $\beta = 0$, then $s_\beta(x;W,\mu)$ is nothing but $s(x;W,\mu)$. For any $\beta > 0$ the estimating equation $\sum_{i=1}^{n} s_\beta(x_i;W,\mu) = 0$ can be viewed as the weighted quasi-likelihood equation with the i-th weight $p^\beta(x_i;W,\mu)$. If x_i is an outlier with a smaller density value $p(x_i;W,\mu)$, then it has only a smaller contribution to the equation. Consequently this supports robustness for the minimum β-divergence method with $\beta > 0$ against outlying. Furthermore, we assume

$$\sup_{z\in\mathbf{R}} q_i(z)\exp(|z|) < \infty$$

for any $i = 1,\ldots,p$, which implies the boundedness of all components of the estimating function $s_\beta(x;W,\mu)$ for any $\beta > 0$. In contrast, taking a limit of β to 0 arises the unboundedness. This redescending property of the influence function is discussed to make an effective learning for a mixture of ICA, see [34].

13.4 Applications to Statistical Pattern Recognition

Pattern recognition aims to identify an attribution of a subject from a feature information with the subject, or more mathematically to predict a class label y based on the feature vector x. Here we assume that x is in a p dimensional feature space $\mathscr{X}$ and that y is in a class-label set $\mathscr{Y}$. See [32] for general discussion in statistical pattern recognition. Thus the methodology of pattern recognition is spontaneously connected with cognitive science in which a classification rule is associated with a process that a biological brain system gives rise to a rational judgment rule through "learning". For example, in a biological population a female individual determines a male individual with optimal gamete based on empirical observations of feature vector x related with the male candidates. In this sense the pattern recognition is a fundamental function of biological brain system. Boosting learning algorithm incorporates artificially the function into combining several different classifiers by linear combination to get a stronger classifier. Recently there is large literature on research for boosting methods discussed from various points of view. For example, see [38] for the motivation of boosting and [15] for AdaBoost. Also see [12, 39] for statistical discussion and [13] for information-geometric discussion.

Consider a mapping h from a feature space $\mathscr{X}$ to a class-label set $\mathscr{Y}$. Then we get a classification rule by $y = h(x)$ in which h is called a classifier. We often consider a function $F(x,y)$ defined on the product space $\mathscr{X} \times \mathscr{Y}$ in which the corresponding classifier is defined by

$$h_F(x) = \operatorname{argmax}\{F(x,y) : y \in \mathscr{Y}\}. \tag{13.26}$$

In this context we call $F(x,y)$ a score function.

In statistical work the objective is to propose an optimal score function F by learning an empirical training data set $E_n = \{(x_1,y_1),\dots,(x_n,y_n)\}$ such that the classifier h_F in (13.26) makes good performance in pattern recognition. The performance of the proposed rule based on the classifier is usually assessed by a test data set $E_m^* = \{(x_1^*,y_1^*),\dots,(x_m^*,y_m^*)\}$, which is organized separately from E_n. Thus, for example the rule is assessed by

$$\operatorname{testerr}(h) = \frac{1}{m}\sum_{i=1}^{m} I(h(x_i^*) \neq y_i^*). \tag{13.27}$$

A lower training error does not always imply a lower test error, which is an essential point in prediction performance in statistical discussion. We say this to overlearning if the gap is excessive.

Consider a space $\mathscr{F}$ of all the score functions and a space $\mathscr{H}$ of all classifiers. Then we point a redundancy when we express a classifier h by a score function F. In fact we consider an equivalent class by

$$\mathscr{F}_h = \{F \in \mathscr{F} : h_F = h\}$$

in which, if F_1 and F_2 are in $\mathscr{F}_h$, then $c_1F_1 + c_2F_2$ is in $\mathscr{F}_h$ for any positive constants c_1, c_2. In this equivalent class $\mathscr{F}_h$ we define a representative element by $F(x,y) = I(h_F(x) = y)$, which has the most economical expression in the sense that F takes only binary values 0 and 1, where I denotes a definition function. In a subsequent discussion we will see that boost learning incorporates this redundancy into effective reinforcement. In this way the redundancy is not a nuisance aspect but helps producing a variety of data learning.

We introduce boosting ideas in the framework established in the preceding section writing a joint data space by $\mathscr{Z} = \mathscr{X} \times \mathscr{Y}$ with the data vector $z = (x,y)$ Consider a subfamily $\mathscr{H}_1$ of $\mathscr{H}$ in which a boosting method is aimed to construct an effective classifier by linearly combining classifiers. The key idea is to embed classifiers $h_1,\dots,h_d$ of $\mathscr{H}_1$ into the score function space $\mathscr{F}$ by

$$\mathscr{F} = \{F(x,y,\alpha) = \sum_{j=1}^{d} \alpha_j I(h_j(x) = y) : \alpha = (\alpha_1,\dots,\alpha_d) \in A\}. \tag{13.28}$$

We expect a stronger classifier h_F than any h_js if F constructed by (13.28) has an optimal linear coefficients α_js. In fact, we will optimize this in a sequential manner as

$$F(x,y,\alpha_1,\ldots,\alpha_{t+1}) = F(x,y,\alpha_1,\ldots,\alpha_t) + \alpha_{t+1}I(h_{t+1}(x) = y).$$

The boosting method iteratively selects the best update to be added the present score function at every step.

13.4.1 U-Loss Function for Score Function

Let us employ the general discussion with a problem of pattern recognition. See [35] for a detailed discussion. Let $p(x,y)$ be a probability density function on the product space of a feature space $\mathscr{X}$ and a class-label set $\mathscr{Y}$ decomposed into as

$$p(x,y) = P(y|x)q(x), \tag{13.29}$$

where $P(y|x)$ is the conditional density function of x given y and $q(x)$ is the marginal density function of x. Then we write a linearly combined classifier given by (13.28)

$$F(x,y) = \alpha^{\mathrm{T}} f(x,y)$$

where $f(x,y) = (I(h_1(x) = y),\ldots,I(h_d(x) = y))$. For this we model

$$\tilde{\mu}_\alpha(y|x) = u(\alpha^{\mathrm{T}} f(x,y) - b(x,\alpha))$$

as a shifted U-model, where $b(x,\alpha) = \sum_{y'\in\mathscr{Y}} \alpha^{\mathrm{T}} f(x,y')p(y'|x)$. This is a special example of the model discussed in (13.17) of Sect. 13.2.4 in the present context. Hence U-loss function is given by

$$\ell_U(\alpha) = \int_{\mathscr{X}} \sum_{y\in\mathscr{Y}} U(\alpha^{\mathrm{T}} f(x,y) - b(x,\alpha))q(x)dx$$

according to the general formula (13.9). An argument similar to that of Sect. 13.2.4 leads to

$$\ell_U(\alpha) - \ell_U(\alpha^*) = D_U(\tilde{\mu}_{\alpha^*}, \tilde{\mu}_\alpha),$$

where $\alpha^* = \operatorname{argmin}_{\alpha\in A} D_U(p,\tilde{\mu}_\alpha)$. The empirical U-loss function is

$$\ell_U^{\mathrm{emp}}(\alpha) = \frac{1}{n}\sum_{i=1}^{n}\sum_{y\in\mathscr{Y}} U(\alpha^{\mathrm{T}}\{f(x_i,y) - f(x_i,y_i)\}), \tag{13.30}$$

where $b(x_i,\alpha) = \alpha^{\mathrm{T}} f(x_i,y_i)$.

Alternatively, the probability constraint leads to a U-probabilistic model

$$\bar{\mu}_\alpha(y|x) = u(\alpha^{\mathrm{T}} f(x,y) - \kappa(x,\alpha)), \tag{13.31}$$

where $\kappa(\alpha)$ is a normalizing constant to have mass 1 by

$$\sum_{y\in\mathscr{Y}} u(\alpha^{\mathrm{T}} f(x,y) - \kappa(x,\alpha)) = 1.$$

Under this constraint U-loss function is

$$\bar{\ell}_U(\alpha) = \int_{\mathscr{X}} \sum_{y\in\mathscr{Y}} \Big[U(\alpha^{\mathrm{T}} f(x,y) - \kappa(x,\alpha)) - P(y|x)\{\alpha^{\mathrm{T}} f(x,y) - \kappa(x,\alpha)\}\Big] q(x)dx.$$

We observe an interesting relation with an equation

$$\bar{\ell}_U(\alpha) - \bar{\ell}_U(\alpha^*) = D_U(\bar{\mu}_{\alpha^*}, \bar{\mu}_{\bar{\alpha}}), \tag{13.32}$$

which is closely related with a Pythagorean relation. The empirical U-loss function is

$$\bar{\ell}_U^{\mathrm{emp}}(\alpha) = \frac{1}{n}\sum_{i=1}^{n}\sum_{y\in\mathscr{Y}} U(\alpha^{\mathrm{T}} f(x_i,y) - \kappa(x_i,\alpha)) + \kappa(x_i\alpha) - \alpha^{\mathrm{T}} f(x_i,y_i). \tag{13.33}$$

Let us look at two variants of U-loss function for the typical case of $U = \exp$ as follows.

$$\ell_{\exp}^{\mathrm{emp}}(\alpha) = \frac{1}{n}\sum_{i=1}^{n}\sum_{y\in\mathscr{Y}} \exp\{\alpha^{\mathrm{T}}\{f(x_i,y) - f(x_i,y_i)\},$$

$$\bar{\ell}_{\exp}^{\mathrm{emp}}(\alpha) = -\frac{1}{n}\sum_{i=1}^{n} \log\frac{\exp\{\alpha^{\mathrm{T}} f(x_i,y_i)\}}{\sum_{y\in\mathscr{Y}}\exp\{\alpha^{\mathrm{T}} f(x_i,y)\}} \tag{13.34}$$

which are called the exponential loss and the log loss functions, respectively. These loss functions lead to AdaBoost and LogitBoost algorithms, see [30] for the interesting understanding for the relation. In particular, the log-loss function is the minimum of the conditional log-likelihood function for a logistic regression model. The exponential loss function was not discussed until [15] did consider in machine learning paradigm. Many other loss functions similar to the log-loss function including the area under the ROC curve was discussed in [12]. See [16] for a close relation with a generalized additive model.

13.4.2 U-Boost

We introduce a form of boost learning algorithms to search a minimizer of a loss function discussed in Sect. 13.4.1. The basic idea is to make a repeated use of m-projection discussed in Sect. 13.1. Thus the present score function $F(x,y)$ updates the next step by embedding a new classifier $h(x)$ by

$$F(x,y) \mapsto F^*(x,y) = F(x,y) + \alpha I(h(x) = y),$$

where we get a solution of (α, h) by

$$(\alpha^*, h^*) = \underset{(\alpha,h)\in\mathbf{R}\times\mathscr{H}}{\text{argmin}} \; \ell_U^{\text{emp}}(F(x,y) + \alpha I(h(x) = y)).$$

This update is implemented by m-projection to the one parameter model parametrized by α, which gives rise to a right triangle satisfying the Pythagorean-like relation. Thus such a procedure is sequentially iterated by association with such a right triangle with respect to U-loss function.

Let us introduce the learning algorithm of U-boost with the shifted model in which the class $\mathscr{H}_1$ of classifiers to be trained for the given example set E_n.

A. fix $w_1(i,y) = \dfrac{1}{n(g-1)} I(y \neq y_i)$ as the initial distribution on E_n, where $g = \text{card}(\mathscr{Y})$.

B. For step $t = 1, \ldots, T$ the weighted error rate is set by

$$\varepsilon_t(h) = \frac{1}{2} \sum_{i=1}^{n} \sum_{y\in\mathscr{Y}} w_t(i,y) I(y \neq y_i)\{f(x_i,y) - f(x_i,y_i) + 1\} \tag{13.35}$$

Next we proceed the following substeps by

(B-1) $h_*^{(t)} = \underset{h\in\mathscr{H}_1}{\text{argmin}}\, \varepsilon_t(h)$.

(B-2) $\alpha_t^* = \underset{\alpha}{\text{argmin}}\, \ell_U^{\text{emp}}(F_{t-1} + \alpha f_*^{(t)})$. where ℓ_U^{emp} is defined in (13.30).

(B-3) Let $F_t(x,y) = \sum_{j=1}^{t} \alpha_j^* I(h_*^{(j)}(x) = y)$. Then, the weight function $w_{t+1}(i,y) \propto u\{F_t(x_i,y) - F_t(x_i,y_i)\}$ leads to update (13.35).

C. Finally we make a decision by $h_{\text{final}}(x) = \underset{y\in\mathscr{Y}}{\text{argmax}}\, F_T(x,y)$, where $F_T(x,y) = \sum_{t=1}^{\text{T}} \alpha_t^* I(h_*^{(t)}(x) = y)$.

We note that U-boost algorithm with probability constraint is similarly implemented just by replacing ℓ_U^{emp} in substep (B-2) into $\overline{\ell}_U^{\text{emp}}$ defined in (13.4.1). In this way U-boost algorithm successfully integrates the knowledge with classifiers in $\mathscr{H}_1$, of which computational cost is rather less than other learning methods such as support vector machines. The most characteristic point is to use a dynamical change of weight functions. We observe that for any t

$$\varepsilon_{t+1}(h_*^{(t)}) = \frac{1}{2}.$$

The proof is immediate because the coefficient vectors α_t^* makes the gradient of $\ell_U^{\rm emp}(F_{t-1}+\alpha f_*^{(t)})$ vanish, see Theorem 3 and the proof in [35] for detailed discussion. In accordance, all U-boost learning algorithms shows the property of least favorable weighting in the sense that The best learning machine $h_*^{(t)}$ with the minimum weighted error rate in the t-step is reduced to the worst machine with respect to the weight $w_{t+1}(i,y)$ updated.

Return the typical case of $U(t)=\exp(t)$. Then the equation in substep (B-2) has a closed solution such that

$$\alpha_t^* = \frac{1}{2}\log\frac{1-\varepsilon_t(h_*^{(t)})}{\varepsilon_t(h_*^{(t)})},$$

which is called AdaBoost M2.

We discuss statistical properties of U-boost methods. It is well known that the Bayes rule is given by

$$h_B(x) = \text{argmax}\{P(y|x) : y \in \mathscr{Y}\}, \tag{13.36}$$

which attains a minimum of the expected error rate among all the classification rules. Almost of classification methods place undue reliance on this fact, in which they reduce to estimating the posterior distribution $P(y|x)$ under a specific model based on example data, cf. [12]. We now explore which relation with the Bayes rule do U-boost methods have. Consider a class of score functions equivalent to the Bayes rule by

$$\mathscr{F}_B = \{F(x,y) : h_F = h_B\}.$$

The U-model of the score function is given by

$$\mathscr{M}_U = \{\mu_F(y|x) = u(F(x,y) - b_F(x)) : F \in \mathscr{F}\}$$

in a nonparametric sense, where the constant $b_F(x)$ is defined by

$$b_F(x) = \sum_{y' \in \mathscr{Y}} F(x,y')p(y'|x). \tag{13.37}$$

Then there exists a F^* of $\mathscr{F}$ such that

$$u(F^*(x,y) - b_{F^*}(x)) = c(x)P(y|x), \tag{13.38}$$

where $c(x)$ is a positive function. This means that F^* is in $\mathscr{F}_B$. In this discussion we claim that

$$F^* = \text{argmin}\{\ell_U(F) : F \in \mathscr{F}\}, \tag{13.39}$$

where

$$\ell_U(F) = \int_{\mathscr{X}} \sum_{y \in \mathscr{Y}} U(F(x,y) - b_F(x))q(x)dx.$$

This is because we observe that

$$\ell_U(F) - \ell_U(F^*) = D_U(\mu_{F^*}, \mu_F)$$

which must be nonnegative by definition of U-divergence as defined in (13.3). In accordance with this, the minimizer F^* belongs to $\mathscr{F}_B$. See [13] for more detailed proof.

We overview that the optimization of the expected U-loss $\ell_U(F)$ is associated with a search of the score function concluding the Bayes rule. Actually, in the practice of the pattern recognition a finite-dimensional vector α of coefficients (13.30) is output by learning finite set of empirical examples in the U-boost learning algorithm. If the distributional assumption for n examples holds asymptotically, then we can conclude the empirical version of the proposition (13.39). However, we here do not proceed a further discussion on asymptotics, which would need more restrictive assumption with several technicalities. See [31] for a rigorous discussion with strict assumptions and also [37] for regularization of AdaBoost.

13.4.3 EtaBoost

We focus on a problem of mislabels in pattern recognition. Let

$$U_\eta(t) = (1-\eta)\exp(t) + \eta t,$$

where η is a constant with $0 \le \eta < 1$. Then we know that

$$u_\eta(t) = (1-\eta)\exp(t) + \eta, \qquad \xi_\eta(u) = \log\frac{u-\eta}{1-\eta},$$

which leads to defining U-divergence by

$$D_\eta(\mu, \nu) = \int_{\mathscr{X}} \sum_{y \in \mathscr{Y}} \left[\nu(x,y) - \mu(x,y) - \{\mu(x,y) - \eta\}\log\frac{\nu(x,y)-\eta)}{\mu(x,y)-\eta)}\right] q(x)dx,$$

which we call Eta-divergence. Hence U_η probabilistic model for a score function $F(x,y) = \alpha^{\mathrm{T}} f(x,y)$ is given by

$$\bar{\mu}_\eta(y|x,\alpha) = (1-\eta)\exp\{\alpha^{\mathrm{T}} f(x,y) - \kappa(x,\alpha)\} + \eta$$

in accordance with the general formula (13.31), where the normalizing constant is

$$\kappa(x,\alpha) = \log\frac{1-\eta}{1-g\eta} + \log\left[\sum_{y' \in \mathscr{Y}} \exp\{\alpha^{\mathrm{T}} f(x,y')\}\right]$$

with g being card($\mathscr{Y}$). It is written that

$$\bar{\mu}_\eta(y|x,\alpha) = \{1-\eta(g-1)\}P_L(y|x,\alpha) + \eta \sum_{y' \neq y} P_L(y|x,\alpha), \tag{13.40}$$

where

$$P_L(y|x,\alpha) = \frac{\exp\{\alpha^{\mathrm{T}} f(x,y)\}}{\sum_{y' \in \mathscr{Y}} \exp\{\alpha^{\mathrm{T}} f(x,y')\}}.$$

Interestingly, this model (13.40) associates with the following insight. As an idealistic situation we assume that the conditional probability $P(y|x)$ of a class-label y given a feature vector follows from a logistic model $P_L(y|x,\alpha)$. However we consider a situation in which a certain cause breaks down this assumption, so that there occurs a mislabel event with probability η. Then the resulting conditional probability is led to $\bar{\mu}_\eta(y|x,\alpha)$. In this way we can view that Eta-divergence D_η associates with mislabels.

We define EtaBoost by Eta-divergence D_η as a special example of U-boost. On account of e probabilistic interpretation of mislabeling as discussed above EtaBoost is generatively a robust learning algorithm. Actually, EtaBoost with probability constraint is closely related with the method in [9] applied to a binary regression with high noise in response variables. See [41] for the robustification of AdaBoost and [42] for a multi-class situation. For a general discussion on outlying including in a feature space, see [28, 29].

13.5 Conclusions

We consider the geometry associated with the U-divergence class which leads to U-loss functions and U-models. These are shown to be connected by Γ-minimaxity. In particular a special choice of the function U yields a robust procedure. The information divergence geometry gives not only a unified look at various applications in statistical machine learning, but also an intuitive understanding for unsupervised and supervised learning algorithms. For the geometric approach, many problems are still unsolved waiting to be challenged in future work.

References

1. Amari, S. (1985). *Differential-Geometrical Methods in Statistics*, volume 28 of *Lecture Notes in Statistics*. Springer, New York.
2. Amari, S. and Nagaoka, H. (2000). *Methods of Information Geometry*, volume 191 of *Translations of Mathematical Monographs*. Oxford University Press, Oxford.
3. Barndorff-Nielsen, O. E. (1978). *Information and Exponential Families in Statistical Theory*. Wiley, Chichester.

4. Basu, A., Harris, I. R., Hjort, N. L., and Jones, M. C. (1998). Robust and efficient estimation by minimising a density power divergence. *Biometrika* 85(3):549–559.
5. Beran, R. (1977). Minimum hellinger distance estimates for parametric models. *Ann. Stat.* 5(3):445–463.
6. Bishop, C. (1995). *Neural Networks for Pattern Recognition*. Clarendon, Oxford.
7. Box, G. E. P. and Cox, D. R. (1964). An analysis of transformations. *J. R. Stat. Soc. B* 26(2):211–252.
8. Csiszar, I. (1967). Information type measures of differences of probability distribution and indirect observations. *Studia Math. Hungarica* 2:299–318.
9. Copas, J. (1988). Binary regression models for contaminated data. *J. R. Stat. Soc. B* 50: 225–265.
10. Eguchi, S. (1983). Second order efficiency of minimum contrast estimators in a curved exponential family. *Ann. Stat.* 11:793–803.
11. Eguchi, S. (1992). Geometry of minimum contrast. *Hiroshima Math. J.* 22:631–647.
12. Eguchi, S. and Copas, J. B. (2002). A class of logistic type discriminant functions. *Biometrika* 89:1–22.
13. Eguchi, S. (2006). Information geometry and statistical pattern recognition. *Sugaku Exposition*, American Mathematical Society, 197–216.
14. Fisher, R. A. (1922). On the mathematical foundations of theoretical statistics. *Philos. Trans. Roy. Soc. London A* 222:309–368.
15. Freund, Y. and Schapire, R. E. (1997). A decision-theoretic generalization of on-line learning and an application to boosting. *J. Comput. Syst. Sci.* 55:119–139.
16. Friedman, J. H., Hastie, T., and Tibshirani, R. (2000). Additive logistic regression: A statistical view of boosting. *Ann. Stat.* 28:337–407.
17. Fujisawa, H. and Eguchi, S. (2005). A new approach to robust parameter estimation against heavy contamination. ISM Research Memo. 947.
18. Fujisawa, H. and Eguchi, S. (2006). Robust estimation in the normal mixture model. *J. Stat. Plan. Inference* 136(11):3989–4011.
19. Good, I. J. (1952). Rational decisions. *J. Roy. Stat. Soc. B* 14:107–114.
20. Grünwald, P. D. and Dawid, A. P. (2004). Game theory, maximum entropy, minimum discrepancy, and robust Bayesian decision theory. *Ann. Stat.* 32:1367–1433.
21. Hampel, F. R., Ronchetti, E. M., Rousseeuw, P. J., and Stahel, W. A. (2005) *Robust Statistics: The Approach Based on Influence Functions*. Wiley, New York.
22. Hastie, T., Tibishirani, R., and Friedman, J. (2001). *The Elements of Statistical Learning*. Springer, New York.
23. Higuchi, I. and Eguchi, S. (1998). The influence function of principal component analysis by self-organizing rule. *Neural Comput.* 10:1435–1444.
24. Higuchi, I. and Eguchi, S. (2004). Robust principal component analysis with adaptive selection for tuning parameters. *J. Machine Learn. Res.* 5:453–471.
25. Huber, P. J. (1981). *Robust Statistics*. Wiley, New York.
26. Hyvarinen, Karhunen, A., and Oja, K. (2001). *Independent Component Analysis*. Wiley, New York.
27. Kamiya, H. and Eguchi, S. (2001). A class of robust principal component vectors. *J. Multivariate Anal.* 77:239–269.
28. Kanamori, T., Takenouchi, T., Murata, N., and Eguchi, S. (2004). The most robust loss function for boosting. Presented by T. Kanamori at Neural Information Processing: 11th International Conference, ICONIP 2004, Calcutta. *Lecture Notes in Computer Science* 3316, 496–501. Berlin: Springer.
29. Kanamori, T., Takenouchi, T., Eguchi, S., and Murata, N. (2007). Robust loss functions for boosting. Robust loss functions for boosting. *Neural Comput.* 19:2183–2244.
30. Lebanon, G. and Lafferty, J. (2002). Boosting and maximum likelihood for exponential models. *Adv. Neural Inform. Process. Syst.* 14.
31. Lugosi, G. and Vayatis, N. (2004). On the Bayes-risk consistency of regularized boosting methods. *Ann. Stat.* 32:30–55.

32. McLachlan, G. J. (2004). *Discriminant Analysis and Statistical Pattern Recognition*. Wiley, New York.
33. Minami, M. and Eguchi, S. (2002). Robust blind source separation by beta-divergence. *Neural Comput.* 14:1859–1886.
34. Mollah, N. H., Minami, M., and Eguchi, S. (2006). Exploring latent structure of mixture ICA models by the minimum beta-divergence method. *Neural Comput.* 18:166–190.
35. Murata, N., Takenouchi. T., Kanamori, T ., and Eguchi. S . (2004). Information geometry of U-Boost and Bregman divergence. *Neural Comput.* 16:1437–1481.
36. Rao, C. R. (1945). Information and accuracy attainable in the estimation of statistical parameters. *Bull. Culcutta Math. Soc.* 37:81–91.
37. Rätsch, G., Onoda, T., and Müller K.-R. (2001). Soft margins for AdaBoost. *Machine Learn.* 42:287–320.
38. Schapire, R. E. (1990). The strength of weak learnability. *Machine Learn.* 5:197–227.
39. Schapire, R. E., Freund, Y., Bartlett, P., and Lee, W. S. (1998). Boosting the margin: A new explanation for the effectiveness of voting methods. *Ann. Stat.* 26:1651–1686.
40. Scott, D. W. (2001). Parametric statistical modeling by minimum integrated square error. *Technometrics* 43:274–285.
41. Takenouchi, T. and Eguchi, S. (2004). Robustifying AdaBoost by adding the naive error rate. *Neural Comput.* 16:767–787.
42. Takenouchi, T., Eguchi, S., Murata N., and Kanamori, T. (2008). Robust Boosting algorithm for multiclass problem by mislabelling model. *Neural Comput.* 20:1596–1630.
43. Vapnik, V. (1995). *The Nature of Statistical Learning Theory*. Springer, New York,

Chapter 14
Model Selection and Information Criterion

Noboru Murata and Hyeyoung Park

Abstract In this chapter, a problem of estimating model parameters from observed data is considered such as regression and function approximation, and a method of evaluating the goodness of model is introduced. Starting from so-called leave-one-out cross-validation, and investigating asymptotic statistical properties of estimated parameters, a generalized Akaike's information criterion (AIC) is derived for selecting an appropriate model from several candidates. In addition to model selection, the concept of information criteria provides an assessment of the goodness of model in various situations. Finally, an optimization method using regularization is presented as an example.

14.1 Introduction

Let us consider a problem of predicting outputs for given inputs based on examples of input–output pairs. This is called a *regression* or *fitting* problem, and it appears in many areas of engineering field, for example, statistics, signal processing, machine learning and so on [4, 11, 13, 17].

Suppose we have a data set

$$\mathscr{D} = \{(x_i, y_i);\ i = 1, 2, \dots, n\}$$

composed of n samples from an unknown probability distribution $P(X,Y)$. Our aim is to construct a model of input–output relation by using data $\mathscr{D}$ and to predict

N. Murata (✉)
Waseda University, Tokyo 169-8555, Japan
e-mail: noboru.murata@eb.waseda.ac.jp

H. Park
Kyungpook National University, Daegu 702-701, Korea
e-mail: hypark@knu.ac.kr

F. Emmert-Streib, M. Dehmer (eds.), *Information Theory and Statistical Learning*,
DOI: 10.1007/978-0-387-84816-7_14,

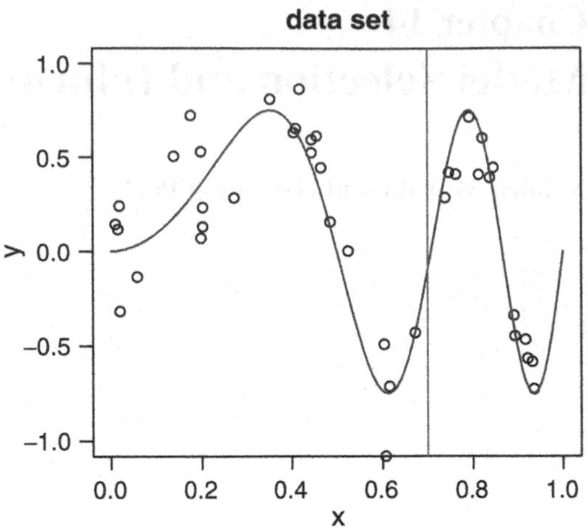

Fig. 14.1 A typical example of the regression problem. The *points* are given data $\mathscr{D}$, and the *vertical solid line* indicates the border above which we are trying to estimate the output for a given input. The *dashed curve* is the conditional expectation $E[Y|X]$ for given X

the corresponding output for a newly given input x. A typical example is depicted in Fig. 14.1, where both, input and output, are one-dimensional. The points are samples of given data, and the vertical solid line indicates the border above which we are trying to estimate the output for a given input. The dashed curve is the conditional expectation of Y for a given X, denoted by $E[Y|X]$ which could give a safe estimate of output Y in most of the cases. Intuitively speaking, we would like to infer the dashed curve based on given data $\mathscr{D}$.

To describe the input–output relation, we may adopt various parametric models, such as a linear combination of monomials

$$f(x;\theta) = a_0 + a_1 x + a_2 x^2 + \cdots + a_d x^d,$$

where the model parameters are summarized in a vector form as $\theta = (a_0, a_1, a_2, \ldots, a_d)^T$. Another example is a so-called neural network [17], and one of the simplest forms is a three-layered perceptron defined as

$$f(x;\theta) = c_1 \varphi(a_1 x + b_1) + \cdots + c_h \varphi(a_h x + b_h),$$

where φ is a nonlinear function (typically sigmoid function), h is the number of hidden units, and the model parameter is $\theta = (a_1, b_1, c_1, \ldots, a_h, b_h, c_h)^T$.

To determine an appropriate parameter θ, we have to establish a measure of the *goodness of fit* for given data. For example, we could employ a sum of squared errors as a criterion

$$\sum_{(x,y)\in\mathscr{D}} (y - f(x;\theta))^2,$$

and choose a parameter θ which minimizes this measure. We denote the optimal parameter chosen so as to minimize such a measure by $\widehat{\theta}$, i.e.

$$\widehat{\theta} = \arg\min_{\theta} \sum_{(x,y)\in\mathscr{D}} (y - f(x;\theta))^2.$$

Then, we can predict the corresponding output for a given input x by plugging the estimated parameter $\widehat{\theta}$ into the model as

$$\widehat{y} = f(x;\widehat{\theta}).$$

In many practical situations, several different candidates are available for modeling the data. In such cases, we would like to select a good model from those candidates, therefore the next problem we have to consider is how to measure the *goodness of model*. Our main aim in the regression problem is to predict the output for a given input as precise as possible. Hence, a natural measure to assess the goodness of model can be the accuracy of prediction.

We can use, for example, the maximum absolute prediction error over all the possible input–output pairs (X,Y)

$$\max_{X,Y} \left| Y - f(X;\widehat{\theta}) \right|,$$

or the average squared prediction error over input–output distribution $P(X,Y)$

$$E\left[(Y - f(X;\widehat{\theta}))^2\right].$$

In this chapter we use the latter measure because it suits the goodness-of-fit criterion, that is, the sum of squared errors.

According to the law of large numbers, it seems that the average prediction error can be well approximated by a sum of squared errors if the data set is sufficiently large,

$$E\left[(Y - f(X;\widehat{\theta}))^2\right] \simeq \frac{1}{n}\sum_{i=1}^{n}\left(y_i - f(x_i;\widehat{\theta})\right)^2.$$

But this is *NOT* true. As shown in Fig. 14.2, the more flexible model, which has a higher degree of freedom, has less prediction error for data $\mathscr{D}$ which are used for estimating the optimal parameter, but has more prediction error for unexperienced data. This is intuitively explained as follows. Since $\widehat{\theta}$ is a function of $\mathscr{D}$, it can be thought that $\widehat{\theta}$ memorizes information of $\mathscr{D}$. Also the more degrees of freedom a model has, the larger its capacity to memorize becomes. Consequently, the above measure directly based on data set $\mathscr{D}$ always overestimates more flexible models.

What we really want to evaluate is the performance of the model for unseen input–output relations (data points), so we could consider the following procedure. Let $\mathscr{D}_{-i}$ be a data set from which an example (x_i, y_i) is excluded

$$\mathscr{D}_{-i} = \mathscr{D} - (x_i, y_i),$$

and $\widehat{\theta}_{-i}$ be an estimated parameter based on $\mathscr{D}_{-i}$

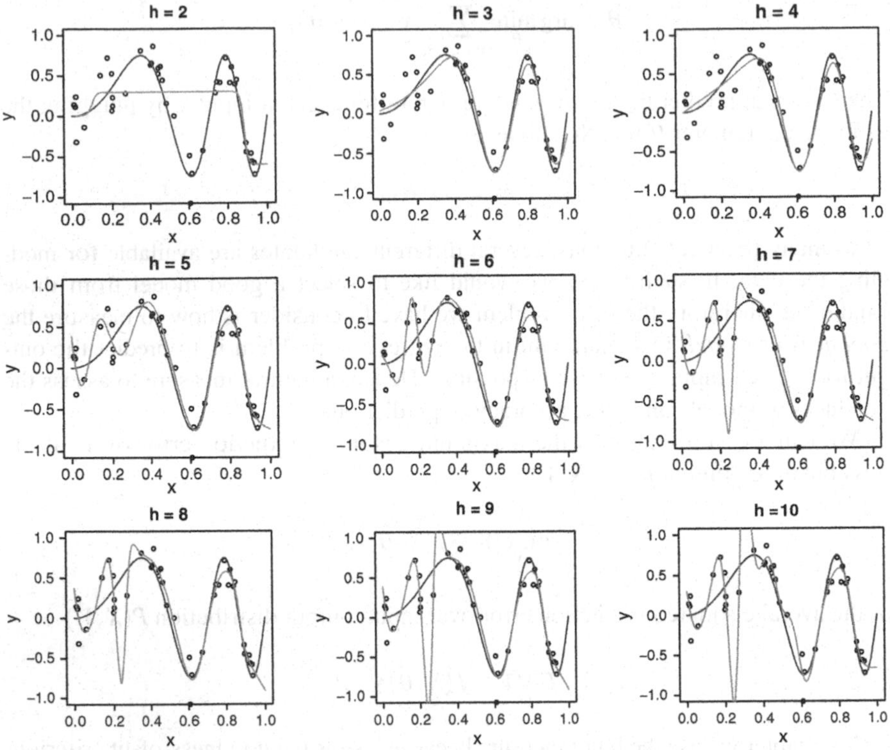

Fig. 14.2 Regression by different models. Fitting results shown in *solid curves* are produced by neural networks with different numbers of hidden units (*top rows*: $h = 2,3,4$, *middle rows*: $h = 5,6,7$, *bottom rows*: $h = 8,9,10$). The *dashed curve* is the conditional expectation of Y for given X. Flexible models, which have a larger number of hidden units, can fit given sample points quite well, but tend to loose generalization ability for predicting unknown input–output relation

$$\widehat{\theta}_{-i} = \arg\min_{\theta} \sum_{(x,y)\in\mathscr{D}_{-i}} (y - f(x;\theta))^2.$$

Hereafter, we call $\widehat{\theta}_{-i}$ a *leave-one-out* estimator. We note that if the number of examples is sufficiently large, $\widehat{\theta}_{-i}$ is close to $\widehat{\theta}$ enough. Also prediction error defined by

$$\left(y_i - f(x_i;\widehat{\theta}_{-i})\right)^2$$

can be a measure of prediction accuracy for an unknown input because (x_i, y_i) is not memorized in $\widehat{\theta}_{-i}$, but the quality strongly depends on (x_i, y_i). Thus, by averaging prediction error of leave-one-out estimates over all data $\mathscr{D}$, we expect to have a good evaluation of the prediction error of $\widehat{\theta}$,

$$E\left[\left(Y - f(X;\widehat{\theta})\right)^2\right] \simeq \frac{1}{n}\sum_{i=1}^{n}\left(y_i - f(x_i;\widehat{\theta}_{-i})\right)^2.$$

This procedure is called *leave-one-out cross-validation*. For a reasonably small data set, this procedure can be exactly carried out, and it is known that the evaluation works well. For a large data set, however, this is sometimes difficult to realize in practice because of the computational complexity of the problem.

To avoid such a difficulty, modified procedures of leave-one-out cross-validation, such as *k-fold cross-validation* [4, 11], are proposed. As yet another measure of the goodness of model, we hereinafter introduce a method of Akaike's information criterion (known as AIC) [1]. There are various points of view to characterize AIC, and we here derive it from leave-one-out cross-validation [23], and analyze its properties by means of statistical tools.

14.2 Mathematical Framework

We start by formulating our problem mathematically. Suppose we have a data set of n samples from a probability distribution $P(X,Y)$, which is denoted by

$$\mathscr{D} = \{(x_i, y_i);\ i = 1, 2, \ldots, n\}. \tag{14.1}$$

We assume $\{(x_i, y_i)\}$ are i.i.d. samples, that means samples are independently generated from an identical distribution $P(X,Y)$. Here we consider a regression problem that means given data are fitted by a parametric model as

$$y = f(x;\theta), \quad \theta \in \Theta \subset R^p \tag{14.2}$$

where θ is a p-dimensional parameter vector included in an open subset Θ in R^p. To measure the goodness of fit, we introduce a loss of the parameter θ at (x,y), which is denoted by

$$l(x,y;\theta). \tag{14.3}$$

As discussed in Sect. 14.1, a simple example of a loss is the squared difference between true output y and modeled output $f(x;\theta)$

$$l(x,y;\theta) = (y - f(x;\theta))^2. \tag{14.4}$$

This squared error can be generalized to an arbitrary power of s as

$$l(x,y;\theta) = |y - f(x;\theta)|^s. \tag{14.5}$$

Another example is a negative log loss. Suppose that output is generated through an additive noise model given by

$$y = f(x;\theta) + \varepsilon, \tag{14.6}$$

and the distribution of noise is given by a density function $q(\varepsilon)$. Then the conditional probability density of output y for given input x is written as

$$p(y|x,\theta) = q(y - f(x;\theta)). \tag{14.7}$$

In this way, f can be naturally connected with a probability distribution, and then a loss is given by

$$l(x,y;\theta) = -\log p(y|x,\theta). \tag{14.8}$$

A special case is given when the noise has the normal distribution with the density

$$q(\varepsilon) = \frac{1}{\sqrt{2\pi\sigma^2}} e^{-\frac{1}{2\sigma^2}\varepsilon^2}, \tag{14.9}$$

where σ^2 is the variance of noise. Then the negative log loss can be written as

$$l(x,y;\theta) = \frac{1}{2\sigma^2}(y - f(x;\theta))^2 + \frac{1}{2}\log(2\pi\sigma^2), \tag{14.10}$$

which is equivalent to the squared error except for constant factors.

The loss $l(x,y;\theta)$ is a point-wise measure for the goodness of fit. In order to evaluate total performance of the model $f(x;\theta)$, we consider an averaged version of loss defined by

$$L(\theta) = E\left[l(X,Y;\theta)\right], \tag{14.11}$$

where the expectation is taken over $P(X,Y)$ from which the data are generated. Then we can define the optimal parameter as the minimizer of the average loss $L(\theta)$ as

$$\theta^* = \arg\min_{\theta} L(\theta). \tag{14.12}$$

Note that the model established with θ^* is considered as optimal in terms of the mean performance over all possible inputs x. In other words, there can be other special, so to say, pathological parameters θ' which gives a more accurate prediction of the output for a specific input than θ^*.

In most of the practical cases, we do not know the distribution $P(X,Y)$ nor $L(\theta)$, and only data points $\mathscr{D}$ are available. Instead of taking an average over the true distribution, we may consider an average over an empirical distribution based on $\mathscr{D}$, that is written as

$$\bar{L}(\theta) = \frac{1}{n} \sum_{(x,y)\in\mathscr{D}} l(x,y;\theta), \tag{14.13}$$

where n is the number of samples in $\mathscr{D}$. When the number of samples is sufficiently large, we can expect that the empirical average loss $\bar{L}(\theta)$ is close enough to the true average loss $L(\theta)$, hence we define the best parameter based on $\mathscr{D}$ by

$$\widehat{\theta} = \arg\min_{\theta} \bar{L}(\theta), \tag{14.14}$$

as a pseudo-optimal estimate of θ^*.

In the next section, we see the validity of $\widehat{\theta}$ by investigating its asymptotic properties in detail.

14.3 Asymptotic Properties

Before describing detailed asymptotic properties of the estimator $\widehat{\theta}$, we summarize two basic theorems concerning sums of independent random variables [5, 12]. First, we define modes of convergence for a sequence of random variables $\{X_n\}$ to a limiting random variable X as follows.

Definition 14.1 (Convergence of random variables).

1. X_n converges to X almost surely (a.s.) if
$$P(\omega | \lim X_n(\omega) = X(\omega)) = 1. \tag{14.15}$$
This is denoted by $X_n \xrightarrow{\text{a.s.}} X$.
2. X_n converges in probability to X if for every $\varepsilon > 0$,
$$P(|X_n - X| > \varepsilon) \to 0 \text{ as } n \to \infty. \tag{14.16}$$
This is denoted by $X_n \xrightarrow{P} X$.
3. X_n converges in law to X if the cumulative distribution function of X_n converges to that of X
$$F_n(x) = P(X_n < x) \to F(x) = P(X < x) \tag{14.17}$$
at all points at which F is continuous. This is denoted by $X_n \xrightarrow{\mathscr{L}} X$, or $X_n \xrightarrow{\mathscr{L}} F$.

Note that all the definitions are given in one-dimensional form, but it can be naturally extended to multi-dimensional cases. Using these notations, two important theorems, the law of large numbers and the central limit theorem, are stated as follows.

Theorem 14.1 (Law of large numbers). *Let $X_1, X_2, \ldots$ be independent and identically distributed random variables. If $E[|X_1|] < \infty$, then*
$$\frac{X_1 + \cdots + X_n}{n} \xrightarrow{\text{a.s.}} E[X_1]. \tag{14.18}$$
If $E[|X_1|] = \infty$, then the above averages diverge almost everywhere.

Since convergence with probability 1 implies convergence in probability, we immediately have a weaken form as follows

Corollary 14.1. *Under the same assumption with Theorem 14.1, if $E[|X_1|] < \infty$, then*
$$\frac{X_1 + \cdots + X_n}{n} \xrightarrow{P} E[X_1]. \tag{14.19}$$

This is the weak law of large numbers, and in the following calculation, we mainly use this version.

Theorem 14.2 (Central limit theorem). *Let $X_1, X_2, \ldots$ be independent and identically distributed random variables with zero mean and finite covariance Σ. Then*

$$\frac{X_1 + \cdots + X_n}{\sqrt{n}} \xrightarrow{\mathscr{L}} \mathscr{N}(0, \Sigma). \tag{14.20}$$

Hereafter $\mathscr{N}(\mu, \Sigma)$ denotes the multivariate normal distribution with mean μ and covariance Σ.

From the law of large numbers, $\bar{L}(\theta)$ converges in probability to $L(\theta)$

$$\bar{L}(\theta) \xrightarrow{P} L(\theta), \tag{14.21}$$

that means, for any parameter $\theta \neq \theta^*$

$$P(\bar{L}(\theta^*) < \bar{L}(\theta)) \to 1 \text{ as } n \to \infty$$

from the definition of the optimal parameter θ^*. Hence, for sufficiently large number of n, we can expect that for any $\theta \neq \theta^*$,

$$\bar{L}(\theta^*) < \bar{L}(\theta)$$

holds with high probability, and $\theta \neq \theta^*$ is hardly chosen as a good parameter. This is formally stated as follows.

Theorem 14.3. *Under mild conditions for loss $l(x, y; \theta)$, estimator $\widehat{\theta}$ converges in probability to the optimal parameter θ^*,*

$$\widehat{\theta} \xrightarrow{P} \theta^*. \tag{14.22}$$

For a proof of this result, see, for example, [12] in which convergence of the maximum likelihood estimator (MLE) are explained in detail. The MLE is equivalent to an estimator derived from the negative log loss in our case, and its proof can be extended to general loss cases. Also, the conditions assumed to loss functions are discussed in [12]. Roughly speaking, loss $l(x, y; \theta)$ is required to be twice-differentiable and bounded with respect to parameter θ.

To describe the convergence of estimator $\widehat{\theta}$ in detail, we prepare some new notations. Let ∂_i be a partial differential operator with respect to the i-th element of parameter θ, that is,

$$\partial_i = \frac{\partial}{\partial \theta_i},$$

and ∇ be a vector of ∂_i's,

$$\nabla = (\partial_1, \ldots, \partial_p)^T.$$

For a function f of θ, ∇f is a p-dimensional vector defined as

$$\nabla f(\theta) = (\partial_1 f(\theta), \ldots, \partial_p f(\theta))^T.$$

In the same way, $\nabla\nabla f$ is a $p \times p$ matrix defined as

$$\nabla\nabla f(\theta) = \begin{pmatrix} \partial_1\partial_1 f(\theta) & \cdots & \partial_1\partial_p f(\theta) \\ \vdots & \ddots & \vdots \\ \partial_p\partial_1 f(\theta) & \cdots & \partial_p\partial_p f(\theta) \end{pmatrix}.$$

The next theorem states the convergence of estimator $\widehat{\theta}$ more precisely than Theorem 14.3.

Theorem 14.4. *Define*

$$G = E\left[\nabla l(X,Y;\theta^*)\nabla l(X,Y;\theta^*)^T\right], \tag{14.23}$$

$$Q = E\left[\nabla\nabla l(X,Y;\theta^*)\right], \tag{14.24}$$

and assume that G and Q are finite and Q is invertible. Then, estimator $\widehat{\theta}$ satisfies

$$\sqrt{n}\left(\widehat{\theta}-\theta^*\right) \xrightarrow{\mathscr{L}} \mathscr{N}\left(0, Q^{-1}GQ^{-1}\right). \tag{14.25}$$

Proof. According to Taylor's theorem, we can expand a differentiable function f of $\theta+\delta$ at θ as

$$f(\theta+\delta) = f(\theta) + \delta^T\nabla f(\theta+\alpha\delta) \quad (0<\alpha<1).$$

Knowing that

$$\nabla\bar{L}(\widehat{\theta}) = 0$$

holds because $\widehat{\theta}$ is the minimizer of $\bar{L}(\theta)$, and expanding $\nabla\bar{L}(\widehat{\theta})$ at θ^* by Taylor's theorem, we have

$$\nabla\bar{L}(\widehat{\theta}) = \nabla\bar{L}(\theta^*) + \nabla\nabla\bar{L}(\theta^* + \alpha(\widehat{\theta}-\theta^*))\left(\widehat{\theta}-\theta^*\right) = 0.$$

Then, assuming that $\nabla\nabla\bar{L}(\theta^*+\alpha(\widehat{\theta}-\theta^*))$ is invertible, the difference between $\widehat{\theta}$ and θ^* can be expressed as

$$\widehat{\theta}-\theta^* = -\nabla\nabla\bar{L}(\theta^*+\alpha(\widehat{\theta}-\theta^*))^{-1}\nabla\bar{L}(\theta^*). \tag{14.26}$$

From the convergence property of estimator $\widehat{\theta}$

$$\widehat{\theta} \xrightarrow{P} \theta^*,$$

and from the law of large numbers

$$\bar{L}(\theta) \xrightarrow{P} L(\theta),$$

we have

$$\nabla\nabla\bar{L}(\theta^*+\alpha(\widehat{\theta}-\theta^*)) \xrightarrow{P} \nabla\nabla L(\theta^*) = Q. \tag{14.27}$$

On the other hand, since θ^* is the minimizer of $L(\theta)$,

$$\nabla L(\theta^*) = E\left[\nabla l(X,Y;\theta^*)\right] = 0$$

holds, and by definition

$$E\left[\nabla l(X,Y;\theta^*)\nabla l(X,Y;\theta^*)^T\right] = G.$$

With the assumption that data $\mathscr{D}$ are i.i.d. samples, $\nabla l(x_i,y_i;\theta^*)$ are zero-mean i.i.d. random variables with finite covariance G, and then by the central limit theorem

$$\sqrt{n}\nabla \bar{L}(\theta^*) = \frac{1}{\sqrt{n}}\sum_{i=1}^{N}\nabla l(x_i,y_i;\theta^*)$$

converges in law to the normal distribution with covariance G,

$$\sqrt{n}\nabla \bar{L}(\theta^*) \xrightarrow{\mathscr{L}} \mathscr{N}(0,G). \tag{14.28}$$

From the above relations (14.26), (14.27) and (14.28), we obtain

$$\begin{aligned}\sqrt{n}\left(\widehat{\theta}-\theta^*\right) &= -\nabla\nabla\bar{L}(\theta^*+\alpha(\widehat{\theta}-\theta^*))^{-1}\sqrt{n}\nabla\bar{L}(\theta^*)\\ &\xrightarrow{\mathscr{L}} \mathscr{N}\left(0,Q^{-1}GQ^{-1}\right).\end{aligned} \tag{14.29}$$

□

Intuitively speaking, estimator $\widehat{\theta}$ differs depending on given data $\mathscr{D}$, and for sufficiently large data set $\mathscr{D}$, $\widehat{\theta}$ has the normal distribution,

$$\widehat{\theta} \sim \mathscr{N}\left(\theta^*,Q^{-1}GQ^{-1}/n\right).$$

Namely:

1. Estimator $\widehat{\theta}$ distributes around the optimal parameter θ^*.
2. The difference between $\widehat{\theta}$ and θ^* is of order $1/\sqrt{n}$ and

$$\widehat{\theta} \to \theta^* \quad \text{as} \quad n \to \infty.$$

The former property is called (asymptotic) unbiasedness and the latter is consistency, both of which are in general thought as desirable properties of estimators.

The average loss $L(\theta)$ and empirical average loss $\bar{L}(\theta)$ can be thought of as distances between the true distribution $P(X,Y)$ and the model $f(\theta)$, and between given data $\mathscr{D}$ and the model, respectively. Strictly speaking, they do not always fulfill all conditions of a distances, especially symmetry, but this point of view allows us to understand the relationship among $P(X,Y)$, $\mathscr{D}$, θ^* and $\widehat{\theta}$ geometrically. Such a geometrical image is illustrated in Fig. 14.3.

From the above results of Theorem 14.4 on asymptotic behaviors of estimates, we also have an asymptotic property of average loss $L(\theta)$ evaluated at pseudo-optimal parameter $\widehat{\theta}$ as follows.

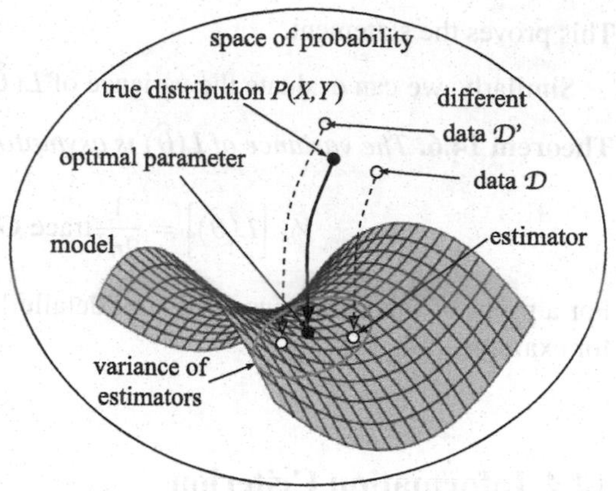

Fig. 14.3 Geometrical image of parameter estimation. Considering an empirical average over given data $\mathscr{D}$ such as (14.13), $\mathscr{D}$ can be regarded as a distribution, therefore data $\mathscr{D}$ and the true distribution $P(X,Y)$ are drawn in the same space. Different data sets give different estimates and estimator $\widehat{\theta}$ distributes around the optimal parameter θ^* depending on given data $\mathscr{D}$

Theorem 14.5. *The expectation of average loss $L(\widehat{\theta})$ with respect to data $\mathscr{D}$ is asymptotically*

$$E_{\mathscr{D}}\left[L(\widehat{\theta})\right] = L(\theta^*) + \frac{1}{2n}\text{trace}\, GQ^{-1}. \tag{14.30}$$

Proof. By expanding $L(\widehat{\theta})$ at θ^* up to the second derivative, we have

$$L(\widehat{\theta}) - L(\theta^*) = \frac{1}{2}(\widehat{\theta}-\theta^*)^T \nabla\nabla L(\theta^* + \alpha(\widehat{\theta}-\theta^*))(\widehat{\theta}-\theta^*),$$

where $0 < \alpha < 1$ and we use

$$\nabla L(\theta^*) = 0.$$

Using a well-known property of the trace operator

$$\text{trace}\, ABC = \text{trace}\, BCA = \text{trace}\, CAB,$$

and previous results

$$\nabla\nabla L(\theta^* + \alpha(\widehat{\theta}-\theta^*)) \xrightarrow{P} \nabla\nabla L(\theta^*) = Q,$$
$$E_{\mathscr{D}}\left[n(\widehat{\theta}-\theta^*)(\widehat{\theta}-\theta^*)^T\right] \longrightarrow Q^{-1}GQ^{-1} \text{ as } n \to \infty,$$

we have

$$\begin{aligned} &E_{\mathscr{D}}\left[2n\left(L(\widehat{\theta}) - L(\theta)\right)\right] \\ &= nE_{\mathscr{D}}\left[\left(\widehat{\theta}-\theta^*\right)^T \nabla\nabla L(\theta^* + \alpha(\widehat{\theta}-\theta^*))\left(\widehat{\theta}-\theta^*\right)\right] \\ &= E_{\mathscr{D}}\left[\text{trace}\left(\nabla\nabla L(\theta^* + \alpha(\widehat{\theta}-\theta^*)) \times n\left(\widehat{\theta}-\theta^*\right)\left(\widehat{\theta}-\theta^*\right)^T\right)\right] \\ &\longrightarrow \text{trace}\left(Q \times Q^{-1}GQ^{-1}\right) \\ &= \text{trace}\, GQ^{-1}. \end{aligned} \tag{14.31}$$

This proves the statement. □

Similarly, we can evaluate the variance of $L(\widehat{\theta})$ as follows.

Theorem 14.6. *The variance of* $L(\widehat{\theta})$ *is asymptotically*

$$V_{\mathscr{D}}\left[L(\widehat{\theta})\right] = \frac{1}{2n^2}\text{trace}\, GQ^{-1}GQ^{-1}. \tag{14.32}$$

For a proof of this result, we need a bit detailed and complicated calculations, see, for example, [14, 15].

14.4 Information Criterion

As discussed in Sect. 14.1, empirical average loss at estimator $\widehat{\theta}$

$$\bar{L}(\widehat{\theta}) = \frac{1}{n}\sum_{i=1}^{n} l(x_i, y_i; \widehat{\theta})$$

is not a good measure for evaluating the goodness of model, because estimator $\widehat{\theta}$ is a function of data $\mathscr{D} = \{(x_i, y_i); i = 1, \ldots, n\}$ and the above measure is apt to over-estimate more flexible models. From a geometrical point of view, empirical average loss $\bar{L}(\widehat{\theta})$ is the distance between given data $\mathscr{D}$ and the estimated model parameter $\widehat{\theta}$ as shown in Fig. 14.4. What we need to measure is, however, the distance between the true distribution $P(X,Y)$ and the estimated model parameter $\widehat{\theta}$, which is not accessible because the true distribution $P(X,Y)$ is usually unknown.

To assess the goodness of model, we introduce the following procedure. Let us define *leave-one-out* data by

$$\mathscr{D}_{-i} = \mathscr{D} - (x_i, y_i), \tag{14.33}$$

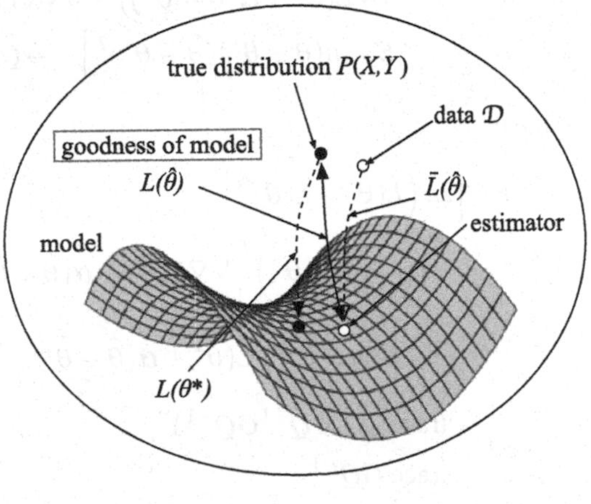

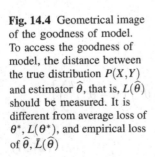

Fig. 14.4 Geometrical image of the goodness of model. To access the goodness of model, the distance between the true distribution $P(X,Y)$ and estimator $\widehat{\theta}$, that is, $L(\widehat{\theta})$ should be measured. It is different from average loss of θ^*, $L(\theta^*)$, and empirical loss of $\widehat{\theta}$, $\bar{L}(\widehat{\theta})$

where the sample (x_i, y_i) is excluded from the original data $\mathscr{D}$. Then, *leave-one-out* estimator is given by

$$\widehat{\theta}_{-i} = \arg\min_{\theta} \sum_{(x,y)\in\mathscr{D}_{-i}} l(x,y;\theta). \tag{14.34}$$

Since $\widehat{\theta}_{-i}$ does not convey any information of sample (x_i, y_i), $l(x_i, y_i; \widehat{\theta}_{-i})$ is expected to be an unbiased evaluation of prediction loss at (x_i, y_i). To evaluate total prediction loss of the model represented by θ, we take an average of point-wise prediction loss over all the data

$$L_{CV} = \frac{1}{n}\sum_{i=1}^{n} l(x_i, y_i; \widehat{\theta}_{-i}), \tag{14.35}$$

in anticipation of

$$L_{CV} \simeq E\left[l(X,Y;\widehat{\theta})\right].$$

This procedure is called *leave-one-out cross-validation.* A geometrical interpretation of cross-validation is depicted in Fig. 14.5.

For a small data set, i.e., reasonably small n, L_{CV} can be easily calculated. On the other hand, for a large data set, obtaining all the leave-one-out estimates is time-consuming, and sometimes the procedure is not practical. Therefore, the following asymptotically equivalent quantity is important [24].

Theorem 14.7. *Estimated average loss by leave-one-out cross-validation L_{CV} is asymptotically equivalent to*

$$L_{IC} = \bar{L}(\widehat{\theta}) + \frac{1}{n}\mathrm{trace}\, GQ^{-1}. \tag{14.36}$$

Proof. By expanding each term of L_{CV} at $\widehat{\theta}$, we have

$$l(x_i, y_i; \widehat{\theta}_{-i}) = l(x_i, y_i; \widehat{\theta}) + \left(\widehat{\theta}_{-i} - \widehat{\theta}\right)^T \nabla l(x_i, y_i; \widehat{\theta} + \alpha_i(\widehat{\theta}_{-i} - \widehat{\theta})),$$

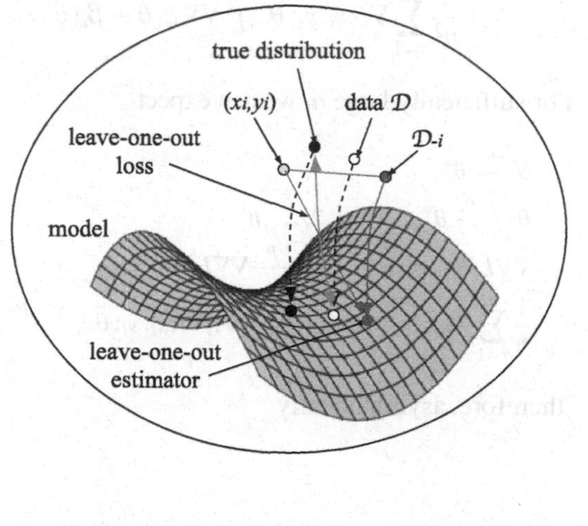

Fig. 14.5 Geometrical image of cross-validation. Data $\mathscr{D}$ is divided into $\mathscr{D}_{-i}$ and (x_i, y_i), and estimator $\widehat{\theta}_{-i}$ based on $\mathscr{D}_{-i}$ is evaluated with (x_i, y_i). This evaluation is repeated over all the data $\mathscr{D}$ and averaged

then L_{CV} is decomposed as

$$L_{CV} = \bar{L}(\widehat{\theta}) + \frac{1}{n}\sum_{i=1}^{n}\left(\widehat{\theta}_{-i} - \widehat{\theta}\right)^T \nabla l(x_i, y_i; \widehat{\theta} + \alpha_i(\widehat{\theta}_{-i} - \widehat{\theta})). \tag{14.37}$$

Knowing that $\widehat{\theta}$ is a solution of $\nabla \bar{L}(\theta) = 0$ because it is the minimizer of $\bar{L}(\theta)$, $\nabla \bar{L}(\widehat{\theta}_{-i})$ is expanded at $\widehat{\theta}$ as

$$\begin{aligned}\nabla \bar{L}(\widehat{\theta}_{-i}) &= \nabla \bar{L}(\widehat{\theta}) + \nabla\nabla \bar{L}(\widehat{\theta} + \beta_i(\widehat{\theta}_{-i} - \widehat{\theta}))\left(\widehat{\theta}_{-i} - \widehat{\theta}\right) \\ &= \nabla\nabla \bar{L}(\widehat{\theta} + \beta_i(\widehat{\theta}_{-i} - \widehat{\theta}))\left(\widehat{\theta}_{-i} - \widehat{\theta}\right).\end{aligned} \tag{14.38}$$

Since $\widehat{\theta}_{-i}$ is the minimizer of

$$\sum_{(x,y)\in\mathscr{D}_{-i}} l(x,y;\theta) = \sum_{(x,y)\in\mathscr{D}} l(x,y;\theta) - l(x_i,y_i;\theta),$$

$\widehat{\theta}_{-i}$ is a solution of $\nabla \bar{L}(\theta) - \nabla l(x_i, y_i; \theta)/n = 0$, that means,

$$\nabla \bar{L}(\widehat{\theta}_{-i}) = \frac{1}{n}\nabla l(x_i, y_i; \widehat{\theta}_{-i}). \tag{14.39}$$

Incorporating both relations (14.38) and (14.39), and assuming $\nabla\nabla \bar{L}$ is invertible, the difference between $\widehat{\theta}_{-i}$ and $\widehat{\theta}$ is written as

$$\begin{aligned}\widehat{\theta}_{-i} - \widehat{\theta} &= \nabla\nabla \bar{L}(\widehat{\theta} + \beta_i(\widehat{\theta}_{-i} - \widehat{\theta}))^{-1}\nabla \bar{L}(\widehat{\theta}_{-i}) \\ &= \frac{1}{n}\nabla\nabla \bar{L}(\widehat{\theta} + \beta_i(\widehat{\theta}_{-i} - \widehat{\theta}))^{-1}\nabla l(x_i, y_i; \widehat{\theta}_{-i}).\end{aligned} \tag{14.40}$$

Then L_{CV} is expressed as

$$\begin{aligned}L_{CV} &= \bar{L}(\widehat{\theta}) \\ &\quad + \frac{1}{n^2}\sum_{i=1}^{n}\nabla l(x_i, y_i; \widehat{\theta}_{-i})^T \nabla\nabla \bar{L}(\widehat{\theta} + \beta_i(\widehat{\theta}_{-i} - \widehat{\theta}))^{-1}\nabla l(x_i, y_i; \widehat{\theta} + \alpha_i(\widehat{\theta}_{-i} - \widehat{\theta})).\end{aligned}$$

For sufficiently large n, we can expect

$$\begin{aligned}&\widehat{\theta} \xrightarrow{P} \theta^*, \\ &\widehat{\theta}_{-i} \xrightarrow{P} \theta^* \quad (i = 1, 2, \ldots, n), \\ &\nabla\nabla \bar{L}(\widehat{\theta} + \beta_i(\widehat{\theta}_{-i} - \widehat{\theta})) \xrightarrow{P} \nabla\nabla L(\theta^*) = Q \quad (i = 1, 2, \ldots, n), \\ &\frac{1}{n}\sum_{i=1}^{n}\nabla l(x_i, y_i; \widehat{\theta} + \alpha_i(\widehat{\theta}_{-i} - \widehat{\theta}))\nabla l(x_i, y_i; \widehat{\theta}_{-i})^T \xrightarrow{P} E[\nabla l(X, Y; \theta^*)\nabla l(X, Y; \theta^*)^T] = G,\end{aligned}$$

therefore, asymptotically

$$L_{CV} \longrightarrow \bar{L}(\widehat{\theta}) + \frac{1}{n}\text{trace } GQ^{-1} = L_{IC}. \tag{14.41}$$

□

In practice, matrices G and Q are replaced by empirical estimates

$$\widehat{G} = \frac{1}{n}\sum_{i=1}^{n} \nabla l(x_i, y_i; \widehat{\theta}) \nabla l(x_i, y_i; \widehat{\theta})^T, \tag{14.42}$$

$$\widehat{Q} = \frac{1}{n}\sum_{i=1}^{n} \nabla\nabla l(x_i, y_i; \widehat{\theta}). \tag{14.43}$$

This kind of measures for model assessment L_{IC} is called *information criterion*, and many kinds of information criteria are proposed so far from various points of view [11, 17]. The most famous one is Akaike's information criterion (AIC) [1], and the above quantity L_{IC} is equivalent to AIC when the negative log loss

$$l(x, y; \theta) = -\log p(y|x, \theta)$$

is employed and the conditional distribution of y given x in $P(X,Y)$ has the density $p(y|x,\theta^*)$ for some unique θ^*. In other words, the probability model constructed from $f(x;\theta)$ is faithful and it includes the true distribution $P(X,Y)$. In this case, the well-known identity

$$G = E[\nabla \log p(y|x,\theta^*) \nabla \log p(y|x,\theta^*)^T] = E[-\nabla\nabla \log p(y|x,\theta^*)] = Q \tag{14.44}$$

holds, which can be easily checked from the identity

$$E[\nabla \log p(y|x,\theta^*)|x] = \int p(y|x,\theta^*) \nabla \log p(y|x,\theta^*)\, dy = 0.$$

Note that conditional expectation $E[\cdot|x]$ is taken over true distribution $p(y|x,\theta^*)$, for example,

$$E[f(Y)|x] = \int f(y) p(y|x,\theta^*)\, dy.$$

Therefore, the second term of L_{IC} is reduced to

$$\frac{1}{n}\text{trace } GQ^{-1} = \frac{1}{n}\text{trace } I_p = \frac{p}{n}, \tag{14.45}$$

where I_p is the $p \times p$ identity matrix.

To assess confidence of information criterion L_{IC}, the next theorem is useful.

Theorem 14.8. *The variance of L_{IC} is asymptotically*

$$V_{\mathscr{D}}[L_{IC}] = \frac{1}{n} V[l(X,Y;\theta^*)]. \tag{14.46}$$

Detailed proof is a bit complicated, so we give a brief one.

Proof. By Taylor's theorem up to the second derivative, we have

$$l(x,y;\widehat{\theta}) = l(x,y;\theta^*) + \left(\widehat{\theta} - \theta^*\right)^T \nabla l(x,y;\theta^*) + \frac{1}{2}\left(\widehat{\theta} - \theta^*\right)^T \nabla\nabla l(x,y;\theta^* + \alpha(\widehat{\theta} - \theta^*))\left(\widehat{\theta} - \theta^*\right).$$

Knowing that $\widehat{\theta} - \theta^*$, a function of $\mathscr{D}$, and $\nabla l(x_i,y_i;\theta^*)$ are correlated, and asymptotically

$$V_{\mathscr{D}}\left[\left(\widehat{\theta} - \theta^*\right)\right] = \frac{1}{n}Q^{-1}GQ^{-1},$$

$$V_{\mathscr{D}}\left[\frac{1}{n}\sum_{(x,y)\in\mathscr{D}} \nabla l(x,y;\theta^*)\right] = \frac{1}{n}G,$$

we have

$$E_{\mathscr{D}}\left[\bar{L}(\widehat{\theta}) - \bar{L}(\theta^*)\right] = O\left(\frac{1}{n}\right). \tag{14.47}$$

Similarly, we can evaluate the mean square of $\bar{L}(\widehat{\theta}) - \bar{L}(\theta^*)$ as

$$E_{\mathscr{D}}\left[\left(\bar{L}(\widehat{\theta}) - \bar{L}(\theta^*)\right)^2\right] = o\left(\frac{1}{n}\right). \tag{14.48}$$

From the above observations (14.47) and (14.48), a dominant order of variance is asymptotically evaluated as

$$\begin{aligned} V_{\mathscr{D}}\left[\bar{L}(\widehat{\theta})\right] &= V_{\mathscr{D}}\left[\bar{L}(\theta^*)\right] \\ &= V_{\mathscr{D}}\left[\frac{1}{n}\sum_{(X,Y)\in\mathscr{D}} l(X,Y;\theta^*)\right] \\ &= \frac{1}{n}V\left[l(X,Y;\theta^*)\right]. \end{aligned} \tag{14.49}$$

□

Actually, more detailed calculation considering higher order terms tells us an interesting fact that the expectation of empirical loss $\bar{L}(\widehat{\theta})$ is asymptotically

$$E_{\mathscr{D}}\left[\bar{L}(\widehat{\theta})\right] = L(\theta^*) - \frac{1}{2n}\operatorname{trace} GQ^{-1}. \tag{14.50}$$

This result shows the expectation of L_{IC} is consistent with the expectation of $L(\widehat{\theta})$, that is,

$$E_{\mathscr{D}}\left[L_{IC}\right] = L(\theta^*) - \frac{1}{2n}\operatorname{trace} GQ^{-1} + \frac{1}{n}\operatorname{trace} GQ^{-1} = E_{\mathscr{D}}\left[L(\widehat{\theta})\right],$$

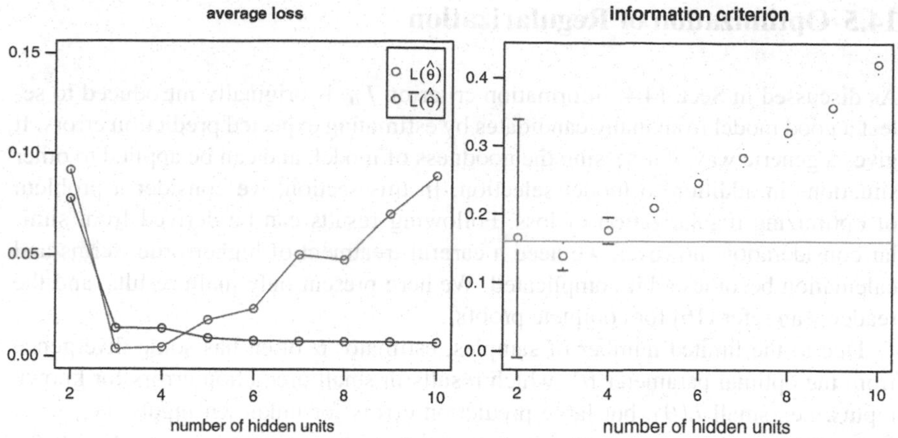

Fig. 14.6 Model selection based on information criterion. *Left*: Average loss $L(\widehat{\theta})$ and empirical average loss $\bar{L}(\widehat{\theta})$ are plotted versus h, the number of hidden units. *Right*: The information criteria L_{IC} are plotted including error bars. It can be seen that the behavior of $L(\theta)$ depending on h is well captured by L_{IC}. The *horizontal solid line* indicates one standard error of the minimum of L_{IC}

and this equivalence supports the validity of L_{IC}. However, we have to be careful with the variance of L_{IC}, which is empirically calculated by

$$\frac{1}{n}V\left[l(X,Y;\theta^*)\right] \simeq \frac{1}{n}\left[\frac{1}{n}\sum_{(x,y)\in\mathscr{D}} l(x,y;\widehat{\theta})^2 - \left(\frac{1}{n}\sum_{(x,y)\in\mathscr{D}} l(x,y;\widehat{\theta})\right)^2\right]$$

and which sometimes has a great influence on the result of model selection. For more detailed proof and discussion, see, for example, [14, 15].

In Fig. 14.6, we show an example of model selection by using the information criterion. Here we consider the regression problem shown in Fig. 14.1, the number of data is 40, and regression curves are modeled by neural networks (three-layered perceptron) with the squared error loss. Due to the non-linearity of neural networks, empirical average loss $\bar{L}(\theta)$ has a lot of local minima, and it is hard to obtain the fully optimal parameter in practice. Therefore, we use the sub-optimal parameter which achieves the lowest loss $\bar{L}(\widehat{\theta})$ among several estimates, each of which is trained from a randomly initialized parameter. In the left plot, average loss $L(\widehat{\theta})$ and empirical average loss $\bar{L}(\widehat{\theta})$ are plotted versus h, the number of hidden units. As h becomes larger, $\bar{L}(\widehat{\theta})$ almost monotonically decreases, while $L(\widehat{\theta})$ once decreases and then increases. This indicates that the over-fit phenomena occur for large h. In the right plot, information criterion L_{IC} for each h is plotted with an error bar (one standard deviation, i.e., $\sqrt{V_{\mathscr{D}}[L_{IC}]}$). We can pick the model which minimizes L_{IC} as the best, or based on *one-standard-error rule* [11], we can select the simplest model within one standard error of the minimum, which is indicated by a horizontal solid line in the plot. In this case, $h = 3$ is the best choice for both strategies.

14.5 Optimization of Regularization

As discussed in Sect. 14.4, information criterion L_{IC} is originally introduced to select a good model from many candidates by estimating expected prediction errors. It gives a generic way of assessing the goodness of model, and can be applied to other situations in addition to model selection. In this section, we consider a problem of optimizing regularization of loss. Following results can be derived from similar consideration, however, we need a careful treatment of higher order terms and calculation becomes a bit complicated. We here present only main results, and the readers can refer [16] for complete proofs.

Due to the limited number of samples, estimator $\widehat{\theta}$ often has some divergence from the optimal parameter θ^*, which results in small prediction errors for known inputs, i.e., small $\bar{L}(\widehat{\theta})$, but large prediction errors for unknown inputs, i.e., large $L(\widehat{\theta})$. This phenomenon is called *over-fitting*. In order to reduce average loss $L(\widehat{\theta})$, the regularization method [4, 11, 17] is widely used.

In the regularization method, we adopt a loss including a regularization term, which is defined by

$$\bar{L}_R(\theta) = \frac{1}{n}\left(\sum_{i=1}^{n} l(x_i, y_i; \theta) + \lambda r(\theta)\right), \tag{14.51}$$

where λ determines the strength of regularization. Typical examples of regularization terms are squared l_2-norm

$$r(\theta) = \|\theta\|_2^2 = \sum_{j=1}^{p} \theta_j^2,$$

and l_1-norm

$$r(\theta) = \|\theta\|_1 = \sum_{j=1}^{p} |\theta_j|.$$

A regularized estimator is defined as the minimizer of regularized loss $\bar{L}_R(\theta)$,

$$\tilde{\theta} = \arg\min_{\theta} \bar{L}_R(\theta). \tag{14.52}$$

By this additional regularization term, estimator $\widehat{\theta}$ is asymptotically modified as

$$\tilde{\theta} - \widehat{\theta} = -\frac{\lambda}{n} Q^{-1} \nabla r(\tilde{\theta}). \tag{14.53}$$

As shown in Sect. 14.3, estimator $\widehat{\theta}$ is asymptotically unbiased, but if we look at higher order terms closely, we can evaluate its asymptotic bias as follows.

Theorem 14.9. *The bias of estimator $\widehat{\theta}$ is asymptotically*

$$E_{\mathscr{D}}\left[\widehat{\theta}\right] = \theta^* + \frac{1}{n}\mathbf{b} + o\left(\frac{1}{n}\right), \tag{14.54}$$

where $\mathbf{b} = (b_1, \ldots, b_p)^T$ *is given by*

$$b_j = \sum_{k,l,m=1}^{p} q^{jk} q^{lm} \left(s_{klm} - \frac{1}{2} \sum_{l',m'=1}^{p} q^{l'm'} t_{kll'} g_{mm'} \right), \tag{14.55}$$

q^{jk} *and* g_{jk} *are the jk-element of* Q^{-1} *and* G*, respectively, and elements of* s *and* t *are given by*

$$s_{klm} = E\left[\partial_k \partial_l l(X,Y;\theta^*) \partial_m l(X,Y;\theta^*)\right], \tag{14.56}$$

$$t_{klm} = E\left[\partial_k \partial_l \partial_m l(X,Y;\theta^*)\right]. \tag{14.57}$$

The bias is thought as one of the sources of over-fits, and the regularization is expected to reduce the bias by modifying estimators as (14.53). Hence, considering average loss of $\tilde{\theta}$, $L(\tilde{\theta})$, the optimal strength of the regularization term is given as follows.

Theorem 14.10. *The optimal regularization strength* λ^**, which asymptotically minimizes average loss* $L(\tilde{\theta})$*, is given by*

$$\lambda^* = \frac{\mathbf{b}^T \nabla r(\theta^*) + \mathrm{trace}\, Q^{-1} G Q^{-1} \nabla\nabla r(\theta^*)}{\nabla r(\theta^*)^T Q^{-1} \nabla r(\theta^*)}. \tag{14.58}$$

In practice, it is not realistic to calculate **b** when the dimension of parameter p is large. To access the optimal strength with computationally feasible way, we propose the following procedure.

First, we define a new information criterion with a slight modification of information criterion L_{IC}.

$$L_{RIC} = \bar{L}(\tilde{\theta}) + \frac{1}{2n} \mathrm{trace}\, \tilde{G} \tilde{Q}^{-1}, \tag{14.59}$$

where $\tilde{G}$ and $\tilde{Q}$ are defined by

$$\tilde{G} = \frac{1}{n} \sum_{(x,y) \in \mathscr{D}} \left(\nabla l(x,y;\tilde{\theta}) + \frac{\lambda}{n} \nabla r(\tilde{\theta}) \right) \left(\nabla l(x,y;\tilde{\theta}) + \frac{\lambda}{n} \nabla r(\tilde{\theta}) \right)^T, \tag{14.60}$$

$$\tilde{Q} = \frac{1}{n} \sum_{(x,y) \in \mathscr{D}} \nabla\nabla l(x,y;\tilde{\theta}) + \frac{2\lambda}{n} \nabla\nabla r(\tilde{\theta}). \tag{14.61}$$

Then, the following theorem justifies the above information criterion.

Theorem 14.11. *The minimizer of* L_{RIC} *with respect to* λ *is asymptotically*

$$\tilde{\lambda} = \frac{\widehat{\mathbf{b}}^T \nabla r(\widehat{\theta}) + \mathrm{trace}\, \widehat{Q}^{-1} \widehat{G} \widehat{Q}^{-1} \nabla\nabla r(\widehat{\theta})}{\nabla r(\widehat{\theta})^T \widehat{Q}^{-1} \nabla r(\widehat{\theta})}, \tag{14.62}$$

where $\widehat{G}$ and $\widehat{Q}$ are the empirical version of G and Q, and $\widehat{\mathbf{b}}$ is defined in the same manner with **b** *by using $\widehat{\theta}$ instead of θ^*.*

Roughly speaking, only the half strength of correction term, trace GQ^{-1}, is used determining the regularization.

With this result, a computationally efficient algorithm is given as follows.

Corollary 14.2. *By simultaneously minimizing the following two losses*

$$\tilde{\theta} = \arg\min_{\theta} \sum_{(x,y)\in\mathscr{D}} l(x,y;\theta) + \tilde{\lambda} r(\theta), \tag{14.63}$$

$$\tilde{\lambda} = \arg\min_{\lambda} \sum_{(x,y)\in\mathscr{D}} l(x,y;\tilde{\theta}) + \frac{1}{2}\text{trace}\, \tilde{G}\tilde{Q}^{-1}, \tag{14.64}$$

we obtain estimator $\tilde{\theta}$ with optimally scaled $\tilde{\lambda}$.

In Fig. 14.7, we show an example of optimizing regularization strength by using the modified information criterion. A neural network with $h = 20$ hidden units, which is a too flexible model and shows over-fit without regularization, is examined for the same regression problem in Fig. 14.6. Here the l_2-norm regularization is used.

In the left plot, the average loss $L(\widehat{\theta})$ and the empirical average loss $\bar{L}(\widehat{\theta})$ are plotted versus λ, the strength of the regularization term. As λ increases, $\bar{L}(\widehat{\theta})$ increases, while $L(\widehat{\theta})$ takes the minimum at appropriately chosen λ. In the right plot, the modified information criterion L_{RIC} is plotted versus λ. We see that L_{RIC} becomes somewhat large value, but estimates a qualitative property of L, and L_{RIC} takes the minimum in the neighborhood of the optimal value of λ.

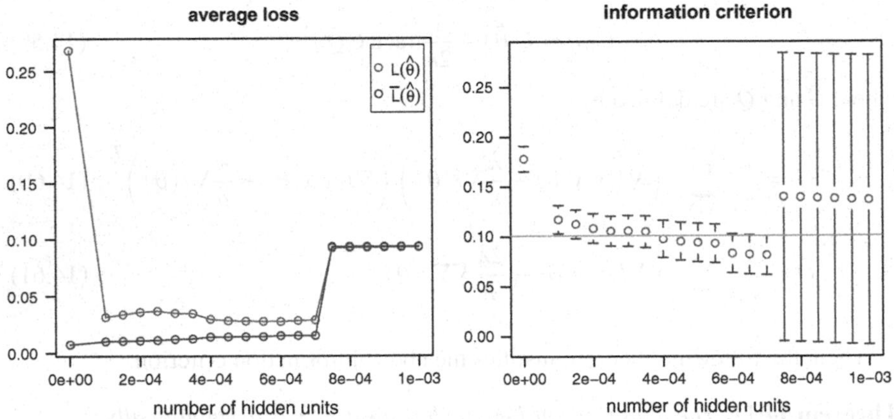

Fig. 14.7 Regularization optimization based on information criterion. In the *left plot*, average loss $L(\widehat{\theta})$ and empirical average loss $\bar{L}(\widehat{\theta})$ are plotted, and in the *right*, the modified information criteria L_{RIC} are plotted versus λ, the strength of the regularization term. A characteristic of L is appropriately estimated by L_{RIC}. The *horizontal solid line* indicates one standard error of the minimum of L_{RIC}

14.6 Conclusions

In this chapter, we have considered a problem of estimating model parameters from data, and by investigating statistical properties of estimated parameters, we have derived the information criteria for assessing the goodness of model and selecting a proper model from many candidates. In general, information criteria are applicable when a reasonably large number of data are available for estimating parameters because their derivation relies on the asymptotic theory. However, even when a small number of data are available, they give an important information of the relationship between the model complexity and the size of data. Here we made an argument in a quite general framework, but there are many variations of information criteria for dealing with specific cases. The reader can refer those works and other types of model selection methods in the literature [2, 3, 6–10, 18–22].

References

1. Akaike H (1974) A new look at the statistical model identification. IEEE Transactions on Automatic Control 19(6):716–723
2. Akaike H (1977) On entropy maximization principle. In: Krishnaia P R (ed) Application of Statistics, 27–41. North-Holland, Amsterdam
3. Allen D M (1974) The relationship between variable selection and data augmentation and a method for prediction. Technometrics 16(1):125–127
4. Bishop C M (2006) Pattern Recognition and Machine Learning. Springer, New York
5. Breiman L (1992) Probability. SIAM, Philadelphia
6. Cover T M, Thomas J A (1991) Elements of Information Theory. Wiley-Interscience, New York
7. Efron B (1979) Bootstrap methods: another look at the Jackknife. Annals of Statistics 7(1): 1–26
8. Efron B, Tibshirani R (1993) An Introduction to the Bootstrap. Chapman and Hall / CRC, Boca Raton
9. Golub G H, Heath M, Wahba G (1979) Generalized cross-validation as a method for choosing a good ridge parameter. Technometrics 21(2):215–223
10. Hannan E J, Quinn B G (1979) The determination of the order of an autoregression. Journal of the Royal Statistical Society, Series B 41(2):190–195
11. Hastie T, Tibshirani R, Friedman J (2001) The Elements of Statistical Learning. Springer, New York
12. Lehmann E L, Casella G (1998) Theory of Point Estimation. Springer, New York
13. MacKay D J C (2003) Information Theory, Inference and Learning Algorithms. Cambridge University Press, Cambridge
14. Murata N, Yoshizawa S, Amari S (1993) Learning curves, model selection and complexity of neural networks. In: Hanson S J, Cowan J D, Giles C L (eds) Advances in Neural Information Processing Systems, 5:607–614, Morgan Kaufmann, San Mateo
15. Murata N, Yoshizawa S, Amari S (1994) Network information criterion – determining the number of hidden units for an artificial neural network model. IEEE Transactions on Neural Networks 5(6):865–872
16. Park H, Murata N, Amari S (2004) Improving generalization performance of natural gradient learning using optimized regularization by NIC. Neural Computation 16(2):355–382

17. Ripley B D (1996) Pattern Recognition and Neural Networks. Cambridge University Press, Cambridge
18. Rissanen J (1978) Modeling by shortest data description. Automatica 14:465–471
19. Rissanen J (1986) Stochastic complexity and modelling. Annals of Statistics 14:1080–1100
20. Rissanen J (1987) Stochastic complexity (with disscussions). Journal of the Royal Statistical Society 49:223–239, 252–265
21. Schwarz G (1978) Estimating the dimension of a model. Annals of Statistics 6:461–464
22. Shibata R (1976) Selection of the order of an autoregressive model by Akaike's information criterion. Biometrika 63:117–126
23. Stone M (1974) Cross-validatory choice and assessment of statistical predictions. Journal of the Royal Statistical Society, Series B 36(2):111–147
24. Stone M (1977) An asymptotic equivalence of choice of model by cross-validation and Akaike's criterion. Journal of the Royal Statistical Society, Series B 39(1):44–47

Chapter 15
Extreme Physical Information as a Principle of Universal Stability

B. Roy Frieden

Abstract The EPI principle for finding a scientific law arises as follows. A coherent universe requires stable scientific laws. Each such law is a dual system AB consisting of a theoretical system A, in a state a, that interacts with a system B that may be an observer. Both A and B are assumed to be real systems. The interaction is via a probe particle, which carries information about a to B, and in doing so perturbs the total system AB in two ways – (1) It perturbs the information flow from A such that a is observed by B as $y = a + x$, where x is an unknown number obeying a frequency law $p(x)$. And (2) the law is likewise perturbed, as $\delta p(x)$. The law, a characteristic of the total system effect AB, is to be found. The above requirement of a stable law requires that the change $I - J$ in information about a from A to B remain invariant to the perturbations, i.e., $\delta(I-J)/\delta p(x) = 0$. This is mathematically equivalent to $I - J =$ *extremum*, and is the EPI principle that may be used to find $p(x)$. Interestingly, the extremum is usually *a minimum*, meaning that, despite the unknown perturbations x and $\delta p(x)$, the output information $I \simeq J$. That is, observation tends to agree with reality, as demanded of a coherent universe. Moreover, the entire observation-interaction procedure has physical reality, meaning that the output physical law is created 'on the spot.' In applications of EPI, information functional I is always of one known form (Fisher's). Also, the epistemic nature of EPI allows a degree of prior knowledge about a to be used to form $J(a)$. In descending order of accuracy in the resulting outputs $p(x)$, these forms of prior knowledge are called (a) abduction (highest quality, with perfect outputs), (b) deduction (next highest) and (c) induction (lowest, giving merely smooth outputs). Numerous applications of EPI are given.

B.R. Frieden
College of Optical Sciences, University of Arizona, Tucson, AZ 85721, USA
e-mail: roy.frieden@optics.Arizona.edu

F. Emmert-Streib, M. Dehmer (eds.), *Information Theory and Statistical Learning*,
DOI: 10.1007/978-0-387-84816-7_15,

15.1 Introduction

Our universe is *coherent*, in the sense that its systems obey *well-defined* laws. These *do not* change randomly during their interactions, even under conditions of maximally complex system structure (Sect. 15.4.7.2). The needed stability (Sect. 15.2) is provided by an abundant available level (15.35) of Fisher information. This *fixes* the probability laws governing all well-defined, i.e. statistically *repeatable*, systems, through a principle of Extreme *physical information* (EPI) [1–14]. These probability laws include those of science and engineering. Depending upon the level of prior knowledge at hand, the exact laws may be found, not merely smooth approximations.

EPI has been used, for example, to find the Lorentz transformation [6], Newton's second law $F = ma$ (Sect. 15.9 or [6]), the wave equations of quantum mechanics (Sect. 15.7 or [1, 6]), quantum gravity [6], classical electrodynamics [6] and classical gravitation [6] as well as the Higgs mass effect [6, 10]. Also found are *non*-quantum effects such as the laws of thermodynamics [1, 2, 6, 7], economic financial investment [7], and the laws governing system transport and population growth in both living and nonliving systems [3, 6]. This includes cancer growth both without [4] and with [8] programs of therapy. Table 15.1 summarizes many such applications. See also the detailed examples in Sects. 15.7–15.10.

EPI itself may be *derived,* in special scenarios of *unitarity*. This is where a unitary transform of any type (Fourier, Laplace, etc.) connects the probability amplitude law of data space with that of an *observable* transform space. The derivation of EPI for a *general* unitary transform is in [14], and for a particular unitary transform, the *Fourier* transform, is in [1, 6, 7].

The unitarity condition applies widely. Unitary spaces occur in quantum problems [1, 6, 7], as well as in non-physical scenarios such as investment economics [7]. Indeed, on the basis of such unitarity, EPI derives the wave equations mentioned above as well as the Wheeler–DeWitt equation of quantum gravity [6] and the Tobin financial investment equation ([7] or Sect. 15.8 of economics).

However, the EPI applications listed in paragraph two includes many scenarios (of mainly classical physics or biology) where unitarity does *not* hold. Hence, for these an entirely different derivation is needed. Such a derivation is given here. It holds for all science, and therefore must be based on a framework that supercedes any one. Hence we ask:

What do all laws of science have in common?

Also, can this unifying property be *quantified?*

15.2 Worldview: Stable Universe

We show next how EPI, a framework for estimating laws of science, arises naturally out of the *defining circumstances for* a law of science. Regard a system generically as either a theoretical or a material body that operates out of a consistent set of rules.

Table 15.1 Past applications of EPI

Source effect	Derived in Ref.	Unitary Trnsf.	J(a)	Type solution
Maxwell–Boltzmann law	[6] Chap. 7	Not used	$\int d\mathbf{x}\, x^2 \sum_n A_n q_n^2,\ A_n = const.$ ($\mathbf{x} \equiv c \cdot$ momenta)	EPI type (b) $\kappa = 1$
Quasi-incompressible turbulence	[6] Chap. 11	Not used	$\int\int d\mathbf{w}\, d\rho\, q^2[\lambda_1 w^2/(2\rho)$ $+H(\rho - \rho_1)(\lambda_2 + \lambda_3 \varepsilon(\rho))]$ ($\mathbf{w}=\rho\mathbf{v}$, $H =$ step funct., $\varepsilon(\rho) =$ internal energy)	EPI type (b)
EPR-spin entanglement (EPR-Bohm effect)	[6] Chap. 12	Not used	$\sum_{ab} \int dx\, q_{ab}^2$ with $x =$ angle betw 2 analyzer orientns **a**,**b**	EPI type (b)
Optimum investment schedule (Tobin q-theory of investment)	[7] Chap. 2	Fourier	$F(K(t)) =$ prodctn fcn of cap. K, $J = \int dt\, F(K(t)) \exp(-\rho t)$ $\equiv \langle F(K(t)) \rangle_t,\ t =$ time	EPI type (a)
(1) Price fluctuations in stocks, via "technical approach" (2) Fluctuations in extrinsic thermodyn. parameters	[5, 6] Chap. 13, [7] Chap. 2 [7] Chaps. 2, 4	Not used	$\sum_m \lambda_m \int dx\, f_m(x) q^2(x)$	EPI type (c)
Population growth	[3, 6] Chap. 14	Not used	$\sum_n p_n (g_n + d_n)^2$ ($g_n, d_n =$ growth, death comps. of fitness coeffs.)	EPI type (b) $\kappa = 1/2$
Molecular replication, RNA molecules	[6] Chap. 14	Not used	$\sum_n p_n \left(\frac{b_n A}{A_n + A} - d_n \right)^2$	EPI type (b) $\kappa = 1/2$
Cancer growth, untreated *in situ*	[6] Chap. 15	Not used	$\int dt\, (q^2/t^2)$	EPI type (b) $\kappa = 4.27$
Quarter-pwr laws biology	[9, 7] Chap. 8	Not used	$\sum_n A_{nj} \cos(4\pi n a),\ j = 0, 1, 2, ..$	EPI type (b)

Consider, first, a real system A that obeys an unknown theory. It is specified by a parameter a. (Example: The system is a real particle moving under the influence of a spring fastened at a point a. The particle moves according to an unknown law whose form is sought.) The parameter a identifies a particular property or state of A.

15.2.1 System Unobserved

Temporarily suppose that the system A remains unobserved. In Kantian terms, the theory behind A is then an unknown and in fact does not yet have definite reality. This observer- (or interaction-) dependent view of physics runs counter to the classical view, whereby the observer merely views passively, as when watching a goldfish in a bowl. This classical view of a passive observer was also that of Einstein – relativity notwithstanding – and essentially why he could never accept the participatory nature [15] of modern quantum theory; also see below.

15.2.2 System Observed

A real system *interacts with* some other system. The interaction can, in particular, be via an *observing* system. This is equivalently Popper's famous criterion [16] for a mathematical statement to be regarded as a scientific law. Here the mathematical statement is that of the unknown law. Hence, the law must be capable of verification or negation *under observation.* For generality we use the word "observation" to include all possible types of real *interaction*. For a law to be real it must at some point interact with some other system law. For brevity, from this point on by system A we mean the unknown law of A that we seek.

Hence, let some aspect of system A be observed by a system B. That aspect must be a, since A is characterized by a. Denote the combined system as AB. This describes both the intrinsic scientific effect A *and* its observation by, or interaction with, B. We assume, a la Popper, that the observation *validates* the existence of the system AB (particle, spring, observing instrument), i.e. the data are consistent with its theory.

Now, in order to observe A, a *probe particle*, such as a photon in a microscope or telescope view, must illuminate A and then travel to the observer (system B). (It is immaterial whether the probe particle was initiated by the observer, as in the microscope case, or by some other particle source, as in a telescope view.) The probe affects both systems A and B, in fact *perturbing both* (possibly through recoil collision with the source). The result is that the ideal parameter value a of A is observed by B as an imperfect datum y, where $y = a + x$ and x obeys some frequency of occurrence law $p(x)$. This law is likewise perturbed. The observer is ignorant of both the particular value of x and the law $p(x)$. Consider x in particular.

15.2.2.1 Nature of x

By its preceding definition, $x \equiv y - a$ represents an error from groundtruth a. In general, x is regarded as any, usually real, number that is unknown to the observer. It is merely *an expression of ignorance* on the part of the observer as to the effect of the perturbation on the net observation. This ignorance can be fundamental – that is, x is (1) intrinsically *random*, as presumably in quantum mechanics, or (2) merely practical – e.g., random only because the observer is using nonideal detection equipment. Or, x is actually deterministic, changing from one reading to the next via a definite trajectory $x(t)$, $t =$ time (for example), but the observer does not know this microlevel truth, and so is content to know merely a 'binned' curve $p(x)$ of the $x(t)$ values over time. (Example: If $x(t) = \cos(t)$, then $p(x) = 1/\left(\pi\sqrt{1-x^2}\right)$). Einstein famously regarded quantum mechanics as being an effect of the latter type.

Thus, in general, x obeys some definite, but unknown, *frequency of occurrence* law $p(x)$. In cases where x is later found to be random, $p(x)$ is as well a *probability* density law. However, since x is *effectively* random, in all the following we treat $p(x)$ *as if it were a probability law*. This simplifies the language.

Note that by this approach no prior physical cause of randomness – such as the second law of thermodynamics or quantum uncertainty – need be invoked. Also, in agreement with the orientations of Einstein and Bohm, in the quantum mechanics that EPI derives the quantum particle can therefore still obey a *classical trajectory on the microlevel*! That is, the 'noise' x can obey some unknown *deterministic* relation $x = x(t)$. Indeed, EPI can often find such relations (e.g., the quarter-power laws of biology, in Sect. 15.5.5.4 or [7, 9]). This is further discussed below (point (5) in Sect. 15.11).

15.2.2.2 Creative Effect of the Perturbation

As we noted, along with perturbation x, the unknown law $p(x)$ – a property of system AB – is likewise perturbed by the probe particle as it travels from A to B. Like x, this second perturbation may be of any size. Also, like x, it does not have to be intrinsically random (see preceding). This second perturbation is a helpful or creative one, since it leads directly to the EPI variational principle, which allows $p(x)$ to be found. Note that since $p(x)$ defines the *total* system effect AB, in Kantian terms it defines the *phenomenon* and not the pure noumenon A (see above). This is the law that the scientist has access to and is, therefore, meaningful.

15.2.2.3 Does Science Exist Independent of the Observer?

Note that in representing AB, $p(x)$ also defines a natural law as having a combined theoretical-observational nature. If the scenario is physical, this is an example of the famous *participatory nature of physics* [15]. Here it generalizes to a *participatory nature of science*.

We found above that the participation by the observer results in perturbing $p(x)$. We will find below that perturbing $p(x)$ leads to the EPI principle, and that EPI implies the laws of science. Thus, the participatory nature of science is not merely a *property of* the laws of science, as presumed in [15] – it also *implies them*.

15.2.3 Information Channel and Variational Principle

The *datum* y from A provides *information* I, in some sense, about the size of a. As we saw, this information $I(a)$ is carried to observer B by the probe particle. Likewise, let $J(a)$ represent the *intrinsic* or *source* level of information about a. It is therefore intrinsic to A. The law of science, then, obeys a general information channel

$$J \underset{x}{\rightarrow} I \tag{15.1}$$

describing the total system AB. Since $p(x)$ represents the system AB, the informations J and I both *depend* upon $p(x)$.

The channel (15.1) indicates that $I(a) - J(a) \equiv K(a)$ represents the change in information about a after the observation. Therefore, $K(a)$ measures the net observed effect (both intrinsic and observed) *on the basis of information transfer* from A to B. Also, ignorance can only reduce information: Since the mean-square error $e^2_{\rm min}$ over observations y goes inversely with information level I (15.10), it is generally nonzero, so that B tends to receive less information I about a than exists in the source. Thus, $I(a) \leq J(a)$, so that $K(a)$ will always be negative or zero (see also 15.11). Since systems A and B, and the law $p(x)$, are all perturbed (by the probe) during the observing process, *so must therefore be the informations I and J* associated with A and B, and consequently their difference $K(a)$. Call these perturbations $\delta p(x)$, δI, δJ and δK. As at the outset, we require the net observed effect $K(a)$ to be stable during its perturbation $\delta K(a)$ with respect to a perturbation $\delta p(x)$,

$$\frac{\delta K(a)}{\delta p(x)} = 0. \tag{15.2}$$

Note that δ is here assumed to be infinitesimal since, for instead a perturbation of *general* size, the solution would depend upon the size of the perturbation and, therefore, be nonunique. As was mentioned at the outset, condition (15.2) should hold even for a system *AB* with *maximum complexity* due to being in a state of maximum Fisher information. This will lead to an information (15.21) ideally characterizing a highly structured system.

Notice that this approach treats the system *AB* as a *knowledge channel,* whose parameter a and law $p(x)$ are sought. Accordingly, the law of science $p(x)$ that so emerges is regarded as *epistemic* in nature, i.e. something to be learned out of prior knowledge and observation. This later allows us to estimate information J below from a limited class of prior knowledge types.

15.2.4 Elementary Example: What Measurement Spaces are Allowed by EPI?

Consider finding the law governing classical particle motion. The system A is the particle, as specified by its ideal position a. To know a, an observation $y = a + x$ is made. Thus, the *data space* is *position* space. (Note that it could not instead be momentum μ – energy E space, which is conjugate to position space, because EPI only permits data spaces of a priori independent degrees of freedom *dof*s. But momentum and energy are not independent *dof*s, instead obeying the relation $E^2 = c^2\mu^2 + m^2c^4$. See Table 1.) Thus the data information $I(a)$ for this problem characterizes position space. In this scenario principle (15.2) gives Newton's second law (15.48) as the scientific law obeyed by *AB*.

Equations (15.1) and (15.2) state unequivocally that the universe is information-oriented, irrespective of the details of its physics. Specifically what *kind* of information is it? A clue is that the space B is the space of the datum, and a datum, by definition, describes a *local* measurement. We can therefore expect *the information to be local*, that is, sensitive to the local x–behavior of $p(x)$.

15.3 Fisher Information

Consider an observation of an unknown a. (As mentioned above, the observation can alternatively be mere interaction between nonliving systems.) The information we seek describes the estimation of a. The system A is observed as N data $\mathbf{y} \equiv y_1, \ldots, y_N$, a generalization of the single-datum situation in Sect. 15.2. The data $\mathbf{y}$ depend on both a and error components $\mathbf{x} \equiv x_1, \ldots x_N$. As in the preceding section, these do not have to be random, but may be regarded so as an expression of ignorance. They could, for example, be actually changing sinusoidally with the time, although the observer does not know this.

Fisher information will result from this derivation even in this general scenario of ignorance of $\mathbf{x}$. (Note that a problem of estimating a still exists, because *how* the $\mathbf{x}$ vary with data set is still regarded as unknown.)

Therefore the data $\mathbf{y}$ unpredictably vary from one set of observations to another, according to some relation $\mathbf{y} \equiv \mathbf{y}(a, \mathbf{x})$ such as $\mathbf{y} = a + \mathbf{x}$ (linear, added case). Then for any fixed a there are many possible sets of data $\mathbf{y}$. Again, these do not have to be randomly changing with each new set of data. However, this unknown $\mathbf{y}$ behavior in the presence of a definite a defines a system likelihood probability law $p(\mathbf{y}|a)$. This describes the behavior of the combined system AB, and is therefore by definition the scientific law that is sought. (This is a generalization of Sect. 15.2, where the simpler problem of finding $p(x)$, x scalar, was considered.) However, it will be useful to first consider how to estimate a.

The estimate will be fashioned from the known data $\mathbf{y}$. If the observer knew as well the prior probability law $p(a)$ for a, that would also help. However, most often this law is lacking. Denote the estimate of a as $\widehat{a}$. Hence, the observer forms a function $\widehat{a}(\mathbf{y})$ purely out of the data, called the "estimator function", which he hopes is a good approximation to a. Suppose he can form estimator functions that are *correct on average*, i.e., that obey $< \widehat{a}(\mathbf{y}) > a$ or, equivalently,

$$< (\widehat{a}(\mathbf{y}) - a) > \equiv \int d\mathbf{y} \, (\widehat{a}(\mathbf{y}) - a) \, p(\mathbf{y}|a) = 0. \tag{15.3}$$

The average is over all possible sets of data $\mathbf{y}$ in the presence of a. Note that the integral exists even if the data sets $\mathbf{y}$ change deterministically. Estimators $\widehat{a}(\mathbf{y})$ obeying (15.3) are called "unbiased," in analogy with "unbiased experiments" of physics. Note that this unbiasedness condition is actually quite weak, since it does not restrict the individual errors in the estimates $\widehat{a}(\mathbf{y})$: they can be quite large. Amazingly, condition (15.3) will *uniquely* lead directly to the information that is needed, Fisher's. (Note that the information even limits the output variability of $\widehat{a}$ in the more general scenario of *biased* estimators [17]). It also will derive the *mean-squared error* as the particular error measure that is meaningful to the problem.

The first step is to differentiate the integral (15.3) $\partial/\partial a$, directly giving

$$\int d\mathbf{y}\,(\widehat{a}(\mathbf{y})-a)\,\partial p/\partial a - \int d\mathbf{y}p = 0, \quad p \equiv p(\mathbf{y}|a). \tag{15.4}$$

Using the identity $\partial p/\partial a = p\,\partial \ln p/\partial a$ and the normalization property $\int d\mathbf{y}p = 1$ in (15.4) gives

$$\int d\mathbf{y}\,(\widehat{a}(\mathbf{y})-a)\,(\partial \ln p/\partial a)\,p = 1.$$

Reversing the equality and factoring the integrand appropriately gives

$$1 = \int d\mathbf{y}\,[(\widehat{a}(\mathbf{y})-a)\sqrt{p}\,]\,[(\partial \ln p/\partial a)\sqrt{p}\,]\,. \tag{15.5}$$

The $\sqrt{p}$ item in each factor is key, as it turns out. Next, we square (15.5) and use the Schwarz inequality. This gives

$$1^2 = 1 \leq \int d\mathbf{y}\,[(\widehat{a}(\mathbf{y})-a)\sqrt{p}\,]^2 \int d\mathbf{y}\left[\frac{\partial \ln p}{\partial a}\sqrt{p}\right]^2.$$

Squaring out the integrands,

$$1 \leq \int d\mathbf{y}\left[(\widehat{a}(\mathbf{y})-a)^2\,p\right] \int d\mathbf{y}\left[\left(\frac{\partial \ln p}{\partial a}\right)^2 p\right]. \tag{15.6}$$

Each of these integrals is now an expectation. The first is

$$\int d\mathbf{y}\left[(\widehat{a}(\mathbf{y})-a)^2\,p\right] \equiv \left\langle(\widehat{a}(\mathbf{y})-a)^2\right\rangle \equiv e^2, \tag{15.7}$$

the mean-squared error in the estimate. The second is

$$\int d\mathbf{y}\left[\left(\frac{\partial \ln p}{\partial a}\right)^2 p\right] \equiv \left\langle\left(\frac{\partial \ln p}{\partial a}\right)^2\right\rangle \equiv I \equiv I(a), \text{ where } p \equiv p(\mathbf{y}|a). \tag{15.8}$$

Quantity I is called the *Fisher information*. Mathematically, it is a *functional*, that is, an integral whose integrand involves a function $p(\mathbf{y}|a)$. Equation (15.8) provides a vital connection between the scientific law $p(\mathbf{y}|a)$ and I that will allow the law to be estimated via principle (15.2).

By the $\partial/\partial a$ operation in (15.8), I effectively measures the gradient content of $p(\mathbf{y}|a)$. Thus, the slower p changes with the value of a the lower is the net information value $I(a)$. This holds to the limit, so that if $p(\mathbf{y}|a) = p(\mathbf{y})$, which is independent of a, then $I(a) = 0$. This makes intuitive sense: If the data are independent of the unknown parameter they certainly carry no information about it.

The use of (15.7) and (15.8) in inequality (15.6) gives the famous "Cramer–Rao inequality,"

$$e^2 I \geq 1. \tag{15.9}$$

This shows that the mean-squared error and the Fisher information obey a complementary relation – if one is small the other tends to be large. More precisely, the *minimum possible* mean-squared error obeys

$$e^2_{\text{min}} = 1/I. \tag{15.10}$$

This indicates that when the information level is high the error is low, and vice-versa. Or, I is a measure of the quality of the estimate. The higher I is the greater is the quality.

The C–R inequality (15.9) has many applications to scientific laws in and of itself. In the special case of intrinsically random (now) errors $\mathbf{x}$ it has been used to derive the *Heisenberg* uncertainty principle [1, 6], uncertainty principles of *population biology* [3, 7], *cancer growth* [4, 7, 8], and *economics* [7], and a decision rule on *predicting population cataclysms* [7]. It is straightforward to show [17, 18] that the minimum error (15.10) can be attained by an estimator $\widehat{a}(\mathbf{y})$ that is the "maximum-likelihood" estimator. By definition, this satisfies a condition $\partial p(\mathbf{y}|a)/\partial a = 0$ with $a \equiv \widehat{a}$.

We now turn to the main problem, to estimate $p(\mathbf{y}|a)$. As noted above, the guiding principle is that $p(\mathbf{y}|a)$ should obey stability (15.2).

15.4 Extreme Physical Information

As we saw at (15.1), the observation process defines an information channel $J \to I$. The amount that pre-exists at, or is 'bound to,' the source effect is denoted as the amount $J \equiv J(a)$. And the amount that is received by the observer is denoted as the amount $I \equiv I(a)$. This is also called the "data information." In effect, J acts as a *source, or reservoir, of information* for the sink of information I in the data. Does this limit the size of I in some way?

15.4.1 EPI Zero-Principle

The information channel (15.1) is a closed system, and hence precludes any further inputs of information into I. Also, the fluctuations x, as effectively *random* inputs to I, can only *subtract* information from the source level J. This is a purely mathematical property of Fisher information I; for example for a normal law $I = 1/\sigma^2$ [6, 14] showing that the more random the system is, the lower is I.

Thus, the observer of I cannot obtain more information than pre-exists at the source, $I \leq J$. Another way of saying this is

$$I - \kappa J = 0, \quad 0 \leq \kappa \leq 1 \tag{15.11}$$

for some constant κ of information efficiency. Equation (15.11) is called the *EPI zero principle*. It can be used *by itself* to find certain probability laws (e.g. in cases of (a) abduction defined below). Examples [6] are the Dirac wave equation and the classical virial theorem. It is also used to form self-consistent EPI solutions ([6] or below) whereby both κ and functional J *are solved fo*r by the use of principles (15.11) and (15.12) below.

The value of κ varies with level of prior knowledge of the source effect. For example, observing an electromagnetic effect on the *classical level* ignores a priori its quantum aspect and, as it turns out (see Table 1), half the existing information, so that $I = J/2$ [6]. In other words, exactly half the source information is wasted when viewing an electromagnetic effect macroscopically.

15.4.2 EPI Variational Principle

Equation (15.2), with $K \equiv I - J$, is directly $\delta(I - J) = 0$. But, if the perturbation of a quantity is zero, that quantity must be at an extreme value,

$$I - J = extremum. \tag{15.12}$$

This is the *EPI variational principle*. It states that the transferred amount of information about a should be stable to perturbation. Note that (15.12) holds whether the scenario is unitary or not (it has *not* been assumed in the development (15.1)–(15.12)). The extremum can be a minimum or a point of inflection, although never a maximum (see Sect.15.4.3). The extremum is to hold even when the system has maximum complexity (Sect. 15.1, 15.2.3, and end of 15.4.7), thereby obeying *strong stability*.

EPI is an evolving approach to science. Previous derivation [1, 6] of the variational principle (15.12) in the absence of unitarity used *the axiom* that $\delta I = \delta J$, which implies that $\delta(I - J) = 0$ and hence gives (15.12) again. *Our approach avoids the need for this axiom and, hence, is a stronger one.*

15.4.3 Nature of Extremum

A particular case of interest is where the extremum in (15.12) is a minimum,

$$I - J = minimum. \tag{15.13}$$

This is in cases where the effect is described by a real coordinate, such as position x in the preceding. By contrast, an imaginary coordinate, such as the famous time coordinate ict in relativity theory, can give rise to a maximum. Assertion (15.13) is confirmed next.

First assume that the coordinates $\mathbf{x},\mathbf{y}$, a of the problem are all real (usual case). For simplicity, we evaluate functional I in a usual case of interest, that of a one dimensional, shift invariant system $\mathbf{y}=y$, $\mathbf{x}=x$ with $p(\mathbf{y}|a) \equiv p(y|a) \equiv p(y-a) = p(x)$, where $x = y - a$. Using this in the defining form (15.8) of I directly gives

$$I = \int dx \frac{1}{p}\left(\frac{dp}{dx}\right)^2, \quad p \equiv p(x). \tag{15.14}$$

We next examine the nature of the resulting extremum. Putting (15.14) into (15.12) and assuming a general form for J

$$J = \int dx j(p,x) \tag{15.15}$$

where the information density j has no explicit dependence upon $dp/dx = p'$, principle (15.12) becomes

$$\int dx \mathscr{L}[p,p',x] = extremum, \quad \mathscr{L} \equiv \frac{1}{p} p'^{\,2} - j(p,x). \tag{15.16}$$

The calculus of variations immediately gives the solution, as

$$\frac{d}{dx}\left(\frac{\partial \mathscr{L}}{\partial p'}\right) - \frac{\partial \mathscr{L}}{\partial p} = 0. \tag{15.17}$$

This is the famous Euler–Lagrange equation. Also, *Legendre's condition of necessity* for the extremum being a *minimum* is that

$$\frac{\partial^2 \mathscr{L}}{\partial p'^{\,2}} > 0 \tag{15.18}$$

(or, conversely, is a maximum if negative). In fact by the second equation (15.16), $\partial \mathscr{L}/\partial p' = 2p'/p$, so that

$$\frac{\partial^2 \mathscr{L}}{\partial p'^{\,2}} = \frac{2}{p} > 0 \tag{15.19}$$

by positivity of any probability p. Note that j not depending upon p' was crucial here. In summary, *the extremum is often a minimum for real system coordinates* and never a maximum.

Thus, (15.12) usually describes a scenario of *minimum* loss of information, that is $I \to J$ in numerical value. Recall that this follows from perturbation of the object and the observing system by the probe particle. Therefore, remarkably, this double perturbation does not simply degrade. True, it caused the existence of the frequency law $p(x)$ that degrades the observation of parameter a. But it evidently also allows, in the face of this degradation, output information approaching *the ideal level* of the source effect. Thus the perturbing effect has both good and bad aspects on the acquired information. In fact it represents *both* players in a 'knowledge game' described below.

We call (15.12) a principle of *Extreme physical information* (EPI), where the *physical information* K is the difference $I - J$. Principle (15.12) is always supplemented by the *zero principle* (15.11). In the opposite case of a purely imaginary coordinate, the same analysis [6] shows that I is now negative, as is J by (15.11), so that $I - J$ now tends toward the negative of a minimum, i.e. a maximum. The nature of the coordinate as to being real or imaginary is a further form of prior knowledge [6].

15.4.4 q-Form of I

The division by p in (15.14) gives an apparent pole at points x where $p \to 0$. However, the pole is only an illusion, which is clarified by working with a probability 'amplitude' q instead, defined as $p(x) = q^2(x)$. Plugging this into (15.14) directly gives the *q-form* of I,

$$I = 4 \int dx q'^2, \; q' \equiv dq/dx. \tag{15.20}$$

The pole in question now no longer exists. However, certain functions $q(x)$ have infinite slopes at isolated points x. To keep I well defined, a *principal value integral* is taken around these points, unless they are physically allowed by the system obeying $q(x)$ [6].

Finally, the elegant form (15.20) is merely a sum of squares, often called an L^2 norm. The analogous L^2 result holds in multidimensional, multicomponent cases $q_n(\mathbf{x})$, $n = 1,...,N$ as well. The L^2 nature of (15.20) is very significant, as taken up below.

15.4.5 Local Information Property

Recall our search for a "local" information measure. In fact (15.14) shows that I depends upon local slope values of the law $p(x)$. Thus, any one large slope value can dominate the answer I regardless of the rest of the curve. This makes I *a local information measure* (The Shannon entropy measure $-\int dx p(x) \ln p(x)$ is, by comparison, well known to be *global* [1, 6, 7].)

15.4.6 Multidimensional Systems

Many systems have multiple *dofs*: they are multidimensional in coordinates $\mathbf{x} \equiv x_1,...x_M$, and have many *complex* components $\psi_n(\mathbf{x})$, $n = 1,...,N/2$. The amplitudes ψ_n are constructed out of N purely real amplitudes q_n via (15.25) below.

As discussed below (15.12), these component amplitudes are regarded as independent, in the sense of maximally adding to information I, so that the principle (15.12) attains a 'strong' extremum and consequently the system obeys strong stability.

15.4.7 Fisher Information Capacity

Consider a system whose component amplitudes $q_n(\mathbf{x})$, $n = 1,...,N$ *do not have overlapping support regions* (for example, N spatially separated triangle functions). The individual contributions (15.20) from each component then *maximally adds* [1, 6] giving a net information

$$I = 4N \int d\mathbf{x} \sum_{n=1}^{N/2} \nabla \psi_n^*(\mathbf{x}) \cdot \nabla \psi_n(\mathbf{x}), \quad \mathbf{x} \equiv x_1,...,x_M, \tag{15.21}$$

$$d\mathbf{x} \equiv dx_1 \cdots dx_M, \; \nabla \equiv \left(\frac{\partial}{\partial x_1}, ..., \frac{\partial}{\partial x_M} \right), \; N \geq 2.$$

(these generally complex amplitudes $\psi_n(\mathbf{x})$ defined at (15.25)). The del ∇ is a generalized gradient operator. Equation (15.21) defines an I that is again an L^2 measure. Also, as a *maximized version* of the actual Fisher information, I is a *Fisher channel capacity* (borrowing terminology from Shannon information theory). This is for the channel (15.1).

15.4.7.1 Mathematical Trace Form

The form (15.21) of the information is also, from its form, the trace of a matrix – the *Fisher information matrix* [6, 14, 17]. It derives as well from the decoherence matrix of quantum physics [6]. In a much earlier work [22], it was shown to derive as a sum of Weizsäcker kinetic energy terms. Most recently, it has been shown [23] that the temperature of a classical statistical system in a generally non-equilibrium state can be represented as the trace of its Fisher information matrix, although not in $\mathbf{x}$ space as here, but in momentum space.

15.4.7.2 Swiss Watch Paradigm

There are two important ramifications of such nonoverlapping support regions. First, since their individual information maximally add, they describe not only maximum *information* but also, by inference, a *maximally structured* system. An example is a classic Swiss watch mechanism with N parts. Here, the non-overlap requirement means that the N individual parts are specified to lie within specific regions within certain tolerances. The tighter the tolerances are the higher the total I becomes (as well as the structural quality of the watch).

Furthermore, such a non-overlapping system of amplitudes $q_n(\mathbf{x})$ can actually be *physically realized* by the EPI output wave equation (e.g. the SWE). These second-order differential equations define amplitudes $\psi_n(\mathbf{x},t)$ in space and time, and *require the user to fix initial-condition wavefunctions* $\psi_n(\mathbf{x},0)$. This permits us to set $\psi_n(\mathbf{x},0) \equiv q_n(\mathbf{x})$, our required separated amplitudes! The proviso is that the phase part of $\psi_n(\mathbf{x},0)$ vary smoothly with $\mathbf{x}$, and this can always be accomplished by the use of generalized functions. For example, if the ideal $\psi_n(\mathbf{x},0)$ are to be separated triangle functions, we choose to convolve these with a Gaussian of variance σ^2. For any small, but finite, value of σ^2 the phase varies smoothly over the complete range of $\mathbf{x}$, as required. In summary, the entire EPI measurement-perturbation description is not merely a convenient 'thought experiment' for purposes of computing physics: it physically 'happens' for a properly prepared initial solution state $\psi_n(\mathbf{x},0)$.

Note that in the special case $M = 1$, $N = 1$, $\psi \equiv q$ (real), (15.21) devolves into the direct (non-channel capacity) form (15.20) of Fisher.

15.4.8 Scope of Solutions

Principles (15.11) and (15.12) result from the worldview taken. They comprise the two facets of a learning based (epistemic) approach to physics called EPI. This approach has been used to derive most of the laws of physics [1, 6, 7], including a key law of quantum gravitation – the Wheeler–DeWitt equation. It has also derived many laws of biology [3, 6, 7, 9] including a new uncertainty principle describing population growth, and laws of economics [5–7], medicine [4, 6–8] and other sciences [6, 7]. These include already-known laws and *new* laws. The premise of an information-oriented universe is thereby verified as well empirically.

What generally are the two information I, J in principles (15.11), (15.12)? Information I is, regardless of application, *of the one basic form* (15.8) or its generalization (15.21). That is, it exists as *a measure of the quality of data*, independent of their particular physical nature. This makes sense since the observation by B is only of output *data* from A. By comparison, the source information J characterizes A and hence the *particular* effect. So far, all we know about J is that it has the general form (15.15). Thus, for any given problem, it is easy to find I, but how is J found?

15.5 Finding Source Information *J*

J represents the information provided by the unknown law. But, a longstanding problem of information theory is:

How much information does a scientific law contain?

Our epistemic viewpoint suggests a route to the answer(s). The route is provided by the nature of the prior knowledge (called *physical insight* or biological insight or etc.) that is at hand. As conceived by the philosopher-physicist Peirce (1839–1914), there are three basic types, giving rise, respectively, to *three levels of accuracy* in the EPI solutions. In descending order of accuracy, these are (a) abduction, (b) deduction, and (c) induction, discussed next. (Examples of J for various scenarios are in the fourth column of Table 15.1).

15.5.1 Exact, Unitary Scenarios: Type (a) Abduction

Cases where there is an observable space that is unitary (say, via Fourier transform) to data space allow for *complete recovery* of the source information, i.e. $I = J$, and consequently a fully accurate estimate of the unknown law. In analogy with Peirce's corresponding term, this is called 'abduction' in the sense of a first, or primary, principle of truth. It is discussed next.

L^2 measures of the form (15.21) are fundamental to all physics. These occur in *observable spaces*, are length-preserving, and hence are *invariant under unitary transformation* U. Two such spaces are space–time and momentum–energy, respectively. As required, these are connected by a unitary *Fourier* transform and are *observable* (why the latter, is discussed below (15.30). This case is well-known to hold in quantum mechanics (15.27) but, interestingly, also holds in *financial investment* (15.36).

In these *unitary scenarios*, by the length-preserving property, $J = I$ so that the zero-property (15.11) of EPI identically holds, with $\kappa = 1$. Also, the EPI variational principle (15.12) may be shown to hold [6, 7, 14]. Indeed, both (15.11) and (15.12) hold *independently*, so that they give generally different (but consistent) solutions $q(\mathbf{x})$. Examples [1, 6] of such solution pairs (15.11), (15.12) are, respectively the Klein–Gordon and Dirac equations of *quantum* mechanics; and [6] Newton's second law $f = ma$ and the Virial theorem of *classical* mechanics.

An exciting prospect is that *new* physical effects can be mathematically defined, and therefore anticipated, by seeking out *new* unitary transformations and applying them to EPI. Again, for this approach to be useful both data space and transform space must be independently *observable*. Finding such transform spaces of course takes insight. Interestingly, one for optimum economic investment was recently found ([7], or Sect. 15.8).

15.5.2 Inexact, Classical Scenarios: Type (b) Deduction

Obviously, exact (type b) solutions are preferred. However, these require the existence of an observable space that is unitary to measurement space. What can be done if such a space is *not* known? Without it, the full amount $J = I$ of information is no longer received. Hence, now $\kappa < 1$, with κ unknown, and an *inexact* solution can only be obtained.

Such solutions, in fact, turn out to describe *classical levels* of physics, for example gravitation and electromagnetics [6]. Here, J and κ *must be solved for*, by simultaneous solution of (15.11) and (15.12). The lower level of prior knowledge now *disallows* finding two distinct solutions (15.11) and (15.12). They are effectively "blurred together" into *one* composite law that obeys both. Therefore, this is called a *self-consistent* EPI solution.

In order to obtain a self-consistent solution $q(x)$ a special form of prior knowledge must be present. This is a *deduction* in Peirce's terminology, since the prior knowledge is generally deduced from other properties of the effect. The deduction is expressed as an *invariance principle*, such as continuity of flow. (In principle an equivalent symmetry given by Noether's theorem could also be used.) Past examples are continuity of flow and gauge invariance [6], used in derivation of the laws of classical gravitation and classical electromagnetics. The efficiency variable κ is found to be exactly $1/2$ in these cases, rather small considering how accurate the solutions are. Table 1 shows a number of examples of these type (b) solutions.

15.5.3 Empirical Scenarios: Type (c) Induction

As one might expect, the lowest level of accuracy follows in scenarios where there is no knowledge of the physics of the source. There is no model for J, and neither κ nor J can be solved for. The only knowledge there is of the source is of an indirect nature: *data* values F_m, $m = 1, \ldots, M$ *from* the unknown law. This occurs, e.g., in econophysics ([5–7] or Sect. 15.10) where the "technical viewpoint" of investment assumes that price data alone suffice as prior knowledge (*effective* invariants). Or, in thermodynamics [2, 7], where the invariants are *extrinsic* measurements of mean values.

But even with no model for J it can be replaced with what *is* known, namely, a sum of data constraint terms. (See one example (c) in Table 15.1.) Principle (15.12) then becomes,

$$I - J = \mathit{extremum}, \text{ where } J = \sum_m \lambda_m \left[\int dx f_m(x) p(x) - F_m \right]. \qquad (15.22)$$

The constants λ_m are called 'undetermined multipliers.' These are found by demanding consistency of the integrals in (15.22) with their corresponding data F_m. This is a generally nonlinear problem, but easy enough to numerically solve, for example by Newton–Raphson methods. However there is an essential departure in the *meaning* of these solutions from those preceding:

Since, as before, J does not depend upon $p'(x)$, for a real coordinate x the extremum in $I - J$ is still a *minimum*. However, the bracketed quantities in (15.22) are now all zero at solution. Therefore, *J is fixed at value zero*. Then *only quantity I is minimized* in (15.22), and it is of course no longer true that $I \simeq J$ at solution. To the contrary, $I \ll J$. This minimized value of I must cause a poor estimate of a, and this is mirrored in poor accuracy for the estimated law $p(x)$ as well. In fact since I measures the gradient content in $p(x)$, *its output shape will merely be as smooth as is possible.* With I so minimized, EPI becomes a principle of minimum Fisher information (MFI) [6, 19]. However, surprisingly, MFI solutions often agree well with ground truth (see [5, 7] or Figs. 15.1 and 15.2). This agreement probably traces from the fact that *groundtruth* frequency laws are often themselves quite smooth, at least approximating a state of minimum Fisher information.

15.5.4 Second Law of Statistical Mechanics

Recall that, at the outset, system variable x could be regarded as generally either random or deterministic. Temporarily consider here a case where x is random.

What if a purely statistical, classical system, as described above by (15.22), is perturbed? In what direction will $p(x)$ go – toward maximum disorder or toward minimum disorder? It is shown [2, 6, 7] that I monotonically decreases

$$dI \leq 0 \tag{15.23}$$

with *the time.* That is, the Fisher information goes down, describing ever *increasing* disorder. This holds under the same statistical conditions for which entropy increases and the second law holds. These conditions are where $p(x|t)$ obeys a Fokker–Planck (general diffusion) equation.

15.5.5 Resulting Information Game

We found in the preceding that EPI-MFI solutions tend to give maximally spread-out or *blurred* $q(x)$ or $p(x)$ curves. This blurring effect also implies that an act of observing is effectively a move in a *mathematical game* [20] with nature [1, 6, 7]. The game is a mnemonic device for capturing the significance of an EPI solution. It is played between the observer and a "demon," both of which try to maximize their information levels, respectively I and J. The conflict of the game arises out of the fact that it is a *zero-sum* game (15.11), whereby any information gain I of the observer is at the expense J of the demon (nature). Hence, the demon minimizes his information payout. This is by maximally blurring the output law. The final move of the game (irrespective of who moves first) is at the EPI *solution* "point" of the playing board.

15.5.5.1 Game Corollary, Finding Unknown Constants

In many EPI derivations the output amplitudes contain *unknown constants*, usually arising out of integration of given Euler–Lagrange equations. We have found [6] that such a constant – call it b – can, in fact, usually be inferred, and by a simple approach that directly traces from the knowledge game. It amounts to a second use of EPI. See also [21].

Assume that the game has been 'played', an EPI-MFI solution $p(x)$ formed. However, the solution contains an unknown constant b. What value should be assigned to it? The answer is obtained by consistently continuing the previous approach that found $p(x)$.

That is, K is evaluated at the EPI solution $p(x)$, the latter containing an unknown constant b. This results in a function $K(b)$. Once again assume that K is stable to variation, now in b. The stability condition becomes $\delta K(b) \equiv \delta(I(b) - J(b)) \equiv 0$. This is a new EPI problem. The source information $J(b)$ here represents prior information about b. However, in most scenarios nothing a priori is known about b. That is, identically $J(b) = 0$, and a type (c) induction solution is the only hope. Then the stability condition simplifies to merely $\delta I(b) = 0$ or, since the only thing left to vary is b,

$$\partial I(b)/\partial b = 0. \tag{15.24}$$

Experience [6, 21] with solutions b to (15.24) is that they *minimize* I. It is as if the demon gets a final move in the knowledge game and, being a "demon," adjusts b such that the output law $p(x)$ is maximally blurred, bequeathing minimal information to the observer.

In summary of the demon's actions to this point, he first optimizes I to be maximally close to the source information J. This can find $p(x)$ to within an arbitrary constant b. Then, to *find* b he minimizes $I(b)$ through choice of the constant, in a *second* knowledge game. In this way a kind of minimax solution is found.

15.5.5.2 Predicting Cancer Growth

Another application of (15.24) is to in situ cancer growth [4, 6, 7]. Here a type (b) EPI solution is sought, giving a growth law $p(t) = Ct^b$, $C = (b+1)T^{-(b+1)}$, b an unknown constant. The corresponding information is computed via (15.14) as $I(b) = T^{-2}b^2(b+1)/(b-1)$. Differentiating as in (15.24) directly gives the solution $b = \frac{1}{2}(1+\sqrt{5}) = 1.618... \equiv \Phi$, the Fibonacci golden mean. This growth law is confirmed by clinical data [4, 6].

15.5.5.3 Finding Unknown Constants and Functions

This approach may be generalized to any other EPI solution type (a), (b) or (c). Also the output $p(x)$ may contain any number of unknown constants and *functions f(x)*.

To find these unknowns, as in the preceding the output $p(x)$ is substituted into the functionals I and J, and the constants and functions are now serially or simultaneously found. This is by varying them so as to again satisfy $\delta(I(f(x)) - J(f(x)) = 0$. Used generally in this way, the Game corollary defines a generalized EPI process for learning nature. The corollary has been widely applied [6, 7, 21]. An example follows.

15.5.5.4 Predicting the Quarter-Power Laws of Biology

Consider the problem of determining the size of a given biological attribute, such as metabolism rate, for an organism of mass x. The *initial* EPI solution is found to be $f(x)^b, b = const.$, $f(x)$ unknown (Chap. 8 of [7]). Using EPI a second time by solving $\delta(I(f(x)) - J(f(x)) = 0$ through variation of $f(x)$ gives the (correct) answer that $f(x) = x$, the mass. Finally, b is found by the Game corollary (15.24) as essentially a *third* variational problem. The solution is $b = n/4, n = 0, \pm 1, \pm 2, \ldots$ The net result $x^{n/4} = p_n(x)$ defines the famous *quarter-power laws of biology.*

15.6 Some Other Past EPI Applications

Table 1 summarizes many past applications of EPI. For each, it lists the source effect to be derived (col. 1), the reference for the derivation (col. 2), the type of unitary transformation that is used (col. 3), the source information functional J (col. 4) and the type of prior knowledge (col. 5).

15.7 Schrodinger Wave Equation: A Type (a) Abduction Solution

We use the EPI approach to derive the nonrelativistic wave equation ([6], Appendix D1). This arises from the simplest possible measurement scenario, a one-dimensional observation $y = a + x$ of the ideal position a of a particle of finite mass m. The particle is moving in a known field of potential $V(x)$ and total energy W, and is in a pure state. The state is unknown, described by N *unknown amplitudes* $q_n(x), n = 1, \ldots, N$. For convenience, these are packed as complex amplitudes

$$\psi_1 \equiv N^{-1/2}(q_1 + iq_2), \psi_2 \equiv N^{-1/2}(q_3 + iq_4), \ldots, \psi_{N/2} \equiv N^{-1/2}(q_{2N-1} + iq_{2N}), \tag{15.25}$$

where $i \equiv \sqrt{-1}$. The ψ_n and N are the new unknowns of the problem. Using these in (15.21) gives

$$I \equiv 4N \int dx \sum_{m=1}^{N} q_m'^2 = 4N \int dx \sum_{n=1}^{N/2} \psi_n^{*\prime} \psi_n'. \tag{15.26}$$

Primes denote derivatives d/dx. Note that each degree of freedom ψ_n represents two consecutive *dofs* q_m, so that N must be even.

A *unitary space* μ exists that is conjugate to measurement space x. It is momentum space μ, and the unitary transform is

$$\psi_n(x) \equiv \frac{1}{\sqrt{2\pi\hbar}} \int d\mu \varphi_n(\mu) \exp(i\mu x/\hbar), \tag{15.27}$$

the Fourier one. Function $\varphi_n(\mu)$ is a corresponding amplitude function in μ space. The constant $\hbar$ is Planck's constant divided by 2π. Here we take (15.27) as a given, but mention in passing that it itself has been recently derived, again using EPI [24].

Cautionary note. A casual look at (15.27) suggests that it, by itself, implies the SWE. This is by merely (1) differentiating (15.27) twice by d/dx, thereby bringing down a factor μ^2 inside the right-side integral; (2) using that the kinetic energy $KE \equiv \mu^2/2m = W - V(x)$; (3) taking this outside the integral sign since $W - V(x)$ is not a function of μ; and (4) using (15.27) as an identity. But in fact $W - V(x)$ equals the KE classically, *but not quantum mechanically*. This is by the following: Coordinates x and μ are not so-called *superselection variables* [25], i.e. do not commute with one another. Therefore, they obey the Heisenberg uncertainty principle. Then, with μ a given value in the integrand of (15.27), and with W and function $V(x)$ assumed known, if $\mu^2/2m = W - V(x)$ held then the value of x would be *precisely known*. But this violates the Heisenberg uncertainty principle. Therefore the casual approach does not work.

Given (15.27), this an EPI approach *(a) of abduction*. Then all the pre-existing information J is obtained as I. Hence, J equals form (15.26) as well. However, if J is so expressed identically, then identically $I - J = 0$ for *all* choices of $\psi_n(x)$. No extremum condition exists. Obviously, J must somehow be re-expressed as a functional that does not just identically equal I. To accomplish this we use the unitary condition (15.27). Substituting it into (15.26) gives (via Parseval's theorem)

$$J = \frac{4N}{\hbar^2} \int d\mu \mu^2 \sum_{n=1}^{N/2} |\varphi_n(\mu)|^2. \tag{15.28}$$

But by (15.25),

$$\sum_{n=1}^{N/2} |\psi_n(x)|^2 = \frac{1}{N} \sum_{n=1}^{N} q_n^2 = p(x). \tag{15.29}$$

Also, by Fourier condition (15.27), the area under the sum $\sum_n |\varphi_n(\mu)|^2$ obeys Parseval's theorem and, hence, equals the area under $\sum_n |\psi_n(x)|^2$. By (15.29), this is the area under $p(x)$, which, by normalization, is unity.

Therefore $\sum_n |\varphi_n(\mu)|^2$ likewise obeys normalization and is a probability law, $p(\mu)$. Using this in (15.28) gives J as essentially a mean-squared momentum value,

$$J = \frac{4N}{\hbar^2} \langle \mu^2 \rangle. \tag{15.30}$$

Note that this quantity physically exists since μ space is observable. This is essentially *why* all such unitary spaces must be observable in EPI applications.

Now, in distinction from the erroneous step in the *Cautionary note* above, *averages* of quantum coordinates *do* obey classical physics (Ehrenfest theorem), so that $\langle \mu^2 \rangle = 2m \langle W - V(x) \rangle$. Using this in (15.30) gives

$$J = \frac{8Nm}{\hbar^2} \langle W - V(x) \rangle \equiv \frac{8Nm}{\hbar^2} \int dx p(x) (W - V(x)) \qquad (15.31)$$

$$= \frac{8Nm}{\hbar^2} \int dx \sum_{n=1}^{N/2} |\psi_n(x)|^2 (W - V(x))$$

by use of (15.29). Then by (15.26) the EPI principle (15.12) becomes

$$I - J = 4N \int dx \sum_{n=1}^{N/2} \left[\psi_n^{*\prime} \psi_n^{\prime} - \frac{2m}{\hbar^2} |\psi_n(x)|^2 (W - V(x)) \right] = extremum. \qquad (15.32)$$

This is a problem (15.16) in the calculus of variations. The net Lagrangian (integrand) is

$$\mathscr{L} = \sum_{n=1}^{N/2} \left[\psi_n^{*\prime} \psi_n^{\prime} - \frac{2m}{\hbar^2} \psi_n^* \psi_n (W - V(x)) \right]. \qquad (15.33)$$

The Euler–Lagrange solution ψ_n is formed as in (15.17), as

$$\psi_n^{\prime\prime} + \frac{2m}{\hbar^2} (W - V(x)) \psi_n = 0, \quad n = 1, \ldots, N/2. \qquad (15.34)$$

This is the SWE without the time, also called the "Schrodinger energy eigenvalue equation." It represents the payoff point of this particular information game.

A remaining problem is to determine N. For this we use the Game corollary (Sect. 15.5.5.1) with $b = N$, according to which N is found through *minimizing I*. Since $I = J$ here, (15.28) represents I as well as J. Every term in the sum (15.28) additively contributes to the total I. Also, we noted before that N must be even. Thus it must be the smallest even number, $N = 2$. Thus, the subscript n may be suppressed, and the measurement scenario obeys one complex degree of freedom $\psi(x)$. This is the usual result.

A fully relativistic analysis [1, 6] of this measurement problem, now with space–time coordinates, yields the Dirac– and Klein–Gordon equations, and

$$I = J = 4N \left(\frac{mc}{\hbar} \right)^2 \qquad (15.35)$$

where c is the speed of light. Thus, *where there is high mass there is high information* about space–time. Since our universe has very high mass, it therefore contains a massive level of information I (even ignoring other information sources such as charge, current, etc., which further add).

15.8 Optimum Investment of Capital: Type (a) Abduction Solution

Can EPI apply to economics? Economics is certainly an example of a *participatory universe*, i.e. where the observer is an investor who affects its laws. A major problem of finance is to find an optimal program $K(t)$ for the investment over time t of capital K in the stock of a firm. This may, in fact, be found as a type (a) abduction solution. The solution is verified by the main result below, the known law of optimum investment (15.40). Since the development is virtually the same as for the SWE, we can be brief here. (Full details are in [7]).

Let program $K(t)$ have a time rate of change $\dot{K}$, and obey an amplitude law $\psi(K(t)) \equiv \psi(K)$ that has a *Fourier conjugate amplitude* $\theta(\mu(t)) \equiv \theta(\mu)$, obeying

$$\psi(K) = (2\pi b)^{-1/2} e^{i\rho K/b} \int d\mu \theta(\mu) e^{i\mu K/b}, \; b = \text{const.}, \; p(\mu) = |\theta(\mu)|^2 \quad (15.36)$$

(cf. (15.27)). The space μ is that of *investment momentum* $\mu \equiv m\dot{K}$. As with *particle* momentum (15.41), m is an inertial *mass*, here of of investment. Also, ρ represents a constant rate of return in the absence of investment, the default rate. As with the quantum problem, a pure state of the system is sought. Let this have a single complex component $\psi(K)$.

Regarding the changes $\dot{K}$, there is *cost* to changing capital stock, described by a function $\varphi(\dot{K})$ called the *adjustment cost function*. By comparison, the economic *source allowing* for investment is the *production function* $F(K)$. This represents the known net *income* as a function of capital stock K. Hence *source* information J must be constructed from $F(K)$. As I is always an *expection* we construct J as one,

$$J \equiv 4 < F(K(t)) >= 4 \int dt F(K(t)) P(t), \; P(t) = \rho \exp(-\rho t) \quad (15.37)$$

(cf. (15.30)). The probability $P(t)$ of investing amount K at time t is taken as exponential, $P(t) = \rho \exp(-\rho t)$. The information I is now

$$I = 4 \int dK |\psi'(K)|^2 = 4b^{-2} \int d\mu |\theta(\mu)|^2 (\mu + \rho)^2 = 4\rho b^{-2} \int dt\, (\mu + \rho)^2 \exp(-\rho t). \quad (15.38)$$

The first equality is by (15.26) with $N = 2$, the second is by (15.36), and the third is by the far-right (15.37) and assuming ergodicity to hold. Combining (15.37) and (15.38) in the EPI principle (15.12) defines a variational problem

$$I - J = \rho \int dt \left[F(K(t)) - b^{-2} (\mu + \rho)^2 \right] \exp(-\rho t) = \textit{extrem}. \quad (15.39)$$

Using $\mu = m\dot{K}$ and varying function $K(t)$ gives the general solution

$$\ddot{K} - \rho \dot{K} + 2^{-1} (b/m)^2 F'(K) - m^{-1} \rho^2 = 0 \quad (15.40)$$

(cf. 15.34). Most often the adjustment cost function is quadratic, $\varphi(\dot{K}) = (m/b)\dot{K}^2$, since this represents a constant cost/time. Then (15.40) becomes the famous Tobin q-theory of optimum investment gain [26]. Thus, optimum *knowledge* in the face of uncertainty equates to optimum gain of *capital.*

15.9 Newton's Second Law: Type (b) Deduction Solution

Suppose we now want to ascertain the *classical equation of motion* for a particle of mass m ([6], Appendix D2). Let it be moving at nonrelativistic speed in a conservative field with potential $V(x)$. Since the particle is classical, it has a definite trajectory $x(t)$ defining position x at time t over a total time interval $(-T,T)$. Its linear momentum obeys

$$\mu = \mu(t) = m\dot{x}(t), \tag{15.41}$$

the dot denoting a time derivative d/dt.

We find the answer as a *deduction* from the preceding SWE derivation. That obeyed $I = J$ and (15.30) with $N = 2$,

$$I = J = \frac{8}{\hbar^2} < \mu^2 > . \tag{15.42}$$

The deduction arises out of taking the classical limit of (15.42). The average in (15.42) is explicitly over all momentum values. Let the particle's momentum values obey *ergodicity*. Then the average is equivalently over *all time* as well,

$$I = \frac{8}{\hbar^2} < m^2\dot{x}^2(t) > = \frac{8m^2}{\hbar^2} \lim_{T\to\infty} \int_{-T}^{T} dt\, p(t)\dot{x}^2(t). \tag{15.43}$$

Here $p(t)$ is the density function ruling time values. Regard time in the usual classical sense, i.e., each time interval $(t, t+dt)$ occurs once and only once, so that these are *uniformly* dense over the total time interval, i.e., $p(t) = (2T)^{-1}$ for $-T \le t \le T$, or 0 for other t. Using this in (15.43) gives

$$I = \frac{8m^2}{\hbar^2} \lim_{T\to\infty} \frac{1}{2T} \int_{-T}^{T} dt\, \dot{x}^2(t). \tag{15.44}$$

A well-known classical limit of quantum mechanics is obtained by letting $\hbar \to 0$. Parameter T is already approaching infinity in (15.44). Therefore, a definite limit for $\hbar^2 T$ in (15.44) can result, by letting $\hbar \to 0$ and $T \to \infty$ such that $\hbar^2 T = A^2$, a *finite* constant. Then (15.44) becomes

$$I = \frac{4m^2}{A^2} \int_{-T}^{T} dt\, \dot{x}^2(t). \tag{15.45}$$

T is regarded as large in this and all further integrals. With I known, we now seek the functional J. Generally represent the bound information as

$$J \equiv \frac{8m}{A^2} \int_{-T}^{T} dt\, j(x), \quad x \equiv x(t) \tag{15.46}$$

with $j(x)$ some unknown information density. As the information source, $j(x)$ defines this particular (Newtonian) problem. Then by (15.45) and (15.46) the EPI extremum condition (15.12) is

$$I - J = \frac{4m^2}{A^2} \int_{-T}^{T} dt \left[\dot{x}^2 - 2\frac{j(x)}{m} \right] = extremum. \tag{15.47}$$

The extremum is through variation of the trajectory $x(t)$. The effective Lagrangian $\mathscr{L}$ for the problem is the integrand of (15.47). The Euler–Lagrange solution (15.17) is directly $m\ddot{x} = -\partial j(x)/\partial x$. This is *Newton/s second law*

$$m\ddot{x} = -\frac{dV(x)}{dx} \equiv F(x), \tag{15.48}$$

for an information density defined as $j(x) \equiv V(x),\ x \equiv x(t)$.

15.10 Financial Laws: Type (c) Induction Solutions

Two extreme strategies for estimating the *financial value* (commonly called *valuation*) of a security such as a stock or bond are the *fundamental* approach and the *technical* approach. The fundamental approach takes into account all details of the underlying business on the microlevel – its yearly sales, debts, profits, etc. This is valuation taking into account all *sources* of valuation. From the perspective of EPI, this implies the existence of a well-defined *source J of information* on valuation. Indeed, a type (a) abduction analysis of optimal *investment* (but not valuation) was made in Sect. 15.8 above out of knowledge of a source function J.

By comparison, a source function J for determining valuation does not seem to exist. Hence a purely *technical* approach to valuation must be taken. This is based entirely on its past *numerical* performance, e.g. its price fluctuation curve. This corresponds to an EPI problem (15.22), rewritten here in terms of amplitude function $q(x)$ as

$$I - J = minimum, \text{ where } I = 4\int dxq'^2(x), \ \ J = \sum_m \lambda_m \left[\int dx f_m(x) q^2(x) - F_m \right]. \tag{15.49}$$

The latter has a Lagrangian and Euler-Lagrange solution

$$\mathscr{L} = 4q'^2(x) - \sum_m \lambda_m f_m(x) q^2(x), \ \ \frac{d}{dx}\left(\frac{\partial \mathscr{L}}{\partial q'}\right) - \frac{\partial \mathscr{L}}{\partial q} = 0. \tag{15.50}$$

Using the first equation in the second gives a solution

$$q''(x) - \frac{q(x)}{4} \sum_m \lambda_m f_m(x) = 0. \tag{15.51}$$

Let us consider the particular problem of estimating $p(t)$, the density function for a perpetual annuity (one that pays interest forever). Here general coordinate $x \equiv t$. Suppose that the knowledge at hand is its "value", defined as its longterm mean interest $\tau \equiv \int dt t p(t)$ (integration limits of 0, ∞). Then the problem (15.51) becomes

$$q''(t) - \frac{q(t)}{4} \lambda t = 0. \tag{15.52}$$

This differential equation has as its solution an Airy function of the second kind $Ai(c_1 t)$, so that the probability is $p(t) = c_1 Ai^2(c_2 t)$. The two constants c_1, c_2 are determined by normalization and the value of τ. The curve is plotted in Fig. 15.1 (labelled as Fisher curve). It is compared with the simple exponential-answer to the same problem using MaxEnt (the principle of maximum entropy) in place of MFI.

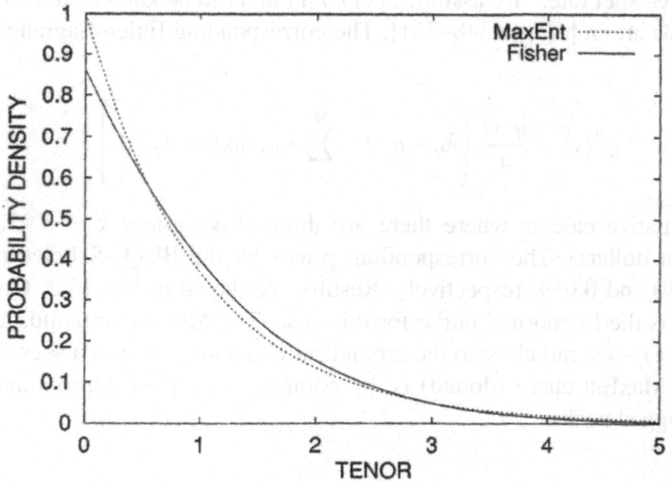

Fig. 15.1 Outputs by EPI (*solid*) and by MaxEnt (*dotted*)

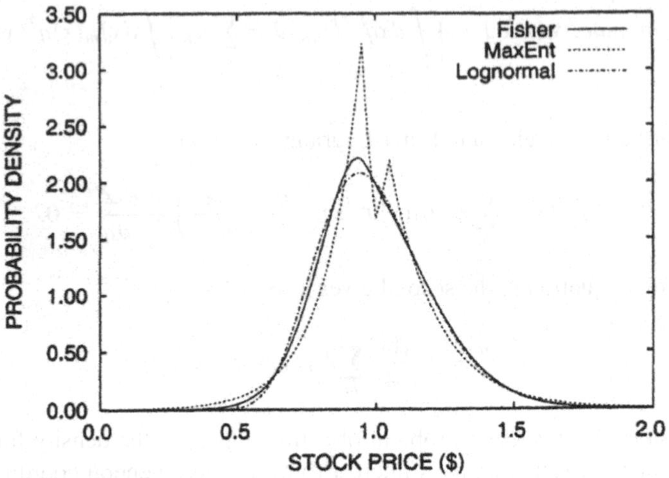

Fig. 15.2 Comparison of MFI, MaxEnt outputs, with Lognormal groundtruth

Both curves are very smooth, and therefore unbiased, although the MFI answer is somewhat smoother. Here the groundtruth curve is not known.

A case where the groundtruth curve *is* known is in options pricing. Here the theoretical answer obeys the Black-Scholes answer of a log-normal curve. Consider the case of a so-called "European call option." This obeys a constraint

$$c(k) = e^{-r(t)t} \int_k^\infty dx p(x), \tag{15.53}$$

where k is the "strike price", t is the time to expiration of the option, and $r(t)$ is the "risk-free spot rate." We assume $c(k)$,k,$r(t)$ and t to be known, and want to find $p(x)$ [details are in [6], pp. 348–351]. The corresponding Euler–Lagrange solution (15.51) is

$$q''(x) = \frac{q(x)}{4} \left[\lambda_0 + e^{-r(t)t} \sum_{m=1}^{M} \lambda_m \max(x - k_m, 0) \right]. \tag{15.54}$$

A representative case is where there are three strike prices $k = 0.95, 1.00$ and 1.05 (all in dollars). The corresponding prices by the Black–Scholes model are $0.105, 0.080$ and 0.059, respectively. Results are shown in Fig. 15.2. Groundtruth (dash-dot) is the Lognormal curve for this case. The MFI curve (solid, labelled as Fisher curve) is overall close to the groundtruth, including required skewness to the right. The MaxEnt curve (dotted) is, by comparison, a poor approximation, with spurious jagged peaks.

15.11 Summary

The following are the main points of the approach:

1. By Popper's criterion [16], a scientific effect is *a body of theory describing a hypothetical system A that is verified under observation* (by a system *B*). The observation is of a required system parameter *a* (Sect. 15.2). Observation results in perturbation of the total system *AB*.
2. The perturbation causes a change in information about *a*. In order for the effect to persist, the information change $I - J$ must be stable (Sects. 15.1 and 15.2.3) in the presence of maximum system structure (Sect. 15.4.7.2). This is equivalent to the EPI principle (15.12), with I obeying (15.21).
3. Nature and the observer thereby play *complementary roles* in the joint system *AB*. This establishes a *participatory nature to science* (Sects. 15.2.1 and 15.2.2.3). This participatory aspect is quantified as the EPI principle.
4. EPI alternatively arises in a scenario of *unitarity* (Sect. 15.1), where a space that is unitary to observation space is likewise an observable. In such scenarios the system *AB* is still assumed to be perturbed, but no assumption 2 (preceding) of stability need be made.
5. The law $p(x)$ is not necessarily a probability law, since *x* is not necessarily random (Sect. 15.2.2.1). It may, e.g., follow a definite - but unknown – "trajectory" $x(t)$ in time. In particular, *the quantum wave equations that EPI derives need not describe a random effect x.* A. Einstein spent the last 20 years of his life seeking a basis for quantum theory that does not assume intrinsic randomness in *x*. The EPI viewpoint effectively accomplishes this on the level of finding $p(x)$, although not $x(t)$. Interestingly, EPI has the further capability of finding *deterministic* effects such as $x(t)$ (e.g., Chap. 8 of [7]) provided an appropriate $J(a)$ can be formed. Thus, it can *in principle* furnish a complete description of quantum mechanics, as both a statistical and (possibly) deterministic theory.
6. The informations *I* and *J* are uniquely derived to be *Fisher* information (Sect. 15.3).
7. In that *J* must be computed for each new observed effect, it represents *the amount of Fisher information that exists in a law of science* (Sect. 15.5), something that has not been defined before.
8. EPI is fundamentally epistemic, and therefore requires some prior knowledge of the state *a*. According to its level of *quality*, $J(a)$ and the output solution are computed at three levels of accuracy. In descending order of quality, these knowledge types are called: Type (a) abduction (Sect. 15.5.1), where a unitary transformation space is known, giving a perfect answer for $p(x)$. Type (b) deduction (Sect. 15.5.2), where an invariance principle is known, giving a slightly imperfect answer, e.g. the *classical* laws of physics: of *classical* mechanics, electromagnetism and gravitation. Type (c) induction (Sect. 15.5.3), where merely empirical data, i.e. samples from the unknown law, are known, so that the EPI solution is merely optimally smooth. As examples, Figs. 15.1

and 15.2 show typical type (c) solutions [6, 7] governing the prices x of economic issues such as stocks (Fig. 15.2), options, etc.

9. I obeys the second law of statistical mechanics (or thermodynamics) if $p(x|t)$ obeys the Fokker–Planck equation (Sect. 15.5.4). Both information I and Shannon entropy thereby predict disorder to rise under the same F–P conditions. Therefore, *entropy is not the sole measure of physical disorder.*
10. The EPI variational principle $I - J = extremum$ has superficially the same form as the famous Lagrangian *action* approach to physics. However, the two approaches differ: first, of course in basic meaning, but more importantly, in how to effect a solution. For example, in type (b) scenarios functional J *is actually solved for* (Sect. 15.5.2). By comparison, the action approach has no zero principle to utilize and, hence, *does not* have a means of solving for the effective J-term (usually, the second) in its Lagrangian.
11. EPI as a process is equivalent to a game of information maximization (Sect. 15.5.5) played by both an observer and a "demon." The observer gains information $I - J$ only at the expense of the demon. Therefore the observer tries to observe data independently, so that their individual informations I maximally add, while the demon tries to blur the data, minimizing the total. The resulting game is zero- and fixed-point, and its solution point is the EPI solution.
12. The Game *corollary* (Sect. 15.5.5.1) describes a 2nd, or further, play of the game for the purpose of evaluating an unknown constant b or function $f(x)$ in the EPI solution $p(x)$ from the *preceding* play. This followup game proceeds as in the first, except that now the output is b or $f(x)$. In most cases nothing is a priori known about b or $f(x)$ so that for this game $J = 0$. It results that $p(x)$ is now maximally blurred through choice of b or $f(x)$. This is a powerful approach for evaluating these unknowns [6, 7, 21]. The game may be played serially any number of times as needed to evaluate unknown quantities from the previous solution. *This provides a sequential framework for learning natural law.*
13. EPI is *itself* a net law AB consisting of a body of theory A that has long-confirmed effects [1–14] observed by physicists B. Thus EPI obeys Popper's criterion for a *scientific law*. In fact, the breadth of phenomena derived by EPI indicates that it is the universal abduction, *inclusive of* all others.

15.12 Some Speculations

The universal physical constants $c, \hbar, e$, etc., have long been regarded as a priori unknown, only capable of empirical evaluation. The Game corollary is a procedure for, in principle, finding these constants. Can they, by the use of suitable models for corresponding levels of prior knowledge $J(c)$, $J(\hbar)$, $J(e)$, etc.?

We found that the received information $I = J \equiv I_{max}$ for quantum systems. That is, nature allows an observer to obtain the full amount of information that exists in a system with non-overlapping wave functions (where the information maximally adds). This implies maximally reliable data on the quantum (finest) level and, hence, *a universe that is maximally condusive to be understood through observation* – a playground of information. These surely are conditions that favor the survival and propagation of intelligent life forms. To what extent is this true, how does it fit into natural selection, and can it be enhanced to promote further human progress?

The Cramer–Rao inequality (15.9) gives rise to the Heisenberg uncertainty principle [1, 6]. It also gives rise to other uncertainty principles – of population biology [3, 7], cancer growth [4, 7] and financial investment [7]. It should be capable of application in like manner to other fields of science as well.

So far, type (a) unitary-transformation cases have been limited to cases of angular rotation and Fourier transformation. But according to the theory, *every other* observable unitary transformation should likewise result in a correct answer for system dynamics. There are in principle an infinity of these. Any of those whose observability is confirmed ought to describe new science in its EPI output.

Going one step further, perhaps what we call 'the laws' of science are but actually *default* laws. In fact, EPI potentially allows one to operate in a 'design mode.' Here the observer works backward from a *desired* answer for system dynamics to the unitary transformation that gives rise to it. This would amount to a kind of nature on demand, a fascinating prospect.

References

1. Frieden, B.R., Soffer, B.H.: Lagrangians of physics and the game of Fisher-information transfer. Phys. Rev. E **52**, 2274–2286 (1995)
2. Frieden, B.R., Plastino, A., Plastino, A.R., Soffer, B.H.: Fisher-based thermodynamics: Its Legendre transform and concavity properties Phys. Rev. E **60**, 48–53 (1999)
3. Frieden, B.R., Plastino, A., Soffer, B.H.: Population genetics from an information perspective. J. Theor. Biol. **208**, 49–64 (2001)
4. Gatenby, R.A., Frieden, B.R.: Application of information theory and extreme physical information to carcinogenesis. Cancer Res. **62**, 3675–3684 (2002)
5. Hawkins, R.J., Frieden, B.R.: Fisher information and equilibrium distributions in econophysics. Phys. Lett. A **322**, 126 (2004)
6. Frieden, B.R.: *Science from Fisher Information, 2nd ed.* (Cambridge Univ. Press, 2004)
7. Frieden, B.R., Gatenby, R.A. eds., *Exploratory Data Analysis using Fisher Information* (Springer, London, 2007)
8. Gatenby, R.A., Frieden, B.R.: Inducing catastrophe in malignant growth. J. Math. Med. Biol. **25**, 267–283 (2008)
9. Frieden, B.R., Gatenby, R.A.: Power laws of complex systems from extreme physical information. Phys. Rev. E **72**, 036101, 1–10 (2005)
10. Frieden, B.R., Plastino, A.: Higgs mass generation from the standpoint of information. Phys. Lett. A **278**, 299–306 (2001)
11. Nagy, A.: Fisher information in density functional theory. J. Chem. Phys. **119**, 9401–9405 (2003)
12. Nalewajski, R.F.: Aspects of the kinetic energy non-additivity in molecular and model subsystems. Mol. Phys. **10**, 2369–2379 (2003)
13. Liu, S.: On the relationship between densities of Shannon entropy and Fisher information for atoms and molecules. J. Chem. Phys. **126**, 191107 (2007)
14. Frieden, B.R.: *Probability, Statistical Optics and Data Testing, 3rd ed.* (Springer, Berlin, 2001) pp. 428–430

15. Wheeler, J.A.: in *Proceedings of the 3rd International Symposium on Foundations of Quantum Mechanics, Tokyo, 1989,* eds. S. Kobayashi, H. Ezawa, Y. Murayama and S. Nomura (Physical Society of Japan, Tokyo, 1990)
16. Popper, K.: *Logik der Forschung* (Julius Springer, Vienna, 1934; 8th German edn., J. C. B. Mohr [Paul Siebeck], Tübingen, 1984). Engl. transl: *The Logic of Scientific Discovery* (Hutchinson, London, 1959)
17. Van Trees, H.L.: *Detection, Estimation and Modulation Theory, Part I* (Wiley, New York 1968)
18. Frieden, B.R.: *Probability, Statistical Optics and Data Testing, 3rd ed.* (Springer, Berlin, 2001) pp. 345, 389–395
19. Huber, P.J.: *Robust Statistics* (Wiley, New York, 1981) pp. 77–86
20. Morgenstern, O., von Neumann, J.: *Theory of Games and Economic Behavior* (Princeton University Press, NJ, 1947). See also an introduction, in H. Lass, *Elements of Pure and Applied Mathematics* (McGraw-Hill, New York, 1957), 368–372
21. Venkatesan, R.C.: "Statistical cryptography using a Fisher-Schrodinger model," in *Foundations of Computational Intelligence, IEEE Symposium FOCI 2007, Honolulu* (IEEE, Honolulu, HI, April 2007) 487–494; also, "Fisher–Schrodinger models for statistical encryption of covert information," in *Quantum Information and Computation V*, eds. E.J. Donkor, A.R. Pirich and H.E. Brandt, in Proceedings of the SPIE **6573** (2007) 65730O
22. Sears, S.B., Parr, R.G., Dinur, U.: On the quantum-mechanical kinetic energy as a measure of the information in a distribution. Israel J. Chem. **19**, 165–173 (1980)
23. Narayanan, K.R., Srinivasa, A.R.: On the Thermodynamic Temperature of a General Distribution. arXiv:0711.1460v2 [cond-mat.stat-mech] 10 Nov 2007
24. Frieden, B.R., Soffer, B.H.: De Broglie's wave hypothesis from a Fisher-based approach to random particle deflection (paper in preparation)
25. Cisneros, C., Martinez, R., Nunez, H., Salas, A.: Limitations on the superposition principle: superselection rules in non-relativistic quantum mechanics. Eur. J. Phys. **19**, 237–243 (1998)
26. Tobin, J.: A General Equilibrium Approach to Monetary Theory. *J. Money, Credit, Bank.* **1**, 15–29 (1969)

Chapter 16
Entropy and Cloning Methods for Combinatorial Optimization, Sampling and Counting Using the Gibbs Sampler

Reuven Rubinstein

Abstract We survey the latest developments in the indicator-based minimum cross-entropy and the MCMC methods for combinatorial optimization (COP's), counting, sampling and rare-event probability estimation as well as we present some new material. The main idea of the indicator-based minimum cross-entropy method, called the *indicator MinxEnt*, or simply IME, is to associate with each counting or optimization problem an auxiliary *single-constrained* convex MinxEnt program of a special type, which has a closed-form solution. The main idea of the MCMC approach is to design a sequential sampling plan, where the "difficult" problem of estimating rare-event probability and counting the cardinality of a set is decomposed into "easy" problems of counting the cardinality of a sequence of related sets. Here we also propose a new algorithm, called the *cloning* algorithm. The main differences between the existing and the proposed algorithm is that the latter one has a special device, called the "cloning" device, which makes the algorithm very fast and accurate. We present efficient numerical results, while solving quite general integer and combinatorial optimization problems as well as counting ones, like SAT and Hamiltonian cycles.

16.1 Introduction

In this paper we survey the latest developments in the cross-entropy combined with *importance sampling* (IS) and the MCMC methods for combinatorial optimization (COP's), counting and rare-event probability estimation, as well as present some new material. In particular, we propose a new algorithm, called the *cloning* algorithm. The main differences between the existing and the proposed algorithm is that

R. Rubinstein
Faculty of Industrial Engineering and Management, Technion – Israel Institute of Technology, Haifa 32000, Israel
e-mail: ierrr01@ie.technion.ac.il

F. Emmert-Streib, M. Dehmer (eds.), *Information Theory and Statistical Learning*,
DOI: 10.1007/978-0-387-84816-7_16,

the latter has a special device, called the "cloning" device, which makes it very fast and accurate for COP's, counting, sampling uniformly on different complex regions, and rare-event probability estimation. In particular, our algorithm is well suited for counting the satisfiability assignments in a SAT problem and solving problems associated with the Boltzmann distribution, like estimating the partition functions in an Ising model and for sampling random variables uniformly distributed on different convex bodies.

16.1.1 The IME Method

The main idea of the indicator-based minimum cross-entropy method, called the *indicator MinxEnt*, or simply IME, is to associate with each counting or optimization problem an auxiliary *single-constrained* convex MinxEnt program of a special type, which has a closed-form solution. We prove that the optimal pdf obtained from the solution of such a specially designed MinxEnt program is a zero variance pdf, provided the "temperature" parameter is set to minus infinity. In addition we prove that the parametric pdf based on the product of marginals obtained from the optimal zero variance pdf coincides with the parametric pdf of the standard *cross-entropy* (CE) method. Thus, we show that originally designed at the end of the 1990s as a heuristic for estimation of rare-events and COP's, CE has a strong connection with the classic MinxEnt, and thus, has a strong mathematical foundation.

The crucial difference between the proposed IME method and its CE counterparts lies in their simulation-based versions: in the latter we always require to generate (via Monte Carlo) a sequence of tuples including the temperature parameter and the parameter vector in the optimal marginal pdf's, while in the former we can fix in advance the temperature parameter (to be set to a large negative number) and then generate (via Monte Carlo) a sequence of parameter vectors of the optimal marginal pdf's alone. In addition, in contrast to CE, neither the elite sample nor the rarity parameter is needed in IME. As result, the proposed IME Algorithm becomes simpler, faster and at least as accurate as the standard CE.

16.1.2 Randomized Algorithms for Counting

The main idea of *randomized algorithms* is to design a sequential sampling plan, where the "difficult" problem of counting $|\mathscr{X}^*|$ is decomposed into "easy" ones of counting the number of elements in a sequence of related sets $\mathscr{X}_1,\ldots,\mathscr{X}_m$. Typically, randomized algorithms explore the connection between counting and sampling problems and in particular the reduction from approximately counting the cardinality of a discrete set to approximately sampling elements of the set, where the sampling is performed by employing the classic MCMC method [15]. A typical randomized algorithm contains the following steps:

1. Formulate the counting problem as the problem of estimating the cardinality of some set $\mathscr{X}^*$.
2. Find sets $\mathscr{X}_0, \mathscr{X}_1, \ldots, \mathscr{X}_m$ such that $|\mathscr{X}_m| = |\mathscr{X}^*|$, and $|\mathscr{X}_0|$ is known.
3. Write $|\mathscr{X}^*| = |\mathscr{X}_m|$ as
$$|\mathscr{X}^*| = |\mathscr{X}_0| \prod_{j=1}^{m} \frac{|\mathscr{X}_j|}{|\mathscr{X}_{j-1}|}, \tag{16.1}$$
4. Develop an efficient estimator $\widehat{\zeta}_j$ for each $\zeta_j = |\mathscr{X}_j|/|\mathscr{X}_{j-1}|$, resulting in an efficient estimator
$$\widehat{|\mathscr{X}^*|} = |\mathscr{X}_0| \prod_{j=1}^{m} \widehat{\zeta}_j, \tag{16.2}$$
for $|\mathscr{X}^*|$

and similar for rare event estimation. Algorithms based on the sequential sampling estimator (16.2) are called in the computer literature [5], *randomized algorithms*.

It is shown in [5, 6, 15] that many interesting counting problems like estimating the volume of a convex body, the permanent of a non-negative matrix, counting the number of independence sets in a graph can be put into the setting (16.1), (16.2).

Example 16.1 (Independent Sets). Consider a graph $G = (V, E)$ with m edges and n vertices. Our goal is to count the number of independent node (vertex) sets of the graph $G = (V, E)$. A node set is called *independent* if no two nodes are connected by an edge, that is, if no two nodes are adjacent.

Consider an arbitrary ordering of the edges. Let E_j be the set of the first j edges and let $G_j = (V, E_j)$ be the associated sub-graph. Note that $G_m = G$, and that G_j is obtained from G_{j+1} by removing an edge. Denoting $\mathscr{X}_j$ the set of independent sets of G_i we can write $|\mathscr{X}^*| = |\mathscr{X}_m|$ in the form (16.1). Here $|\mathscr{X}_0| = 2^n$, since G_0 has no edges and thus every subset of V is an independent set, including the empty set. Note that here $\mathscr{X}_0 \supset \mathscr{X}_1 \supset \cdots \supset \mathscr{X}_m = \mathscr{X}^*$.

Example 16.2 (Knapsack Problem). Given items of sizes $a_1, \ldots, a_m > 0$ and a positive integer $b \geq \min_i a_i$, find the numbers of vectors $\mathbf{x} = (x_1, \ldots, x_n) \in \{0,1\}^n$, such that
$$\sum_{i=1}^{n} a_i x_i \leq b.$$
The integer b presents the size of the knapsack, and x_i indicates whether or not item i is put into the knapsack. Let $\mathscr{X}^*$ denote the set of all feasible solutions, that is, all different combinations of items that can be placed into the knapsack without exceeding its size. The goal is to determine $|\mathscr{X}^*|$.

To put the knapsack problem into the frame work (16.1), assume without loss of generality that $a_1 \leq a_2 \leq \cdots \leq a_n$ and define $b_j = \sum_{i=1}^{j} a_i$, with $b_0 = 0$. Denote $\mathscr{X}_j$ the set of vectors $\mathbf{x}$ that satisfy $\sum_{i=1}^{n} a_i x_i \leq b_j$, and let m be the largest integer such that $b_m \leq b$. Clearly, $\mathscr{X}_m = \mathscr{X}^*$. Thus, (16.1) is established again.

Quite often randomized algorithms deal with estimation of a partition function $Z(\lambda)$ of the system at some designed temperature via generation samples from the Boltzmann distribution. This can be achieved by computing the ratios of the partition functions similar to (16.1) for a carefully desired sequence of temperatures $0 = \lambda_0 < \lambda_1 < ... < \lambda_T = \lambda$, also called cooling schedule, where Z(0) is trivial to compute and the ratios $Z(\lambda_{i+1})/Z(\lambda_i)$ are easy to estimate by sampling from the distribution corresponding to $Z(\lambda_i)$ [15]. The challenging problem is to define a "smart" cooling schedule. The best known is an adaptive one due to [15], which has length $T = O^*(\sqrt{\ln Z(0)})$. In particular for some well-studied problems such as estimating the partition function of the Ising model, or approximating the number of colorings or matchings of a graph, their [15] cooling schedule is of length $O^*(\sqrt{n})$, where n is the size of the problem.

Here we propose a new algorithms, called the *cloning* algorithm. The main differences between the existing and the proposed algorithm is that the latter has a special device, called the "cloning" device, which makes it very fast and accurate. In particular, our algorithm is well suited for counting the satisfiability assignments in a SAT problem and solving problems associated with the Boltzmann distribution, like estimating the partition functions in an Ising model and for sampling random variables uniformly distributed on different convex bodies.

The rest of our paper is organized as follows. In Sect. 16.2 we show how rare-event probability estimation and counting can be treated using the MinxEnt the classic MinxEnt program. In particular, we establish the connections between counting, rare-events and MinxEnt, and we present our new MinxEnt method, which involves indicator functions in the MinxEnt programs and is called *indicator-based MinxEnt* or simply the *IME* program. We also establish the relationship of the proposed IME program to the earlier CE and MinxEnt ones treated in [9, 10]. In particular we show that the optimal pdf obtained from the IME program coincides with *zero variance* importance sampling (IS) pdf, provided the temperature parameter $\lambda = -\infty$. In Sect. 16.2.2 we present our main IME algorithms for counting. Here we also show how to count the cardinality of the set of feasible solutions of LIP's (linear integer programs) using the IME algorithm. Section 16.2.3 deals with IME for combinatorial optimization. Section 16.3 deals with the cloning method, for rare events, counting, combinatorial optimization and uniform sampling on complex convex bodies, involving sampling from the optimal zero variance IS pdf, rather than sampling from the parametric distributions as CE and IME do. Our inspiration and motivation comes from Botev and Kroese paper [1], to which we devote Sect. 16.3.3. Here we also have a Sect. 16.3.1 on the work of Diaconis and Holmes [3] and Ross [7], from which we adopted some basic ideas. Section 16.4 presents numerical results. In particular, Sect. 16.4.1 gives numerical results with the IME method, while Sect. 16.4.4 gives numerical results with the cloning method. It follows from our numerical results that for COP's both methods work well, while for counting problems the cloning method often outperforms IME. Finally, in Sect. 16.5 conclusions and some final remarks are given.

16.2 The MinxEnt Method for Rare Events, Counting, and Optimization

Here we establish the connection between rare-event probabilities, MinxEnt and counting. In particular we discuss how to employ MinxEnt for estimating the following rare-event probability

$$\ell = \mathbb{E}_u \left[I_{\{S(X) \geq b\}} \right], \tag{16.3}$$

were $S(X)$ is quite an arbitrary sample function, $X \sim f(\mathbf{x}; u)$, where $f(\mathbf{x}; u)$ is a fixed distribution parameterized by a vector u, and b is large number, so ℓ is a very small probability.

To estimate ℓ we can use the following non-parametric IS estimator

$$\widehat{\ell} = \frac{1}{N} \sum_{k=1}^{N} \left[I_{\{S(X_k) \geq b\}} \frac{f(X_k; u)}{g(X_k)} \right], \tag{16.4}$$

or using a parametric one

$$\widehat{\ell} = \frac{1}{N} \sum_{k=1}^{N} \left[I_{\{S(X_k) \geq b\}} \frac{f(X_k; u)}{f(X_k; p)} \right], \tag{16.5}$$

respectively. Here $X_1, \ldots, X_N$ in (16.4) and (16.5) is a random sample from $g(\mathbf{x})$ and from $f(\mathbf{x}; p)$, respectively. Note that p is a parameter vector, that is, typically different from u. At this point, it is crucial to note that in order to obtain a low-variance estimator $\widehat{\ell}$ we shall use below $g(\mathbf{x})$ and p obtained from the MinxEnt program.

If not stated otherwise, we assume below that $f(\mathbf{x}; u)$ is a *uniform* distribution. Since any counting quantity can be derived using the probability (16.3) (see [10]), we shall use the *same* IS pdfs $g(\mathbf{x})$ and $f(\mathbf{x}; p)$ given in (16.4) and (16.5) to estimate the counting quantity, denoted by $|\mathscr{X}^*|$. It is readily seen [10] that using $g(\mathbf{x})$ and $f(\mathbf{x}; p)$, the estimator of $|\mathscr{X}^*|$ can be written as

$$\widehat{|\mathscr{X}^*|} = \frac{1}{N} \sum_{k=1}^{N} I_{\{S(X_k) \geq b\}} \frac{1}{g(X_k)}, \tag{16.6}$$

and as

$$\widehat{|\mathscr{X}^*|} = \frac{1}{N} \sum_{k=1}^{N} I_{\{S(X_k) \geq b\}} \frac{1}{f(X_k; p)}, \tag{16.7}$$

respectively, provided again $f(\mathbf{x}; u)$ is a uniform distribution and $X_1, \ldots, X_N$ is a random sample either from $f(\mathbf{x}; p)$ or from g.

16.2.1 The MinxEnt Method for Rare Events

To establish the connection between MinxEnt and rare-events (see [9]), note that while estimating rare events probabilities using in (16.5) $g(\mathbf{x})$ as IS pdfs, it is common to take the ones obtained from the solution of the following *single constrained* MinxEnt program

$$\begin{aligned} \min_g \mathcal{D}(g,h) = \min_g \int \ln \frac{g(\mathbf{x})}{f(\mathbf{x})} \quad g(\mathbf{x})\mathrm{d}\mathbf{x} &= \min_g \mathbb{E}_g \left[\ln \frac{g(X)}{f(\mathbf{x})}\right] \\ \text{s.t. } \mathbb{E}_g[S(X)] &\geq b\,, \\ \int g(\mathbf{x})\mathrm{d}\mathbf{x} &= 1\,. \end{aligned} \tag{16.8}$$

In other words, one typically takes the optimal MinxEnt pdf $g(\mathbf{x})$ derived from (16.8), that is [13]

$$g(\mathbf{x}) = \frac{f(\mathbf{x})\exp\{-\lambda S(\mathbf{x})\}}{\mathbb{E}_f\left[\exp\{-\lambda S(X)\}\right]}\,, \tag{16.9}$$

where λ is obtained from the solution of the following system of equations

$$\frac{\mathbb{E}_f\left[S(X)\exp\{-\lambda S(X)\}\right]}{\mathbb{E}_f\left[\exp\{-\lambda_j S(X)\}\right]} = b \tag{16.10}$$

as the importance sampling pdf in (16.4). Alternatively, if $g(\mathbf{x})$ is a complex pdf (which is typically the case), one can *approximate* $g(\mathbf{x})$ by the product of its *marginal* pdf's $g_i(x_i) = f_i(x_i, p_i)$, $i = 1,\ldots,n$ [9], that is, use (16.5) instead of (16.4), where $f(\mathbf{x},p)$ is a parametric pdf, that differs from the prior pdf $f(x) = f(\mathbf{x},u)$ only in p.

We shall explain now how to derive the optimal parameter vector p while considering for simplicity the single constrained case (16.8) and assuming that $X = (X_1,\ldots,X_n)$ is a binary random vector with probabilities with independent components that is $X \sim \mathsf{Ber}(u)$, $u = (u_1,\ldots,u_n)$.

Indeed, applying the CE method to

$$Z(\lambda) = \mathbb{E}_u\left[\exp\{\lambda S(X)\}\right]\,, \tag{16.11}$$

which is called the *partition function* we immediately obtain

$$p_j = \frac{\mathbb{E}_u[X_j \exp\{-S(X)\lambda\}]}{\mathbb{E}_u[\exp\{-S(X)\lambda\}]}\,,\; j = 1,\ldots,n \tag{16.12}$$

with λ satisfying (16.10) (for $m = 1$).

The diagram connecting rare-events, MinxEnt and counting can be represented as

$$\{\mathbf{x} \in \mathbb{R}^n : S(\mathbf{x}) \geq b\} \longrightarrow \mathbb{E}_u I_{[}\{S(X) \geq b\}] \longrightarrow \text{MinxEnt (16.12)} \longrightarrow \text{Count as (16.7)}. \tag{16.13}$$

Note that the p_j's in (16.12) were extensively used in [9] for rare-event estimation and COP's while updating the parameter vector p using simulation. In this paper we shall use a different approach for deriving $g(\mathbf{x})$ and the associated parameter vector p.

16.2.1.1 Multiple Events with the Indicator-Based MinxEnt Method

Consider counting on the set

$$\mathscr{X}^* = \{\mathbf{x} \in \mathbb{R}^n : S_i(\mathbf{x}) \geq b_i,\ i = 1,\ldots,m\}\,, \tag{16.14}$$

where $S_i(\mathbf{x})$, $i = 1,\ldots,m$ are arbitrary functions. In this case we can associate with (16.14) the following *multiple-event* probability

$$\ell = \mathbb{P}_u \left\{ \bigcap_{i=1}^{m} [S_i(X) \geq b_i] \right\} = \mathbb{E}_u \left[\prod_{i=1}^{m} I_{\{S_i(X) \geq b_i\}} \right]. \tag{16.15}$$

Note that (16.15) extends (16.3) in the sense that it involves simultaneously an *intersection* of m events $\{S_i(X) \geq b_i\}$, that is, *multiple* events rather than a *single* one $\{S(X) \geq b\}$. Note also that some of the constraints may be equality ones, that is $\{S_i(X) = b_i\}$, that is the entire set $\mathscr{X}^*$ can be defined as

$$\begin{aligned} S_i(\mathbf{x}) &= b_i, \quad i = 1,\ldots,m_1\,, \\ S_j(\mathbf{x}) &\leq b_j, \quad j = m_1 + 1,\ldots,m\,. \end{aligned} \tag{16.16}$$

We assume that each individual event in (16.16) is *not rare*, that is each probability of the type $\mathbb{P}_u\{S_i(X) \geq b_i\}$ is not a rare-event probability, say $\mathbb{P}_u\{S_i(X) \geq b_i\} \geq 10^{-4}$, but their intersection forms a *rare-event probability* ℓ. Similar to the single-event case in (16.3) we are interested in efficient estimation of ℓ defined in (16.15). As before, we shall use the IS estimators (16.5) and (16.7). The crucial issue is how to approximate efficiently $g(\mathbf{x})$ and in particular how to estimate efficiently the parameter vector p in $f(\mathbf{x}, p)$.

The main idea here is to design an IS pdf $g(\mathbf{x})$ such that under $g(\mathbf{x})$ all constraints $\{S_i(\mathbf{x}) \geq b_i,\ i = 1,\ldots,m\}$ are fulfilled. This is equivalent of saying that the rare-event probability ℓ in (16.15) becomes certain under $g(\mathbf{x})$, that is,

$$\mathbb{E}_g \left[\prod_{i=1}^{m} I_{\{S_i(X) \geq b_i\}} \right] = 1\,. \tag{16.17}$$

In other words, (16.17) states that under such an *ideal* IS pdf $g(\mathbf{x})$ all m indicators must be equal to unity with probability 1. This can also be written as

$$\mathbb{P}_g \left\{ \left(\sum_{i=1}^{m} C_i(X) \right) = m \right\} = \mathbb{E}_g \left[I_{\{\mathscr{C}(X) = m\}} \right] = 1\,, \tag{16.18}$$

where

$$\mathscr{C}(X) = \sum_{i=1}^{m} C_i(X) = \sum_{i=1}^{m} I_{\{S_i(X) \geq b_i\}} , \tag{16.19}$$

and $C_i(X) = I_{\{S_i(X) \geq b_i\}}$. Similar to (16.17), formula (16.18) states that under $g(\mathbf{x})$ the probability of the sum of m indicator random variables $C_i(X)$ to be equal m (m is the number of indicators $C_i(X)$) must be equal 1.

We shall call the indicators $C_i(X) = I_{\{S_i(X) \geq b_i\}}$, $i = 1, \ldots, m$, the *local* ones to distinguish them from $\mathscr{C}(X) = \sum_{i=1}^{m} C_i(X) =$, which is called the *global* indicator.

It follows from the above that in order to satisfy all the multiple events $\{S_i(X) \geq b_i,\ i = 1, \ldots, m\}$ we can consider the following *single* constrained MinxEnt program

$$\mathbb{E}_g \left(\sum_{i=1}^{m} C_i(X) \right) = m. \tag{16.20}$$

That is, similar to (16.8) we define the following single-constrained MinxEnt program

$$\begin{aligned} min_g \mathcal{D}(g,h) = \min_g \int \ln \frac{g(\mathbf{x})}{f(\mathbf{x})} \quad g(\mathbf{x}) d\mathbf{x} = \min_g \mathbb{E}_g \left[\ln \frac{g(X)}{f(\mathbf{x})} \right] \\ \text{s.t.} \quad \mathbb{E}_g \left[\sum_{i=1}^{m} C_i(X) \right] = m \\ \int g(\mathbf{x}) d\mathbf{x} = 1 . \end{aligned} \tag{16.21}$$

In other words, in order to estimate the rare-event probability ℓ given in (16.15) and to count the cardinality of the set (16.14) we can use the single-constrained MinxEnt program (16.21). The solution of (16.21) is

$$g(\mathbf{x}) = Z^{-1}(\lambda) f(\mathbf{x}, u) \exp\left\{ -\lambda \sum_{i=1}^{m} C_i(\mathbf{x}) \right\} \tag{16.22}$$

where $Z(\lambda) = \mathbb{E}_u [\exp\{-\sum_{i=1}^{m} \lambda C_i(X)\}]$ is the partition functions and, as usual λ is the "temperature" obtained from the solution of the following equation

$$Z^{-1}(\lambda) \mathbb{E}_u \left[\sum_{i=1}^{m} C_i(X) \exp\left\{ -\lambda \sum_{j=1}^{m} C_j(X) \right\} \right] = m . \tag{16.23}$$

It is important to note that if (16.23) has no solution then the set $\{S_i(\mathbf{x}) \geq b_i,\ i=1, \ldots, m\}$ is empty.

It is crucial to note that the classic *multi-constrained* MinxEnt program involves *expectations of* $S_i(X)$, while the proposed single-constrained one (16.21) is based on the *expectations of the indicators of* $S_i(X)$, so the name indicator MinxEnt program or simply IME program.

For $m = 1$ the IME program (16.21) reduces to

$$\min_g \mathcal{D}(g,h) = \min_g \mathbb{E}_g \left[\ln \frac{g(X)}{f(\mathbf{x})} \right]$$
$$\text{s.t. } \mathbb{E}_g[C(X)] = 1 \qquad (16.24)$$
$$\int g(\mathbf{x})d\mathbf{x} = 1 ,$$

where $C(X) = I_{\{S(X) \geq b\}}$.

Observe also that in this case the single-constrained programs (16.24) and (16.8) do not coincide: in the former case we use an expectation of the indicator of $S(X)$, that is $\mathbb{E}\{I_{\{S(X)\geq b\}}\}$, while in the later case we use an expectation of $S(X)$, that is, $\mathbb{E}\{S(X)\}$. We shall treat the program (16.24) in more details in Sect. 16.2.3.1.

The following Lemmas 16.1–16.3 are taken from [11].

Lemma 16.1. *The optimal λ of the IME program (16.21) satisfying (16.23) is $\lambda = -\infty$. This means that in practice there is no need to solve (16.23) since one can set in (16.22) λ to a big negative number, like $\lambda = -100$*

Lemma 16.2. *The optimal pdf $g(\mathbf{x})$ in (16.22) corresponds to a* uniform *pdf over the set $\{\mathbf{x} \in \mathbb{R}^n : S_i(\mathbf{x}) \geq b_i,\ i = 1,\ldots,m\}$.*

Lemma 16.2 automatically implies that the optimal $g(\mathbf{x})$ is a zero-variance IS. Thus, solving the MinxEnt program (16.21) we obtained a zero variance IS sampling pdf $g(\mathbf{x},\lambda)$ with $\lambda = -\infty$.

Lemma 16.3. *For $\lambda = -\infty$ the optimal IME pdf $g(\mathbf{x})$ in (16.22) coincides with the zero-variance IS pdf (see [12])*

$$g^*(\mathbf{x}) = \frac{f(\mathbf{x},u) I_{\{\mathscr{C}(\mathbf{x})=m\}}}{\mathbb{E}_{\mathbf{u}}\left[I_{\{\mathscr{C}(\mathbf{x})=m\}}\right]}. \qquad (16.25)$$

Remark 16.1. Efficient sampling from the Boltzmann distribution of the form (16.9) (and that of (16.22)) with $\lambda = -\infty$ is essential in many applications and in particular in computer science. As mentioned in Sect. 16.1 the common approach [15] is to use a sequence of temperature parameters λ_t, $t = 1,\ldots,T$ in the MCMC-based methods with adjustable (adaptive) choice of λ_t and such that eventually $\lambda_T = -\infty$. Since both the Boltzmann distribution (16.9) with $\lambda = -\infty$ and the optimal zero variance IS pdf

$$g^*(\mathbf{x}) = \ell^{-1} f(\mathbf{x},u) I_{\{S(\mathbf{x})\geq b\}}, \qquad (16.26)$$

where $\ell = \mathbb{E}_{\mathbf{u}}\left[I_{\{S(\mathbf{x})\geq b\}}\right]$, *coincide*, it would be interesting to investigate situations for which the IS pdf $g^*(\mathbf{x})$ (16.26) is more beneficial than the MinxEnt pdf $g(\mathbf{x},\lambda)$ (16.9) and vice versa. Note that while using (16.26) one deals directly with the rare event probability $\ell = \mathbb{E}_{\mathbf{u}}\left[I_{\{S(\mathbf{x})\geq b\}}\right]$, rather than with the partition function $Z(\lambda) = \mathbb{E}_u\left[\exp\{\lambda S(X)\}\right]$. In particular, as far as the MCMC-based methods are concerned, one could use (in the former case) the sequence b_t, $t = 1,\ldots,T$ with $b_T = b$ rather than the sequence λ_t, $t = 1,\ldots,T$ with $\lambda_T = -\infty$.

It is also interesting to note that the zero variance IS pdf $g^*(\mathbf{x})$ is derived from the following variance minimization program [13]

$$\min_g \mathrm{Var}_g \left[I_{\{S(X)\geq b\}} \frac{h(X,u)}{g(X)} \right],$$

while the zero variance Boltzmann pdf $g(\mathbf{x})$ is obtained from the Kullback–Leibler program (16.8).

It is important to note that $g(\mathbf{x})$ (derived from the solution of the IME program (16.21) and based on the indicators of the sample functions $S_i(X)$) is an optimal *zero variance pdf*, while the optimal pdf $g(\mathbf{x})$ in the classic MinxEnt program, like (16.8) (based on the sample functions $S_i(X)$ by themselves), *is not a zero-variance pdf*, (it leads only to variance reduction).

Observe again that generating samples from a multidimensional Boltzmann pdf, like $g(\mathbf{x})$ in (16.22) is a difficult task. The existing MCMC (Markov Chain Monte Carlo) algorithms [13] are typically quite slow, in particular when the "temperature" parameter λ is high. Recently several efficient speed up MCMC procedures to sample efficiently from the Boltzmann pdfs have been proposed. Among these is the dynamic programming approach in [4] and adaptive cooling schedule in [15].

Returning back to $g(\mathbf{x})$ in (16.22) and taking into account that generating samples from $g(\mathbf{x})$ is difficult we shall approximate by the *product of its marginal pdf's* $g_i(x_i) = f_i(x_i, p_i)$, $i = 1,\dots,n$, that is, we shall write the components p_i, $i = 1,\dots,n$ of the optimal vector p as

$$p_i = \frac{\mathbb{E}_u\left[X_i \exp\{-\lambda \sum_{i=1}^m C_i(X)\}\right]}{\mathbb{E}_u\left[\exp\{-\sum_{i=1}^m \lambda C_i(X)\}\right]}, \quad i = 1,\dots,n, \tag{16.27}$$

which coincides with (16.12) up to the notations. Note that when each component of X is an arbitrary r-point discrete random variable then (16.27) extends to

$$p_{ij} = \frac{\mathbb{E}_u\left[I_{X_i=j} \exp\{-\lambda \sum_{i=1}^m C_i(X)\}\right]}{\mathbb{E}_u\left[\exp\{-\sum_{i=1}^m \lambda C_i(X)\}\right]}, \quad i = 1,\dots,n;\ j = 1,\dots,r. \tag{16.28}$$

It is important to note that formula (16.28) is similar to the corresponding CE one [12]

$$p_j^* = \frac{\mathbb{E}_u[X_j I_{\{\sum_{i=1}^m C_i(X)=m\}}]}{\mathbb{E}_u[I_{\{\sum_{i=1}^m C_i(X)=m\}}]} \tag{16.29}$$

with one main difference: the indicator function $I_{\{\sum_{i=1}^m C_i(X)=m\}}$ in the CE formula has been replaced by $\exp\{-\lambda \sum_{i=1}^m C_i(X)\}$.

In summary, to estimate efficiently ℓ and the associated counting quantity $|\mathscr{X}^*|$, we shall use again the IS estimator (16.5), where p in $f(\mathbf{x};p)$ is given in (16.27) and it is obtained from the solution of the IME program (16.21).

The diagram explaining the connection between the rare-events, IME and counting is similar to (16.13) and it can be presented as

$$\{S_i(\mathbf{x}) \geq b_i,\ i = 1,\dots,m\} \longrightarrow \mathbb{E}_u\left[\prod_{i=1}^m I_{\{S_i(X)\geq b_i\}}\right] \longrightarrow \text{IME (16.21), (16.27)}$$
$$\longrightarrow \text{Count via (16.7).} \tag{16.30}$$

Table 16.1 The vector p for several values of the vector $(b^{(1)}, b^{(2)})$

$b^{(1)}$	$b^{(2)}$	p_{11}	p_{12}	p_{13}	p_{21}	p_{22}	p_{23}	p_{31}	p_{32}	p_{33}	$\mathscr{S}(p)$	$\widehat{\lvert\mathscr{X}^*\rvert}$
(0,0,0)	(0,0,0)	0	0	0	0	0	0	0	0	0	0	1
(0,0,1)	(1,0,0)	0	0	1	0	0	0	0	0	0	0	1
(0,2,2)	(2,2,0)	0	1	1	0	1	1	0	0	0	0	1
(1,1,2)	(2,1,1)	0.6	0.6	0.8	0.2	0.2	0.6	0.2	0.2	0.6	5.19	5.07
(1,1,3)	(2,2,1)	0.5	0.5	1	0.5	0.5	1	0	0	1	2.77	1.97
(2,2,2)	(3,1,2)	1	1	1	0.33	0.33	0.33	0.66	0. 66	0.66	3.82	3.015
(3,3,3)	(3,3,3)	1	1	1	1	1	1	1	1	1	0	1

Example 16.3 (Counting 0–1 Tables with Fixed Margins). The set $A\mathbf{x} = b$ is given as

$$\sum_{i=1}^{n} x_{ij} = b_j^{(1)}, \quad j = 1,\ldots,m$$

$$\sum_{j=1}^{m} x_{ij} = b_i^{(2)}, \quad i = 1,\ldots,n \tag{16.31}$$

$$X_{ij} \in \{0,1\}, \quad \forall i,j.$$

Table 16.1 presents such a summary of the results.

Note that for the extreme values of the vectors $b^{(1)}$ and $b^{(2)}$, namely for $b^{(1)} = b^{(2)} = (0,0,0)$ and $b^{(1)} = b^{(2)} = (3,3,3)$, we obtain degenerated solutions (all components of p are either zeros or all ones, respectively). In this case, we also have $\mathscr{S}(p) = 0$ and $\widehat{|\mathscr{X}^*|} = 1$, as expected. Similar, for $b^{(1)} = (1,1,3)$ and $b^{(2)} = (2,2,1)$ we obtain two feasible solutions with the corresponding values of X: $X_1 = (1,0,0,0,1,0,1,1,1)$ and $X_2 = (0,1,0,1,0,0,1,1,1)$, respectively. In this case the estimate of $|\mathscr{X}^*|$ (based on a sample $N = 1{,}000$) is $\widehat{|\mathscr{X}^*|} = 1.9712$ and similarly for the other values of $(b^{(1)}, b^{(2)})$.

Consider finally the extreme case where m and n are even, $b_j^{(1)} = \frac{n}{2}, j = 1,\ldots,m$, and $b_i^{(2)} = \frac{m}{2}, i = 1,\ldots,n$. In this case it is readily seen that the optimal IME vector $p = (1/2,\ldots,1/2)$, that is, it coincides with the original one u and, thus the IME based on $f(\mathbf{x},p)$ is useless.

16.2.1.2 The Connection Between CE and IME

To establish the connection between CE and IME we need the following

Theorem 16.1. *For $\lambda = -\infty$ the optimal parameter vector p in (16.27), associated with the marginal pdfs of the optimal IME pdf $g(\mathbf{x},\lambda)$ in (16.22), coincides with the optimal parameter vector p in (16.29), which is associated with the CE method. The latter is obtained from the solutions of the following cross-entropy program [13]*

$$\min_p \mathbb{E}_{g^*}\left[\ln \frac{g^*(X)}{f(X,p)}\right], \tag{16.32}$$

where $g^*(\mathbf{x})$ is the zero variance IS pdf defined in (16.25).

Proof. The proof is given in [11]. □

Theorem 16.1 is crucial for the foundations of the CE method. Indeed, designed originally in [8] as a heuristics for rare-event estimation and COP's, Theorem 16.1 states that CE has strong connections with the IME program (16.21) and, thus, has strong mathematical foundation since the latter is so. The important connection between CE and IME is that the optimal IME parametric pdf $f(\mathbf{x},p) = f(\mathbf{x},p,\lambda)$ (with p in (16.27) and $\lambda = -\infty$) and the CE pdf (with p as in (16.29)) obtained heuristically from the solution of the cross-entropy program (16.32) are the same.

It follows from above that as an alternative to the original CE program (16.32) one can consider the following one

$$\min_p \mathbb{E}_{g}\left[\ln \frac{g(X,\lambda)}{f(X,p)}\right], \tag{16.33}$$

where $g(\mathbf{x},\lambda)$ (with $\lambda = -\infty$) *is the zero variance IME pdf* (16.22).

Clearly, solving, say for the Bernoulli random variables the original CE program (16.32) and the one in (16.33) we immediately obtain the optimal parameter vectors p in (16.29) and (16.27), respectively.

16.2.2 The IME Counting Algorithm

Here we present the indicator-based Minx-Ent (IME) counting algorithm for rare-events and counting the number of feasible solutions on the set $\mathscr{X}^*$ defined (16.16).

We call our method, the indicator-based MinxEnt (IME), to distinguish it from the cross-entropy (CE) for the following reasons. As we shall see below

1. CE
   - Generates iteratively a sequence of tuples $\{\widehat{p}_t, \widehat{m}_t\}$, where $\widehat{p}_t$ and $\widehat{m}_t$, denote the estimates of the optimal parameter vector in the parametric pdf $f(\mathbf{x},p)$ and the approximation of m at the t-th iteration, respectively.
   - Involves a *rarity parameter* ρ and *elite sampling*, while generating the sequence $\{\widehat{p}_t, \widehat{m}_t\}$.

2. IME
   - Generates only a sequence of vectors $\{p_t\}$, since, as mentioned earlier, the temperature parameter λ can be fixed in advance (as a large negative number).
   - Neither the rarity parameter ρ, nor the elite samples are involved in IME.

For this reason, the indicator-based IME is much simpler than CE and the classic MinxEnt and, as it follows from our numerical results, it is faster and at least as accurate as its counterparts CE and MinxEnt.

It is also important to keep in mind that the IME method is based on the MinxEnt program (16.21), which integrates the set (multiple events) defined in (16.16) into a single (stochastic) one given as $\mathbb{E}_g[\sum_{i=1}^m C_i(X)] = m$.

As mentioned, since we cannot sample from $g(\mathbf{x})$ we approximate it by the product of its marginal pdf's $g_i(x_i) = f_i(x_i, p_i)$, $i = 1, \ldots, n$, where the components p_i, $i = 1, \ldots, n$ of the optimal vector p are given in (16.27). As soon as an estimate of p is derived we estimate ℓ and $|\mathscr{X}^*|$ using the IS estimator (16.5) and (16.7), respectively.

We proceed below with the standard CE and the IME algorithms.

16.2.2.1 The Standard CE Algorithm

In the standard CE, one uses a *multi-level* approach, that is, one generates simultaneously a sequence of the parameter vector p_t of the parametric pdf's $f(\mathbf{x}, p_t)$ and levels $\{m_t\}$. Starting with $f(\mathbf{x}, p_0) = f(\mathbf{x}, u)$, that is, taking the prior $f(\mathbf{x}, u) = f(\mathbf{x}, p_0)$, one

1. Updates m_t as
$$m_t = \mathbb{E}_{g_{t-1}}[\mathscr{C}(X) \mid \mathscr{C}(X) \geq q_t],$$
where q_t is the $(1-\rho)$-quantile of $\mathscr{C}(X)$ under g_t and as before $\mathscr{C}(X) = \sum_{i=1}^m C_i(X)$.
2. Updates g_t as the solution to the above MinxEnt program for level m_t, rather than m.

Specifically, m_t can be estimated from a random sample $X_1, \ldots, X_N$ of g_{t-1} as the average of the $N_e = \lceil \rho N \rceil$ elite sample performances:

$$\widehat{m}_t = \frac{\sum_{i=N-N_e+1}^{N} \mathscr{C}_{(i)}}{N_e}, \tag{16.34}$$

where $\mathscr{C}_{(i)}$ denotes the i-th order-statistics of the sequence $\mathscr{C}(X_1), \ldots, \mathscr{C}(X_N)$.

The updating of p in CE is performed according to

$$\widehat{p}^{*}_{t,j} = \frac{\sum_{k=1}^{N} X_{kj}\, I_{\{\mathscr{C}(X_k) \geq \widehat{m}_t\}}\, W(X_k; u, \widehat{p}_{t-1})}{\sum_{k=1}^{N} I_{\{\mathscr{C}(X_k) \geq \widehat{m}_t\}}\, W(X_k; u, \widehat{p}_{t-1})}. \tag{16.35}$$

Note that since the prior pdf $f(\mathbf{x}, u)$ is uniform (in our case $f(\mathbf{x}, u) = \text{Ber}\,(u)$, $u = (1/2, \ldots, 1/2)$ we can write for convenience $W(X_k; u, \widehat{p}_{t-1})$ as $\frac{1}{f(X_k; \widehat{p}_{t-1})}$.

16.2.2.2 The IME Counting Algorithm

Since in IME λ is fixed (λ is a large negative number), the components of p can be updated according to the following formula

$$\widehat{p}_{t,j} = \frac{\sum_{k=1}^{N} X_{kj} \exp\{-\lambda \mathscr{C}(X_k)\} \, W(X_k; u, \widehat{p}_{t-1})}{\sum_{k=1}^{N} \exp\{-\lambda \mathscr{C}(X_k)\} \, W(X_k; u, \widehat{p}_{t-1})}. \tag{16.36}$$

For application purposes we not only set λ to a large negative number, like $\lambda = -100$, but we also use in (16.36) instead of $\mathscr{C}(X_k)$, its so-called normalized value

$$\mathscr{C}^{(n)}(X_k) = \frac{\mathscr{C}(X_k)}{\max_{k=1,\dots,N} \mathscr{C}(X_k)} \,. \tag{16.37}$$

Using (16.37) the resulting updating of $\widehat{p}_t$ can be written as

$$\widehat{p}_{t,j} = \frac{\sum_{k=1}^{N} X_{kj} \exp\{-\lambda \mathscr{C}^{(n)}(X_k)\} \, W(X_k; u, \widehat{p}_{t-1})}{\sum_{k=1}^{N} \exp\{-\lambda \mathscr{C}^{(n)}(X_k)\} \, W(X_k; u, \widehat{p}_{t-1})}, \tag{16.38}$$

The main reason for using $\mathscr{C}^{(n)}(X)$ instead of $\mathscr{C}(X)$ is for convenience only; to make sure that $\lambda \mathscr{C}^{(n)}(X_k)$ is a large negative number, say $\lambda \mathscr{C}^{(n)}(X_k) = -100$, when $\mathscr{C}^{(n)}(X_k) = 1$.

Algorithm 16.2.1 (IME Algorithm for Counting)

1. *Define* $\widehat{p}_0 = u$. *Set* $\lambda = M$, *say* $M = -100$. *Set* $t = 0$ *(iteration = level counter).*
2. $t \leftarrow t+1$. *Generate a sample* $X_1, \dots, X_N$ *from the density* $f(\mathbf{x}; \widehat{p}_{t-1})$ *and compute* $\widehat{p}_t$ *according to (16.38).*
3. *Smooth out the vector* $\widehat{p}_t$ *according to*

$$\overline{p}_t = \alpha \widehat{p}_t + (1-\alpha) \widehat{p}_{t-1} \,, \tag{16.39}$$

   where α, $(0 < \alpha < 1)$ *is called the* smoothing parameter.
4. *If* $\mathscr{C}(X) < m$, *reiterate from step 2. Else proceed with step 5.*
5. *Reiterate steps 2–3. for 2–4 more iterations. Estimate the counting quantity* $|\mathscr{X}^*|$ *as*

$$\widehat{|\mathscr{X}^*|} = \frac{1}{N} \sum_{k=1}^{N} I_{\{\mathscr{C}(X_k)=m\}} \frac{1}{f(X_k; \widehat{p}_t)} \,. \tag{16.40}$$

Our numerical results of Sect. 16.4.1 clearly show that the IME Algorithm 16.2.1 is quite robust with respect to λ, provided λ is a large negative number, say $-50 \geq \lambda \geq -1{,}000$. To see this, let $\lambda = -100$ and assume for simplicity that

$\mathscr{C}^{(n)}(X_k)\}$ obtains values from the set $\{1,\ 0.9,\ldots,0.1\}$. In this case, the updating of the parameter vector p according to (16.38) will be based on the following *exponential* sequence $\{\exp(100),\ \exp(90),\ldots,\exp(10)\}$. Clearly, the dominating term is $\exp(100)$, while the remaining ones are negligible. Similar conclusions hold for some other large negative values of λ, like $-50 \geq \lambda \geq -1{,}000$.

Remark 16.2 (Convergence of Algorithm 16.2.1). Since for fixed λ Algorithm 16.2.1 updates only the single parameter vector $\widehat{p}$, the convergence and the speed of the convergence of $\widehat{p}$ to the true optimal parameter vector p^* with the components

$$p_j = \frac{\mathbb{E}_u X_j \exp\{-\lambda\, \mathscr{C}^{(n)}(X)\}}{\mathbb{E}_u \exp\{-\lambda\, \mathscr{C}^{(n)}(X)\}} \tag{16.41}$$

follows from Theorems A1 and A2 of [14].

16.2.2.3 IME for Counting the Number of Feasible Solutions in Integer Programs

Here we show how to apply Algorithm 16.2.1 to count the set of feasible solutions

$$\begin{aligned} &\sum_{k=1}^{n} a_{ik} x_k = b_i\,, \quad i = 1,\ldots,m_1\,,\\ &\sum_{k=1}^{n} a_{jk} x_k \geq b_j\,, \quad j = m_1+1,\ldots,m_1+m_2\,,\\ &\mathbf{x} \geq 0\,,\ x_k \text{ integer } \forall k = 1,\ldots,n \end{aligned} \tag{16.42}$$

of the following integer program containing both equality and inequality constraints

$$\begin{aligned} &\min c'\mathbf{x},\\ \text{s.t.}\ &\sum_{k=1}^{n} a_{ik} x_k = b_i\,, \quad i = 1,\ldots,m_1\,,\\ &\sum_{k=1}^{n} a_{jk} x_k \geq b_j\,, \quad j = m_1+1,\ldots,m_1+m_2\,,\\ &\mathbf{x} \geq 0\,,\ \ x_k \text{ integer } \forall k = 1,\ldots,n\,. \end{aligned} \tag{16.43}$$

Here c and $\mathbf{x}$ are n-dimensional vectors.

It is readily seen that in this case Algorithm 16.2.1 is directly applicable, provided the first m_1 terms $C_i(X)$ in (16.20) (out of the total of $m = m_1 + m_2$ ones) are defined as

$$C_i(X) = I_{\{\sum_{k=1}^{n} a_{ik} X_k = b_i\}}\,, \quad i = 1,\ldots,m_1, \tag{16.44}$$

while the remaining m_2 ones are defined as

$$C_i(X) = I_{\{\sum_{k=1}^{n} a_{ik} X_k \geq b_i\}}\,, \quad i = m_1+1,\ldots,m_1+m_2. \tag{16.45}$$

In short, in order to count the number $|\mathscr{X}^*|$ of feasible solutions of the set (16.42) we associate with it the following rare-event probability

$$\ell = \mathbb{P}_u\{X \in \mathscr{X}^*\} = \mathbb{E}_u \left[\prod_{i=1}^{m_1} I_{(\sum_{k=1}^{n} a_{ik}X_k = b_i)} \prod_{j=m_1+1}^{m_1+m_2} I_{(\sum_{k=1}^{n} a_{jk}X_k \geq b_j)} \right] \tag{16.46}$$

and then apply to it Algorithm 16.2.1.

Example 16.4 (SAT Example). As a simple example consider the following SAT assignment

$$(x_1 + \bar{x}_2)(\bar{x}_1 + \bar{x}_2 + x_3)(x_2 + x_3) .$$

In this case we have the following system of linear constraints

$$\begin{aligned} x_1 + (1 - x_2) &\geq 1 \\ (1 - x_1) + (1 - x_2) + x_3 &\geq 1 , \\ x_2 + x_3 &\geq 1 , \end{aligned}$$

where each $x_1, x_2, x_3 \in \{0, 1\}$, where $\bar{x}_i = 1 - x_i$.

Proceeding with our example, we can write ℓ as $\ell = \mathbb{P}_u(C_1 + C_2 + C_3 = 3)$, where $C_1 = I_{\{X_1 - X_2 \geq 0\}}$, $C_2 = I_{\{X_1 + X_2 - X_3 \leq 1\}}$ and $C_3 = I_{\{X_2 + X_3 \geq 1\}}$.

16.2.3 IME for Combinatorial Optimization

We consider here both unconstrained and constrained optimization.

16.2.3.1 Unconstrained Combinatorial Optimization

Consider the following non-smooth (continuous or discrete) unconstrained optimization program.

$$\max_{\mathbf{x} \in \mathbb{R}^n} S(\mathbf{x}) .$$

Denote by b^*, the optimal function value.

In this case the MinxEnt program becomes

$$\begin{aligned} &\min_g \mathcal{D}(g, h) = \min_g \mathbb{E}_g \left[\ln \frac{g(X)}{f(\mathbf{x})} \right] \\ &\text{s.t. } \mathbb{E}_g\{I_{\{S(X) \leq b\}}\} = 1 \\ &\int g(\mathbf{x}) \mathrm{d}\mathbf{x} = 1 . \end{aligned} \tag{16.47}$$

The corresponding updating of the component of the vector $\widehat{p}_t$ can be written as

$$\widehat{p}_{t,j} = \frac{\sum_{k=1}^{N} X_{kj} \exp\{-\lambda I_{\{S(X_k) \le \widehat{b}_t\}}\}}{\sum_{k=1}^{N} \exp\{-\lambda I_{\{S(X_k) \le \widehat{b}_t\}}\}}, \tag{16.48}$$

where λ is a big negative number.

Algorithm 16.2.2 (IME Unconstrained Optimization Algorithm)

1. *Define $\widehat{p}_0 = u$, say choose, $f(\mathbf{x}, u)$ uniformly distributed over $\mathscr{X}$. Set λ to a big negative number, say $\lambda = -100$. Set $t = 1$ (iteration = level counter).*
2. *Generate a sample $X_1, \dots, X_N$ from the density $f(\mathbf{x}; \widehat{p}_{t-1})$ and compute the elite sampling value $\widehat{b}_t$ of $S(X_1), \dots, S(X_N)$.*
3. *Use the same sample $X_1, \dots, X_N$ and compute $\widehat{p}_t$, according to (16.48).*
4. *Smooth out the vector $\widehat{p}_t$ according to ((16.39)).*
5. *If the stopping criterion is met, stop; otherwise, set $t = t+1$ and return to Step 2.*

As a stopping criterion one can use for example: if for some $t \ge d$, say $d = 5$,

$$\widehat{b}_{t-1,(N)} = \widehat{b}_{t,(N)} = \cdots = \widehat{b}_{t-d,(N)} \tag{16.49}$$

then stop.

16.2.3.2 Constrained Combinatorial Optimization: The Penalty Function Approach

Consider the particular case of the problem (16.43) with inequality constraints only, that is

$$\begin{aligned} &\max_{\mathbf{x}} \sum_{k=1}^{n} c_k x_k \\ \text{s.t. } &\sum_{k=1}^{n} a_{ik} x_k \ge b_i\,, \quad i = 1, \dots, m\,, \\ &\mathbf{x} \ge 0\,,\; x_k \text{ integer } \forall\, k = 1, \dots, n\,. \end{aligned} \tag{16.50}$$

Assume in addition that the vector $\mathbf{x}$ is binary and all components b_i and a_{ik} are positive numbers. Using the penalty method approach we can reduce the original constraint problem (16.50) to the following unconstrained one

$$\min_{\mathbf{x}}\{S(\mathbf{x}) = \sum_{k=1}^{n} c_k x_k + M(\mathbf{x})\}\,, \tag{16.51}$$

where the penalty function is defined as

$$M(\mathbf{x},\beta) = \beta \sum_{i=1}^{m} \min\left\{ \sum_{k=1}^{n} a_{ik}x_k - b_i, 0 \right\}. \tag{16.52}$$

Here

$$\beta = \frac{a + \sum_{k=1}^{n} c_k}{\min_{ik}(a_{ik} - b_i)}, \tag{16.53}$$

where a is a positive number. If not stated otherwise, we assume that $a = 1$. Note that the penalty parameter β is chosen such that if $\mathbf{x}$ satisfies all constraints in (16.50), then $M(\mathbf{x}) = 0$ and $S(\mathbf{x}) \geq 0$. Alternatively, if $\mathbf{x}$ does not satisfy all constraints in (16.50), then $M(\mathbf{x}) \leq -(a + \sum_{k=1}^{n} c_k x_k)$ and $S(\mathbf{x}) \leq -a$. Clearly the optimization program (16.50) can again be associated with the rare-event probability estimation problem, where $m \in (0, \sum_{k=1}^{n} c_k)$ and X is a vector of iid Ber(1/2) components.

16.3 The Cloning Method for Rare Events, Counting, Sampling and Optimization

Consider estimation of the following rare event probability

$$\ell(m) = \mathbb{E}_f\left[I_{\{S(X) \geq m\}}\right]. \tag{16.54}$$

Here $S(X)$ is the sample performance, $X \sim f(\mathbf{x})$, m is fixed.

To estimate $\ell(m)$ we shall use the well known chain rule (nested events) according to which the desired probability $\ell(m)$ can be written either as

$$\ell(m) = \mathbb{E}_f[I_{\{S(X) \geq m_0\}}] \prod_{t=1}^{T} \mathbb{E}_f[I_{\{S(X) \geq m_t\}} | I_{\{S(X) \geq m_{t-1}\}}] = c_0 \prod_{t=1}^{T} c_t, \tag{16.55}$$

or as

$$\ell(m) = \mathbb{E}_f[I_{\{S(X) \geq m_0\}}] \prod_{t=1}^{T} \mathbb{E}_{g^*_{t-1}}[I_{\{S(X) \geq m_t\}}] = c_0 \prod_{t=1}^{T} c_t, \tag{16.56}$$

where

$$c_t = \mathbb{E}_f[I_{\{S(X) \geq m_t\}} | I_{\{S(X) \geq m_{t-1}\}}] = \mathbb{E}_{g^*_{t-1}}[I_{\{S(X) \geq m_t\}}] \tag{16.57}$$

and $c_0 = \mathbb{E}_f[I_{\{S(X) \geq m_0\}}]$. Here $\{m_t,\ t = 0, 1, \ldots, T\}$ is a fixed grid satisfying $-\infty < m_0 < m_1 < \cdots < m_T = m$; f denotes the proposal pdf $f(\mathbf{x})$; and $g^*_{t-1} = g^*(\mathbf{x}, m_t)$ denotes the zero variance importance sampling (IS) pdf at iteration t that is

$$g^*(\mathbf{x}, m_t) = \ell_t^{-1} f(\mathbf{x}) I_{\{S(\mathbf{x}) \geq m_t\}}, \tag{16.58}$$

where $\ell_t = \mathbb{E}_f\left[I_{\{S(X) \geq m_t\}}\right]$ is the normalization constant. Thus, an estimator of $\ell(m)$ can be obtained by taking the product of conditional expectations (probabilities)

$$\mathbb{E}_f[I_{\{S(X) \geq m_t\}} | I_{\{S(X) \geq m_{t-1}\}}],\ t = 0, 1, \ldots, T$$

under f, or as the product of the unconditional ones

$$c_t = \mathbb{E}_{g^*_{t-1}}[I_{\{S(X)\geq m_t\}}]$$

under g^*_{t-1}, that is it can be written as

$$\widehat{\ell}(m) = \prod_{t=0}^{T} \widehat{c}_t = \frac{1}{N^T} \prod_{t=0}^{T} N_t, \tag{16.59}$$

where

$$\widehat{c}_t = \frac{1}{N} \sum_{i=1}^{N} I_{\{S(X_i)\geq m_t\}} = \frac{N_t}{N}. \tag{16.60}$$

Note also that g^*_{t-1} in $c_t = \mathbb{E}_{g^*_{t-1}}[I_{\{S(X)\geq m_t\}}]$ means that we use the zero variance IS pdf g^*_{t-1} of level m_{t-1} to calculate c_t at level m_t. *So, the estimator $\widehat{c}_t$ of $c_t = \mathbb{E}_{g^*_{t-1}}[I_{\{S(X)\geq m_t\}}]$ is not a zero variance IS estimator, since $m_{t-1} \neq m_t$.* In fact, we will see that in most interesting counting models g^*_{t-1} is *uniformly* distributed on the set $\{\mathbf{x} : S(\mathbf{x}) \geq m_t\}$. Note finally that *the sequence $\{c_t\}$ will be required only for rare events and counting, but not for optimization problems.*

For such an estimator to be useful, the levels m_t should be chosen such that each quantity $\mathbb{E}_f[I_{\{S(X)\geq m_t\}} | \, I_{\{S(X)\geq m_{t-1}\}}]$ is not too small, say approximately equal to 10^{-2}. Note only we assume that the levels m_t are chosen such that each c_t is not to small, but we shall also require *existence of at least one sequence* $\{m_0, m_1, \ldots, m_T = m\}$ to insure that all c_t, $t = 0, 1, \ldots, T$ values are not to small, that is they all are bounded, say by what we call the *rarity parameter* ρ, $\rho = 10^{-2}$.

In both counting and optimization problems we shall generate an adaptive sequence of tuples

$$\{(m_0, g^*(\mathbf{x}, m_{-1})),\ (m_1, g^*(\mathbf{x}, m_0)),\ (m_2, g^*(\mathbf{x}, m_1)), \ldots, (m_T, g^*(\mathbf{x}, m_{T-1}))\}, \tag{16.61}$$

where as before $g^*(\mathbf{x}, m_{-1}) = f(\mathbf{x})$. This is in contrast to CE where we generate a sequence of tuples

$$\{(m_0, v_0),\ (m_1, v_1), \ldots, (m_T, v_T)\}, \tag{16.62}$$

where $\{v_t,\ t = 1, \ldots, T\}$ is a sequence of parameters in the parametric family of fixed distributions $f(\mathbf{x}, v_t)$. The crucial difference is, of course, that in our approach, $\{g^*(\mathbf{x}, m_{t-1}) = g^*_{t-1},\ t = 0, 1, \ldots, T\}$ is a sequence of *zero variance pdfs or their approximation, rather than is a sequence $\{v_t,\ t = 1, \ldots, T\}$ of parameter vectors.* Otherwise the CE and the cloning methods are similar. In particular we shall see that the cloning algorithm for optimization coincides with the CE one, provided the updating of v and sampling from $f(\mathbf{x}, v)$ is replaced by updating of $g^*(\mathbf{x}, m_{t-1}) = g^*_{t-1}$ and sampling from g^*_{t-1}. It is also important to note that regardless of the fact that $\{g^*(\mathbf{x}, m_{t-1}),\ t = 1, \ldots, T\}$ will be not explicitly available, we still use the tuple representation (16.61) since, as we shall see below in order to update the parameters c_t and m_t at iteration t we will be still able to *sample* (using the MCMC machinery) from the optimal IS pdf $g^*(\mathbf{x}, m_{t-1}) = g^*_{t-1}$ corresponding to iteration $t-1$. Finally, we will see that the cloning method typically works better than its CE counterpart, especially for rare-events and counting. The main reason is that, while sampling from an optimal sequence of pdfs g^*_{t-1} (or even from their approximations) is more beneficial than sampling from sequence of a parametric family, like $f(\mathbf{x}, v_t)$. In other words the sequence 16.61 is more informative than the one in 16.62.

As mentioned, the chain rule approach (16.1), (16.55) has been extensively used in randomized algorithms [5, 6] for estimating counting quantities associated with some graphs. Their sampling mechanism is, however, completely different from our.

16.3.1 The Method of Diaconis–Holmes–Ross

In the method of Diaconis–Holmes [3] and Ross [7], called the *Diaconis–Holmes–Ross* (DHR) method, each quantity $\mathbb{E}_f[I_{\{S(X)\geq m_t\}}|I_{\{S(X)\geq m_{t-1}\}}]$ is estimated *separately and independently* by using the combination of the MCMC, in particular, the Gibbs sampler, provided the set of levels $\{m_t,\ t = 0,\ldots,T\}$ is fixed in advance. Ross [7] presents several interesting applications using the Gibbs sampler. In the Appendix we recapitulate the original Ross Algorithm for calculating $\widehat{c}_t$.

The main idea of the DHR [3] algorithm, while estimating each conditional probability

$$\mathbb{E}_f[I_{\{S(X)\geq m_t\}}|I_{\{S(X)\geq m_{t-1}\}}]$$

is to run *Markov Chain Monte Carlo* (MCMC) and count the proportion of values satisfying $S(\mathbf{x}) \geq m_t$.

More formally, their algorithm can be written as

Algorithm 16.3.1 (DHR Algorithm) *Given a sequence of levels $m_0 < m_1 < \cdots < m_T = m$ and the sample size N, execute the following steps:*

1. *Acceptance-Rejection. Set a counter $t = 1$. Initialize by generating a sample $X_1,\ldots,X_N$ from the proposal density $f(\mathbf{x})$. Let $\widetilde{\mathscr{X}}_0 = \{\widetilde{X}_1,\ldots,\widetilde{X}_{N_0}\}$ be the largest subset of the population $\{X_1,\ldots,X_N\}$ (elite samples) for which $S(X_i) \geq m_0$. Note that $\widetilde{X}_1,\ldots,\widetilde{X}_{N_0} \sim g^*(\mathbf{x}, m_0)$ and that*

$$\widehat{\ell}(m_0) = \widehat{c}_0 = \frac{1}{N}\sum_{i=1}^{N} I_{\{S(X_i)\geq m_0\}} = \frac{N_0}{N} \tag{16.63}$$

is an unbiased *estimator of $\ell(m_0)$.*

2. *MCMC step. Find a feasible point X such that $S(X) \geq m_{t-1}$. Starting from X, run the MCMC sampler, such that after some* burn-in *period, each vector $X = (X_1,\ldots,X_n)$, of the* new *population denoted as $\mathscr{X} = \{X_1,\ldots,X_N\}$ is* approximately *distributed as $g^*(\mathbf{x}, m_{t-1})$.*
3. *Let $\widetilde{\mathscr{X}}_t = \{\widetilde{X}_1,\ldots,\widetilde{X}_{N_t}\}$ be the subset of the population $\{X_1,\ldots,X_N\}$ for which $S(X_i) \geq m_t$. Take $\widehat{c}_t$ in (16.60) as an estimator of c_t in (16.57). Note that $\widetilde{X}_1,\ldots,\widetilde{X}_{N_t}$ is distributed only* approximately *$g^*(\mathbf{x}, m_t)$. Note also that as a*

feasible point X satisfying $S(X) \geq m_t$ one can take, for example, any point from the subset $\{\widetilde{X}_1, \ldots, \widetilde{X}_{N_t}\}$.

4. *If $t = T$ go to step 5, otherwise, set $t = t+1$ and repeat from step 2.*
5. *Deliver $\ell(m)$ in (16.59) as an estimator of $\ell(m)$.*

As we shall see below our cloning algorithm adopts the basic steps of DNR Algorithm 16.3.1, except its main MCMC one, which is totaly different in our method.

For convenience we also present below the

16.3.2 The Gibbs Sampler

There are two basic versions of the Gibbs sampler [13]: *systematic* and *random*. In the former one, the components of the vector $X = (X_1, \ldots, X_n)$ are updated in a fixed, say increasing order: $1, 2, \ldots, n, 1, 2, \ldots$ while in the latter, they are chosen randomly, that is according to a discrete uniform n-point pdf. Below we present the systematic Gibbs sampler algorithm. In a systematic Gibbs sampler, for a given vector $X = (X_1, \ldots, X_n) \sim g(\mathbf{x})$, one generates a *new* vector $\widetilde{X} = (\widetilde{X}_1, \ldots, \widetilde{X}_n)$ with the same distribution $\sim g(\mathbf{x})$ as follows.

Algorithm 16.3.2 (Gibbs Sampler)

1. Draw $\widetilde{X}_1$ from the conditional pdf $g(x_1 | X_2, \ldots, X_n)$.
2. Draw $\widetilde{X}_i$ from the conditional pdf $g(x_i | \widetilde{X}_1, \ldots, \widetilde{X}_{i-1}, X_{i+1}, \ldots, X_n)$, $i = 2, \ldots, n-1$.
3. Draw $\widetilde{X}_n$ from the conditional pdf $g(x_n | \widetilde{X}_1, \ldots, \widetilde{X}_{n-1})$.

Note that in Gibbs sampler it is assumed that generating samples from the conditional pdfs

$$g(x_i | X_1, \ldots, X_{i-1}, X_{i+1}, \ldots, X_n),\ i = 1, \ldots, n$$

is simple.

Example 16.5 (Sum of Independent Random Variables).
Consider estimation of ℓ with $S(\mathbf{x}) = \sum_{i=1}^n X_i$, that is

$$\ell = \mathbb{E}_f \left[I_{\{\sum_{i=1}^n X_i \geq m\}} \right] . \tag{16.64}$$

In this case, generating random variables X_i, $i = 1, \ldots, N$ for a fixed value m can be easily performed by using the Gibbs sampler based on the following conditional pdf

$$g^*(x_i, m | \mathbf{x}_{-i}) = c_i(m) f_i(x_i) I_{\{x_i \geq m - \sum_{j \neq i} x_j\}} , \tag{16.65}$$

where $|\mathbf{x}_{-i}$ denotes conditioning on all random variables but *excluding* the i-th component and $c_i(m)$ denotes the normalization constant.

Note also that each of the n conditional pdfs $g^*(x_i, m | \mathbf{x}_{-i})$ presents a truncated version of the proposal marginal pdf $f_i(x_i)$ with the truncating point at $m - \sum_{j \neq i} x_j$.

In short, the random variable $\widetilde{X}$ from $g^*(x_i, m|\mathbf{x}_{-i})$ presents a shifted original random variable $X \sim f_i(x_i)$. Generation from such truncated single dimensional pdf $g^*(x_i, m|\mathbf{x}_{-i})$ is easy and can be typically performed by using the inverse-transform method, provided the inverse-transform method can be applied to $f_i(x_i)$.

To proceed with the inverse-transform method denote by $b_i = m - \sum_{j \neq i} x_j$. It is readily seen that following the way of the construction of the elite sampling, we will always have that $b \geq 0$ and the range of b_i will be always the *same* as the range of the proposal random variable X_i.

Note that the inverse-transform algorithm for generating shifted random variables $Y = X + b$ from $X \sim f(x)$ with a fixed shifting (location) parameter $b = m - \sum_{j \neq i} x_j$ is based on the following relationship

$$\mathbb{P}(Y \leq x) = \mathbb{P}(X + b \leq x) = \mathbb{P}(X \leq x - b) = F(x - b).$$

It is important to note that sampling a Bernoulli random variable $\widetilde{X}_i$ from (16.65) using the Gibbs sampler can be performed as follows. Generate $Y \sim \text{Ber}\,(1/2)$. If

$$I_{\{Y \geq m - \sum_{j \neq i} X_j\}},$$

then set $\widetilde{X}_i = Y$, otherwise set $\widetilde{X}_i = 1 - Y$.

16.3.3 The Method of Botev and Kroese

The main drawback of DHR [3] Algorithm 16.3.1 is that for each fixed level m_t it starts basically from scratch. That is, at each iteration, starting at some feasible point X, it runs a single Markov chain before the entire sample $X_1, \ldots, X_N$ becomes distributed approximately stationarity, that is approximately $g_t^*(\mathbf{x}, m_{t-1})$. The time of reaching the stationarity (the burn-in period) might be quite long.

To overcome this difficulty, Botev and Kroese [1] introduced several important enhancements into the DHR Algorithm 16.3.1.

The main enhancement of Botev and Kroese [1] is that their algorithm has an additional step, called the *bootstrap resampling step*. It reuses iteratively all the elite samples $\widetilde{X}_1, \ldots, \widetilde{X}_{N_t} \sim g^*(\mathbf{x}, m_t)$ from the previous Markov chain runs and, thus to run in parallel many Markov chains. By doing so, the elite samples at different levels become dependent, but the stationarity in terms of sampling from the optimal importance sampling $g_t^*(\mathbf{x}, m_t)$ is preserved. To define the level sets $\{\widehat{m}_t\}_{t=0}^T$, Botev and Kroese make an additional (pilot) run. Note that in Botev and Kroese the level sets $\{\widehat{m}_t\}_{t=0}^T$ are defined adaptively, while in the DHR Algorithm 16.3.1 they are assumed to be fixed in advance. Clearly, in the former and the latter cases, T is a deterministic and a random variable, respectively. Note that the level sets $\{\widehat{m}_t\}_{t=0}^T$ in the latter case are chosen similarly to the CE method, in the sense that they involve a rarity parameter ρ . The adaptive choice of $\widehat{m}_t$ seems to be more natural and more flexible than the fixed one. Note, finally that since DHR Algorithm 16.3.1

always starts with a single (elite) sample and the samples between different levels are independent, it requires quite a long burn-in period for the samples $X_1,\dots,X_N$ to become at least approximately stationary. As a result, $\widehat{c}_t$ is typically a biased estimator of the true parameter c_t.

It is crucial to note that in contrast to the CE and the MinxEnt algorithms [13], both algorithms, Botev–Kroese [2] and DHR [3]

1. Sample from the *optimal or approximately optimal zero variance IS* nonparametric distribution $g^*(\mathbf{x},\widehat{m}_t)$ rather than from the parametric one $f(\mathbf{x},\widehat{p}_t)$. The latter is associated with the original (proposal) pdf $f(\mathbf{x},u)$.
2. Do not involve any optimization procedure, like the MinxEnt program, which minimizes the Kulback–Leibler divergence, subject to some constraints, They are based solely on the samples from $g^*(\mathbf{x},\widehat{m}_t)$, $t=0,1,\dots,T$, or their approximations.

Before presenting the Botev–Kroese algorithm we summarize its main features.

- It requires a pilot run to define a sequence $\{\widehat{m}_t\}$ such that $\widehat{m}_0 < \widehat{m}_< \dots < \widehat{m}_T = m$
- It samples recursively from the sequence of *zero variance* IS pdfs: $\{g_t^*\} = \{g^*(\mathbf{x},\widehat{m}_t)\}$, where each pdf $g^*(\mathbf{x},\widehat{m}_t)$ is associated with a sequence $\widehat{m}_0 < \widehat{m}_< \dots < \widehat{m}_T = m$.
- The exact sampling from $g_0^* = g^*(\mathbf{x},\widehat{m}_0)$ is obtained from the original distribution $f(\mathbf{x})$ by using the acceptance-rejection method (with the acceptance probability ρ). The goal of the sample from g_0^* (or from an associated kernel density approximation based on that sample) is to help generate exact samples from g_1^*.
- The recursive process of sampling from $g^*(\mathbf{x},m_t)$ is continued until eventually the level m is reached and, thus one can generate from the desired optimal IS pdf $g^*(\mathbf{x}) = g^*(\mathbf{x},m)$.
- The estimators of c_t are dependent and thus the entire estimator of $\ell(m)$ given in (16.59), which is based on the product of dependent random variables $\widehat{c}_t$ is biased.

The resulting Botev–Kroese [1] algorithm for rare events estimation can be written as

Algorithm 16.3.3 (Botev–Kroese Algorithm for Rare Events) Given a sequence of levels $\widehat{m}_0 < \widehat{m}_1 < \cdots < \widehat{m}_T = m$ and the sample size N, execute the following steps

1. *Acceptance-Rejection.* Set a counter $t=1$. Initialize by generating a sample $X_1,\dots,X_N$ from the proposal density $f(\mathbf{x})$. Let $\widetilde{\mathscr{X}}_0 = \{\widetilde{X}_1,\dots,\widetilde{X}_{N_0}\}$ be the largest subset of the population $\{X_1,\dots,X_N\}$ (elite samples) for which $S(X_i) \geq \widehat{m}_0$. Note that $\widetilde{X}_1,\dots,\widetilde{X}_{N_0} \sim g^*(\mathbf{x},\widehat{m}_0)$ and that $\widehat{\ell}(\widehat{m}_0)$ in (16.63) is an *unbiased* estimator of $\ell(m_0)$.
2. *Bootstrap step.* Sample *uniformly with replacement* N times from the population $\widetilde{\mathscr{X}}_{t-1} = \{\widetilde{X}_1,\dots,\widetilde{X}_{N_{t-1}}\}$ to obtain a new (bootstrap) population $\{X_1^*,\dots,X_N^*\}$. Note that $X_1^*,\dots,X_N^* \sim g^*(\mathbf{x},\widehat{m}_{t-1})$.

3. *MCMC step*. For each vector $X^* = (X_1^*, \dots, X_n^*)$ of the population $\{X_1^*, \dots, X_N^*\}$ generate, say by using the Gibbs sampler, a new vector $\widetilde{X}^* = (\widetilde{X}_1^*, \dots, \widetilde{X}_n^*)$. Note that the *new* population $\{\widetilde{X}_1^*, \dots, \widetilde{X}_N^*\}$ of $\widetilde{X}^*$'s is distributed again $\sim g^*(\mathbf{x}, \widehat{m}_{t-1})$. Denote the new population thus obtained by $\{X_1, \dots, X_N\}$.
4. Let $\widetilde{\mathscr{X}}_t = \{\widetilde{X}_1, \dots, \widetilde{X}_{N_t}\}$ be the subset of the population $\{X_1, \dots, X_N\}$ for which $S(X_i) \geq \widehat{m}_t$. Take $\widehat{c}_t$ in (16.60) as an estimator of c_t given in (16.57). Note again that $\widetilde{X}_1, \dots, \widetilde{X}_{N_t}$ is distributed $g^*(\mathbf{x}, \widehat{m}_t)$.
5. If $t = T$ go to step 6, otherwise set $t = t+1$ and repeat from step 2.
6. Deliver $\ell(m)$ given in (16.59) as an estimator of $\ell(m)$.

The pilot run algorithm of Botev–Kroese for selection of the levels m_t, $t = 1, \dots, T$ is as follows

Algorithm 16.3.4 (Pilot Run Algorithm for Levels Selection) Given the rarity parameter $\rho \in (0,1)$ and the sample size N_p execute the following steps:

1. *Acceptance-Rejection*. Set a counter $t = 1$. Generate a sample $X_1, \dots, X_{N_p}$ from the proposal density $f(\mathbf{x})$. Let $\widehat{m}_0$ be the $(1-\rho)$ sample quantile of $S(X_1), \dots, S(X_{N_p})$. Let $\widetilde{\mathscr{X}}_0 = \{\widetilde{X}_1, \dots, \widetilde{X}_{N_0}\}$ be the largest subset of the population $\{X_1, \dots, X_{N_p}\}$ (elite samples) for which $S(X_i) \geq \widehat{m}_0$.
2. *Bootstrap step*. Sample *uniformly with replacement* N times from the population $\widetilde{\mathscr{X}}_{t-1} = \{\widetilde{X}_1, \dots, \widetilde{X}_{N_{t-1}}\}$ to obtain a new (bootstrap) population $\{X_1^*, \dots, X_{N_p}^*\}$. Note that $X_1^*, \dots, X_{N_p}^* \sim g^*(\mathbf{x}, \widehat{m}_{t-1})$.
3. *MCMC step*. For each vector $X^* = (X_1^*, \dots, X_n^*)$ of the population $\{X_1^*, \dots, X_{N_p}^*\}$ generate, say by using the Gibbs sampler, a new vector $\widetilde{X}^* = (\widetilde{X}_1^*, \dots, \widetilde{X}_n^*)$. Note that the *new* population $\{\widetilde{X}_1^*, \dots, \widetilde{X}_{N_p}^*\}$ of $\widetilde{X}^*$'s is distributed again $\sim g^*(\mathbf{x}, \widehat{m}_{t-1})$. Denote the new population thus obtained by $\{X_1, \dots, X_{N_p}\}$.
4. Set
$$\widehat{m}_t = \min\{m, \widehat{a}\}\,, \tag{16.66}$$
where $\widehat{a}$ is the $(1-\rho)$ sample quantile of $S(X_1), \dots, S(X_{N_p})$. Let $\widetilde{\mathscr{X}}_t = \{\widetilde{X}_1, \dots, \widetilde{X}_{N_t}\}$ be the subset of the population $\{X_1, \dots, X_N\}$ for which $S(X_i) \geq \widehat{m}_t$. Note again that $\widetilde{X}_1, \dots, \widetilde{X}_{N_t}$ is distributed $g^*(\mathbf{x}, \widehat{m}_t)$
5. If $\widehat{m}_t = m$, set $t = T$ and go to step 6; otherwise, set $t = t+1$ and reiterate from step 2.
6. Deliver the sequence of estimated levels $\widehat{m}_0, \widehat{m}_0, \dots, \widehat{m}_{T-1}, m$.

Remark 16.3 (Reducing Dependency). In order to reduce the dependence between the vectors $X_1, \dots, X_N$ at different iterations t, it is suggested in [1] to use (similar to the DHR) some burn-in periods. More specifically, given the bootstrap sample $X_1^*, \dots, X_N^*$,

1. Generates at the MCMC step instead of $\widetilde{X}_1^*, \dots, \widetilde{X}_N^*$ a larger new population, namely $\{\widetilde{X}_1^*, \dots, \widetilde{X}_{rN}^*\}$, $r > 1$.
2. Takes only the last N samples from $\{\widetilde{X}_{1r}^*, \dots, \widetilde{X}_{rN}^*\}$ (discarding the first $(r-1)N$ ones) and denotes these, as before, by $X_1, \dots, X_N$.

It is not difficult to see that if the size of the elite sample $\widetilde{X}_1,\ldots,\widetilde{X}_{N_t}$ equals 1 at every iteration t, and if these single elites are independent for all t, $(t = 1,\ldots,T)$, then we automatically obtain the DHR Algorithm 16.3.1.

In the following two sections we deal with the cloning algorithms and their applications to multiple events, in particular for counting the number of feasible solutions in an integer program.

16.3.4 Cloning Algorithms

Here we present two alternative algorithms, called the *cloning algorithms* for rare-events and counting assuming that $\mathscr{X}$ is a discrete space. Both cloning algorithms are somewhat closer to DHR Algorithm 16.3.1 rather than to Botev–Kroese Algorithm 16.3.3. In particular, the similarity to the former is that we do not use bootstrapping, while the similarity to the latter is that we use elite samples. The main differences between the proposed and the existing algorithms are that in both our algorithms we

- Introduce a new mechanism, called the *cloning* mechanism.
- No pilot run is used in our algorithms.

The main difference between our two algorithms is that the first is based on the classic zero variance IS pdf (16.58) and is called the *cloning algorithm*, while the second one uses the Boltzmann zero variance distribution instead of the classic one and is called the *Boltzmann cloning algorithm.*

Before proceeding further, we need the following

Remark 16.4. For the discrete random variables formula

$$\mathbb{P}(S(X) \geq m_t)|S(X) \geq m_{t-1}) = \rho$$

is not valid anymore because of the discretization. We can instead generate a sequence of $\{\rho_t\}$ by arguing as follows. For fixed ρ, m_{t-1} and a given sequence of the sample functions $S(X_i)$, $i = 1,\ldots,N$ we choose m_t to be the closest (integer valued) $(1-\rho)$-th empirical value of the elite sample of $S(X_i)$, $i = 1,\ldots,N$. In short, we include in the elite sample of the next population all points for which $\{S(X_i) \geq m_t\}$ holds.

For example, assume that $\rho = 0.5$, $N = 6$ and that the ordered values of the $S(X_i)$, $i = 1,\ldots,N$ for iteration t are $\{1,1,2,2,2,3\}$. Then we have $m_t = 2$ and the modified $\rho = 0.5$ becomes $\rho_t = N_t/N = 5/6$.

16.3.4.1 The Main Cloning Algorithm

We start by introducing what we call (1) the simplified, (2) the basic and the (3) enhanced versions of our algorithm.

1. *The simplified version.* Let as before N, ρ_t and N_t be the fixed sample size, the actual rarity parameter (see Remark 16.4) and the number of elites at iteration t, respectively. At the simplified version we apply to each of the N_t elites a *burn-in period of length* ρ_t^{-1}. Bo doing so we generate $\rho_t^{-1} N_t = N$ samples at each level m_t. The rationale of this is based on the fact that if ρ is not small, say $\rho = 0.1$, then we have enough stationary elite samples and the goal of the Gibbs sampler is merely to continue with these stationary N_t elites and thus to generate N *new* stationary samples for the next level.

Below we present a simplified version of our main algorithm for counting, which provides a good insight to it.

Algorithm 16.3.5 (The Simplified Algorithm for Multiple Events) Given the rarity parameter $\rho \in (0,1)$ and the sample size N execute the following steps:

1. *Acceptance-Rejection.* Set a counter $t = 1$. Generate a sample $X_1, \ldots, X_N$ from the proposal density $f(\mathbf{x})$. Let $\widetilde{\mathscr{X}_0} = \{\widetilde{X}_1, \ldots, \widetilde{X}_{N_0}\}$ be the largest subset of the population $\{X_1, \ldots, X_N\}$ (elite samples) for which $S(X_i) \geq \widehat{m}_0$. Take $\widehat{c}_0 = \widehat{\ell}(\widehat{m}_0)$ in (16.63) is an *unbiased* estimator of c_0. Note that $\widetilde{X}_1, \ldots, \widetilde{X}_{N_0} \sim g^*(\mathbf{x}, \widehat{m}_0)$.
2. *MCMC.* For each vector $\widetilde{X}_k = (\widetilde{X}_{1k}, \ldots, \widetilde{X}_{nk})$, $k = 1, \ldots, N_{t-1}$ of the elite sample $\{\widetilde{X}_1, \ldots, \widetilde{X}_{N_{t-1}}\} \sim g^*(\mathbf{x}, \widehat{m}_{t-1})$ obtained at the $t-1$-th iteration apply ρ_t^{-1} burn-in periods, while using the MCMC (and in particular the Gibbs) sampler and, thus generate ρ_t^{-1} new vectors $(X_{k1}, \ldots, X_{k\rho_t^{-1}})$. Note that the *new entire* population $\{(X_{s1}, \ldots, X_{s\rho_t^{-1}}),\ s = 1, \ldots, N_{t-1}\}$ of length N, which is denoted as $\{X_1, \ldots, X_N\}$, is distributed approximately $g^*(\mathbf{x}, \widehat{m}_{t-1})$.
3. Let $\widetilde{\mathscr{X}_t} = \{\widetilde{X}_1, \ldots, \widetilde{X}_{N_t}\}$ be the subset of the population $\{X_1, \ldots, X_N\}$ for which $S(X_i) \geq \widehat{m}_t$. Take $\widehat{c}_t$ in (16.60) as an estimator of c_t given in (16.57). Note again that $\widetilde{X}_1, \ldots, \widetilde{X}_{N_t}$ is distributed approximately $g^*(\mathbf{x}, \widehat{m}_t)$.
4. If $m_t = m$ go to step 5, otherwise set, set $t = t + 1$ and repeat from step 2.
5. Deliver $\widehat{\ell}(m)$ given in (16.59) as an estimator of $\ell(m)$ and $|\widehat{\mathscr{X}^*}| = \widehat{\ell}(m)|\mathscr{X}|$ as an estimator of $|\mathscr{X}^*|$.

Note that Algorithm 16.3.5 can be also viewed as a particular case of Algorithm 16.3.3 with both the bootstrap step and the pilot run being omitted, and the length of the burn-in period being equal to ρ^{-1}.

Our numerical experience with all three algorithms: 16.3.1, 16.3.3, 16.3.5 clearly indicate that neither perform satisfactorily for multiple events. In particular, they often stop without reaching the final level m. To overcome this difficulty we turn next to our

2. *Basic version.* As compared to Algorithm 16.3.5 this version contains an additional step for an adaptive choice of ρ. As we shall see below this additional step will prevent from stopping all three algorithms before reaching the target level m.

To proceed note first that the draw back of the approach for updating ρ based on Remark 16.4 is that one can often run into a situation, where $N_t = N$ and thus $\rho_t = 1$. This is the main reason that all three algorithms often stop before reaching the desired level m. The following example provides details. Assume that $N = 9$

and let the proposal (nOn-adaptive) $\rho = 1/3$. Consider the following two sample scenarios of the ordered values of $S(X_i)$, $i = 1, \ldots, 9$:

(1) $S(X_i) = (1,1,1,2,2,2,2,3,3)$ and (2) $S(X_i) = (1,1,1,1,1,1,1,2,3)$. Following Remark 16.4 it is readily seen that in cases (i) and (ii) the actual values of ρ are $\rho_1 = 6/9$ and $\rho_2 = 1$, respectively. They are based on the elite sequences $\{2,2,2,2,3,3\}$ and $\{1,1,1,1,1,1,1,2,3\}$, respectively. Note that both $\rho_1 > \rho$ and $\rho_2 > \rho$. Note, however, that we can not use $\rho_2 = 1$ (corresponding to $S(X_i) = 1$), since as soon as we obtain $\rho = 1$ our algorithm will stop and, thus the level m will be never reached. To prevent this we modify ρ_2 by moving from the level corresponding to $S(X_i) = 1$ to the next level of $S(X_i)$, that is to the level $S(X_i) = 2$. This corresponds to the elite sequence $\{2,3\}$ with the new modified $\rho_2 = N_t/N = 2/9 < \rho$. Based in this we shall further require that ρ should be in some fixed interval (a_1, a_2), say $(a_1, a_2) = (0.01, 0.25)$. This means that when the number of elites $N_t > a_2 N$ we automatically switch from a lower elite level to a higher one; if $a_1 N \le N_t \le a_2 N$ $(\rho \in (a_1, a_2))\rho \in (a_1, a_2)$, and thus $a_1 \le \rho_t \le a_2$, we accept N_t as the size of the elites sample; and if $N_t < a_1 N$, we proceed sampling until $N_t = a_1 N$, that is until we obtain at least $a_1 N$ elites.

We summarize this as

Remark 16.5. Adaptive choice of ρ For fixed N, ρ and for fixed interval $a_1 \le \rho \le a_2$, say $0.01 \le \rho \le 0.25$, let $S_{\lceil 1-\rho \rceil} = S_{min}$ be the smallest elite value of the ordered sample $S(X_i)$, $i = 1, \ldots, N$. The adaptive choice of ρ is performed as follows:

- *Include into the elite sample all additional values* $S_{\lceil 1-\rho \rceil} = S_{min}$ (see Remark 16.4), provided that at the iteration t the number of elite samples $N_t \le a_2 N$. Denote the adaptive ρ as ρ_1. Clearly, $\rho_1 \ge \rho$.
- *Remove from the elite all values* $S_{\lceil 1-\rho \rceil} = S_{min}$, provided the number of elite samples $N_t > a_2 N$. Note that by doing so we switch from a lower elite level to a higher one. If $N_t \ge a_1 N$, accept N_t as the elite sample and denote the modified ρ as ρ_2. If $N_t < a_1 N$, we proceed sampling until $N_t \ge a_1 N$, that is until we obtain at least $a_1 N$ samples.

  Note that the main reason that both Algorithms 16.3.3 and 16.3.5 *fail* to reach the target value m is that we used ρ based on Remark 16.4, rather than the adaptive ρ satisfying $a_1 < \rho \le a_2$.

  Note finally, that if $N_t = N$ and if there is only a *single elite value left* then we automatically obtain DHR Algorithm 16.3.1. Since Algorithm 16.3.1 is typically less efficient than the proposed one, it is desirable either to reject such an elite sample and start a new one, or enlarge the current one until we get several elites.

We call Algorithm 16.3.5 with the additional step, called *adaptive choice of* ρ , the *Basic Algorithm.* It is important to note that adding the above step to all three Algorithms (16.3.1, 16.3.3 and 16.3.5), improves essentially their performance in the sense that all three have reached the desired level m in most of our experiments. In particular, while estimating rare-events for the sum of iid Bernoulli random variables we found that the Basic Algorithm outperforms Algorithm 16.3.3. It is our understanding that the bootsraping step Algorithm 16.3.3 is the one which causes strong deviation of the resulting sample $X_1,\ldots,X_N$ from the uniform, and thus it should be removed.

To increase further the accuracy of the Basic Algorithm, we turn next to the (3) *Enhanced version*, where we introduce two additional steps to the Basic Algorithm.

1. *Screening step*. Since the optimal zero variance pdf $g^*(\mathbf{x},m_t)$ must be *uniformly distributed* for each fixed m_t, our algorithm checks at each iteration whether or not *all elite vectors* $\widetilde{X}_1,\ldots,\widetilde{X}_{N_t}$ *are different*. If this is not the case, we screen out (clean) all redundant elite samples. We denote the resulting elite sample as $\widehat{X}_1,\ldots,\widehat{X}_{N_t}$ and call it, *the truly uniform sample*. Observe that this procedure prevents the empirical pdf associated with $\widehat{X}_1,\ldots,\widehat{X}_{N_t}$ from deviation from the uniform one.
2. *Cloning*. The goal of cloning is to find a good balance (in terms of bias-variance) between the number of elite and the burn-in periods, denoted by b, in the Gibbs sampling. Note that DHR Algorithm 16.3.1 and Botev–Kroese Algorithm 16.3.3 correspond to $b=N$ and $b=1$, respectively.
   For fixed N and b, the adaptive *cloning* parameter η at iteration $t-1$ is defined as
   $$\eta_{t-1} = \left\lceil \frac{N}{bN_{t-1}} \right\rceil - 1 = \left\lceil \frac{N_{cl}}{N_{t-1}} \right\rceil - 1\,. \tag{16.67}$$
   Here $N_{cl} = N/b$ is called the *cloned sample size* and, as before, N_{t-1} denotes the number of truly uniform elites (left after the screening) at iteration $t-1$. Note that $\lceil\cdot\rceil$ denotes rounding to the largest integer. The goal of η is to reproduce η times the N_{t-1} truly uniform elites. Note also that because of its adaptive nature $(\rho_2 \le \rho \le \rho_1)$, the parameter ρ is not directly involved in the calculation of η.
   As an example, let $N=1{,}000$, $b=10$. Consider two cases: $N_{t-1}=21$ and $N_{t-1}=121$. We obtain $\eta=4$ and $\eta=0$ (no cloning).
   Our numerical studies show that it is quite reasonable to choose $3\le b\le 10$ to have manageable bias-variance balance. In this case, the Gibbs sampler is applied b times to each vector X of the cloned samples of size $N_{cl}=b^{-1}N$.

Remark 16.6 (An alternative for b and η). Denote $\pi_{t-1} = \frac{N}{N_{t-1}}$. Then as an alternative to (16.67) one can use the following strategy in defining b and η: find b and η from $b\eta \approx \pi_{t-1}$ and take $b\approx\eta$. In short

$$b\approx\eta\approx\left(\frac{N}{N_{t-1}}\right)^{1/2}. \tag{16.68}$$

Consider again the two cases: $N_{t-1}=21$ and $N_{t-1}=121$ and let as before $N=1{,}000$. We have $b\approx\eta=7$ and $b\approx\eta=3$, respectively.

If not stated otherwise we shall use (16.67). Note, finally, that we always keep all samples generated by the Gibbs sampler.

With this at hand we next present the

Cloning step. Given the number of burn-in periods b and the size N_{t-1} of truly uniform elites at iteration $t-1$, find the cloning parameter $\eta_{t-1} = \left\lceil \frac{N}{bN_{t-1}} \right\rceil - 1$. Reproduce η_{t-1} times each vector $\widehat{X}_k = (\widehat{X}_{1k}, \ldots, \widehat{X}_{nk})$ of the truly elite sample $\{\widehat{X}_1, \ldots, \widehat{X}_{N_{t-1}}\}$, that is, take η identical copies of each vector $\widehat{X}_k$ obtained at the $t-1$-th iteration. Denote the entire new population (ηN_{t-1} cloned vectors plus the original truly elite sample $\{\widehat{X}_1, \ldots, \widehat{X}_{N_{t-1}}\}$) by $\mathscr{X}_{cl} = \{(\widehat{X}_1, \ldots, \widehat{X}_1), \ldots, (\widehat{X}_{N_{t-1}}, \ldots, \widehat{X}_{N_{t-1}})\}$. To each of the cloned vectors of the population $\mathscr{X}_{cl}$ apply the MCMC (and in particular the Gibbs sampler) for b burn-in periods. Denote the *new entire* population by $\{X_1, \ldots, X_N\}$. Observe that each component of $\{X_1, \ldots, X_N\}$ is distributed approximately $g^*(\mathbf{x}, \widehat{m}_{t-1})$. We call such an MCMC procedure involving η cloning and b burn-in periods – the *cloning procedure*.

Recall that because of low dependence between the elite samples *neither bootstrap step nor pilot run* is used in our algorithms. Note that the pilot run in Algorithm 16.3.3 typically takes nearly the same amount of time as the main algorithm, provided one wants to estimate the levels m_t reliably. Note, however, that since the levels $\widehat{m}_t$'s are random variables, our algorithm will generate only from approximately zero variance IS pdfs. For large samples we can neglect the randomness of $\widehat{m}_t$.

Figures 16.1 and 16.2 demonstrate how cloning works for $\ell(m) = \mathbb{E} I_{\{X_1 + X_2 \geq m\}}$ with $\eta = 0$ (no cloning) and $\eta_1 = \eta_2 = 2$ (cloning), respectively. The sample size was $N = 10$. We have 2 and 3 elite samples (mice) at levels m_0 and m_1 at Fig. 16.1. The counting quantity at the final level m is $|\mathscr{X}^*| = 3$. Similarly we have 2 and 3 elite samples at levels m_0 and m_1, respectively at Fig. 16.2. Since in this case $\eta_1 = \eta_2 = 2$ after the cloning we obtain 6 and 8 samples (mice), respectively. At the final level m we obtain again $|\mathscr{X}^*| = 3$.

Below we present our main cloning algorithm, which presents an enhanced version of Algorithm 16.3.5 for rare events and counting with discrete distributions.

Algorithm 16.3.6 (Main Cloning Algorithm for Multiple Events) Given the proposal rarity parameter ρ, say $\rho = 0.1$, the parameters a_1 and a_2, say $a_1 = 0.01$ and $a_2 = 0.25$, such that $\rho \in (a_1, a_2)$, the sample size N, say $N = m \times n$, the burn in period b, say $3 \leq b \leq 10$ execute the following steps:

1. *Acceptance-Rejection.* Set a counter $t = 1$. Generate a sample $X_1, \ldots, X_N$ from the proposal density $f(\mathbf{x})$. Let $\widetilde{\mathscr{X}}_0 = \{\widetilde{X}_1, \ldots, \widetilde{X}_{N_0}\}$ be the largest subset of the population $\{X_1, \ldots, X_N\}$ (elite samples) for which $S(X_i) \geq \widehat{m}_0$. Note that $\widetilde{X}_1, \ldots, \widetilde{X}_{N_0} \sim g^*(\mathbf{x}, \widehat{m}_0)$ and that $\widehat{\ell}(\widehat{m}_0)$ in (16.63) is an *unbiased* estimator of $\ell(m_0)$.
2. *Adaptive choice of* ρ. For each iteration use the adaptive choice of $a_1 \leq \rho \leq a_2$, with a_1 and a_2 defined in Remark 16.5.
3. *Screening.* Denote the elite sample obtained at iteration $t-1$ by $\{\widetilde{X}_1, \ldots, \widetilde{X}_{N_{t-1}}\}$. Screen out the redundant elements of the subset $\{\widetilde{X}_1, \ldots, \widetilde{X}_{N_{t-1}}\}$ and denote the resulting one as $\{\widehat{X}_1, \ldots, \widehat{X}_{N_{t-1}}\}$.

Fig. 16.1 Mice without cloning

4. *Cloning*. Given the number of burn-in periods b and the size N_{t-1} of truly uniform elites at iteration $t-1$, find the cloning parameter η_{t-1} according to $\eta_{t-1} = \left\lceil \frac{N}{bN_{t-1}} \right\rceil - 1$. Reproduce η_{t-1} times each vector $\widehat{X}_k = (\widehat{X}_{1k}, \ldots, \widehat{X}_{nk})$ of the truly elite sample $\{\widehat{X}_1, \ldots, \widehat{X}_{N_{t-1}}\}$, that is, take η identical copies of each vector $\widehat{X}_k$ obtained at the $t-1$-th iteration. Denote the entire new population (ηN_{t-1} cloned vectors plus the original truly elite sample $\{\widehat{X}_1, \ldots, \widehat{X}_{N_{t-1}}\}$) by $\mathscr{X}_{cl} = \{(\widehat{X}_1, \ldots, \widehat{X}_1), \ldots, (\widehat{X}_{N_{t-1}}, \ldots, \widehat{X}_{N_{t-1}})\}$. To each of the cloned vectors of the population $\mathscr{X}_{cl}$ apply the MCMC (and in particular, the Gibbs sampler) for b burn-in periods. Denote the *new entire* population by $\{X_1, \ldots, X_N\}$. Observe that each component of $\{X_1, \ldots, X_N\}$ is distributed approximately $g^*(\mathbf{x}, \widehat{m}_{t-1})$.
5. *Estimating* c_t. Let $\widetilde{\mathscr{X}_t} = \{\widetilde{X}_1, \ldots, \widetilde{X}_{N_t}\}$ be the subset of the population $\{X_1, \ldots, X_N\}$ for which $S(X_i) \geq \widehat{m}_t$. Take $\widehat{c}_t$ in (16.60) as an estimator of c_t given in (16.57). Note again that $\widetilde{X}_1, \ldots, \widetilde{X}_{N_t}$ is distributed approximately $g^*(\mathbf{x}, \widehat{m}_t)$.
6. *Stopping rule*. If $t = T$, go to step 7, otherwise set $t = t+1$ and repeat from step 2.

Fig. 16.2 Mice with cloning, $\eta = 2$

7. *Final Estimator*. Deliver $\widehat{\ell}(m)$ given in (16.59) as an estimator of $\ell(m)$ and $|\widehat{\mathscr{X}^*}| = \widehat{\ell}(m)|\mathscr{X}|$ as an estimator of $|\mathscr{X}^*|$.

Remark 16.7 (The direct estimator). As an alternative to the estimator $|\widehat{\mathscr{X}^*}|$ obtained by Algorithm 16.3.6 we can use the one based on *direct counting* of the number of the truly uniform samples obtained just after crossing the level m. Such counting estimator, denoted by $|\widehat{\mathscr{X}^*_{dir}}|$, is associated with the *empirical* distribution of the optimal zero variance uniform distribution $g^*(\mathbf{x},m)$. We found numerically that $|\widehat{\mathscr{X}^*_{dir}}|$ is extremely useful and very accurate. Note that it can be applied only for counting problems with $|\mathscr{X}^*|$ being not too large. In particular, $|\mathscr{X}^*|$ should be less than the sample size N, that is $|\mathscr{X}^*| < N$. Note also that counting problems with values small relative to $|\mathscr{X}|$ are known as the must difficult ones and in many problems one is indeed interested to count only if $|\mathscr{X}|^*$ is no greater then some fixed quantity, say $\mathscr{N}$. Clearly, this is possibly only if $N \geq \mathscr{N}$.

It is important to note that $|\widehat{\mathscr{X}^*_{dir}}|$ is typically much more accurate than its counterpart, the standard estimator $|\widehat{\mathscr{X}^*}| = \widehat{\ell}|\mathscr{X}|$. The reason is that $|\widehat{\mathscr{X}^*_{dir}}|$ is obtained *directly* by counting all distinct values $S(X_i)$, $i = 1,\ldots,N$, satisfying $S(X_i) \geq m$, that is $|\widehat{\mathscr{X}^*_{dir}}|$ can be written as $|\widehat{\mathscr{X}^*_{dir}}| = \widehat{\ell}_{dir}|\mathscr{X}^*|$, where

$$\widehat{\ell}_{dir} = \frac{1}{N} \sum_{i=1}^{N_{dir}} I_{\{S(X_i) \geq m\}},$$

and where $N_{dir} = |\widehat{\mathscr{X}^*_{dir}}|$ is the number of different (distinguishable) uniform samples of $S(X_i)$, $i = 1, \ldots, N$ at level m.

To increase further the accuracy of $|\widehat{\mathscr{X}^*_{dir}}|$ one can take a larger sample at the last step of Algorithm 16.3.6, that is while estimating $c_T = c_m$.

Table 16.2 presents comparison of performance of Algorithm 16.3.6 for different parameter configurations of as well as with the DHR and Botev–Kroese (BK) Algorithms while estimating

$$\ell = \mathbb{E}_f \left[I_{\{\sum_{i=1}^n X_i \geq m\}} \right],$$

where the X_i's are iid Ber(1/2), $n = 20$ and $m = 19$. Table 16.3 presents similar data for $n = 100$ and $m = 99$. In both experiments, we set $N = 1{,}000$ and $\rho = 0.05$ and the results were averaged over 100 independent runs.

Here RE and RE_{dir} denote the relative error of the estimators $|\widehat{\mathscr{X}^*}|$ and $|\widehat{\mathscr{X}^*_{dir}}|$, respectively. Note that in the version N_0 1 of Algorithm 16.3.6 we set $a_1 = 0.01$ and $a_2 = 0.5$, while we did not restrict the remaining ones, that is, they were $a_1 = 0$ and $a_2 = 1$. It follows from these tables that Algorithm 16.3.6 with the burn-in period $b = 10$ is the best. Similar performance was obtained for $3 \leq b \leq 10$. Note that the version N_0 2 and N_0 3 are somewhat similar to the version N_0 4 and N_0 5 of DHR and BK, respectively, in the sense that the burn-in periods are $b = N$ and $b = 1$, respectively. Note also that for the case $n = 100$ and $m = 99$, Algorithm 16.3.3 did

Table 16.2 Comparative performance of the algorithms for the sum of $n = 20$ iid Bernoulli random variables for $m = 19$, $N = 1{,}000$ and $\rho = 0.05$

N_0	Algorithm	b	η	$\lvert\widehat{\mathscr{X}^*}\rvert$	RE	$\lvert\widehat{\mathscr{X}^*_{dir}}\rvert$	RE_{dir}	CPU
1	New	10	$\eta_t = \frac{N}{bN_t} - 1$	21.20	0.047	21.0	0	0.3
2	New	N/N_t	0	21.92	0.147	21.0	0	0.9
3	New	1	N/N_t	21.78	0.163	21.0	0	0.7
4	BK	1	N/N_t	27.32	0.305	21.0	0	0.3
5	DHR	N	0	20.22	0.137	21.0	0	14.5

Table 16.3 Comparative performance of the algorithms for the sum of $n = 100$ iid Bernoulli random variables for $m = 99$, $N = 1{,}000$ and $\rho = 0.05$

N_0	Algorithm	b	η	$\lvert\widehat{\mathscr{X}^*}\rvert$	RE	$\lvert\widehat{\mathscr{X}^*_{dir}}\rvert$	RE_{dir}	CPU
1	New	10	$\eta_t = \frac{N}{bN_t} - 1$	101.7	0.041	101.0	0	33
2	New	N/N_t	0	122.6	0.206	100.7	0.01	27
3	New	1	N/N_t	118.2	0.172	101.0	0	26
4	BK	1	N/N_t	-	-	-	-	∞
5	DHR	N	0	207.3	0.354	101	0.0	288

Table 16.4 Dynamics of Algorithm 16.3.6 for the sum of 100 Bernoulli random variables with $N = 1{,}000$, $\rho = 0.05$ and $b = 5$

t	$\lvert\mathscr{X}^*\rvert$	$\lvert\mathscr{X}^*_{dir}\rvert$	N_t	$N_t^{(s)}$	m_t^*	m_{*t}	ρ_t
1	5.4e+027	0.0	66	66	65	57	0.07
3	3.7e+025	0.0	70	70	71	67	0.07
5	2.9e+023	0.0	96	96	77	73	0.06
7	2.1e+021	0.0	105	105	81	77	0.10
10	1.1e+018	0.0	212	212	87	83	0.05
12	3.8e+016	0.0	375	375	88	85	0.19
15	1.2e+014	0.0	515	515	90	88	0.15
17	1.7e+012	0.0	395	395	94	90	0.13
20	1.4e+009	0.0	159	159	95	93	0.09
22	4.1e+006	0.0	77	77	96	95	0.06
25	4,188.1	27.0	31	31	99	98	0.03
26	99.1	101.0	26	22	99	99	0.02
27	99.1	101.0	1,078	99	100	99	1.00

not converge to $m = 99$. It is interesting to note that while increasing the sample size from $N = 1{,}000$ to $N = 10{,}000$ for the Bernoulli model with $n = 100$ and $m = 99$, while all the rest of the data remained the same, we found that the performance of all 5 versions substantially improved and this case Algorithm 16.3.3 also reaches the level $m = 99$. In particular the relative error *RE* was decreased approximately by a factor of 5. Note that if the data would be independent, one would obtain a decrease in relative error by a factor of $10^{1/2} \approx 3$ only. Still the relative efficiencies of the 5 versions were similar to the case of $N = 1{,}000$. Finally, it follows from the tables that, in spite of the high relative error *RE* and substantial bias of some of the versions of $\lvert\widehat{\mathscr{X}^*}\rvert$, the direct estimator $\lvert\widehat{\mathscr{X}^*_{dir}}\rvert$ is very accurate for all 5 cases. The explanation for such nice behavior of $\lvert\widehat{\mathscr{X}^*_{dir}}\rvert$ is given in Remark 16.7.

Table 16.4 presents the dynamics for one of the runs of Algorithm 16.3.6 for the sum of 100 Bernoulli random variables for $m = 99$ with $N = 1{,}000$, $\rho = 0.05$ and $b = 5$. We used the following notations

1. N_t and $N_t^{(s)}$ denotes the actual number of elites and the one after screening, respectively.
2. m_t^* and m_{*t} denotes the upper and the lower elite levels reached, respectively.
3. $\rho_t = N_t/N$ denotes the adaptive rarity parameter.

We also run Algorithm 16.3.6 for the sum of Bernoulli random variables with $n = m$ and obtained similar results. Clearly, in this case $\lvert\mathscr{X}^*\rvert = 1$.

We finally define the most naive acceptance-rejection version of Algorithm 16.3.6, called the *($N = 1$)-policy algorithm*. According to ($N = 1$)-policy algorithm at each fixed level m_{t-1} we use the acceptance-rejection single trial method, until for the first time we reach a higher level $m_t > m_{t-1}$. Using the ($N = 1$)-policy algorithm we always managed to reach any level $m \leq n$, even for large n, like $n = 1{,}000$. It is not difficult to understand that the ($N = 1$)-policy algorithm performs as a *randomized "bisection" method.*

Clearly, the total number of Bernoulli samples M to reach the level $m = n$ increases with n. For example, for $n = m = 10{,}100,\ 1{,}000$ we found that in average $M = 14,\ 250,\ 8{,}000$, respectively.

For the continuous case Algorithm 16.3.6 simplifies substantially. In particular its steps 2 and 3 can be omitted. Also, one has to take into account that, since there is no screening N_{t-1} in the cloning step (step 4) presents the number of elites rather than what we call the number of truly uniform elites.

Below we present for completeness the continuous version of Algorithm 16.3.6.

Algorithm 16.3.7 (Cloning Algorithm for the Continuous Case) Given the rarity parameter ρ, say $\rho = 0.1$, the sample size N, say $N = m \times n$, the burn-in period b, say $3 \le b \le 10$, execute the following steps:

1. *Acceptance-Rejection*. Set a counter $t = 1$. Generate a sample $X_1, \ldots, X_N$ from the proposal density $f(\mathbf{x})$. Let $\widetilde{\mathscr{X}_0} = \{\widetilde{X}_1, \ldots, \widetilde{X}_{N_0}\}$ be the largest subset of the population $\{X_1, \ldots, X_N\}$ (elite samples) for which $S(X_i) \ge \widehat{m}_0$. Note that $\widetilde{X}_1, \ldots, \widetilde{X}_{N_0} \sim g^*(\mathbf{x}, \widehat{m}_0)$ and that $\widehat{\ell}(\widehat{m}_0)$ in (16.63) is an *unbiased* estimator of $\ell(m_0)$.
2. *Cloning*. Given the number of burn-in periods b and the elite size N_{t-1} at iteration $t-1$, find the cloning parameter $\eta_{t-1} = \left\lceil \frac{N}{bN_{t-1}} \right\rceil - 1$. Reproduce η_{t-1} times each vector $\widehat{X}_k = (\widehat{X}_{1k}, \ldots, \widehat{X}_{nk})$ of the elite sample $\{\widehat{X}_1, \ldots, \widehat{X}_{N_{t-1}}\}$, that is, take η identical copies of each vector $\widehat{X}_k$ obtained at the $t-1$-th iteration. Denote the entire new population (ηN_{t-1} cloned vectors plus the original elite sample $\{\widehat{X}_1, \ldots, \widehat{X}_{N_{t-1}}\}$) by $\mathscr{X}_{cl} = \{(\widehat{X}_1, \ldots, \widehat{X}_1), \ldots, (\widehat{X}_{N_{t-1}}, \ldots, \widehat{X}_{N_{t-1}})\}$. To each of the cloned vectors of the population $\mathscr{X}_{cl}$ apply the MCMC (and in particular, the Gibbs sampler) for b burn-in periods. Denote the *new entire* population by $\{X_1, \ldots, X_N\}$. Observe that each component of $\{X_1, \ldots, X_N\}$ is distributed approximately $g^*(\mathbf{x}, \widehat{m}_{t-1})$.
3. *Estimating* c_t. Let $\widetilde{\mathscr{X}_t} = \{\widetilde{X}_1, \ldots, \widetilde{X}_{N_t}\}$ be the subset of the population $\{X_1, \ldots, X_N\}$ for which $S(X_i) \ge \widehat{m}_t$. Take $\widehat{c}_t$ in (16.60) as an estimator of c_t given in (16.57). Note again that $\widetilde{X}_1, \ldots, \widetilde{X}_{N_t}$ is distributed approximately $g^*(\mathbf{x}, \widehat{m}_t)$.
4. *Stopping rule*. If $m_t = m$ go to step 5, otherwise set $t = t + 1$ and repeat from step 2.
5. *Final Estimator*. Deliver $\widehat{\ell}(m)$ given in (16.59) as an estimator of $\ell(m)$ and $|\widehat{\mathscr{X}^*}| = \widehat{\ell}(m)|\mathscr{X}|$ as an estimator of $|\mathscr{X}^*|$.

16.3.4.2 The Boltzmann Cloning Algorithm

Many problems including the classic Ising model are based on the Boltzmann distribution

$$g(\mathbf{x}) = Z^{-1}(\lambda) f(\mathbf{x}) \exp\{-\lambda S(\mathbf{x})\}, \tag{16.69}$$

where $Z(\lambda) = \mathbb{E}_f\left[\exp\{-\lambda S(X)\}\right]$ is called the partition function.

Note that in analogy to (16.55)–(16.56) we can write $Z = Z(m)$ as

$$\begin{aligned} Z(m) &= \mathbb{E}_f\left[\exp\{-\textstyle\sum_{i=1}^{m_0}\lambda C_i(X)\}\right] \\ &\quad\times\textstyle\prod_{t=1}^{T}\mathbb{E}_f\left[\exp\{-\sum_{i=1}^{m_t}\lambda C_i(X)\}\,|\exp\{-\sum_{i=1}^{m_{t-1}}\lambda C_i(X)\}\right] \\ &= \mathbb{E}_f\left[\exp\{-\textstyle\sum_{i=1}^{m_0}\lambda C_i(X)\}\right]\prod_{t=1}^{T}\mathbb{E}_{g_{t-1}}\left[\exp\{-\sum_{i=1}^{m_t}\lambda C_i(X)\}\right] \\ &= z_0\textstyle\prod_{t=1}^{T} z_t, \end{aligned} \tag{16.70}$$

were

$$\begin{aligned} z_t &= \mathbb{E}_f\left[\exp\{-\textstyle\sum_{i=1}^{m_t}\lambda C_i(X)\}\,|\exp\{-\sum_{i=1}^{m_{t-1}}\lambda C_i(X)\}\right] \\ &= \mathbb{E}_{g_{t-1}}\left[\exp\{-\textstyle\sum_{i=1}^{m_t}\lambda C_i(X)\}\right], \end{aligned} \tag{16.71}$$

$z_0 = \mathbb{E}_f\left[\exp\{-\sum_{i=1}^{m_0}\lambda C_i(X)\}\right]$ and as usual, the sequence $\{m_t,\ t = 0, 1, \ldots, T\}$ satisfies $0 < m_0 < m_1 < \ldots < m_T = m$ and it is chosen on-line (adaptively), f denotes the original pdf $f(\mathbf{x})$, and $g_{t-1} = g(\mathbf{x}, m_t, \lambda = -\infty)$ denotes the zero variance (Boltzmann) pdf at iteration t.

The estimator of $Z(m)$ can be written in analogy to (16.59) and (16.60) as

$$\widehat{Z}(m) = \prod_{t=0}^{T}\widehat{z}_t, \tag{16.72}$$

where

$$\widehat{z}_t = \frac{1}{N}\sum_{j=1}^{N}\exp\left\{-\sum_{i=1}^{m_t}\lambda C_i(X_j)\right\} \tag{16.73}$$

and $X_j \sim g(\mathbf{x}, \lambda, m_{t-1})$.

Here we present a modified version of Algorithm 16.3.6 for rare events, counting and estimation of the partition function $Z(\lambda)$.

To proceed, note [11] that (16.69) can be derived from the solution of the following single-constrained MinxEnt program

$$\begin{aligned} &\min_g \mathcal{D}(g,h) = \min_g \textstyle\int \ln\frac{g(\mathbf{x})}{f(\mathbf{x})}\, g(\mathbf{x})\mathrm{d}\mathbf{x} = \min_g \mathbb{E}_g\left[\ln\frac{g(X)}{f(X)}\right] \\ &\text{s.t. } \mathbb{E}_g[\textstyle\sum_{i=1}^{m} C_i(X)] = m \\ &\textstyle\int g(\mathbf{x})\mathrm{d}\mathbf{x} = 1. \end{aligned} \tag{16.74}$$

It is also important to note (see [11]) that for $\lambda = -\infty$ the optimal pdf $g(\mathbf{x})$ in (16.69) coincides with the zero-variance IS pdf (16.58), that is with

$$g^*(\mathbf{x}, m) = \ell^{-1} f(\mathbf{x}) I_{\{S(\mathbf{x}) \geq m\}}. \tag{16.75}$$

Because of this relation one can easily *switch* from the Boltzmann pdf (16.69) to the IS zero-variance pdf (16.75) and thus to apply the original Algorithm 16.3.6 instead of the Boltzmann type algorithm below and vise-versa, provided $\lambda = -\infty$ in (16.69). This, for example, means that when $\lambda = -\infty$ the classic Ising model can be treated by using the IS zero-variance pdf (16.75) rather than via the original Boltzmann pdf (16.69).

To proceed note that similar to the parametric MinxEnt algorithm (see [11]) no acceptance-rejection step is explicitly involved in the Boltzmann cloning algorithm our algorithm below. As result neither the *rarity parameter* ρ nor *elite sampling* are explicitly involved either. In fact, the elite sampling will be hidden some how in the sense that it will be explicitly available only after the sequence $\{\widehat{m}_t,\ t = 1,\ldots,T\}$ is generated. Observe that here $\widehat{m}_t$ corresponds to the *maximum* level reached so for at iteration t.

To clarify, consider a rare events probability ℓ associated with the sample performance $S(X)$ being the sum of iid Bernoulli random variables, that is $S(X) = \sum_{i=1}^{m} X_i$. Assume for concreteness that we took a sample $N = 1{,}000$ and while running the simulation we obtained (at some iteration t) that all 1,000 sample values $S_k \in (a,b)$, $k = 1,\ldots,1{,}000$, where, for concreteness, say $a = 20$ and $b = 50$. Since $\widehat{m}_t$ corresponds to the maximum level reached at iteration t, we clearly have that $\widehat{m}_t = b = 50$ and since $\lambda = -\infty$, all the terms of S_k which are less than b are negligible in the exponent $\exp\{-\sum_{i=1}^{m} \lambda X_{ik}\}$. Assume, finally that the number of terms, which we also call here, the number of elite samples, say N_t for which $S_k = 50$ equals 25. So, as before, we can define $\widehat{\rho}_t = N_t/N$ and call it *adaptive inexplicit* ρ, (for our case it is $\widehat{\rho}_t = \frac{25}{1{,}000} = 1/40$).

Note that typically we expect $\widehat{\rho}_t > 10^{-2}$, since we take at each iteration a sample size $N > m/\rho$, where say $\rho = 10^{-1}$. If however, at some iteration t, we occasionally obtain that $\widehat{\rho}_t < \rho$, we can reject the largest value $\widehat{m}_t$ obtained so far and take instead the value $\widehat{m}_t - 1$, provided $N > (\widehat{m}_t - 1)/\rho$ and so far. In our example $\widehat{m}_t - 1 = 49$.

The Boltzmann cloning algorithm basically coincides with Algorithm 16.3.6. Here we sample from the zero variance Boltzmann pdf $g(\mathbf{x}, m_t, \lambda = -\infty)$ (see (16.69)) rather then from the zero variance IS pdf $g^*(\mathbf{x}, m_t)$ (see (16.58)) and the step 2 of Algorithm 16.3.6 for chosing the adaptive $\rho \in (a_1, a_2)$ is replaced by the corresponding *adaptive inexplicit* one defined above. Also, as in Algorithm 16.3.6 we use the term *truly uniform elites*. This is regardless of the fact that the elite samples are obtained only in an inexplicit way.

Algorithm 16.3.8 (Cloning Algorithm with the Boltzmann Distribution) Given the sample size N, execute the following steps:

1. *Acceptance-Rejection.* Set a counter $t = 1$. Generate a sample $X_1,\ldots,X_N$ from the proposal density $f(\mathbf{x})$. Let $\widetilde{\mathscr{X}}_0 = \{\widetilde{X}_1,\ldots,\widetilde{X}_{N_0}\}$ be the largest subset of the population $\{X_1,\ldots,X_N\}$ (elite samples) for which $S(X_i) \geq \widehat{m}_0$. Take $\widehat{z}_0 = \widehat{z}(\widehat{m}_0)$ in (16.73) is an *unbiased* estimator of z_0. Note that $\widetilde{X}_1,\ldots,\widetilde{X}_{N_0} \sim g(\mathbf{x}, \lambda, \widehat{m}_0)$.

2. *Screening*. Denote the elite sample obtained at iteration $t-1$ by $\{\widetilde{X}_1,\ldots,\widetilde{X}_{N_{t-1}}\}$. Screen out the redundant elements of the subset $\{\widetilde{X}_1,\ldots,\widetilde{X}_{N_{t-1}}\}$ and denote the resulting one as $\{\widehat{X}_1,\ldots,\widehat{X}_{N_{t-1}}\}$.
3. *Adaptive inexplicit choose of* ρ. Estimate ρ according to the adaptive inexplicit rule $\widehat{\rho}_t = N_t/N$, where N_t includes all elite samples $S(X_i)$ corresponding to the maximum level $\widehat{m}_t$ of the entire sample $S(X_i)$, $i = 1,\ldots,N$.
4. *Cloning*. Given the size N_{t-1} of truly uniform elites at iteration $t-1$, find the cloning and the burn-in parameters η_{t-1} and b_{t-1} according to (16.68). Reproduce η_{t-1} times each vector $\widehat{X}_k = (\widehat{X}_{1k},\ldots,\widehat{X}_{nk})$ of the truly elite sample $\{\widehat{X}_1,\ldots,\widehat{X}_{N_{t-1}}\}$, that is, take η identical copies of each vector $\widehat{X}_k$ obtained at the $t-1$-th iteration. Denote the entire new population (ηN_{t-1} cloned vectors plus the original truly elite sample $\{\widehat{X}_1,\ldots,\widehat{X}_{N_{t-1}}\}$) by $\mathscr{X}_{cl} = \{(\widehat{X}_1,\ldots,\widehat{X}_1),\ldots,(\widehat{X}_{N_{t-1}},\ldots,\widehat{X}_{N_{t-1}})\}$. To each of the cloned vectors of the population $\mathscr{X}_{cl}$ apply the MCMC (and in particular the Gibbs sampler) for b_{t-1} burn-in periods. Denote the *new entire* population by $\{X_1,\ldots,X_N\}$. Observe that each component of $\{X_1,\ldots,X_N\}$ is distributed approximately $g(\mathbf{x},\lambda,\widehat{m}_{t-1})$.
5. *Estimating* z_t. Let $\widetilde{\mathscr{X}_t} = \{\widetilde{X}_1,\ldots,\widetilde{X}_{N_t}\}$ be the subset of the population $\{X_1,\ldots,X_N\}$ for which $S(X_i) \geq \widehat{m}_t$. Take $\widehat{z}_t$ in (16.73) as an estimator of z_t given in (16.71). Note again that $\widetilde{X}_1,\ldots,\widetilde{X}_{N_t}$ is distributed approximately $g(\mathbf{x},\lambda,\widehat{m}_t)$.
6. *Stopping rule*. If $m_t = m$ go to step 6, otherwise set, set $t = t+1$ and repeat from step 2.
7. *Final Estimator*. Deliver $\widehat{Z}(m)$ given in (16.72) as an estimator of $Z(m)$ given in (16.70).

16.3.5 Applications: Counting the Number of Feasible Solutions in an Integer Program

Here we show how to use the Gibbs sampler for counting on the set (16.42), that is on

$$\sum_{k=1}^{n} a_{ik}x_k = b_i,\ i = 1,\ldots,m_1,$$

$$\sum_{k=1}^{n} a_{jk}x_k \geq b_j,\ j = m_1+1,\ldots,m_1+m_2,$$

$$\mathbf{x} \geq 0,\ x_k \text{ integer } \forall k = 1,\ldots,n\,.$$

While using Gibbs sampler, we take into account the additivity properties of the functions $S_i(\mathbf{x}) = \sum_{k=1}^{n} a_{ik}x_k$ in (16.42) and also formula (16.46)

Example 16.6 (SAT Example 16.4 Continued). We shall show how to apply Gibbs sampler for counting the number of assignments in the SAT Example 16.4, that is,

$$(x_1 + \bar{x}_2)(\bar{x}_1 + \bar{x}_2 + x_3)(x_2 + x_3) .$$

Recall that

1. The set of the associated linear integer constraints in this case is as

$$x_1 + (1 - x_2) \geq 1$$

$$(1 - x_1) + (1 - x_2) + x_3 \geq 1,$$

$$x_2 + x_3 \geq 1 ,$$

   where each $x_1, x_2, x_3 \in \{0, 1\}$.
2. The associated probability ℓ is $\ell = \mathbb{P}_u(C_1 + C_2 + C_3 = 3)$, where $C_1 = I_{\{X_1 - X_2 \geq 0\}}$, $C_2 = I_{\{X_1 + X_2 - X_3 \leq 1\}}$ and $C_3 = I_{\{X_2 + X_3 \geq 1\}}$.

We have

$$\begin{aligned}
g^*(x_1, \widehat{m}_{t-1} | \mathbf{x}_{-1}) &= f_1(x_1) I_{\{x_1 \geq \widehat{m}_{t-1} - I_{\{-x_2 \geq 0\}} - I_{\{x_2 - x_3 \leq 1\}} - C_3\}}, \\
g^*(x_2, \widehat{m}_{t-1} | \mathbf{x}_{-2}) &= f_2(x_2) I_{\{x_2 \geq \widehat{m}_{t-1} - I_{\{x_1 \geq 0\}} - I_{\{x_1 - x_3 \leq 1\}} - I_{\{x_3 \geq 1\}}\}}, \\
g^*(x_3, \widehat{m}_{t-1} | \mathbf{x}_{-3}) &= f_3(x_3) I_{\{x_1 \geq \widehat{m}_{t-1} - C_1 - I_{\{x_1 + x_2 \leq 1\}} - I_{\{x_2 \geq 1\}}\}},
\end{aligned} \tag{16.76}$$

where $f_i(x_i)$, $i = 1, 2, 3$ are independent $\text{Ber}(p = 1/2)$ distributions.

Note that $I_{\{X_1 \geq 0\}} = 1$ in $g^*(x_2, \widehat{m}_{t-1} | \mathbf{x}_{-2})$.

It readily follows from the above example, that in order to count on a quite general set of linear (integer or continuous) constraints (16.42) with given matrix $\mathbf{A} = \{a_{ij}\}$, one only needs to apply the Gibbs method, while sampling from the following simple one-dimensional conditional pdfs $g^*(x_i, \widehat{m}_{t-1} | \mathbf{x}_{-i})$

$$g^*(x_i, \widehat{m}_{t-1} | \mathbf{x}_{-i}) = \text{Ber}(1/2) I_{\{\sum_{r \in R_i} C_r(X) \geq \widehat{m}_{t-1} - \sum_{r \notin R_i} C_r(X)\}} , \tag{16.77}$$

where $R_i = \{j : \ a_{ij} \neq 0\}$.

Recall that sampling a random variable $\widetilde{X}_i$ from (16.77) using the Gibbs sampler can be performed as follows. Generate $Y \sim \text{Ber}\ (1/2)$. If

$$\sum_{r \in R_i} C_r(x_1, \ldots, x_{i-1}, Y, x_{i+1}, \ldots, x_n) \geq \widehat{m}_{t-1} ,$$

then set $\widetilde{X}_i = Y$, otherwise set $\widetilde{X}_i = 1 - Y$.

Note also that before performing the simulation, one has to store the corresponding set of indexes R_i associated with each conditional marginal pdf $g^*(x_i, \widehat{m}_{t-1} | \mathbf{x}_{-i})$.

It follows from the above that counting multiple events and in particular the cardinality of the set (16.42) of the integer program (16.43) can be efficiently performed using the cloning Algorithm 16.3.6, that is one can estimate $|\mathscr{X}^*|$ via $|\widehat{\mathscr{X}^*}| = \widehat{\ell}|\mathscr{X}|$, where $\widehat{\ell}$ by itself is an estimator of ℓ given in (16.59). Similar arguments can be applied for estimating the volume of bodies given by the set

$$\mathscr{X}^* = \{\mathbf{x} \in \mathbb{R}^n : S_i(\mathbf{x}) = b_i,\ i = 1,\ldots,m_1; S_j(\mathbf{x}) \geq b_j,\ j = 1,\ldots,m_2\}\,. \quad (16.78)$$

In particular, for a polyhedron, $\mathscr{X}^*$ reduces to the set associated with the following linear programming constraints

$$\begin{aligned} &\sum_{k=1}^{n} a_{ik} x_k = b_i,\ i = 1,\ldots,m_1, \\ &\sum_{k=1}^{n} a_{jk} x_k \geq b_j,\ j = 1,\ldots,m_2, \\ &x_k \in (a,b),\ \forall k = 1,\ldots,n\,. \end{aligned} \quad (16.79)$$

It also follows that, when possible, it is desirable to use the estimator $|\widehat{\mathscr{X}^*_{dir}}|$ defined in Remark 16.7, since it is much more accurate than the standard one $|\widehat{\mathscr{X}^*}|$.

16.3.6 Sampling Uniformly on Different Regions

Since the sequences produced by Algorithm 16.3.6 a truly uniform, so it should be suitable for uniform sampling on $\mathscr{X}^*$. This means, for example, that while counting the number of feasible solutions defined on the set of the linear integer program with the constraints (16.42), one can sample according to a *discrete uniform* pdf inside the corresponding region $\mathscr{X}^*$. As for another example, if we have linear programming constraints (16.79) instead of (16.42), then the continuous version of Algorithms 16.3.6 can be used to sample uniformly inside the corresponding polyhedron.

Since to sample on $\mathscr{X}^*$ only the elites at level $m_T = m$ matter, we can run the cloning Algorithms 16.3.6 at all intermediate levels $m_0, m-1,\ldots,m_{T-1}$ with less samples, while at the final level, $m_T = m$, we can take a larger sample. The corresponding elites could be further employed to generate an approximate uniform sampling on the entire region $\mathscr{X}^*$.

As a simple example, consider sampling on the region

$$\mathscr{X}^* = \{X : \sum_{i=1}^{n} X_i \geq m\},$$

where the X_i's are iid each distributed Ber (1/2). Let $n = 100$ and $m = 98$. We have $|\mathscr{X}^*| = 10{,}001$. For this example we employed Algorithm 16.3.6 with $N = 100$,

$\rho = 0.1$ and $b = 5$ for the first $T-1$ iterations. By doing so we obtain at each iteration, on average, 10 truly uniform elites. At the last iteration we increased the sample to 1,000 using the same $\rho = 0.1$ and the same $b = 5$ and thus obtained on average, 100 truly uniform elites. We finally run each of these 100 Markov chains (in steady-state) for a long burn-in, say $b = 500$ and, thus generating a total of $N = 50{,}000 = 100 \times 500$ Gibbs samples, for which statistics was collected. For this simple example we found that

1. The direct estimator $|\widehat{\mathscr{X}^*_{dir}}|$ with the above sample $N = 50{,}000 = 100 \times 500$ found all 10,001 different Bernoulli points.
2. The resulting sample of the size 50,000 is distributed very close to the uniform on the set $\mathscr{X}^* = \{X : \sum_{i=1}^{100} X_i \geq 98\}$, that is the histogram over these 10,001 points is close to the uniform.

More research on uniform sampling on different regions is under way.

16.3.7 Integer Programming: The Penalty Function Approach

The cloning Algorithm 16.3.6 can be readily modified for unconstrained optimization as follows

Algorithm 16.3.9 (Cloning Algorithm for Integer Programming) Given the rarity parameter ρ, say $\rho = 0.1$, the parameters a_1 and a_2, say $a_1 = 0.01$ and $a_2 = 0.25$, the sample size N, say $N = m \times n$, the burn-in period b, say $3 \leq b \leq 10$, execute the following steps:

1. *Acceptance-Rejection.* Set a counter $t = 1$. Generate a sample $X_1, \ldots, X_N$ from the proposal density $f(\mathbf{x})$. Let $\widetilde{\mathscr{X}_0} = \{\widetilde{X}_1, \ldots, \widetilde{X}_{N_0}\}$ be the largest subset of the population $\{X_1, \ldots, X_N\}$ (elite samples) for which $S(X_i) \geq \widehat{m}_0$.
2. *Adaptive choice of* ρ. For each iteration use the adaptive choice of $a_1 \leq \rho \leq a_2$, with a_1 and a_2 defined in Remark 16.5.
3. *Screening.* Denote the elite sample obtained at iteration $t-1$ by $\{\widetilde{X}_1, \ldots, \widetilde{X}_{N_{t-1}}\}$. Screen out the redundant elements of the subset $\{\widetilde{X}_1, \ldots, \widetilde{X}_{N_{t-1}}\}$ and denote the resulting one as $\{\widehat{X}_1, \ldots, \widehat{X}_{N_{t-1}}\}$.
4. *Cloning.* Given the number of burn-in periods b and the size N_{t-1} of truly uniform elites at iteration $t-1$, find the cloning parameter η_{t-1} according to $\eta_{t-1} = \left\lceil \frac{N}{bN_{t-1}} \right\rceil - 1$. Reproduce η_{t-1} times each vector $\widehat{X}_k = (\widehat{X}_{1k}, \ldots, \widehat{X}_{nk})$ of the truly elite sample $\{\widehat{X}_1, \ldots, \widehat{X}_{N_{t-1}}\}$, that is take η identical copies of each vector $\widehat{X}_k$ obtained at the $t-1$-th iteration. Denote the entire new population (ηN_{t-1} cloned vectors plus the original truly elite sample $\{\widehat{X}_1, \ldots, \widehat{X}_{N_{t-1}}\}$) by $\mathscr{X}_{cl} = \{(\widehat{X}_1, \ldots, \widehat{X}_1), \ldots, (\widehat{X}_{N_{t-1}}, \ldots, \widehat{X}_{N_{t-1}})\}$. To each of the cloned vectors of the population $\mathscr{X}_{cl}$ apply the MCMC (and in particular the Gibbs sampler) for b burn-in periods. Denote the *new entire* population by $\{X_1, \ldots, X_N\}$.

5. *Estimating* m_t. Let $\widetilde{\mathscr{X}_t} = \{\widetilde{X}_1, \ldots, \widetilde{X}_{N_t}\}$ be the subset of the population $\{X_1, \ldots, X_N\}$ for which $S(X_i) \geq \widehat{m}_t$. Deliver $\widehat{m}_t$.
6. *Stopping*. If for some $t \geq d$, say $d = 5$,

$$\widehat{m}_t = \cdots = \widehat{m}_{t-d} \tag{16.80}$$

then stop and deliver $\widehat{m}_{t,N}$ as the estimator of the optimal solution; otherwise, set $t = t + 1$ and return to Step 2.

16.4 Numerical Results

16.4.1 Numerical Results with IME

We present here numerical results with the proposed algorithms for counting and optimization. For counting problems we found that the variance minimization (VM) algorithms [13] are more robust than its IME and CE counterparts. In contrast, for combinatorial optimization, like TSP, we found that all proposed algorithms perform similarly. The main reason for that is, *there is no need to use LR's in optimization*. If not stated otherwise, we set the rarity parameter $\rho = 0.001$ and the smoothing parameter $\alpha = 0.7$. Note that $\rho = 0.001$ applies to the elite samples only for the intermediate states of our algorithms, that is, when $\widehat{m}_t < m$. For $\widehat{m}_t = m$, we accumulate all elite samples.

A huge collection of instances (including real-world) is available on OR-LIB site: http://people.brunel.ac.uk/ mastjjb/jeb/orlib/scpinfo.html; for multiple-knapsack http://hces.bus.olemiss.edu/tools.html or http://elib.zib.de/pub/Packages/mp-testdata/ip/sac94-suite/index.html. Knapsack instances generator is given on: http://www.diku.dk/ pisinger/codes.html. A huge collection of SAT problems is given on SATLIB website www.satlib.org.

To study the variability in the solutions, we run each problem 10 times and report our statistics based on these 10 runs of our algorithms. In the following tables, the quantities are defined as follows (for each iteration t):

1. "Mean, max and min $|\widehat{\mathscr{X}^*}|$" denote the sample mean, maximum and minimum and minimal values of the 10 estimates of $|\mathscr{X}^*|$.
2. "Mean, max and min Found" denote the sample mean, maximum and minimum of values found in each of the 10 samples of size N. Note that the maximum value can be viewed as the lower bound of the true unknown quantity $|\mathscr{X}^*|$.
3. PV denotes the proportion of generated values, averaged over 10 replications.
4. RE denotes the mean relative error for $|\widehat{\mathscr{X}^*}|$, averaged over the 10 runs.
5. λ denotes the mean λ, averaged over the 10 runs.
6. $\mathscr{S}$ denotes the mean entropy averaged over the 10 runs.
7. m denotes the mean number of satisfied constraints at t-th iteration and averaged over the 10 runs.

In all counting problems, we compared the performance of the standard CE and the IME Algorithm 16.2.1 (with fixed $\lambda = -100$) with their VM counterparts. While running the algorithms we found that for some particular instances, all CE-based algorithms produce incorrect estimators, while their counterpart, the VM-based algorithm, always delivers correct (unbiased) ones. This undesirable phenomenon of CE-based algorithms has not yet been fully understood and it is under investigation.

In all our numerical studies, we generated the matrices $A = (a_{ij})$ randomly and made sure that they are sparse. The sparsity insures that the counting quantity $|\mathscr{X}^*|$ is small (is associated with rare-event probability, that is, the most difficult cases), while random matrices generation insures the diversity of the cases. All cases have been checked first on small randomly generated models, such that $|\mathscr{X}^*|$ is relatively small, say $0 \leq |\mathscr{X}^*| \leq 100$, and such that their exact solution via full enumeration is available. Only, after that have larger models been tested.

To speed up the convergence, we implemented the following.

- We set $\lambda = -10$ for its first 2–3 iterations and for the remaining ones we set $\lambda = -100$ the IME Algorithm 16.2.1.
- In many counting problems involving rare-events, the elements of $\widehat{p}_t$ are approaching either 0's or 1's as t increases. We set them automatically either to 0's or 1's as soon as they reach, say 0.01 and 0.99, respectively. By doing so, at iterations $t+1,\dots,T$ one needs to generate and update only a very small portion of p's, namely those which remain in the interval (0.01, 0.99).

Below we consider separately counting, and combinatorial optimization.

16.4.1.1 Counting

Recall that in all our experiments with the IME Algorithm 16.2.1, we set $\lambda = -100$. We also present the performance of our algorithms to count the number of optimal solutions in some constrained optimization problems, where the optimal solution was either obtained via full enumeration (for small models) or (the best known solution) was taken from the web site.

16.4.2 Counting Hamiltonian Cycles

Similar, to the *random K-PERM* we define the so-called *random K-Hamiltonian matrix*, denotes as *K-HAM matrix*, where, as before, K, $(K < n)$ denotes the number of independent uniformly distributed Bernoulli random variables at each row of the randomly generated matrix A. We found empirically that in order for $|\mathscr{X}|^*$ to be very small relative to $|\mathscr{X}|$, the parameter K should be chosen as $K \leq 0.15n$.

Table 16.5 presents the performance of the IME Algorithm 16.2.1 for a 4-HAM randomly generated (30×30) matrix using $N = 100{,}000$ samples. The trajectories

Table 16.5 Performance of the IME algorithm for the HC problem for a 4-HAM matrix $A = (30 \times 30)$ and $N = 100,000$

t	$\lvert\mathscr{X}^*\rvert$			Found			PV	RE
	Mean	Max	Min	Mean	Max	Min		
0	36.27	283	0	0.20	1	0	0.0000	2.3608
1	62.84	109	32	22.10	26	17	0.0005	0.3272
2	66.09	76	55	55.10	60	49	0.0065	0.0885
3	62.75	68	56	55.30	62	49	0.0344	0.0496

Table 16.6 Performance of the IME Algorithm 16.2.1 for the random 3-SAT for the instance matrix $A = (40 \times 160)$ and $N = 100,000$

t	$\lvert\mathscr{X}^*\rvert$			*Found*			PV	RE	S	m
	Mean	Max	Min	Mean	Max	Min				
0	0.0	0.0	0.0	0	0	0	0.00	NaN	13.86	151
1	0.0	0.0	0.0	0	0	0	0.00	NaN	13.25	153
2	96.0	960.3	0.0	0	1	0	0.00	3.000	12.28	155
3	88.9	328.6	0.0	2	9	0	0.00	1.164	11.13	157
4	93.9	120.1	0.0	42	106	0	0.00	0.387	8.46	159
5	111.0	134.6	45.9	98	113	12	0.05	0.207	6.52	160
6	113.0	123.1	105.9	109	113	98	0.22	0.038	4.49	160
7	109.5	113.3	104.5	109	111	105	0.38	0.025	3.55	160
8	109.7	113.5	105.1	109	111	105	0.49	0.021	3.16	160
9	109.9	114.6	104.5	109	111	105	0.53	0.025	3.04	160
10	111.7	116.9	104.8	109	111	105	0.54	0.027	3.00	160

(tours) were generated using the *node transition algorithm* (see Algorithm 4.7.1 of [12]). The results are self-explanatory.

16.4.3 The SAT Problem

Table 16.6 presents the performance of the IME Algorithm 16.2.1 for a random 3-SAT problem with an instance matrix $A = (40 \times 160)$ for $N = 100,000$. The results are self-explanatory. Note that running the same problem with the standard CE method, the results were worse, in particular in terms of relative error.

Figure 16.3 presents a typical dynamics of the IME Algorithm 16.2.1 with the instance matrices $A = (40 \times 160)$.

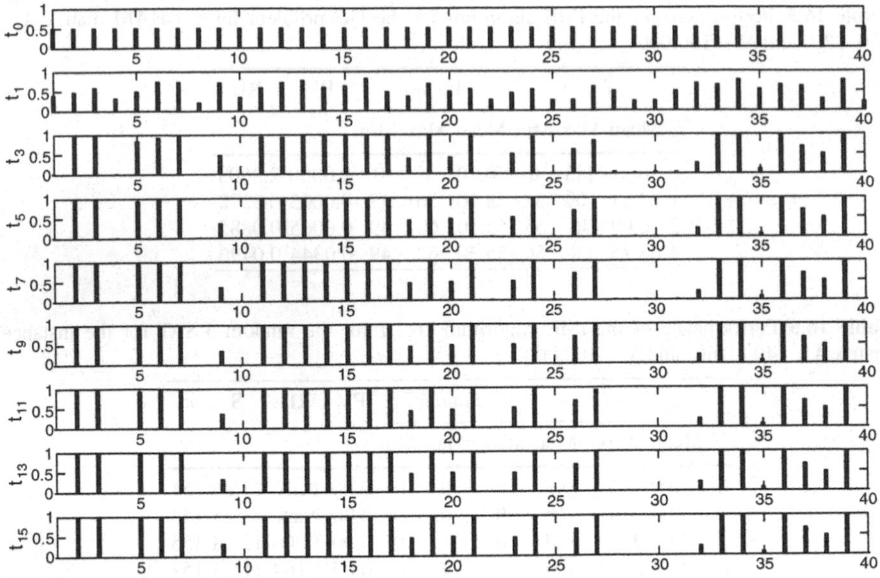

Fig. 16.3 Typical dynamics of the IME Algorithm 16.2.1 for the random 3-SAT with an instance matrix $A = (40 \times 160)$ and $N = 100{,}000$

16.4.3.1 Optimization

In this section we present performance of CE, VM, MinxEnt and IME for unconstrained optimization. Table 16.7 presents comparative studies of the following 4 methods: CE, VM, MinxEnt and IME for a couple of TSP models taken from http://www.iwr.uni-heidelberg.de/groups/comopt/software/ TSPLIB95/atsp/.

In all numerical results we use the same CE parameters as for the `ft53` problem, that is, $\rho = 10^{-2}$, $N = 10n^2$, $\alpha = 0.7$ and $d = 5$ (see (16.49)). To study the variability in the solutions, each problem was repeated 10 times. In Table 16.7, n denotes the number of nodes of the graph, $\bar{T}$ denotes the average total number of iterations needed before stopping, $\widehat{b}_1$ and $\widehat{b}_T$ denote the average initial and final estimates of the optimal solution, b^* denotes the best known solution, $\bar{\varepsilon}$ denotes the average relative experimental error based on 10 replications, ε_* and ε^* denote the smallest and the largest relative error among the 10 generated shortest paths, and finally CPU denotes the average CPU time in seconds.

It follows that all 4 methods work reasonable well and it is difficult to give priority to any of them.

We apply next the CE algorithm to solve the Knapsack problem using the penalty function approach (see (16.51)–(16.53).

As a particular problem consider the Sento2.dat knapsack problem given in http://people.brunel.ac.uk/ mastjjb/jeb/orlib/files/mknap2.txt. The problem has 30 constraints and 60 variables. We ran the CE method for 10 independent runs with $\rho = 0.01$, $N = 50{,}000$, $\alpha = 0.9$ and $a = 1$. In all our experiments, CE always found

Table 16.7 Comparative studies for TSP

File	n	b^*	Alg.	$\widehat{b}_1$	$\widehat{b}_T$	$\bar{\varepsilon}$	ε_*	ε^*	$\bar{T}$	CPU
ftv33	34	1,286	CE	3,248	1,333	0.0365	0.0000	0.0684	17.8	56
			VM	3,366	1,286	0.0000	0.0000	0.0000	23.8	127
			PME	3,296	1,308	0.0171	0.0000	0.0412	19.8	173
			IME		1,305	0.0154	0.0000	0.0435	18.30	76
ry48p	48	14,422	CE	40,254	14,840	0.0289	0.0133	0.0579	31.2	424
			VM	42,208	14,960	0.0373	0.0162	0.0597	61.7	935
			PME	41,041	14,952	0.0367	0.0228	0.0537	34.0	992
			IME		14,888	0.0323	0.0160	0.0461	30.60	731

Table 16.8 A typical dynamics of the CE algorithm with $a = 1$ for the problem Sento2

t	$\widehat{m}_t$	$\widehat{m}_t^*$
1	5,742.0	7,685.0
2	7,461.0	8,392.0
3	8,052.0	8,395.0
4	8,297.0	8,522.0
5	8,427.0	8,606.0
6	8,509.0	8,668.0
7	8,562.0	8,704.0
8	8,604.0	8,713.0
9	8,624.4	8,722.0
10	8,640.0	8,722.0
11	8,649.0	8,722.0

the optimal solution $S(\mathbf{x}^*)$= 8,722. Table 16.8 presents a typical dynamics of the CE algorithm. Here $\widehat{m}_t$ and $\widehat{m}_t^*$ denote the elite sample value at iteration t and the best solution found during the first t iterations, respectively.

16.4.4 Numerical Results with the Cloning Method

We present here numerical results with the proposed algorithms for counting and optimization. If not stated otherwise, we set the rarity parameter $\rho = 0.1$ for $N =$ 1,000 and $\rho = 0.01$ for $N =$ 1,000.

A huge collection of instances (including real-world) is available on OR-LIB site: http://people.brunel.ac.uk/mastjjb/jeb/orlib/scpinfo. html; for multiple-knapsack http://hces.bus.olemiss.edu/tools.html or http://elib.zib.de/pub/Packages/mp-testdata/ip/sac94-suite/index.html. Knapsack instances generator is given on: http://www.diku.dk/ pisinger/codes.html.

To study the variability in the solutions, we run each problem 10 times and report our statistics based on these 10 runs of our algorithms. In the following tables, the quantities are defined as follows (for each iteration t):

1. "Mean, max and min $|\widehat{\mathscr{X}^*}|$" denote the sample mean, maximum and minimum and minimal values of the 10 estimates of $|\mathscr{X}^*|$.
2. "Mean, max and min $|\widehat{\mathscr{X}^*_{dir}}|$" denote the sample mean, maximum and minimum values of the *empirical estimator* (see Remark 16.7) found in each of the 10 samples of size N. Note that the maximum value of the "direct estimator" can be viewed as the lower bound of the true unknown quantity $|\mathscr{X}^*|$.
3. RE denotes the mean relative error for $|\widehat{\mathscr{X}^*}|$, averaged over the 10 runs.
4. CPU denotes the mean relative error for $|\widehat{\mathscr{X}^*}|$, averaged over the 10 runs.

16.4.5 The SAT Problem

Table 16.9 presents the performance of Algorithm 16.3.6 for a random 3-SAT problem with an instance matrix $A = (20 \times 80)$ for $N = 1{,}000$ and $\rho = 0.05$. We found that the average relative error is RE = 0.08 and the average CPU time is 67 s.

The results are self-explanatory. Table 16.10 presents the dynamics for one of the runs of Algorithm 16.3.6 for the same model.

As before, we used the following notations

1. N_t and $N_t^{(s)}$ denote the actual number of elites and the one after screening, respectively.
2. m_t^* and m_{*t} denote the upper and the lower elite levels reached, respectively.
3. $\rho_t = N_t/N$ denotes the adaptive rarity parameter.

Table 16.9 Performance of Algorithm 16.3.6 for 3-SAT with the matrix $A = (20 \times 80)$

t	$\lvert\widehat{\mathscr{X}^*}\rvert$			$\lvert\widehat{\mathscr{X}^*_{dir}}\rvert$			m_t
	Mean	Max	Min	Mean	Max	Min	
1	8,500.5	10,223.6	7,401.7	0.2	2.0	0.0	75
2	1,740.5	2,075.3	1,565.7	1.7	4.0	0.0	77
3	208.3	232.9	167.3	5.9	8.0	3.0	78
4	14.3	16.3	11.8	14.7	15.0	14.0	79
5	14.3	16.3	11.8	15.0	15.0	15.0	80

Table 16.10 Dynamics of Algorithm 16.3.6

t	$\lvert\widehat{\mathscr{X}^*}\rvert$	$\lvert\widehat{\mathscr{X}^*_{dir}}\rvert$	N_t	$N_t^{(s)}$	m_t^*	m_{*t}	ρ_t
1	8,500.5	0.2	71	71	77	75	0.06
2	1,740.5	1.7	110	109	79	77	0.12
3	208.3	5.9	206	191	80	78	0.21
4	14.3	14.7	109	87	80	79	0.10
5	14.3	15.0	70	15	80	80	0.07

Tables 16.11 and 16.12 present data similar to Tables 16.9 and 16.10, respectively, for the random 3-SAT problem with the instance matrix $A = (75 \times 325)$ taken from www.satlib.org. We set $N = 10{,}000$ and $\rho = 0.1$ for all iterations until Algorithm 16.3.6 reached the desired level 325. After that we switched to $N = 100{,}000$ for the last iteration. The results are self-explanatory.

We found that the average relative error is RE = 0.08 and the average CPU time is 25 min for each run. It is readily seen that at iteration 21 we obtained $\rho = 1$ for $m_t = 324$; after that Algorithm 16.3.6 switches automatically from $\rho = 0.05$ to $\rho_2 = 0.02$. This in turn results in switching from $m_t = 324$ to $m_t = m = 325$. Note again that without the adaptive mechanism Algorithm 16.3.6 would terminate at $m_t = 324$ without reaching the final destination $m = 325$.

Similar to the CE algorithms, we also applied the cloning Algorithm 16.3.9 to solve the knapsack problem Sento2. We ran the algorithm for 10 independent runs with $\rho = 0.1$, $N = 1{,}000$ and $b = 5$. We found that Algorithm 16.3.9 always converged to the optimal solution = 8,772. We also ran different problems from the same side and found that the results with Algorithm 16.3.9 were exact.

Table 16.11 Performance of Algorithm 16.3.6 for the random 3-SAT with the clause matrix $A = (75 \times 325)$, $N = 10{,}000$ and $\rho = 0.1$

	$\lvert\mathscr{X}^*\rvert$			$\lvert\widehat{\mathscr{X}^*_{dir}}\rvert$			
t	Mean	Max	Min	Mean	Max	Min	m_t
1	5.4e+020	5.8e+020	5.2e+020	0.0	0.0	0.0	292
2	5.6e+019	6.0e+019	5.3e+019	0.0	0.0	0.0	297
3	6.4e+018	7.0e+018	6.0e+018	0.0	0.0	0.0	301
4	1.2e+018	1.3e+018	1.1e+018	0.0	0.0	0.0	304
5	1.7e+017	1.9e+017	1.6e+017	0.0	0.0	0.0	306
6	1.9e+016	2.0e+016	1.8e+016	0.0	0.0	0.0	308
7	5.9e+015	6.3e+015	5.5e+015	0.0	0.0	0.0	310
8	1.7e+015	1.8e+015	1.6e+015	0.0	0.0	0.0	311
9	4.5e+014	4.7e+014	4.2e+014	0.0	0.0	0.0	312
10	1.1e+014	1.1e+014	1.0e+014	0.0	0.0	0.0	313
11	2.4e+013	2.6e+013	2.1e+013	0.0	0.0	0.0	314
12	4.9e+012	5.5e+012	4.2e+012	0.0	0.0	0.0	315
13	9.2e+011	1.0e+012	7.8e+011	0.0	0.0	0.0	316
14	1.5e+011	1.7e+011	1.3e+011	0.0	0.0	0.0	317
15	2.4e+010	2.7e+010	2.0e+010	0.0	0.0	0.0	318
16	3.2e+009	3.6e+009	2.6e+009	0.0	0.0	0.0	319
17	3.7e+008	4.3e+008	3.2e+008	0.0	0.0	0.0	320
18	3.4e+008	4.3e+008	3.7e+007	0.5	2.0	0.0	321
19	3.6e+007	4.2e+007	3.0e+007	7.5	12.0	4.0	322
20	2.7e+006	3.2e+006	2.2e+006	76.9	89.0	59.0	323
21	1.3e+005	1.6e+005	1.0e+005	1,167.3	1,214.0	1,113.0	324
22	2,226.4	2,728.8	1,780.7	2,255.4	2,258.0	2,250.0	325
23	2,226.4	2,728.8	1,780.7	2,255.4	2,258.0	2,250.0	325

Table 16.12 Dynamics of Algorithm 16.3.6 for the random 3-SAT with the clause matrix $A = (75 \times 325)$

t	$\lvert\mathscr{X}^*\rvert$	$\lvert\widehat{\mathscr{X}^*_{dir}}\rvert$	N_t	$N_t^{(s)}$	m_t^*	m_{*t}	ρ_t
1	5.4e+020	0.0	1,036	1,036	306	292	0.10
2	5.6e+019	0.0	1,300	1,300	308	297	0.14
3	6.4e+018	0.0	973	973	308	301	0.10
4	1.2e+018	0.0	1,086	1,086	311	304	0.11
5	1.7e+017	0.0	1,807	1,807	313	306	0.19
6	1.9e+016	0.0	1,271	1,271	314	308	0.15
7	5.9e+015	0.0	1,013	1,013	314	310	0.11
8	1.7e+015	0.0	2,842	2,842	317	311	0.30
9	4.5e+014	0.0	2,450	2,450	317	312	0.28
10	1.1e+014	0.0	2,616	2,616	317	313	0.26
11	2.4e+013	0.0	1,932	1,932	318	314	0.24
12	4.9e+012	0.0	2,154	2,154	320	315	0.22
13	9.2e+011	0.0	1,776	1,776	320	316	0.21
14	1.5e+011	0.0	1,652	1,652	320	317	0.19
15	2.4e+010	0.0	1,656	1,656	321	318	0.16
16	3.2e+009	0.0	1,488	1,488	321	319	0.15
17	3.7e+008	0.0	1,209	1,209	323	320	0.13
18	3.4e+008	0.5	1,111	1,111	323	321	0.12
19	3.6e+007	7.5	9,815	9,811	325	322	0.09
20	2.7e+006	76.9	7,338	7,334	325	323	0.07
21	1.3e+005	1,167.3	4,708	4,622	325	324	0.05
22	2,226.4	2,255.4	1,718	1,182	325	325	0.02
23	2,226.4	2,255.4	99,288	2,250	325	325	1.00

The advantage of the cloning Algorithm 16.3.6, as compared to CE, is that it can count the number of all optimal solutions with quite high accuracy simultaneously with optimization.

16.5 Conclusions and Further Research

In this paper we presented the latest developments in *importance sampling* (IS) combined with MinxEnt and in the MCMC methods for counting, rare-event probability estimation, sampling and combinatorial optimization problems (COP's). In particular, we discuss how in the former the optimal parameter vector in the *parametric* IS density can be efficiently obtained using the IME method and how in the latter one can sample from the optimal zero-variance nonparametric IS density.

The main idea of the indicator-based minimum cross-entropy method, called the *indicator MinxEnt*, or simply IME, is to associate with each counting or optimization problem an auxiliary *single-constrained* convex MinxEnt program of a special type, which has a closed-form solution. We proved that the optimal pdf obtained from the solution of such a specially-designed MinxEnt program is a zero variance pdf, provided the "temperature" parameter is set to minus infinity. In addition, we proved that the parametric pdf based on the product of marginals obtained from the optimal zero variance pdf coincides with the parametric pdf of the standard *cross-entropy* (CE) method. The crucial difference between the proposed IME method and its standard CE counterparts is in their simulation-based versions: in the latter we always require to generate (via Monte Carlo) a sequence of tuples including the temperature parameter and the parameter vector in the optimal marginal pdf's, while in the former we can *fix* in advance the temperature parameter (to be set at large negative number) and then generate (via Monte Carlo) a sequence of parameter vectors of the optimal marginal pdf's only. In addition, in contrast to CE, neither the elite sample nor the rarity parameter is needed in IME. As a result, the proposed IME Algorithm becomes simpler, faster and at least as accurate as the standard CE.

The main idea of the MCMC approach is to design a sequential sampling plan, where the "difficult" problem of estimating rare-event probability and counting the cardinality of a set is decomposed into "easy" problems of counting the cardinality of a sequence of related sets. Such sequential sampling algorithms are called *randomized algorithms*. Typically the randomized algorithms involve estimation of a partition function at some desired temperature via generation samples from the Boltzmann distribution. They are based on computing the ratios of the partition functions for a carefully desired sequence of temperatures, also called a cooling schedule.

Here we proposed a new algorithm, called the *cloning* algorithm. The main differences between the existing and the proposed algorithm is that the latter has a special device, called the "cloning" device, which makes it very fast and accurate. In particular, our algorithm is well suited for counting the satisfiability assignments in a SAT problem and solving problems associated with the Boltzmann distribution, like estimating the partition functions in an Ising model and for sampling random variables uniformly distributed on different convex bodies.

We presented numerical results with both the IME and the cloning algorithms and we show that both methods work well for COP's, while for counting problems, like SAT and Hamiltonian cycles, the cloning one typically outperforms the IME.

Further Research

As for further research, we consider the following issues:

1. Use the large deviation theory to prove polynomial convergence and speed of convergence of the IME Algorithm 16.2.1 for rare-event probability estimation and thus, for estimation of the counting quantity $|\mathscr{X}^*|$ according to (16.7).
2. Establish rigorous mathematical foundations for the cloning algorithms.
3. Apply the cloning algorithms to a broad variety of optimization and counting problems, like Hamiltonian cycles, counting 0–1 Tables, self-avoiding walks, counting problems associated with graph coloring, cliques and counting the number of multiple extreme in a multi-extremal function.
4. Use the cloning algorithms to generate samples uniformly distributed on different regions $\mathscr{X}^*$.

Appendix

Ross' Algorithm for Estimating c_t is given as follows.

Algorithm 16.5.1 (Ross' Algorithm)

1. Set J=N=0.
2. Choose a vector **x** such that $S(\mathbf{x}) \geq m_{t-1}$.
3. Generate a random vector $U \sim U(0,1)$ and set I=Int$(nU)+1$.
4. If $I = k$, generate X given the conditional distribution of X_k, given that $X_j = x_j$, $j \neq k$.
5. If $S(x_1,\ldots,x_{k-1},X,x_{k+1},\ldots,x_n) < m_{t-1}$, return to 4.
6. $N = N+1$, $x_k = X$.
7. If $S(\mathbf{x}) \geq m_t$, then $J = J+1$.
8. Go to 3.

References

1. Z. I. Botev, D. P. Kroese: An Efficient Algorithm for Rare-event Probability Estimation, Combinatorial Optimization, and Counting. *Methodology and Computing in Applied Probability*, (in press)
2. Z. I. Botev, D. P. Kroese, T. Taimre: Generalized Cross-Entropy Methods. *Proceedings of RESIM06*, 1–30 (2006)
3. P. Diaconis, S. Holmes: Three Examples of the Markov Chain Monte Carlo Method. *Discrete Probability and Algorithms*, D. Aldous et al. (eds.), 43–56 (Springer, New York, 1994)
4. A. Ghate, R. L. Smith: A Dynamic Programming Approach to Efficient Sampling from Boltzmann Distribution (accepted for publication)
5. M. Mitzenmacher, E. Upfal: *Probability and Computing: Randomized Algorithms and Probabilistic Analysis* (Cambridge University Press, New York, 2005)
6. R. Motwani, R. Raghavan: *Randomized Algorithms* (Cambridge University Press, New York, 1997)
7. Sh. M. Ross: *Simulation*, 3rd edn. (Academic, New York, 2002)
8. R. Y. Rubinstein: Optimization of Computer Simulation Models with Rare Events. *European Journal of Operations Research* 99, 89–112 (1997)
9. R. Y. Rubinstein: A Stochastic Minimum Cross-Entropy Method for Combinatorial Optimization and Rare-event Estimation. *Methodology and Computing in Applied Probability*, 1, 1–46 (2005)
10. R. Y. Rubinstein: How Many Needles Are in a Haystack, or How to Solve #P-Complete Counting Problems. *Methodology and Computing in Applied Probability*, 1, 1–42, 2007
11. R. Y. Rubinstein: Semi-Iterative Minimum Cross-Entropy Algorithms for Rare-Events, Counting, Combinatorial and Integer Programming. *Methodology and Computing in Applied Probability* (in press)
12. R. Y. Rubinstein, D. P. Kroese: *The Cross-Entropy Method: a Unified Approach to Combinatorial Optimization, Monte-Carlo Simulation and Machine Learning* (Springer, Berlin, 2004)
13. R. Y. Rubinstein, D. P. Kroese: *Simulation and the Monte Carlo Method*, 2nd edn. (Wiley, New York, 2007)
14. R. Y. Rubinstein, A. Shapiro: *Discrete Event Systems: Sensitivity Analysis and Stochastic Optimization*, (Wiley, New York, 1993)
15. D. Stefankovic, S. Vempala, E. Vigoda: Adaptive Simulated Annealing: A Near-Optimal Connection between Sampling and Counting. *FOCS 2007*

Index